T0192540

Clinicopathologic principles for veterinary medicine

Clinicopathologic principles for veterinary medicine

Edited by

WAYNE F. ROBINSON and
CLIVE R. R. HUXTABLE

*School of Veterinary Studies, Murdoch University
Murdoch, Western Australia*

CAMBRIDGE UNIVERSITY PRESS
Cambridge
New York New Rochelle Melbourne Sydney

PUBLISHED BY THE PRESS SYNDICATE OF THE UNIVERSITY OF CAMBRIDGE
The Pitt Building, Trumpington Street, Cambridge, United Kingdom

CAMBRIDGE UNIVERSITY PRESS
The Edinburgh Building, Cambridge CB2 2RU, UK
40 West 20th Street, New York NY 10011–4211, USA
477 Williamstown Road, Port Melbourne, VIC 3207, Australia
Ruiz de Alarcón 13, 28014 Madrid, Spain
Dock House, The Waterfront, Cape Town 8001, South Africa

http://www.cambridge.org

First published 1988
First paperback edition 2003

A catalogue record for this book is available from the British Library

Library of Congress cataloguing in publication data

Clinicopathologic principles for veterinary medicine / edited by Wayne
F. Robinson and Clive R. R. Huxtable.
p. cm.
Includes index.
ISBN 0 521 30883 6 hardback
1. Veterinary clinical pathology. I. Robinson, Wayne F.
II. Huxtable, Clive R. R.
[DNLM: 1. Pathology, Veterinary. SF 769 C641]
SF772.6.C57 1988
636.089′607–dc 19
DNLM/DLC
for Library of Congress 87–32006 CIP

ISBN 0 521 30883 6 hardback
ISBN 0 521 54813 6 paperback

Contents

Contributors

John R. Bolton, B.V.Sc., Ph.D., M.A.C.V.Sc. *Senior Lecturer in Large Animal Medicine*

Leonard K. Cullen, B.V.Sc., M.A., M.V.Sc., Ph.D., D.V.A., F.A.C.V.Sc. *Senior Lecturer in Anesthesiology*

Clive E. Eger, B.V.Sc., M.Sc., Dip. Sm. An. Surg. *Senior Lecturer in Small Animal Medicine and Surgery*

John Grandage, B.Vet.Med., D.V.R., M.R.C.V.S. *Associate Professor of Anatomy*

John McC. Howell, B.V.Sc., Ph.D., D.V.Sc., F.R.C Path., M.A.C.V.Sc., F.A.N.Z.A.A.S., M.R.C.V.S. *Professor of Pathology*

Clive R. R. Huxtable, B.V.Sc., Ph.D., M.A.C.V.Sc. *Associate Professor of Pathology*

Jennifer N. Mills, B.V.Sc., M.Sc., Dip. Clin. Path. *Senior Lecturer in Clinical Pathology*

David A. Pass, B.V.Sc., M.Sc., Ph.D., Dip. Am. Coll. Vet. Path. *Associate Professor of Pathology*

W. John Penhale, B.V.Sc., Ph.D., Dip. Bact., M.R.C.V.S. *Associate Professor of Microbiology and Immunology*

David W. Pethick, B.Ag.Sc., Ph.D. *Lecturer in Biochemistry*

Wayne F. Robinson, B.V.Sc., M.V.Sc., Ph.D., Dip. Am. Coll. Vet. Path, M.A.C.V.Sc. *Associate Professor of Pathology*

Susan E. Shaw, B.V.Sc., M.Sc., F.A.C.V.Sc., Dip. Am. Coll. Int. Med. *Senior Lecturer in Small Animal Medicine*

V. E. O. Valli,* D.V.M., M.Sc., Ph.D., Dip. Am. Coll. Vet. Path. *Professor of Veterinary Pathology*

Sheila S. White, B.V.M.S., Ph.D., M.R.C.V.S. *Senior Lecturer in Anatomy*

Peter E. Williamson, B.V.Sc., Ph.D. *Senior Lecturer in Reproduction*

Robert S. Wyburn, B.V.M.S., Ph.D., D.V.R., F.A.C.V.Sc., M.R.C.V.S. *Associate Professor of Veterinary Medicine and Surgery (Radiology)*

* Department of Veterinary Pathology, *University of Guelph, Guelph, Ontario, Canada N1G 2WI*

Except where otherwise stated, all contributors are faculty members of the School of Veterinary Studies, Murdoch University, Murdoch WA 6155, Australia.

Preface

This book is written for veterinary medical students as a primer for their clinical years and should also be of benefit beyond graduation.

As the title suggests, our aim is to highlight the essential relationship between tissue diseases, their pathophysiologic consequences and clinical expression. The book is designed to emphasize the principles of organ system dysfunction, providing a foundation on which to build.

The basis of the book is an integrated course in systemic pathology and medicine taught at this school, and it is a source of satisfaction that all but one of the contributors teach in the course. The approach taken is similar in many respects to the pattern followed in other schools throughout the world. Our experience and no doubt that of many others is that the two disciplines of pathology and medicine are enriched by such integration, a merger rather than a polarization. We have endeavoured to encapsulate these views in the first chapter of the book entitled 'The relationship between pathology and medicine'.

To our co-authors we extend our heartfelt thanks. Their contributions of time and expertise are greatly appreciated.

January 1987

W. F. Robinson
C. R. R. Huxtable
Perth, Australia

Acknowledgements

We are indebted to a number of dedicated helpers who do not appear in name elsewhere. Sue Lyons with her trusty word processor has typed and corrected numerous chapter drafts with dedication, speed and accuracy. Hers was a most onerous task carried out with cooperation and willingness. Pam Draper and Diane Surtees were also of immense help with some of the chapter typing. The creativity and expertise of Gaye Roberts, whose line drawings and diagrams are of the highest quality, are evident throughout the book. Geoff Griffiths lent his able photographer's eye to the printing of the graphic artwork and Jennifer Robinson dealt swiftly with the split infinitive and other grammatical transgressions. To all, our profound gratitude is extended.

We also wish to express our deep appreciation to the publisher, Cambridge University Press, and especially to Dr Simon Mitton, the editorial director, who enthusiastically supported the initial idea and helped throughout the writing and production phases. Finally, we would like to thank both the School of Veterinary Studies and Murdoch University for grants to complete the graphic artwork.

Wayne F. Robinson and
Clive R. R. Huxtable

1 The relationship between pathology and medicine

The aim of this book is to assist the fledgling clinician to acquire that 'total view' of disease so essential for the competent diagnostician. The typical veterinary medical student first encounters disease at the level of cells and tissues, amongst microscopes and cadavers and then proceeds rather abruptly to a very different world of lame horses, vomiting dogs, panting cats, scouring calves, stethoscopes, blood counts, electrocardiographs and anxious owners. In this switch from the fundamental to the business end of disease, the link between the two is often obscured. It is easy to forget that all clinical disease is the result of malfunction (hypofunction or hyperfunction) within one or several organ systems, and that such malfunction springs from some pathologic process within living tissues.

Although some disease processes are purely functional, in most instances the pathologic events involve structural alteration of the affected organ, which may or may not be reversible or repairable. At least one of the basic reactions of general pathology, such as necrosis, inflammation, neoplasia, atrophy or dysplasia, will be present.

The expert clinician, having recognized functional failure in a particular organ as the cause of a clinical problem, is easily able to conjure up a mental image of the likely underlying lesion and take effective steps to characterize it. This characterization of the underlying disease opens the way for establishing the etiology and appropriate prognosis and management. By contrast, the novice tends to stop short at the stage of identifying organ malfunction, neglecting the important step of characterizing and comprehending the nature of the tissue disease. A good example is provided by the clinical state of renal failure, recognized by a number of characteristic clinical findings. This failure may result from a diversity of pathologic states, some readily reversible, some relentlessly progressive. The need to accurately characterize the tissue disease is appreciated by the expert, but frequently neglected by the novice.

The diagnostic process must therefore combine clinical skills with a sound understanding of pathology. Lesions causing tissue destruction will only become clinically significant when the functional reserve of the affected organ has been exhausted. This fact clearly establishes the important principle that tissue disease does not necessarily induce clinical disease, and that many quite spectacular structural lesions have no functional significance. The critical factor is the erosion of functional reserve capacity or, conversely, the stimulation of significant hyperfunction.

Modern veterinary medicine provides an expanding battery of clinical diagnostic aids, by which organ function may be assessed and tissue disease processes characterized. This happy situation catalyzes the fusion of the clinical sciences and tissue pathology. Whilst we cannot promise diamonds, we hope that the veterinary student will find a crystalline and easily digestible fusion in the chapters of this book.

These introductory remarks pave the way for the enunciation of some general principles.

The limited nature of clinical and pathologic responses

The clinical signs resulting from malfunction of a particular organ may be likened to the themes and variations of a particular musical composition. Regardless of variations induced by different etiology and pathogenesis, the thread of the basic theme is always apparent to the thoughtful investigator. In the case of renal failure, for example, two basic themes – failure of urinary concentration and elevation of non-protein nitrogenous compounds in the plasma – are always present. Variations are provided by items such as large or small urine output, large or small urinary protein concentration and few or numerous inflammatory cells in the urine. Particular patterns of variations based on the common theme provide opportunity for differentiating types of disease processes.

Pathologic responses are limited in scope and modified by the differing characteristics of various organs. Ultimately all lesions can only fall into those basic categories defined in general pathology, such as inflammation/repair, proplasia/retroplasia, neoplasia, developmental anomaly, degeneration/infiltration, circulatory malfunction or non-structural biochemical abnormality. The most important modifying factors are the developmental age of the affected tissue and its intrinsic regenerative ability.

The progression of the diagnostic process

The clinician's initial contact with a patient usually occurs when the owner reports the **recognition** of an abnormality. Through further questioning and a physical examination of the animal, the recognition of abnormality is further refined to a localization of the problem to a particular organ or tissue, and often the 'single' problem may prove to be a plethora of problems. The next step is usually **confirmation** of suspicions by the use of appropriate clinical aids such as radiography and the taking of blood and tissue samples. Then follows **characterization**, directly or by inference, of the underlying pathologic process. This is ideally accompanied by **identification** of the specific cause, by further testing or by inference from previous experience. The culmination of all these steps and procedures is the **prediction** of the outcome of the process. This method of investigation has widespread acceptance and again demonstrates the inextricable link between the clinical appearance of the disease and the underlying pathology. Recognition, localization and confirmation are the essence of clinical skill, whereas characterization and identification involve knowledge of tissue reactions. The last and most important step of prediction is a combination of the two disciplines.

Disease versus failure

The prevalence of disease far outweighs the prevalence of tissue or organ failure. A certain threshold must be reached before an organ system fails. This varies greatly from organ to organ and the interpretation of failure must necessarily be broad. The concept of organ failure applies well to the heart, lungs, kidneys, liver, exocrine pancreas and some endocrine organs. In these organs, failure implies an inability to meet the metabolic needs of the body. Organ failure in this sense cannot be applied so strictly to organs such as the brain, muscle, bone, joints and skin. These rarely fail totally, but rather produce severe impediments to normal function when focally damaged.

However, the overriding concept remains, that disease does not necessarily equate with failure. A lesion may be visible grossly in an organ, leaving no doubt that disease is present, but organ function may not be impaired. Conversely, comparatively small lesions may be of great clinical significance when they are critically located, or have a potent metabolic effect. The skilled and experienced observer will be able to assess the type and character of any lesion and decide if it has nil, moderate or marked effect on organ function.

Reversible versus irreversible disease

One of the central features of the clinician's skill is the ability to estimate the outcome of a disease process. While a number of factors need to be considered, the two most important are the conclusions reached about the nature of the disease process and the inherent ability of a particular tissue to replace its specialized cells.

The nature of the disease process may, for example, be a selective degeneration and necrosis of specialized cells. This may be caused by a number of agents and may be accompanied by an inflammatory process. If the offending cause is removed or disappears and the architectural framework remains, a number of organs have the capacity to replace the lost cells. Prominent in this regard are the skin, liver, kidney, bone, muscle and most mucosal lining cells. However, tissues such as the brain, spinal cord and heart muscle have little or no capacity for regeneration.

Sometimes, when a disease process is highly destructive, it matters little if the organ has the capacity to regenerate and the only savior in the circumstances is the ability of some systems to compensate. The remaining unaffected tissue undergoes hypertrophy or hyperplasia and to some extent increases its efficiency. An example of this is the ability of one kidney to enlarge and compensate when the other is lost because of a disease such as chronic pyelonephritis.

Another factor that needs to be taken into account is the potential reversibility of the disease process itself. There are numerous examples of chronic diseases in which there is little hope of reversal. A number of the inherited or familial diseases fit this pattern, as do many malignant neoplastic diseases. In these cases, a disease may be recognized in its early stages, but there is an inexorable progression. It is important to characterize the nature of the disease as quickly as possible so that suffering by the animal and emotional and monetary costs to the owner can be minimized.

W. John Penhale

2 The immune system

Knowledge of immunology has now become essential for the comprehension of many disease processes. In addition to the awareness of an expanding spectrum of diseases which have at their core immunologic mechanisms, basic information is also required on the cells of the immune system and their interactions and effector mechanisms.

The immune system is extremely complex, performing a variety of activities directed towards maintaining homeostasis. It consists of an intricate communications network of interacting cells, receptors and soluble factors. As a consequence of this complex organization, it is immensely flexible and is able greatly to amplify or markedly to diminish a given response, depending upon the circumstances and momentary needs of the animal. A normally functioning immune system is an effective defense against the intrusion of noxious foreign materials such as pathogenic microbial agents, toxic macromolecules and to some extent against endogenous cells which have undergone neoplastic transformation. However, by virtue of its inherent complexity, the system has the potential to malfunction and, since it also has the ability to trigger effector pathways leading to inflammation and cell destruction, may then cause pathologic effects ranging from localized and mild to generalized and life threatening.

The intensity of a particular immune response depends on many factors, including genetic constitution, and hormonal and external environmental influences. Amongst these, it is now becoming clear that genetic background plays a highly influential role, and to a significant extent, therefore, immunopathologic events are a reflection of genetically determined aberrations in immune regulation.

This chapter is designed to bridge the interface between immunology and disease and will be concerned largely with the involvement of immunologic processes in disease pathogenesis. Accordingly, emphasis will be placed on the effector pathways and regulating mechanisms and detailed accounts will not be given of the organization of the system as a whole or of its primary role in host defense.

The organization and regulation of the immune system

In the absence of immune function, death from infectious disease is inevitable. In order to counteract infectious agents, the system has evolved to recognize molecular conformations foreign to the individual (antigenic determinants) and to promote their elimination. To accomplish this effectively, the system is ubiquitously distributed throughout body tissues and has as basic operational features: **molecular recognition**, **amplification and memory**, together with a range of **effector pathways** by which foreign material may be eliminated. The last of these can be divided broadly into the humoral and cell-mediated immune responses.

In addition, such a system requires precise regulation in order to avoid excessive and hence wasteful responses, and also potentially dangerous reactivity to self components.

These diverse activities are performed by a limited number of morphologically distinct cell types which are capable of migrating through the organs and tissues, performing their functions remote from their sites of origin and maturation. In this section, the chief features and interactions of these cells where considered germane to the main theme of this chapter will be reviewed briefly.

Cells of the immune system

The ability of the individual to recognize and respond to the intrusion of foreign macromolecules resides in cells of the lymphoid series. Lymphoid cells are distributed throughout the body both in circulating fluids and in solid tissues. In the latter, they occur either diffusely or in aggregates of varying degrees of organization. In strategic regions of the body, they collectively form discrete encapsulated lymphoid organs such as the spleen and lymph nodes.

The central cell of lymphoid tissues is the immunocompetent lymphocyte. These cells have receptor molecules on their cytoplasmic membranes which enable them to recognize, and to interact with, complementary antigenic, as well as endogenously derived physiologic molecules.

Lymphocytes are activated by contact with appropriate antigenic determinants and then undergo transformation, proliferation and further differentiation (Fig. 2.1). Ultimately, one or more effector pathways are initiated and the antigen concerned may then be eliminated. Activated cells secrete a variety of biologically active effector molecules which are responsible both for cellular regulation and effector functions. In addition, a proportion of the expanded cell population remains dormant as memory cells and accounts for the augmented secondary response on re-exposure to the same antigen.

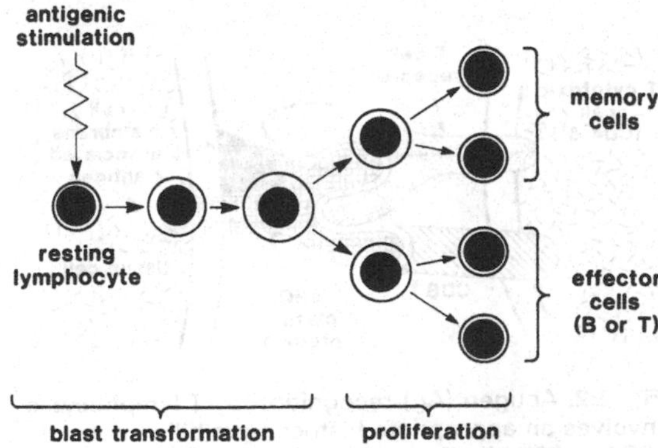

Fig. 2.1. Resting lymphocytes following contact with an appropriate antigen undergo blast transformation followed by proliferation and further differentiation.

Lymphocytes are divided into B and T cell classes on the basis of ontogeny and function. Functionally, B lymphocytes are responsible for humoral, and T lymphocytes for cell-mediated immune responses. These cells also differ in their distribution within lymphoid tissues and in their expression of cell surface molecules (markers). Thus the immune system can be regarded as a system composed of dual but interacting compartments.

The B lymphocyte

Cells of this lineage are the progenitors of antibody-secreting plasma cells and in mammals develop initially from stem cells situated in the bone marrow by a process of antigen-independent maturation. Subsequently, after migration to peripheral lymphoid tissues, they undergo further differentiation induced by antigen contact and mature to plasma cells. Depending on the nature of antigen concerned, B cell activation may require the cooperation of a subpopulation of T cells (T helper cells). Generally, small asymmetric molecules such as polypeptides will not stimulate B cells directly, and require T cell cooperation, whilst many polysaccharides are capable of causing a direct (but limited) B cell response. The antibodies generated may exist in several different molecular types or classes (immunoglobulins (Ig) A, D, E, G and M). The first antibodies generated are often of IgM class and later, particularly after restimulation, a switch in production to IgG, and less frequently to IgA and IgE classes, occurs.

The functional activities of B cells depend on an array of cell surface receptor molecules, including Ig receptors for antigen, histocompatibility markers, receptors for the Fc region of IgG and for complement (C3b component).

The T lymphocyte

T lymphocytes which undergo maturation in the thymus are key cells in the expression of many facets of immunity, where they perform a variety of functions essentially concerned with immune regulation and the elimination of abnormal cells.

T cells orchestrate the immune response by modulating the activities of both B and other T cells. Regulation may be either positive or negative. So, T cells are involved in initiating immune responses (T helper cells) and also terminating them (T suppressor cells). T cells are also the principal cells involved in initiating cellular immune events which include such phenomena as delayed hypersensitivity reactions and allograft rejection.

Another facet of cell-mediated immunity is cytotoxicity, executed by T cells having the capacity to kill other cells, as exemplified in the destruction of virus-infected cells and in the rejection of allografts.

These various functions are performed by major subsets of T lymphocytes which have distinctive surface markers and which appear to belong to different T cell lineages. Two major subsets are now well defined both functionally and phenotypically. T helper/inducer cells cooperate in the production of antibodies by B cells and with other T cells in cellular immune reactions. They also act as inducers of cytotoxic/suppressor cells. Helper/inducer cells may be identified serologically by the presence of the **CD4** marker (defined by a monoclonal antibody) on their surfaces. Cytotoxic/suppressor T lymphocytes are involved in the suppression of immune responses and in the killing of virus-infected and other abnormal cells. They also express a specific cell marker, **CD8**, on their cell membrane.

Antigen recognition by T lymphocytes

In major contrast to B cells, T cells recognize antigen only when it is presented on a cell surface. Furthermore, the antigen-presenting cell *must* be of histocompatibility type identical with that of the T cell concerned. Thus, in this instance, antigen recognition is restricted and can only be accomplished in the context of an appropriate histocompatibility molecule. The latter occurs in several different classes and it is now clear that the major subsets of T cells described above, recognize antigen in association with *different* histocompatibility classes. Thus helper/inducer cells are restricted to the recognition of antigen on cells bearing the class II molecules (immune-associated anti-

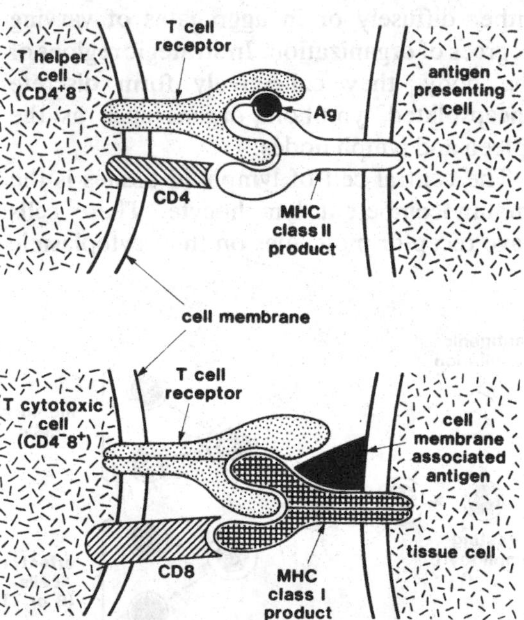

Fig. 2.2. Antigen (Ag) recognition by T lymphocytes involves an appropriate histocompatibility molecule (CD4 and CD8 in diagram), and a combination of the T cell receptor, the antigen and an appropriate histocompatibility product (class I or class II) on the presenting cell. MHC, major histocompatibility complex.

gen, **Ia**) and suppressor/cytotoxic cells are similarly restricted to antigen recognition on cells bearing class I. Furthermore, it now appears that the CD4 and CD8 markers found mutually exclusively on different subsets of the two major T cell types act as the respective binding sites for the two classes of histocompatibility molecules. So CD4 in T helper cells links to the non-variant part of class II antigens and CD8 to class I. A speculative arrangement is shown diagramatically in Fig. 2.2.

The T cell antigen receptor

Recent studies have shown that this is a two-chain structure with domains, some of which bear considerable homology in amino acid sequence to those of immunoglobulin light chains. In this regard it therefore resembles a number of other important cell surface molecules such as class I and II histocompatibility antigens and is evidently a member of the immunoglobulin supergene family (Fig. 2.3).

Soluble factors secreted by T cells

Following activation, T lymphocytes manufacture and secrete an as yet undetermined number of biologically important soluble substances commonly called lymphokines. These substances affect the behavior of other cells and play a prominent role in immunologically induced inflammatory change as well as in

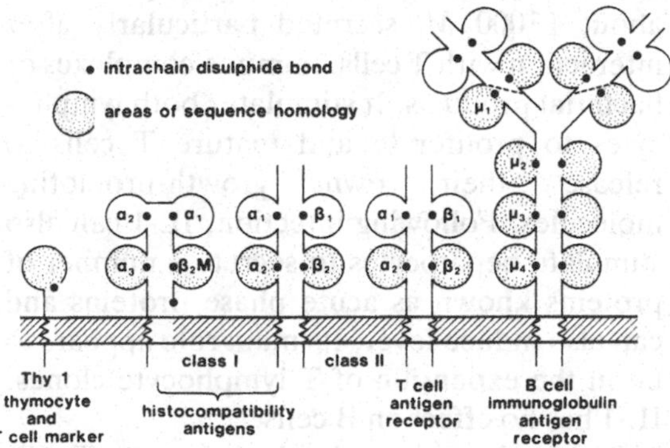

Fig. 2.3. The cell membrane and glycoprotein molecules of the immunoglobulin supergene family. α_{1-3}, β_{1-2}, μ_{1-4}, domains within supergene.

Table 2.1. *Factors produced by activated lymphocytes (lymphokines)*

Factors affecting macrophages	
Migration inhibitory factor	(MIF)
Macrophage-activating factor	(MAF)
Macrophage chemotactic factor	(MCF)
Ia antigen-inducing factor	
Factors affecting polymorphonuclear leukocytes	
Leukocyte inhibitory factor	(LIF)
Leukocyte chemotactic factor	(LCF)
Factors affecting lymphocytes	
T cell growth factor (TCF) or interleukin-2 (IL-2)	
Factors affecting antibody production: B cell growth factor 1 (BCGF-1) – now IL-4	
Transfer factor	
Specific and non-specific suppressor factors	
Interferon	
Factors affecting other cell types	
Lymphotoxin	
Growth inhibitory factor	
Interferon	
Osteoclast-activating factor	
Colony-stimulating activity	

various stages of the immune response itself. At present, at least 60 of these factors have been described and it has proved to be difficult to isolate and to characterize them biochemically. Consequently, at present, it is not known how many distinct lymphokines are produced but they are generally small polypeptides (15000–60000 M_r) which have very short half lives *in vivo*. Those characterized can be divided into four groups according to the target cell they affect (Table 2.1).

'Null' lymphocytes

Although the majority of lymphocytes bear surface markers of either T or B cells, a small number do not and are termed 'null' cells. Null lymphocytes probably encompass a number of cell lineages in various stages of differentiation. Among them are included killer (K cells) and natural killer (NK) cells. K cells are characterized by membrane receptor molecules for the Fc portion of the IgG molecule and can consequently bind to antibody-coated cells. These cells may be subsequently destroyed and this phenomenon is termed antibody-dependent, cell-mediated cytotoxicity (ADCC).

NK cells can similarly bind and kill some types of tumors and virus-infected cells, but in the absence of antibody. The molecular basis of the binding and recognition of diverse cellular targets in this instance is not clear presently. In contrast to B cells, these cells do not express surface IgM or IgD molecules. Their exact lineage is not established but they appear to share at low level some of the early differentiation antigens occurring on both macrophages and T cells.

Non-lymphoid cells involved in immune reactions

Macrophages

Mononuclear phagocytes are widely distributed throughout body tissues and form an important component of the defense mechanism by removing micro organisms from blood and tissues. Their most important characteristic is their ability to pinocytose soluble molecules and phagocytose particles. Certain types have the ability also to process and present this internalized foreign material to immunocompetent lymphocytes. In addition, they provide factors necessary for lymphocyte activation and proliferation. They play a crucial role in the early inductive events of the immune response. Macrophages also respond to external stimuli emanating from activated lymphocytes and are important effector cells in cell-mediated immune reactions.

Particulate antigens are taken up via phagocytosis, soluble antigens by pinocytosis. Aggregated material is ingested much more rapidly than is non-aggregated, with the bulk of ingested foreign material rapidly degraded by lysosomal enzymes. The remainder (approximately 10%) is only partially degraded and persists in macromolecular form associated with the cell membrane or in special vacuoles inaccessible to lysosomal enzymes. In this latter situation it can survive within cells in which intense phagocytosis and catabolic activities are in progress. Some undegraded antigen may eventually be released but most is attached to the cell membrane, where it lies in close proximity to membrane-bound major histocompatibility complex (MHC) molecules. Such membrane-associated material fulfils the arrangement required by T cells for effective antigen recognition (surface antigen associated with MHC class II markers) once it reappears on the cell surface following fusion of the vacuole and cell membranes.

Macrophages, by synthesizing and secreting a great many substances, have the potential to exert a regulatory influence on their surrounding environment in inflammation, tissue repair and the critical inductive steps of immunity. The secreted substances may be grouped into three categories:

1 Products involved in defense processes such as complement components and interferon.
2 Enzymes capable of affecting extracellular proteins which are of importance in generating inflammation, such as hydrolytic enzymes, plasminogen activators and collagenase.
3 Factors which modulate the function of surrounding cells. Most of these have not been characterized biochemically, but included in this category are those factors which influence immune function and only these will be discussed in depth in this section.

Interleukin I (IL-1), also known as lymphocyte-activating factor (LAF) is a protein of about 15000 M_r secreted particularly after interaction with T cells, immune complexes or bacterial products. It stimulates both lymphocytes to proliferate and mature T cells to release their own growth-promoting molecules. Following infection, IL-1 can also stimulate hepatocytes to secrete a number of proteins known as acute phase proteins and can also induce fever. Its main role appears to be in the expansion of T lymphocyte clones. IL-1 has no effect on B cells.

B lymphocyte activating factor (BAF) affects only B cells and enhances the production of antibodies. Its production is influenced by some macrophage-activating stimuli such as endotoxin.

In addition to the above, factors affecting other cells are also generated during the course of macrophage activation. One such factor stimulates bone marrow stem cells to differentiate into monocytes and granulocytes. This factor is a glycoprotein with a molecular weight between 45000 and 65000. Another soluble factor stimulates fibroblast growth and probably plays a role in wound healing.

Other cells involved in antigen presentation
Dendritic cells, which take their name from their tree-like appearance, are present in the spleen, where they comprise about 1% of the total nucleated cell population. They are in smaller numbers in lymph nodes and Peyers patches and occupy a strategic position within the lymphoid follicles. These cells lack many of the markers of both lymphocytes and macrophages, although they carry surface MHC class I and II antigens. These bone-marrow-derived cells are thought to present antigen to lymphocytes.

Langerhans cells are bone marrow derived and appear to be of macrophage lineage. They resemble dendritic cells morphologically, but differ in surface markers and are distributed through the epidermis. They are believed to function in the immune response in the skin by taking up antigens and presenting them to T cells.

Although cells of the immune system, B cells are activated by presenting antigen to T helper cells in association with class II MHC molecules in a manner analogous to that of macrophages and other antigen-presenting cells.

Effector cells of immune reactions
A number of leukocytes and connective tissue cells participate as effector cells in immunologic reactions. These reactions will be detailed later, but a brief reference is appropriate here. They include polymorphonuclear leukocytes (granulocytes) and mast cells. Neutrophils are involved in reactions mediated by antigen–antibody–complement complexes, and basophils in inflammatory reactions mediated by IgE antibodies. Eosinophils are frequent participants in allergic reactions involving IgE antibodies. Mast cells are similarly involved in IgE-mediated reactivity and like basophils carry surface receptors for these immunoglobulins. However, in contrast to basophils, these are connective-tissue cells which are not found in the blood.

Regulation of the immune response

The precise regulation of the immune system is crucial to the health of the individual for reasons given on p. 22. The regulation of this complex system is dependent on a number of interacting mechanisms which are as yet not fully understood. Ultimately, the extent of regulation of a particular immune response depends to a significant degree on genetic make-up, which is discussed in detail later.

Three essential regulatory interactions take place between the various cells of the system:

1 The activation of T helper/inducer cells by antigen presented by macrophages.
2 The T helper/inducer cell-driven differentiation of B cells to produce antibodies.
3 The activation of suppressor mechanisms to restrict antibody- and cell-mediated immunity.

Macrophage/lymphocyte interactions
An essential step in the initiation of immunity to all polypeptide antigens is the activation of T helper/inducer cells, a process first requiring the interaction of helper T cells with macrophages. As previously discussed, T helper cells recognize only antigen presented on the surface of macrophages in conjunction with the appropriate glycoprotein histocompatibility molecules. Antigen presentation requires physical contact between T lymphocytes and macrophages. The macrophages then secrete IL-1, which promotes T cell proliferation (Fig. 2.4). Under macrophage influence, the T cell expresses interleukin-2 (IL-2) receptors on its

surface and also secretes this factor. IL-2 production is necessary for the proliferation of all T cells. In this way macrophages exert a very important positive regulatory influence on the early stages of the immune response and to a large extent may determine its character, as, for example, whether the response will be predominantly of the T cell or B cell type or to what extent antibodies or memory cells are generated. Once lymphocytes are activated, they in turn influence macrophage behavior by secreting a variety of soluble mediators, as previously described. Although much is still uncertain concerning macrophage/lymphocyte interaction, it is clear that the macrophage is highly influential in both normal and abnormal immunologic reactivity.

B–T cell collaboration

Helper T cells interact with B cells promoting their growth and differentiation. In these interactions the B and T cells do not need to recognize the same antigenic determinants, provided that both of these are present on the one molecule. While the T cell-macrophage interaction is the main event resulting in clonal expansion of the T helper clones, the interaction of the T helper cells with B cells has a similar effect on B cells. This interaction leads to the clonal expansion of the B cells and their ultimate differentiation into antibody-secreting plasma cells. Although there are many unresolved issues in this collaboration the following three main stages are recognized (Fig. 2.4).

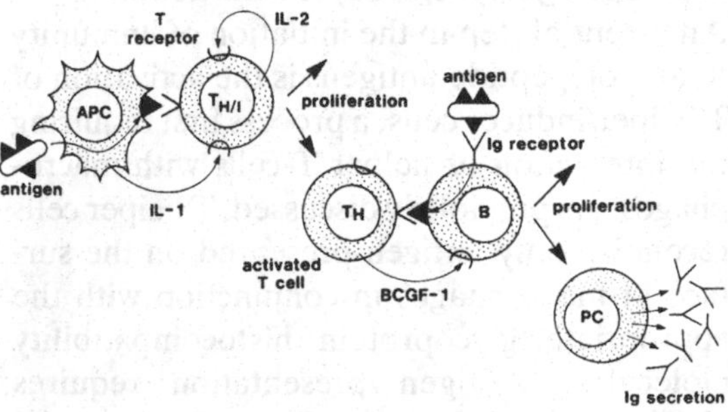

Fig. 2.4. Cellular interactions leading to the generation of antibodies. APC, antigen-presenting cell; $T_{H/I}$, T helper/inducer cell; B, B cell; PC, plasma cell; IL-1, interleukin 1; IL-2, interleukin 2; BCGF-1, B cell growth factor 1 (or IL-4); Ig, immunoglobulin.

1 Recognition of antigen by the B cell via surface immunoglobulin receptors.
2 B cells present antigen fragments to T cells, the cells interacting in a process modulated by class II MHC glycoproteins.
3 T cells undergo expansion under IL-2 influence and secrete lymphokines that promote B cell growth and differentiation and lead ultimately to antibody production by plasma cells.

In the first stage, the B cell binds antigen by way of its Ig receptor and then internalizes it. Following this, the immunogenic determinant reappears on the cell surface and in stage 2 the T helper cell recognizes and binds to the B cell. Thus, B cells serve as antigen-presenting cells to T cells in much the same way that macrophages do.

T–T cell interactions

T cell–T cell interactions occupy a key position in the regulation of the immune response. These interactions center around the generation of T cells of the cytotoxic/suppressor lineage by T suppressor/inducer cells, following the latter's activation by macrophage-presented antigen. Evidence suggests that both antigen-specific and non-specific suppressor cells may be generated under these circumstances and that this suppressor circuit is capable of down-regulating an ongoing antibody response or even inducing a state of specific unresponsiveness or tolerance, depending on circumstances (Fig. 2.5). T helper and T suppressor cells can be regarded as opposing cell types and the response to an antigen may be the result of a critical balance between these cells. Suppressor cells have also been shown to be capable of specifically suppressing other immune phenomena, such as delayed-type hypersensitivity, contact sensitivity and target cell killing by cytotoxic cells.

T suppressor cells are generated concurrently with the appearance of T helper cells and the development of the response to antigen. The physiologic development of T suppressors can therefore be regarded as a cellular mechanism that inhibits and controls the expansion and continuation of the immunologic process. A number of conditions have been found to favor the generation of suppressor T cells. These include:

1 Very high or very low concentrations of antigen.
2 The nature of antigen – in particular highly soluble antigen, which can escape phagocytosis.
3 Repeated exposure to antigen.
4 Route of antigen entry – in particular the intravenous route.
5 Age – very young individuals have a tendency to develop strong T suppressor activity, which declines with age.

(*a*) T_H cell stimulation (T_S absent)

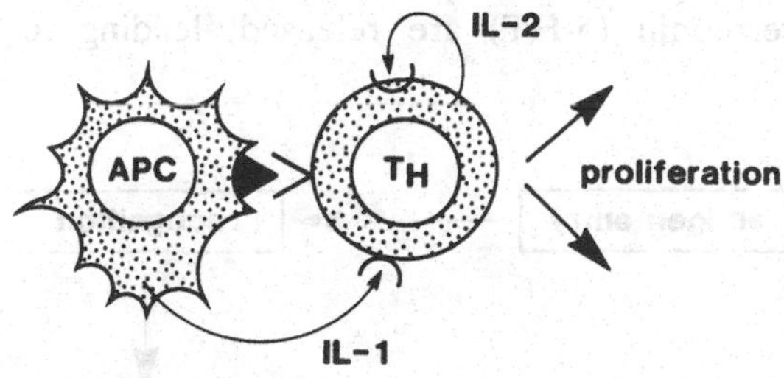

(*b*) T_H cell inhibition by T suppressor cells

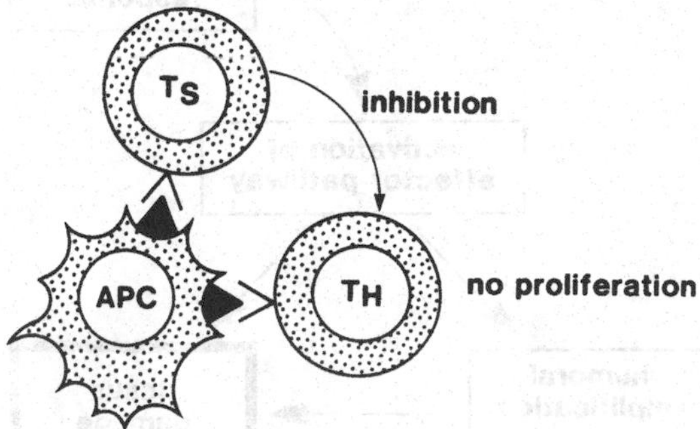

Fig. 2.5. T suppressor cell regulation of T helper cell activity. APC, antigen-presenting cell; T_H, T helper cell; T_S, T suppressor cell; IL-1, interleukin 1; IL-2, interleukin 2.

The termination of the immune response

Several distinct mechanisms are thought to act in concert to halt an immune response, thereby conserving resources.

1 **The elimination of antigen.** The persistence of antigen in immunogenic form in macrophages is relatively short lived. Once antigen disappears, the impetus of the response decreases.
2 **The presence of antibody.** Antibody can itself inhibit further generation by binding circulating antigen and promoting its elimination. In addition, immune complexes are known to inactivate B cells by binding to their Fc receptors. Thus, antibody generation acts as an important feedback regulatory mechanism.
3 **The emergence of suppressor T cells.** As discussed, these cells are a significant regulatory component normally activated during the immune response.
4 **Anti-idiotype antibody generation.** The unique molecular configuration of the antibody receptor site (the 'idiotype') can itself act as an immunologic determinant and may thus stimulate the production of anti-idiotypic antibodies. This has led to the concept that immunoregulation may be at least partly accomplished by the existence of functional regulatory networks of interacting lymphocytes.

Immunologic aspects of inflammation and tissue injury

Although the initiation of the immune response generally provides protection against microorganisms that threaten the welfare of the host it can also prove to be deleterious. The immune response to an infecting microorganism may lead to its elimination, but the same response may produce significant pathologic or even lethal effects in the host. Even more inappropriate immune reactions, giving rise to pathologic changes, may be induced by inert non-toxic environmental antigens or, indeed, self-components.

A number of distinct immunologic mechanisms can result in inflammation (Fig. 2.6) and frequently a particular disease may involve a combination of these pathways. The factors which condition these reactions are complex and not clearly evident in all situations but include the type of antigen, and its route of entry, the quantity and duration of exposure, and the tissue wherein the reaction takes place. Also involved are those factors which influence the immune system in general.

Furthermore, both the type of immune reaction and the associated clinicopathologic phenomena may be further complicated by the subsequent activation of one or more of the non-specific enzyme cascades, for example, the blood clotting mechanism. These will be collectively referred to as the humoral amplification systems and their close interrelationship frequently leads to their joint activation after initiation of the immune process. The sequence of immune-triggered events leading to inflammation and tissue injury is summarized in Fig. 2.7.

It can now be appreciated that the immune system is able to orchestrate a spectrum of pathologic changes resulting from mild local inflammation to severe and widespread tissue necrosis or even circulatory collapse. These immune effector mechanisms, together with the humoral amplification systems will be discussed below.

type I
anaphylactic reactions
type II
cytotoxic reactions
IgE
mast cells
basophils
IgM, IgG, C
macrophages
K cells
inflammation
T cells
macrophages
immune complex C
neutrophils
type IV
delayed hypersensitivity
type III
immune complex mediated

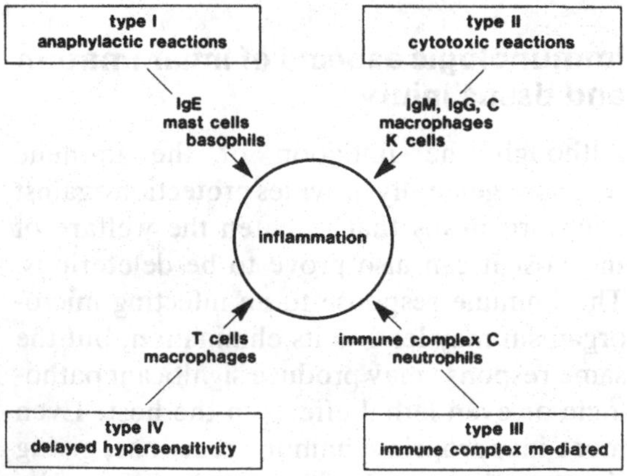

Fig. 2.6. Immunologic mechanisms in the generation of inflammation. C, complement.

Immune effector mechanisms involved in disease production

The various immune mechanisms involved in the production of damaging reactions have been classified into four basic types and this classification will be used in the present discussion.

Type I (anaphylactic) reaction

Essentially, this involves the rapid degranulation of mast cells or basophils previously sensitized by antibodies of the IgE class following contact with the corresponding antigen (Fig. 2.8 (*a*) and (*b*)).

Only antigens which are polyvalent are able to cause mast cell degranulation. Triggering of degranulation requires that adjacent IgE molecules on the cell surface are cross-linked by antigen. With degranulation, various chemical mediators such as histamine and serotonin (5-HT) are released, leading to

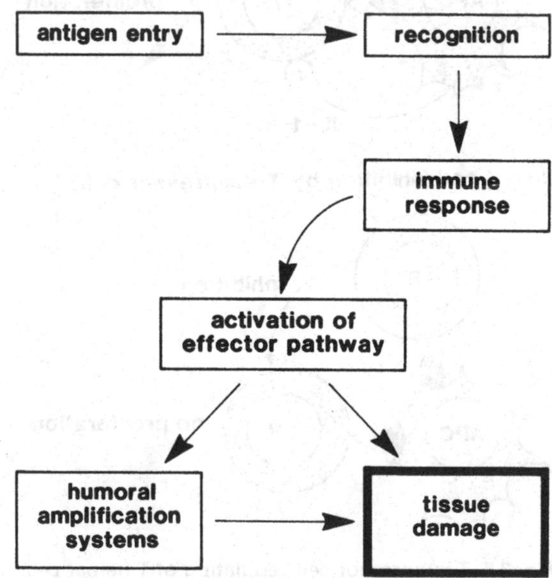

Fig. 2.7. Sequence of events leading to immune-mediated tissue injury.

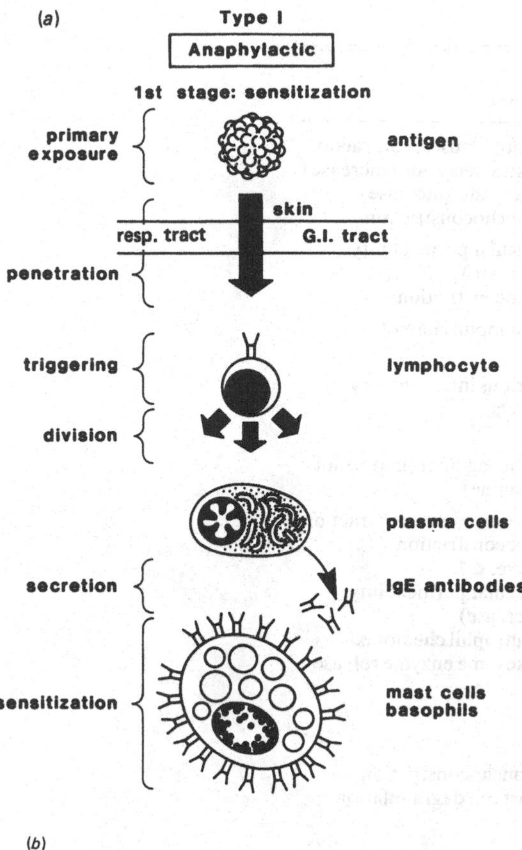

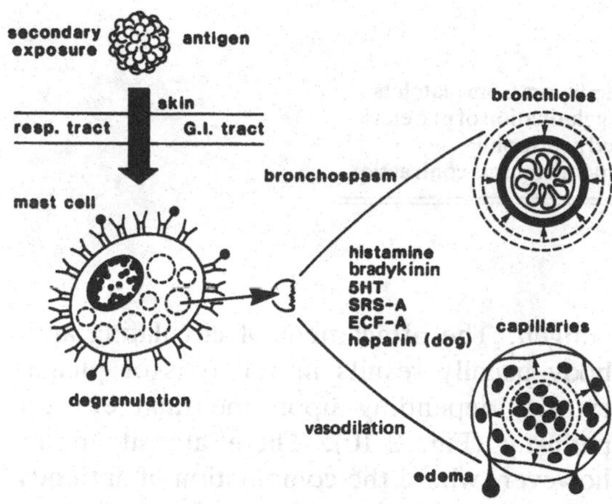

Fig. 2.8. Type I hypersensitivity. (*a*) Sequence of events ultimately leading to the sensitization of mast cells and basophils. (*b*) Events following secondary exposure to the antigen. G.I., gastrointestinal; 5HT, 5-hydroxytryptamine; SRS-A, slow-releasing substance of anaphylaxis; ECF-A, eosinophil chemotactic factor of anaphylaxis.

contraction of smooth muscle and an increase in the permeability of small blood vessels.

Mediators of anaphylactic reactions

There are two classes of chemical mediator responsible for anaphylactic reactions. The preformed or primary mediators, such as histamine and 5-HT, are stored in mast cell or basophil granules and are released within seconds of antigen contact. The secondary mediators are molecules synthesized following interaction with antigens. The principal secondary mediators are lipid derivatives mobilized by enzymatic action from cell membrane phospholipids (Fig. 2.9) and include the leukotrienes, prostaglandins and platelet-activating factor. The various mediators generated and their properties are summarized in Table 2.2.

In essence, it is apparent that the binding of antigen to IgE surface receptors results in the release and production of potent molecules by mast cells, basophils and perhaps other cells. These molecules are especially important pathologically when their large scale production gives rise to systemic circulatory and respiratory effects. The precise manner of their interaction in the production of all type I manifestations is not clear. Fortunately, the

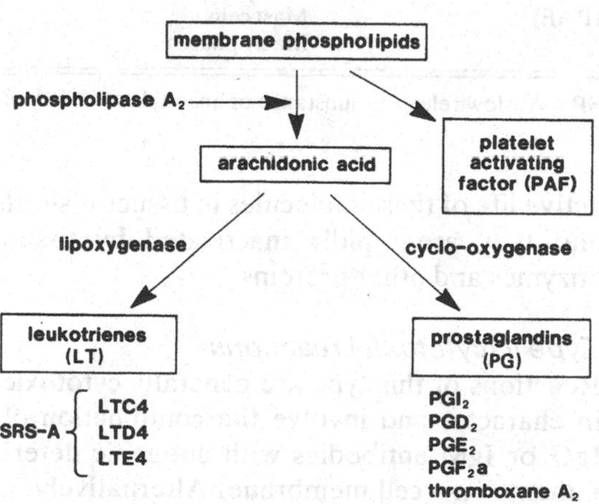

Fig. 2.9. Major secondary mediators in the anaphylactic reaction. a, activated; SRS-A, slow-releasing substance of anaphylaxis.

Table 2.2. *Biologic mediators of type I reactions*

Preformed mediators	Nature and origin	Action
Histamine	Decarboxylation of histidine Man and guinea pig	Smooth muscle contraction Gastric secretion (increase) Heart rate (increase) Bronchoconstriction
Serotonin	5-Hydroxytryptamine Mouse and rat	Vascular permeability (increase) Vasoconstriction
Eosinophil chemotactic factor	Tetrapeptide	Eosinophil chemotaxis
Enzymes	Varies with species, e.g. chymotrypsin and glucuronidase	Various inflammatory effects
Heparin	Proteoglycan	Anticoagulant (important in canine)
Leukotrienes (LT)	Cell membrane of: Basophils Mast cells Macrophages via lipoxygenase action on arachidonic acid	Smooth muscle contraction Vasoconstriction (increase) Vascular permeability (increase) Neutrophil chemotaxis Lysosome enzyme release
SRS-A	LTC4 LTD4 LTE4	
Prostaglandins and thromboxane	Cell membrane of: basophils mast cells macrophages via cyclo-oxygenase action on arachidonic acid Stimulated by LT5	Bronchoconstriction Mast cell degranulation
Platelet-activating factor (PAF)	Cell membranes of Basophils Mast cells Macrophages	Mediators from platelets Agglutination of platelets and neutrophils Smooth muscle contraction

SRS-A, slow releasing substance of anaphylaxis; LT, leukotrienes.

active life of these molecules in tissues is short and they are rapidly inactivated by tissue enzymes and other proteins.

Type II (cytotoxic) reactions

Reactions of this type are generally cytotoxic in character and involve the combination of IgG or IgM antibodies with antigenic determinants on a cell membrane. Alternatively, a free antigen or hapten may be adsorbed on to a tissue component or cell membrane and antibody subseqneutly binds with this adsorbed antigen. The attachment of circulating antibody usually results in cell lysis or phagocytosis, depending upon the final effector pathway (Fig. 2.10). There are situations, however, where the combination of antibody with cell-bound determinants does not result in cytotoxicity but causes a pathologic effect by blocking and inactivating physiologically important cell surface molecules such as hormone receptors.

The target for cytotoxic reactions may be either a specific cell type within a tissue or the circulating blood, or a variety of cell types carrying similar surface determinants (exogenously or endogenously derived).

The attachment of antibody to cells targets them for attack by either the complement sequence or by various effector cell types. Complement-fixing antibody is not required for the latter activity, but the cells involved require receptor sites for the Fc portion of the IgG molecule. By this means, bringing of effector cells into close proximity of the targets initiates the final attack phase. In some instances, cells of the monocyte/macrophage series engulf and phagocytose the antibody-coated target cells. However, controversy still surrounds the identity of the main cell type responsible for non-phagocytic cytotoxicity. It is generally accepted that cells in the monocyte/macrophage series can lyse target cells by this mechanism, but the identity of lymphoid cells which also have this ability is still uncertain. The term killer or K cell has been introduced because of this characteristic.

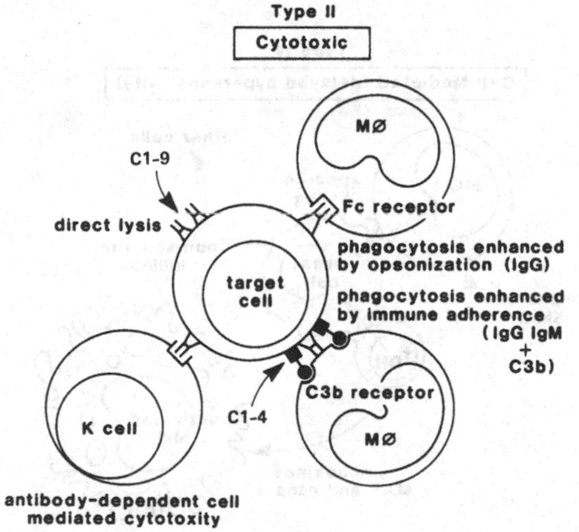

Fig. 2.10. Type II hypersensitivity (cytotoxic). Effector mechanisms are depicted, but the common factor is the binding of specific antibody to the target cell. Mϕ, macrophage; C1–9, complement factors.

Type III (immune complex-mediated) reaction

In this type of reaction immune-mediated injury results from the deposition of immune complexes within tissues and has inflammation as its main feature. Immune complexes formed with IgG antibody (and to a lesser extent IgM) can fix complement and, therefore, have the potential to cause tissue injury by means of complement-induced inflammation. The sequence of events leading to type III tissue damage following immune complex deposition is shown in Fig. 2.11.

A variety of factors are involved in the deposition of complexes in vulnerable tissue sites, particularly the subendothelial regions of small blood vessels.

1 Size of complex

The outcome of the formation of immune complexes *in vivo* depends not only on the absolute concentration of antigen and anti-

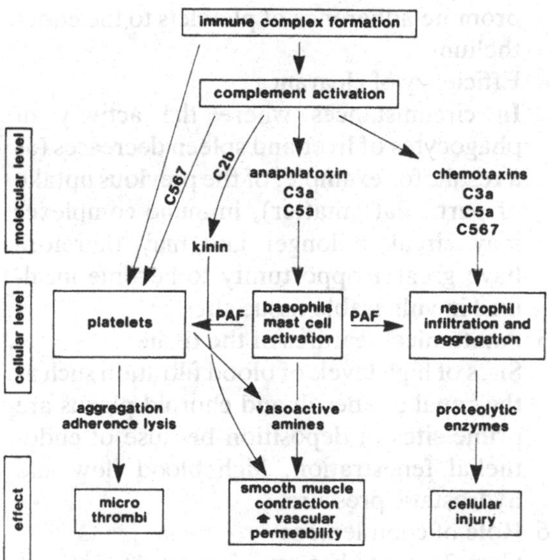

Fig. 2.11. Type III hypersensitivity (immune complex). PAF, platelet-activating factor; C, complement components.

body, which determines the intensity of the reaction, but also on their relative proportions, which govern the nature of the complexes and hence their distribution within the body. Between antibody excess and mild antigen excess the complexes are rapidly precipitated and tend to be localized at the site of introduction of antigen, whereas in moderate to gross antigen excess, soluble complexes are formed which circulate. Small soluble complexes tend to escape phagocytosis in the liver, spleen and elsewhere, and by circulating freely have the opportunity to penetrate vascular endothelium. They may cause systemic reactions by being widely deposited in such sites as the kidneys, synovia, skin and choroid plexus.

2 Vasoactive amines
The penetration of endothelia by immune complexes requires the production of vasoactive amines. These may be supplied by activation of mast cells, basophils and platelets (see Fig. 2.11).

3 Hemodynamic factors
Complexes tend to become localized in vessels where there is an increase in blood pressure and/or turbulence which tends to promote adherence of platelets to the endothelium.

4 Efficiency of clearance
In circumstances where the activity of phagocytes of liver and spleen decreases (as a result, for example, of the previous uptake of particulate matter), immune complexes may circulate longer and may therefore have greater opportunity to become localized in vulnerable tissue sites.

5 Anatomical features of the tissue
Sites of high levels of blood filtration such as the renal glomeruli and choroid plexus are prime sites of deposition because of endothelial fenestration, high blood flow and hydrostatic pressure.

6 Role of complement
Complement has an important role in modulating the size and facilitating the removal of immune complexes, and in the case of C2 and C4 deficiencies the incidence of immune complex disease is increased, possibly because of increased persistence of complexes.

7 Persistence of antigens
Long-lasting disease is only seen when antigen persists in the system over an extended period, such as, for example, in chronic infections and autoimmune disease.

8 Host response
Immune complex disease may occur only in certain individuals who produce moderate amounts of antibody of moderate affinity. Those generating high antibody titers of good affinity tend to eliminate antigen more effectively and therefore give less opportunity for immune complex deposition.

Type IV (cell-mediated) reactions

Cell-mediated reactions result from interactions between sensitized T lymphocytes and their corresponding antigen. They occur without involvement of antibody or complement and are mediated by the release of lymphokines, by direct cytotoxicity, or both. The sequence of events shown in this form of immune reactivity is shown in Fig. 2.12. The first stage in the reaction is the binding of antigen by small numbers of antigen-specific T

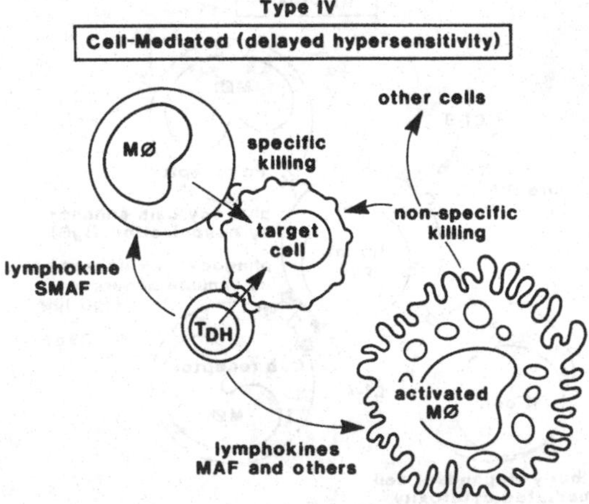

Fig. 2.12. Type IV hypersensitivity (cell mediated). T_{DH}, T lymphocyte (delayed hypersensitivity); Mϕ, macrophage; MAF, macrophage-activating factor; SMAF, specific macrophage-arming factor.

lymphocytes. This initial stage is followed by cellular proliferation and the production and release of soluble mediators with a wide variety of biologic activities. These lymphokines have various effects on macrophages, polymorphonuclear leukocytes, lymphocytes and others. Their overall effect is to amplify the initial cellular response by recruitment of other lymphocytes, polymorphonuclear leukocytes and, in particular, to attract, localize and activate macrophages at the site of the lesion. In addition, the recruited lymphocytes (both B and T) are induced to undergo mitogenesis.

Because the reaction depends upon both cell infiltration and proliferation, the generation of inflammatory changes is relatively slow as compared to type I and II reactions and generally does not reach its full magnitude until 24–48 hours after the challenging exposure to antigen.

There are distinct mediators for some of the functions that have been described. However, it is not yet clear whether there are a small number of molecules with multiple functions at different concentrations or whether a different lymphokine molecule is specific for each function. The clotting system may also be involved in the early stages of the reaction. Activated macrophages give rise to tissue damage and this may then activate the clotting system via factor VII. Such macrophages may become surrounded by a fibrin net and this is subsequently lysed by the action of plasmin. The kinin system as well as the clotting and fibrinolytic mechanisms may also be involved in modulating the extent and duration of inflammation.

There are normal control mechanisms that lead to resolution of such a lesion but these have not yet been clarified. In the situation where prolonged exposure to the antigen occurs, the lesions may progress to the stage of local necrosis or granuloma formation.

Humoral amplification systems

As previously indicated, the various interrelationships between these systems often lead to their involvement after initial activation of immune processes. Each is composed of a series of protein substrates, inhibitors and enzymes. They include the complement, coagulation, kinin and fibrinolytic systems (Fig. 2.13).

The complement system

This complex system of twenty distinct serum proteins is outlined in Fig. 2.14. The involvement of the complement system can be initiated by a wide variety of stimuli along either the classic or alternative pathways of activation. Activation of the system leads to a variety of biologic consequences apart from the classic function of cell lysis. Cleavage products C3a and C5a, termed anaphylatoxins, induce the release of histamine from the granules of mast cells, thereby producing increased capillary permeability, edema and smooth muscle contraction. Both C3a and C5a, together with the trimolecular complex C567, also have chemotactic activity for polymorphonuclear leukocytes. These products

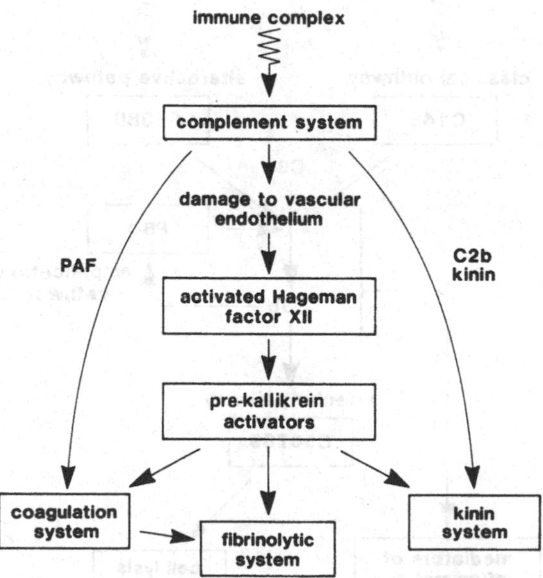

Fig. 2.13. The interrelations between immune reactions and enzyme cascade systems (humoral amplification systems). PAF, platelet-activating factor.

are amongst the most powerful inflammatory agents liberated within tissues and are key contributors to the degree of inflammation occurring at the site of antigen–antibody combination involving complement activation.

Other amplification systems are also involved with the complement system. For example, the fibrinolytic enzyme plasmin can directly attack C1, C3 and C5; the plasma proteolytic enzyme thrombin (which converts fibrinogen to fibrin) can attack C3; and a fragment of C2 has a kinin-like activity in causing increased vascular permeability and contraction of smooth muscle.

The coagulation system

Although the complement and coagulation mechanisms do not have a common means of activation, they interact at a number of levels. As mentioned above, thrombin formed during activation of the later stages of the coagulation cascade has the ability to act on various complement components. The complement system may also activate coagulation pathways indirectly via effects on platelets. These may include platelet adherence, aggregation and lysis by binding of the trimolecular complex C567, and more indirectly by complement-induced damage to the endothelium of small blood vessels, leading to the activation of the coagulation pathway by Hageman factor (factor XII). Factor XII is activated by exposure to collagen and this leads to the activation of subsequent stages in the coagulation system (Fig. 2.13).

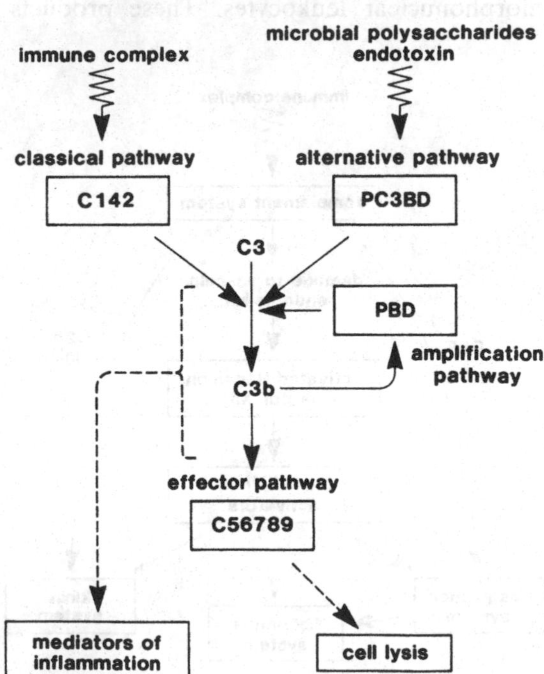

Fig. 2.14. Pathways of the complement system. C, complement components; B, factor B; D, factor D; P, properdin.

The kinin system

This is also known as the kallikrein system. It is initiated by the activation of factor XII and is completed eventually by the formation of kallikrein, which acts on an α-globulin substrate, kininogen, to form bradykinin. Bradykinin is a nonapeptide which produces marked and prolonged slow contractions of smooth muscle as well as dilation of peripheral arterioles and increased capillary permeability. The pathway of formation of bradykinin can be inhibited in at least three stages by C1 inactivator (C1 esterase inhibitor). Further involvement of this system in inflammatory effects comes from the chemotactic effect of kallikrein for polymorphonuclear leukocytes. Immunologic triggering of this system could occur via factor XII activation following immune injury to the vascular endothelium.

The fibrinolytic system

This system is also initiated by the activation of factor XII and then proceeds through intermediate stages to the formation of plasmin from its precursor plasminogen. Plasmin is a proteolytic enzyme of broad specificity which can digest not only fibrin but also fibrinogen, factor XIIa, clotting factors V and VIII, prothrombin, C1 inactivator, C1, C3 and C5. Clearly both factor XII and plasmin have several actions relevant to different humoral amplification systems.

In summary, these four systems are involved in several mechanisms which serve to amplify and control an initial small stimulus. They are particularly suited to modifying the vascular reaction and cellular events in immune as well as non-immune reactions in terms of inflammation, thrombosis and tissue necrosis, and in hemostasis and tissue repair.

Factors affecting the immune system and the induction of immune-mediated disease

Many factors may exert an influence on the immune response and consequently on the occurrence and/or severity of immune-mediated disease. Major constitutive influences include genetic composition, sex and age. Superimposed on these are a variety of external factors such as stress, nutrition and infectious disease. Of these, individual factors or combinations of these predominate in the etiology of each type of immune-mediated disease. Moreover, since several forms of immune-mediated disease may occur simultaneously in an individual, it follows that they must either have common predisposing factors, or that the development of one disease may predispose to the second. Figure 2.15 summarizes some of the important inter-

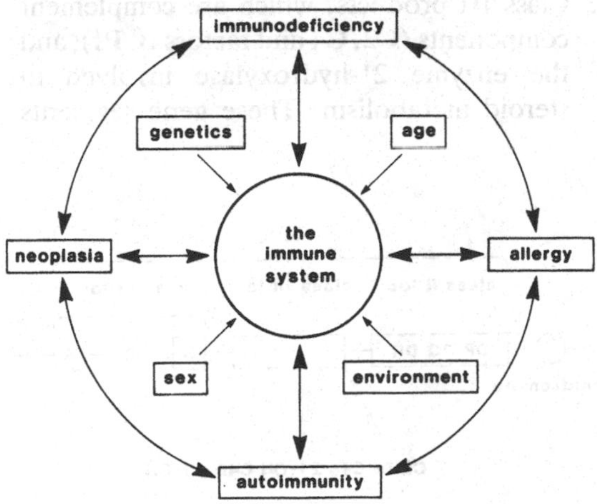

Fig. 2.15. Factors affecting the immune system and the induction of the immune-mediated response.

relationships which will be discussed in this section. Environmental factors giving rise to severe immunosuppression will be given further consideration in the section on Secondary immunodeficiency (p. 26).

Genetic factors

All immune function is ultimately genetically predetermined and, in the main, the genetic repertoire effectively covers all the responses required to counteract the hostile elements in the organism's environment. Nevertheless, certain genetic combinations may confer on the individual a subtle inability to respond to a particular infectious agent by leaving a 'hole' in the repertoire. Furthermore, the rare occurrence of grossly deleterious genes or gene deletions can lead to more drastic malfunctions of the immune system, manifesting as primary immunodeficiency.

There are several gene systems which are directly involved in determining the immune capability of the individual.

1 Genes which encode the variable regions of immunoglobulins; that is, the antigen-combining site of antibodies and B cell receptors.
2 Genes which similarly encode the variable region of the antigen receptor of the T cell.
3 Genes which encode the class I and II major histocompatibility antigens.

In addition, other genes may contribute less directly to immune competence, for example by influencing the function of antigen-processing and other accessory cells.

Since genes in the first two categories directly encode the specificity of the two forms of antigen receptors, their primary involvement is obvious and in the context of disease it may be envisaged that random genetic recombination, deletion and mutation may give rise to inappropriate receptor configurations which might be autoreactive or, conversely, fail to recognize a significant environmental antigen.

In the third category, major histocompatibility genes are grouped in the major histocompatibility complex (MHC) and are widely distributed on the surfaces of lymphoid and other cells. Moreover, MHC genes in particular exert a regulatory influence on immunologic reactivity and possess important disease associations, particularly with those of the immune-mediated type. This influence most likely arises from the requirement previously mentioned, that, in the case of T cells, antigenic determinants must be recognized in close association with MHC gene-encoded molecules on cell surfaces. In effect, MHC molecules determine whether presentation of an antigenic determinant will take place, since the association depends upon compatible charge and spatial configurations of the two molecules concerned. Thus not *all* antigenic determinants can be presented in the context of a given MHC molecule. Since class II molecules are involved in antigen presentation to T helper cells which cooperate with B cells, MHC class II genes will, for the reason given above, determine the immune response. In consequence, they have been called immune response (Ir) genes and their cell surface products, immune-associated (Ia) antigens. Because of this major contribution to immune reactivity and the well-documented disease associations, further description of this area is warranted here.

Histocompatibility and the immune response

This important area developed from early tissue-grafting studies which showed that tissue rejection was an immunologic mechanism involving the recognition of donor tissue graft antigens by the recipient's cytotoxic T cells. The antigens concerned are called histocompatibility antigens and the very high degree of polymorphism of these antigens accounts for the virtually total tissue incompatibility of non-related individuals. Subsequently, immune response studies in inbred strains of laboratory animals of a particular histocompatible type indicated that a major group of these antigens and their determinant genes also had an important physiologic role in the cell interactions that govern immune responses.

The major histocompatibility gene complex

The first MHC to be studied was that of the mouse. In this species, the complex, known as H-2, is located on chromosome 17. An indication of the significance of this locus is the finding that its arrangement is very similar in all mammalian species investigated. In man, this complex is located on chromosome 6 and has also been extensively investigated. By serologic and other laboratory techniques it has been possible to 'map' the disposition of the various loci on these chromosomes. Similar studies are now underway for most of the major species of domestic animals. The arrangement of the human MHC (HLA) is shown in Fig. 2.16. This complex contains a series of multiple-allelic genes which broadly encode products of three types:

1 Class I histocompatibility (transplantation) antigens found on all tissue cells apart from erythrocytes and encoded by A, B and C loci.
2 Class II histocompatibility (immune associated, Ia) antigens, which are restricted to cells of the immune system, principally macrophages, B cells and some T cells and encoded by loci DP, DQ and DR.
3 Class III products, which are complement components (C2, C4 and factors B, Bf), and the enzyme 21-hydroxylase involved in steroid metabolism. These gene segments

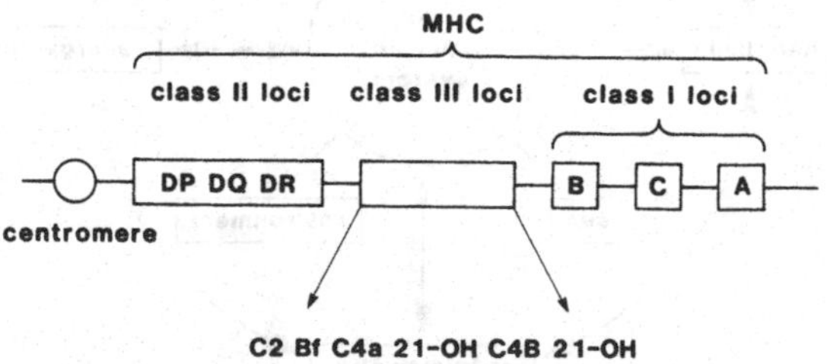

Fig. 2.16. Human major histocompatibility complex (MHC). Arrangement of loci on chromosome 6. For lettered loci, see the text.

are located between those for class I and II products. The reason for this juxtaposition is unknown.

Because of the multiple-allelic nature of the genes, many antigenic variants are possible (20–40 per locus). Furthermore, since the genes are co-dominant, the products of each class I or II locus are expressed on the cell surface. Thus, in man, each tissue cell will display up to six class I antigens and this accounts for the enormous number of possible combinations and hence tissue diversity.

Structure of class I and II antigens: Both class I and II antigens are transmembrane glycoproteins with intrachain disulfide bonding creating characteristic folding of the chains into 'domains'. These bear considerable homology to immunoglobulin domains in amino acid sequencing and it is apparent that, like the T receptor and other important surface molecules, they belong to the immunoglobulin supergene family presumably derived by the evolutionary duplication and diversification of a common gene (Fig. 2.3). Figure 2.3 shows that class I and II molecules, apart from differences in tissue distribution, also differ structurally in that class I molecules consist of a single chain of three domains with which the serum protein β_2M is non-covalently associated, whilst class II molecules consist of a pair of two-domain chains.

MHC disease associations: In man, statistical analysis has shown that susceptibility to certain diseases is associated with particular HLA antigens. Several broad groups of disease associations are recognized including: autoimmune diseases and diseases with a suspected autoimmune etiology, for example rheumatoid arthritis, autoimmune thyroiditis and juvenile diabetes mellitus; diseases of unknown etiology such as multiple sclerosis; non-immune diseases such as congenital adrenal hyperplasia; and infectious diseases such as leprosy.

These associations are important because they provide new insights and approaches to the investigation of the pathogenesis of particular diseases, and, in some instances, are of value in diagnosis. The reasons for these associations are currently unknown although several mechanisms have been postulated including:

1 The similarity of MHC determinants and those of infectious agents ('mimickry').
2 MHC antigens may act as receptors for microorganisms.
3 Particular MHC antigens may have differing effects on the efficiency of cellular recognition of antigen determinants of pathogens.
4 MHC genes may be in close linkage with 'disease susceptibility' genes within the MHC complex.

Influence of sex

In general, females of all mammalian species are known to be more responsive immunologically than their male counterparts. This difference is due to the influence of sex hormones, which have been shown markedly to affect the immune system at several points, although the precise mechanism(s) of action at the cellular level is unknown. Steroid sex hormones are known to affect the epithelial cells of the thymus and avian bursa, macrophages and lymphocytes. The principal cell affected appears to be the T lymphocyte and there is evidence to suggest that sex hormones can alter balances between T helper and suppressor cells. It is as a consequence of differential steroid hormone production in the respective sexes and their influence on lymphoid tissues that females are generally better responders than males. They are thus more resistant to infectious agents but conversely are more prone to the development of immune-mediated disease of the autoimmune type.

The effect of age

Immunologic responsiveness is known to vary considerably with age. For example, the neonatal and the aged tend to have poorer immunologic reactivity than young adults. In particular, the decline of suppressor cells with age is well documented and is likely to be an important contributor to the increasing incidence of aberrant immune responses and autoimmune disease in older age groups.

The influence of environmental factors

The nutritional status of the individual can strongly influence immune capability. Deficiency of proteins and essential vitamins, as in starvation, can profoundly depress cellular function and can consequently lead to reduced immune capability and eventually secondary immunodeficiency. At a more subtle level, diets rich in saturated fatty acids have been shown experimentally to increase the incidence of experimental autoimmune disease.

Many infectious agents are known to modulate immune function in one way or another. Some microorganisms, particularly a number of viruses, have immunosuppressive properties and this activity is particularly common in viruses with tropisms for lymphoid tissues. Where severe destruction of lymphoid tissues follows infection, general immunodeficiency may be the consequence. However, more subtle infections of these tissues may cause other effects. Thus, oncogenic viruses, by infecting particular lymphoid cell types, may partially subvert the immune system and cause aberrant responses including self-reactivity. At an even more subtle level, viral infection, particularly in the prenatal or neonatal period, may lead to specific tolerance induction to the antigens of the virus involved, but leave general responsiveness unimpaired. On the other hand, infectious agents or their products are capable of non-specifically *stimulating* lymphoid tissues and thus give rise to heightened immune responses with potential autoimmune consequences. This type of effect is particularly the property of endotoxins from Gram-negative bacteria and cell wall constituents of mycobacteria.

Finally, it is possible that microorganisms with cross-reactivity for self-components may specifically trigger immunologic responses to these components with autoimmune consequences.

The spectrum of immune-mediated disease

In the vast majority of animals, cells of the immune system, acting alone or in combination with the other defense mechanisms of the body can be expected effectively to combat or to limit disease. However, there are occasions when disease is enhanced or initiated by a over- or underreaction of the immune system. Such diseases are broadly referred to as **immune-mediated disease**. In man, there is a wide spectrum of well-documented examples of immune-mediated disease. Although much less is known about this subject in domestic animals, a comparable diversity is likely to occur.

Immune-mediated disease may be classified broadly into two major categories: immune hypoactivity and immune hyperactivity. A third category which may reflect either type of reactivity is the consequence of neoplasia of the immune cells. As already mentioned, these conditions may in some instances be linked (Fig. 2.17). For example, the commitment to the production of neoplastic lymphoid cells can render the individual immuno-

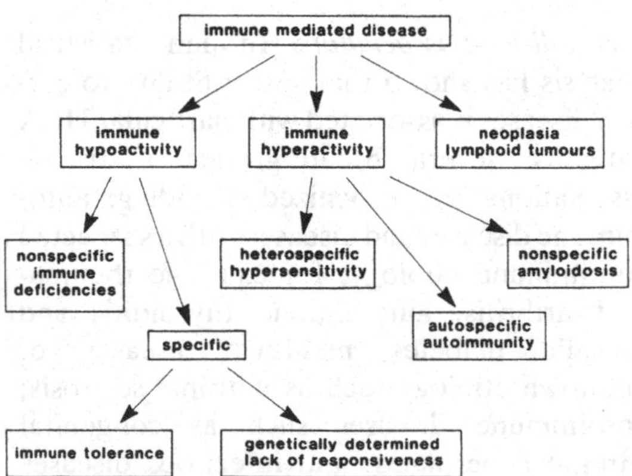

Fig. 2.17. The spectrum of immune-mediated disease.

deficient, and conversely, immune deficiency may lead to neoplasia.

The diagnosis of immune-mediated disease may present a considerable challenge for the following reasons: the diversity of disease types involving many organ systems; the chronic and often insidious nature of many of these conditions, with accompanying clinical signs that may be vague and difficult to define; the not infrequent concurrence of two or more of these conditions as a consequence of their interrelatedness and common predisposing factors; and the similarity of clinical signs and pathology of certain immune-mediated diseases with diseases of other etiology.

For these reasons there may be failure to appreciate that the condition observed represents the consequence of a primary aberration within the immune system. In view of these difficulties, immune-mediated disease should be suspected in the following circumstances.

- In all chronic disease of unknown origin, particularly those characterized by periods of remission and relapse.
- In all chronic disease restricted to a particular breed.
- When there are infections with unusual agents, such as commensal and normally non-pathogenic microorganisms.
- When repeated infections fail to respond to appropriate treatment.
- When individuals succumb to vaccination with live organisms.
- In infectious disease in the neonatal animal.
- When there is chronic leukopenia or leukocytosis.

Immune hypoactivity (failure)

Under this broad category may be grouped a variety of defects ranging from a gross deficiency resulting from generalized immune failure, to a subtle inability to respond to a particular antigen.

Immunodeficiencies

Immunodeficiency diseases are the consequence of a failure of one or more components of the immune system which generally result in reduced resistance to infectious agents and hence are usually manifest as infectious disease. Infections with particular microorganisms are, to a certain extent, characteristic of individual types of immunodeficiency. In addition to infectious disease, immunodeficiency may also underlie autoimmunity or neoplasia.

Immunodeficiency may arise as a primary impairment during the course of fetal development or as a secondary result of an environmental insult to some component(s) of the fully developed system in the mature animal. In consequence, primary immunodeficiency problems are generally observed in the neonatal and young animal, whilst secondary immunodeficiency may occur at any time, and is the more common.

Primary immunodeficiency disease

Primary failure of the immune system is an inherited or developmental defect which can occur at any of the maturational stages of the immune system and may give rise to characteristic clinical problems. The extent of failure depends on the stage of ontogeny at which the defect occurs. Figure 2.18 outlines the overall development pathways of the system and indicates points where defects have been identified, principally in man. For example, a defect occurring at the point of lymphoid precursor differentiation, 2 in Fig. 2.18, may lead to failure of both arms of the lymphoid system, with disastrous consequences, since both cell- and antibody-mediated responses will be affected. A defect that occurs in thymic development alone at point 3 will be reflected in an inability to mount a cell-mediated response. Similarly, a lesion restricted to the B cell system at point 4 will only affect antibody-mediated responses.

As a broad generalization, impairment of the humoral system alone leads to **enhanced susceptibility to Gram-negative and pyogenic bacterial infections**, whilst that of the cell-mediated arm enhances **susceptibility to intracellular pathogenic agents** such as viruses,

Table 2.3. *Humoral immunodeficiency*[a]

Disease	Species	Notes
Selective deficiency of IgM	Horse (all breeds) Dog	*Klebsiella* infection Doberman (unsubstantiated)
Selective IgG deficiency	Cow (Red Danish) Horse Chicken	Selective IgG2 deficiency 1 case recorded UCD 140 line of chicken
Selective IgA deficiency	Dog Chicken	German Shepherd dogs associated with chronic gastrointestinal tract infection (possible) Hypothyroid OS strain
Agammaglobulinemia	Horse	Total B cell deficiency 1 case, thoroughbred
Transient hypogamma-globulinemia	Horse	Delayed onset neonatal Ig synthesis
Failure of colostral transfer	All species	Neonate or dam fault
Dysgammaglobulinemia	UCD chicken	T_s deficiency: IgG (decreased), IgM, IgA (increased)

Ig, immunoglobulin; T_s, T suppressor cell.
[a] This refers to selective deficiency of one or more classes of immunoglobulin (B cell) and may be associated with variable T cell deficiency and infectious or autoimmune disease. Some patients may be clinically normal.

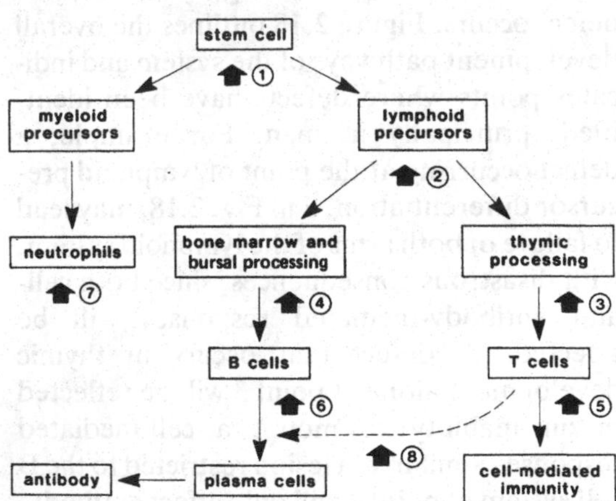

Fig. 2.18. Developmental pathways of the immune system and developmental blocks leading to immune failure. The defects are classified as (1) recticular dysgenesis, (2) severe combined immunodeficiency, (3) thymic aplasia, (4) agammaglobulinemia, (5) lymphokine deficiency, (6) deficiency in individual immunoglobulins, (7) neutrophil defects, (8) T suppressor defect (dysgammaglobulinemia).

some bacteria (e.g. *Mycobacterium*, *Brucella*) and fungi. In addition, defects may also occur in the development of the major accessory components which act in concert with the immune system, such as in production of the individual components of the complement sequence or cells of the phagocytic series. Tables 2.3 and 2.4 list many of the identified humoral and cell-mediated immunodeficiencies recognized in animals.

Genetic defects are known in many species for most of the complement proteins, all of which are inherited as autosomal recessive traits. Some of these are recorded in domestic animal species. The common manifestations associated with defects of early acting components (C1, C2, C4) are those of immune-complex disease, particularly non-organ specific types of autoimmunity. In man, late component (C5–C8) defects have been associated with recurrent Neisserial infection. C3

Table 2.4. *Cell-mediated immunodeficiency*

Disease	Species	Notes
Pneumocystis pneumonia	Dog (miniature Dachshund)	Australia *Pneumocystis carinii* (possible)
Demodecosis	Dog	Worldwide T cell dysfunction shown
Recurrent infections	Dog (Weimeraner)	USA and Australia Thymic atrophy
Lethal trait A-46	Black Pied Danish cattle	Scandinavia Defective Zn metabolism?
Athymic/thymic hypoplasia	Experimental 'nude' mice and rats Hairless/immunodeficient guinea pig	
Severe combined immune deficiency (SCID)	Arab horses	Worldwide Inherited Neonates Decreased IgM and CMI *in vitro* Decreased lymphoid tissue: nodes and thymus

Table 2.5. *Phagocytic defects*

Disease	Species	Defect
Cyclic neutropenia	Collie dog (Gray Collie syndrome)	Decreased neutrophil production Cyclic – 12 days
Chediak–Higashi syndrome	Hereford cattle Aleutian mink Persian cat White tiger Killer whale Man	Decreased chemotaxis Decreased intracellular killing
Irish Setter granulopathy	Dog	Decreased intracellular killing

deficiency leads to pyogenic infections, as also does C3b inactivator deficiency, since this defect causes C3 deficiency due to its excessive consumption. C1 inhibitor deficiency is associated with hereditary angioneurotic edema due to over activity of C1 and consequent liberation of C2b kinin fragments.

These clinical associations point out the importance of complement in the elimination and/or solubilization of immune complexes and also in bactericidal and opsinization effects.

Several classes of congenital deficiency syndromes associated with phagocytic failure have been reported in man, including failure in intracellular killing of bacteria, failure in opsonization, defective chemotaxis and defective phagocyte production. These defects render the affected individual highly susceptible to infectious agents, particularly pyogenic and Gram-negative bacteria. Those which have been reported in domestic animals are listed in Table 2.5.

The most subtle form of immunologic

'failure' is genetically determined lack of responsiveness to a particular infectious agent. The individual concerned has not inherited the necessary genetic programming to respond immunologically to a particular infectious agent. In consequence, such an individual is susceptible to this agent even though capable of mounting effective responses to other pathogens. This highly restricted hypoactivity is likely to account for a proportion of the individuals within a population who respond poorly to a particular vaccine or succumb during the course of an outbreak of infection. As indicated above (p. 20), a number of gene families contribute to the immune repertoire and could be involved in this phenomenon. The Ir genes of the MHC are likely to be most significant.

Secondary immunodeficiency

Common causes of secondary immunodeficiency include infectious agents, neoplasia, senility, drugs, nutritional status and failure of colostral transfer.

Infection with particular microorganisms is among the most important causes of secondary immunodeficiency and several viruses are particularly involved. These may bring about immunosuppression in several ways. Firstly, many lymphotropic viruses cause severe and widespread lymphoid destruction. In this category are the viruses causing canine distemper, feline panleukopenia, feline leukemia, African swine fever, bovine virus diarrhoea (BVD), equine herpes I and rinderpest.

Other viruses are less destructive but nevertheless may still be immunosuppressive by involving the primary lymphoid organs. For example, in mice, a herpesvirus infection can cause thymic atrophy and in chickens the virus of infectious bursal disease causes necrosis of the bursa of Fabricius. In both cases, lymphopenia and immunosuppression follow infection. In addition to causing extensive lymphoid damage, some viruses, for example BVD virus, may also exert generalized immunosuppressive effects through the stimulation of interferon production. Viruses may also impair immune function when they infect accessory cells such as neutrophils and macrophages and cause defective leukocyte degranulation and phagocytosis. Secondary bacterial invasion is a usual sequel to virus-induced immunosuppression.

Immunosuppression may also accompany infections with other agents including parasites such as *Demodex*, *Toxoplasma*, *Trypanosoma* and *Trichinella*.

Under particular conditions an individual may be rendered specifically hyporesponsive to a given antigen whilst retaining full immune competence to others. This phenomenon, called **immune tolerance**, can occur naturally during the course of certain infectious diseases and when it develops it renders the animal incapable of eliminating the agent concerned. This state is most frequently the result of virus infections which occur very early in life or *in utero*. For this reason, viruses transmitted vertically are most likely to induce tolerance, chronic infection and persistent viremia. Examples of such infections are feline leukemia in kittens and BVD in calves.

Secondary immunodeficiency in the neonatal animal may arise from failure to acquire maternal immunoglobulins. This is the most commonly occurring immunodeficiency problem of the domestic animal species and particularly affects calves, foals, piglets and lambs. These young animals depend upon the acquisition of maternal immunoglobulins, via colostrum, to tide them over the crucial neonatal stage until they are able to develop their own active immunity. Animals which fail to acquire sufficient maternal immunoglobulins are highly susceptible during the first week of life to septicemia and enteric infection with Gram-negative bacteria.

The most common occurrence is inadequate colostral intake. This can occur for a variety of reasons such as poor mothering qualities of the dam, weakness or physical defects in the offspring, or poor husbandry practices.

Immune hyperactivity

Immune-mediated disease caused by excessive immunologic activity can be grouped into three categories according to the specificity of the response. Thus in hypersensitivity the response is to external heterologous antigens, in autoimmune diseases to autologous antigens, whilst amyloidosis appears to result from non-specific over activity.

Hypersensitivity

Hypersensitivity reactions, also loosely called allergies, are essentially situations in which heterologous antigen (allergen) interacts with components of the immune system producing a reaction that is detrimental to the host. In some instances, a beneficial aspect can be identified, but this is outweighed by the adverse effects. As detailed previously, hypersensitivity reactions have been classified into four types on the basis of the mechanisms involved. Because it is possible for two or more of these mechanisms to be activated simultaneously, the etiology of numerous inflammatory lesions of this type is multifactorial.

Type I hypersensitivity (anaphylactic)

This type of hypersensitivity, also variously termed immediate hypersensitivity, atopy, allergy or anaphylaxis, is the most rapidly developing and dramatic of all the adverse immune reactions. Because it tends to cause irritation, discomfort, severe distress or even death, the underlying beneficial action in promoting rapid antigen removal is often overlooked at the clinical level.

Type I hypersensitivities are inflammatory reactions mediated by certain immunoglobulins, especially IgE but also some IgG subclasses. Because such antibodies are bound to mast cells and basophils, cross-linking of the immunoglobulin molecules with antigen leads to the rapid release of pharmacologically active substances (see Fig. 2.8 (*a*) and (*b*)).

It is not entirely clear under which conditions IgE is preferentially produced, but antigens that are well-known potential stimulators of this type of response include proteins of pollen grains, insect venoms, and some helminth antigens. An important factor is an hereditary predisposition to produce antibodies of this class, and this has been observed in many species including man and the dog. Those individuals, having an above average tendency to mount an IgE response, are said to be **atopic**. This tendency is thought to affect approximately 1–2% of the dog population of western countries. Inheritance of the trait is probably via a recessive gene and there is an apparent breed disposition involving Terriers, Dalmatians and Irish Setters.

The clinical manifestations of type I hypersensitivity relate to the release of vasoactive substances. The severity of the reaction depends on the number of mast cells stimulated, and is therefore a function of the dose of antigen. The location of the reaction relates to the sites of mast cell activation. The most severe form is systemic anaphylactic shock, in which a rapidly delivered intravenous dose of antigen triggers widespread mast cell degranulation, with potentially fatal results. The clinical signs of systemic anaphylaxis vary across the species, presumably because of differences in the distribution of mast cells, the types and quantities of mediators induced and the sensitivity of particular organs. In cattle, therefore, pulmonary signs predominate, whilst, in the dog, engorgement of the hepatic portal system is the major pathologic change.

Type I reactions of lesser severity are a much more common clinical problem and the major organ of involvement is the skin (see Chapter 11). The culpable antigens are either inhaled or ingested and are taken via the circulation to combine with mast-cell-bound antibodies in the dermis. The intradermal release of mast cell products initiates an intensely pruritic dermatitis. The reaction may also be triggered in the respiratory or alimentary mucosa to provoke sneezing and coughing or diarrhea, respectively. Table 2.6 lists some of the specific clinical conditions associated

Table 2.6. *Clinical type I hypersensitivities*

Condition	Allergen (if known)	Species
Generalized		
Systemic anaphylaxis	Insect venoms Penicillin, etc. Vaccines	All
'Milk allergy'	Casein (self-produced)	Dairy cow
Localized		
Atopic or allergic Rhinitis Inhalant dermatitis Asthma Urticaria	Pollens	Dog
Food allergy	Protein-rich foods	
	Milk, meat, wheat, milk, fish, eggs	Dog
	Fish meal, alfalfa	Pig
	Oats, clover, alfalfa	Horse
Chronic obstructive pulmonary disease	Spores	Horse

with type I hypersensitivity in a number of animal species.

Type II hypersensitivity (cytotoxic)

This type of reaction is termed cytotoxic because antibody binding to cell surfaces initiates cellular destruction. The latter is accomplished by several mechanisms including complement activation, phagocytosis or K cell-mediated lysis (see Fig. 2.10).

Type II hypersensitivity has been implicated in a range of pathologic conditions (Table 2.7), the most notable of which is isoimmune hemolytic anemia of neonates (see Chapter 3). This condition is seen in neonates which have received preformed antibody to red cells, via their maternal colostrum. It occurs most commonly in multiparous horses, occasionally in cattle and pigs and rarely in small animals. The generation of anti-red-cell antibody requires exposure of the dam to foreign red cell antigens of the same group as the fetus. In the mare, this appears to happen spontaneously as a result of the transplacental leakage of fetal red cells, but sensitization may also be induced inadvertently by the administration of vaccines or blood transfusions.

Antigens or haptens which modify cell surfaces *in vivo* are capable of initiating type II reactions against the cells concerned. This may, in consequence, result in the elimination of the antigen. In this way, for example, aspirin and its derivatives may cause hemolytic anemia, and sulfonamides may induce agranulocytosis. A similar mechanism can account for the destruction of circulating cells in infectious diseases in which modifying antigens are shed by microorganisms. This is exemplified by *Salmonella* infections in which precocious destruction of red blood cells is a feature.

Finally, this mechanism is the basis of the damage wrought in certain forms of autoimmune disease such as autoimmune hemolytic anemia.

Type III hypersensitivity (immune complex)

Formation of immune complexes *in vivo*, through antigen–antibody combination, can lead to a sequence of pathologically significant

Table 2.7. *Clinical type II hypersensitivities*

Clinical type II hypersensitivities
1 Isoimmune reactions
Neonatal hemolytic anemia
Incompatible blood transfusion
2 Drug-induced cytotoxicty
Red cell modifying
Penicillin
Quinine
L-Dopa[a]
Aminosalicylic acid
Phenacetin
Granulocyte modifying
Sulfonamides
Phenylbutazone
Aminopyrine
Phenathiazine
Chloramphenicol
Platelet modifying
Phenylbutazone
Quinine
Apronalide
Sulfonamides
Chloramphenicol
3 Infectious diseases
Salmonellosis (particularly avian)
Equine infectious anemia virus
Aleutian disease virus
Anaplasmosis
Trypanosomiasis
4 Autoimmune diseases
Autoimmune hemolytic anemia
Idiopathic thrombocytopenia
Automimmune glomerulonephritis (Goodpasture's syndrome)

[a]3,4-Dihydroxy-L-phenylalanine.

events under particular circumstances. One of the most important of these is activation of the complement cascade. When complement-bound immune complexes are deposited within tissues, the subsequent generation of chemotactic factors leads to a local accumulation of neutrophils which discharge their hydrolytic enzymes to cause local inflammation and tissue destruction (Fig. 2.19). The extent of tissue destruction and the severity of the condition depend on the quantity of complex generated and its sites of deposition, and it would generally appear that large quantities must be deposited before clinically significant disease occurs.

Two main types of immune complex disease are recognized: local and systemic. In the former case, complexes are formed within localized tissue sites after large quantities of antigen are introduced directly at the sites. In systemic reactions, disease is produced when antigen gains access to the circulation. Immune complexes form within the circulation and are carried with it to lodge at vulnerable sites, most notably within the walls of glomerular capillaries (see Chapter 9). In either case, prior sensitization with antigen must have occurred for immune complexes to be formed or, alternatively, antigen must persist, unsequestered by phagocytic cells, until antibody is generated.

The archetype of local type III reactions is the Arthus reaction, which occurs when antigen is injected subcutaneously into an animal that possesses circulating homologous antibody of the precipitating type. This reaction commences as an erythematous swelling and proceeds to local hemorrhage and thrombosis, culminating in necrosis. Maximum intensity is reached by 6–8 hours, and, histologically, at this time the damaged blood vessels are densely infiltrated with neutrophils.

A similar local reaction is seen in a number

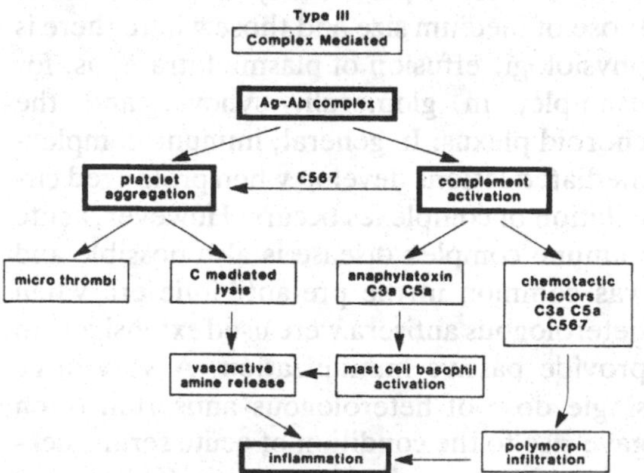

Fig. 2.19. Immune complex formation results in a number of deleterious effects including the formation of microthrombi, increased vascular permeability and injury or destruction of cells. C, complement or its components; Ag, antigen; Ab, antibody.

of natural disease conditions. For example hypersensitive or allergic pneumonitis is an acute alveolitis and vasculitis, with exudation of fluid into the alveolar spaces. It is seen in the lungs of cattle housed during the winter and exposed to high dust levels from moldy hay. This condition is analogous to farmer's lung of man and is caused by hypersensitivity to the inhaled spores of the mold *Micropolyspora faeni*, which are generated in profusion in the hay under appropriate conditions.

Chronic obstructive pulmonary disease in horses ('heaves') is a somewhat similar hypersensitivity pneumonitis, although with a more complex etiology, as type I reactions are probably also involved. The offending antigens are also likely to be derived from fungal spores in moldy hay (see Chapter 5).

Staphylococcal hypersensitivity in dogs is a chronic pruritic, multifocal dermatitis induced by type III reactions to bacterial products (see Chapter 11). The animals show evidence of hypersensitivity by skin testing and have a characteristic neutrophilic dermal vasculitis.

Generalized type III hypersensitivity is likely to occur under conditions of antigen excess when circulating immune complexes are soluble and hence poorly phagocytosed. These complexes may become deposited in blood vessel walls under certain circumstances. Vessels particularly involved include those of medium size and those where there is physiologic effusion of plasma filtrate, as, for example, in glomeruli, synovia and the choroid plexus. In general, immune complex-mediated lesions develop when prolonged circulation of complexes occurs. However, acute immune complex disease is also possible and was common in the pre-antibiotic era when heterologous antisera were used extensively to provide passive immunization. A very large single dose of heterologous antiserum often gave rise to the condition of acute serum sickness, with generalized vasculitis. This causes erythema, edema and urticaria of the skin, neutropenia, lymph node enlargement, joint swelling and proteinuria. The latter was the consequence of immune complex deposition within glomeruli. Fortunately, these effects were usually of short duration and subsided within a few days.

Prolonged systemic exposure to antigen may lead to chronic type III hypersensitivity, with more serious consequences. The primary site of injury in this instance is the kidney, where continued deposition of immune complexes may lead to glomerular disease (see Chapter 9). By the use of appropriate techniques, the aggregates of immune complex can be demonstrated within glomerular capillary walls or in the mesangium (Fig. 2.20).

Since immune complex-mediated lesions occur when prolonged antigenemia persists in the presence of antibodies, glomerulonephritis is a characteristic of a number of chronic infectious diseases. Table 2.8 lists conditions in which this mechanism is considered to play an important role.

Finally, it should be pointed out that in many cases of glomerulonephritis and arteritis the antigens responsible for the formation of immune complexes are unknown. It is most likely that these diseases involve a range of antigens which could be viral, bacterial or autologous constituents.

Type IV hypersensitivity

As detailed previously, certain antigens when deposited in tissue sites provoke cellular rather than antibody-mediated responses. Since the cellular responses may take many hours to develop, they are referred to as

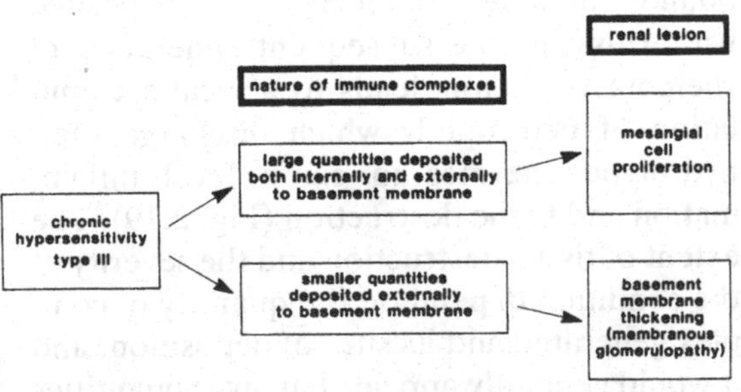

Fig. 2.20. The renal effects of chronic type III hypersensitivity.

Table 2.8. *Diseases involving type III hypersensitivity*

Agent or disease	Pathology
Infectious	
Erysipelas	Arthritis
Leptospirosis	Opthalmia
Streptococcus equi	Purpura
Staphylococcus aureus	Dermatitis
Canine adenovirus I	Uveitis, GN[a]
Feline leukemia	GN
Feline infectious peritonitis	Peritonitis, GN
Aleutian disease	GN arthritis
Hog cholera	GN
Bovine viral diarrhea	GN
Equine viral arteritis	Arteritis
Equine infectious anemia	Anemia, GN
Dirofilariasis	GN
Autoimmune disease	
Systemic lupus erythematosis	GN dermatitis arthritis
Rheumatoid arthritis	Arthritis, arteritis
Others	
Pyometra	GN
Chronic pneumonia	GN
Bacterial endocarditis	GN
Acute pancreatic necrosis	GN
Malignant tumors	GN
Finnish landrace glomerulonephropathy	GN, choroid plexus
Polyarteritis nodosa	Arteritis (particularly renal and ophthalmic arteries)

[a]GN, glomerulonephropathy.

delayed hypersensitivity reactions. The classical reaction of this type is the response to intradermally injected tuberculin in the animal infected with *Mycobacterium tuberculosis*. Hypersensitivity is also largely responsible for the chronically progressive lesion known as the tubercle, which develops during the course of tuberculosis. Owing to the persistence within tissues of intracellular mycobacteria, whose cell walls contain large quantities of poorly metabolized waxes, a chronic form of delayed hypersensitivity is generated. In consequence, large numbers of macrophages accumulate in the lesions, many of which die attempting to ingest invading bacteria, whilst others fuse to form multinucleated giant cells. The developing lesion thus consists of a mass of necrotic material containing both living and dead microorganisms surrounded by a layer of macrophages (in this situation called epithelioid cells) and some giant cells. Persistent tubercules become relatively well-organized granulomas and develop a fibrous tissue wall. Collagen formation by fibroblasts in this situation is thought to be T-cell-mediated via lymphokines.

A further example of a type IV reaction is **allergic contact dermatitis** (see also Chapter 11). This condition arises as the result of the absorption into the epidermis of certain chemicals which have the ability to bind to epidermal proteins. Once bound to the carrier protein the chemical acts as a hapten and the fully immunogenic complex stimulates a cell-mediated immune reaction. The chemicals that induce allergic contact dermatitis are usually relatively simple; they include such compounds as formaldehyde, picric acid, aniline dyes, plant resins, organophosphates and even salts of metals such as nickel, chromium and beryllium. The resulting lesions may vary greatly in severity, ranging from a mild erythema to severe erythematous vesiculation.

Yet another illustration of a type IV reaction is arthropod hypersensitivity. The feeding habits of some parasitic arthropods effectively promote the establishment of hypersensitivity. Most such reactions occurring in the domestic species tend to be complex with a strong type I component, but may also include type IV reactivity. For example, the reaction which develops around the hair follicles in Demodectic mange may be type IV hypersensitivity, since it is mononuclear in character and may proceed to granuloma formation. There is also evidence to suggest that animals with the generalized demodicosis condition are immunosuppressed. Similarly, in flea bite dermatitis intense mononuclear infiltrations occur in the vicinity of bites, and sensitized animals also respond to provocative skin tests. In this instance, low molecular weight compounds of flea saliva can act as haptens by binding to dermal collagen.

Autoimmunity

As the term implies, autoimmunity is the occurrence of immunologic reactivity to autologous constituents and can be viewed as a breakdown in self-tolerance. This aberrant response to self may lead to the activation of one, or possibly more, of the tissue-damaging effector pathways and eventually to the appearance of clinical disease. Diseases of this type, with a common underlying basis of immunologic self-reactivity, are called autoimmune diseases. Owing to the large number and wide distribution of body constituents which are potentially autoantigenic, a broad range of such diseases is possible, involving virtually all organ systems and, in consequence, autoimmune disease parallels infectious disease in its diversity.

The mechanism of self-tolerance and its breakdown

It is now clear that for many body constituents, immunologic self-tolerance is maintained by an active regulatory mechanism which is responsible for inhibiting the activation of pre-existing potentially autoreactive T/B lymphocytes. Although the detail is far from clarified, suppressor T lymphocytes are believed to play a key role in this process. Conversely, breakdown in self-tolerance may be caused by the bypassing of suppressor down-regulation (Fig. 2.21). This is thought to occur in two ways; firstly by loss and/or reduced function of T suppressor cells (Fig. 2.22), or secondly by the presentation of the self-antigen in modified form which can be recognized as 'foreign' by other non-suppressed T inducer cells, and a self-response initiated accordingly (Fig. 2.23). In these regards, suppressor cell function is known to decline generally with age and may also be affected by stress and infectious agents. The latter may also play an important part by producing cross-reactive antigen or modifying body components during the course of infection (particularly viral infection). In these ways, infectious agents may act as important triggers of autoimmunity which, once initiated, may be self-sustaining in the absence of effective immune regulation.

One other possible cause of autoimmunity is the release of sequestered antigens either by traumatic injury, or by tissue injury during infection. Since these antigens are not normally exposed to immunocompetent cells, there may be absence of immunoregulation

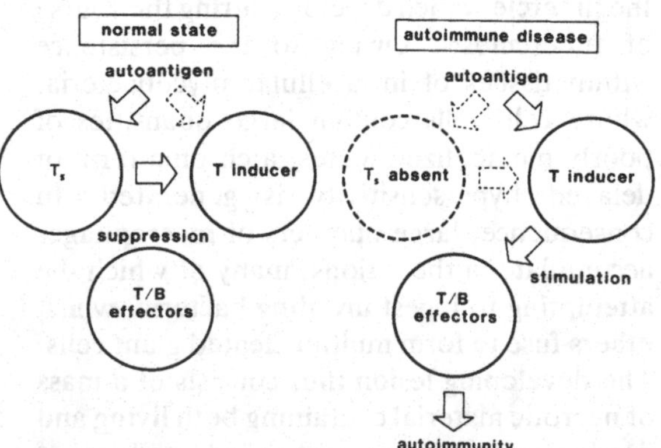

Fig. 2.21. The mechanism of self-tolerance and its breakdown in autoimmunity. T_s, T suppressor cell.

and an immune response may be initiated. An important example of this mechanism may be autoimmunity to sperm.

Pathogenetic mechanisms in autoimmunity

The immunologic mechanisms causing inflammation and tissue destruction during the course of autoimmune disease are thought to be essentially those involved in hypersensitivity reactions, except that in these situations the antigens involved are autologous constituents. In some types of disease, the pathogenetic mechanisms are well defined, but in others there is still controversy concerning the respective importance of cellular as opposed to humoral mechanisms. In addition, in certain types of autoimmunity, a unique immunologic injury occurs which involves the generation of auto-antibodies to the physiologically sensitive receptor molecules on tissue cells. This results in characteristic malfunctions, but without structural tissue changes. In this sense the latter will be referred to as anti-receptor autoimmune diseases.

Autoimmune disease involving type I hypersensitivity is rare. It is, however, seen in the condition of milk allergy of cattle in which casein, normally only found in the udder, gains access to the general circulation. This stimulates an immune response which is largely of the IgE type. If milking is delayed, casein may gain access to the circulation in sufficient concentration to cause clinical signs of acute systemic anaphylaxis. This condition is also seen occasionally in other domestic animals such as the mare.

Autoimmune reactions involving type II hypersensitivity are seen in a number of conditions in which cytotoxic antibodies are generated to cell membrane components. In consequence, the cells involved are destroyed by one or more of the several cytotoxic effector pathways such as antibody-mediated cytolysis or phagocytosis. Examples of diseases in which this mechanism is involved are autoimmune hemolytic anemia and idiopathic thrombocytopenia (see Chapter 3).

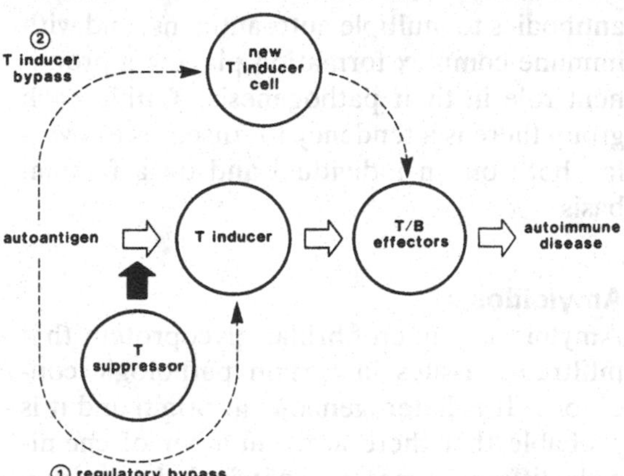

Fig. 2.22. Loss of self-tolerance with declining T suppressor function (regulatory bypass).

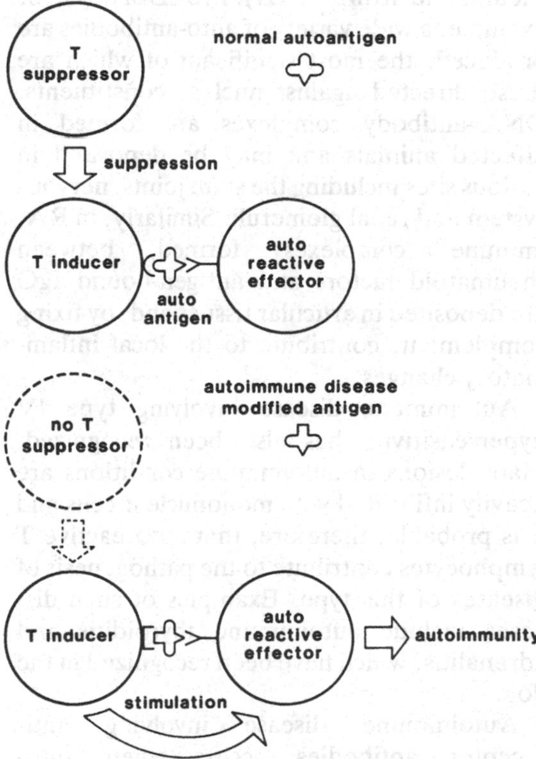

Fig. 2.23. Tolerance breakdown by auto-antigen modification (T inducer bypass).

Table 2.9. *Classification of autoimmune disease*

Organ specific type	Systemic type
Thyroiditis	Systemic lupus erythematosis
Hypoparathyroidism	Rheumatoid arthritis
Pernicious anemia	Sjogrens syndrome (Sicca)
Addison's disease	Dermatomyositis
Addison's-like disease	Idiopathic thrombocytopenia
Diabetes mellitus (some forms)	Hemolytic anemia
Autoimmune oophoritis	Autoimmune leukopenia

Since autoimmune disease involving type III hypersensitivity is chronic in nature, the autoantigen(s) concerned are constantly available. They provide appropriate conditions for the formation and persistence of pathologic immune complexes. Thus type III hypersensitivity reactions occur in a number of autoimmune diseases and are particularly important in non-organ-specific conditions such as systemic lupus erythematosus (SLE) and rheumatoid arthritis (RA). In SLE of dogs, for example, a wide variety of auto-antibodies are produced, the most significant of which are those directed against nuclear constituents. DNA–antibody complexes are formed in affected animals and may be deposited in various sites including the skin, joints, nervous system and renal glomeruli. Similarly, in RA, immune complexes formed between rheumatoid factors and antigen-bound IgG are deposited in articular tissues and, by fixing complement, contribute to the local inflammatory changes.

Autoimmune disease involving type IV hypersensitivity has also been recognized. Many lesions in autoimmune conditions are heavily infiltrated with mononuclear cells and it is probable, therefore, that autoreactive T lymphocytes contribute to the pathogenesis of diseases of this type. Examples of such diseases include autoimmune thyroiditis and adrenalitis, which have been recognized in the dog.

Autoimmune disease involving anti-receptor antibodies occurs when auto-antibodies are generated and bind to receptor molecules for physiologic mediators on cell surfaces, thus impeding their function. The bound antibody either blocks the cell's physiologic response to the mediator, or, alternatively, acts as an unregulated cell signal. The former effect is seen in myasthenia gravis in dogs and man, in which auto-antibodies to the acetylcholine receptor of the muscle end plate block nervous transmission at the neuromuscular junctions. The clinical consequence is rapid fatigue and profound muscular weakness (see also Chapter 13).

Autoimmune diseases of animals

This diverse group of immune-mediated diseases can be divided into two broad groups on the basis of similarities in pathogenesis and occurrence (Table 2.9). Organ-specific diseases are confined to one organ and only a limited range of auto-antibody specificities are observed. In contrast, the second group are systemically distributed, with auto-antibodies to multiple auto-antigens, and with immune complex formation playing a prominent role in their pathogenesis. Within each group there is a tendency for diseases to overlap both on an individual and on a familial basis.

Amyloidosis

Amyloid is a microfibrillar glycoprotein that infiltrates tissues in certain pathologic conditions. It is heterogeneous in origin and it is probable that there are a number of chemically different types, two of which derive from hyperactivity of the immune system. It may be classified as immunocytic, if it is associated with a myeloma or other lymphoid tumor. Immunocytic amyloid is a proteolytic digestive product of immunoglobulin light chains. Amyloid is classified as reactive, if it is associated with chronic suppurative conditions such as mastitis, osteomyelitis, abscessation, traumatic pericarditis or tuberculosis. The amyloid of the reactive form is not immunoglobulin but is probably derived from the proteolytic digestion of a serum globulin derived from the liver, which is deposited in close association with plasma cells. Reactive amyloidosis is a major problem in animals undergoing hyperimmunization for commercial antiserum production. It is also commonly associated with autoimmune disorders.

All forms of amyloid have a similar basic chemical structure which consists of protein fibrils having their polypeptide chains arranged in the form known as β pleated sheets. This is a uniquely stable molecular conformation that renders the material extremely insoluble and resistant to proteolytic digestion. Amyloid accumulation is therefore irreversible, leading to gradual tissue atrophy and loss of functions.

Tumors of the immune system

Neoplastic transformation may occur in lymphoid cells of both branches of the immune system at almost any stage in their maturation process. Such cells remain thereafter in arrested maturation and form a clone which proliferates in an uncontrolled manner, unresponsive to normal regulatory mechanisms. Eventually, they may swamp the organism, displacing normal cells spatially and functionally. Lymphoid tumors are both the consequence and cause of immunosuppression and, as a broad generalization, tumors of T cell lineage interfere with the cell-mediated immune system and B cell tumors interfere with the humoral arm of the immune system. In consequence, affected individuals are predisposed to a variety of infectious diseases particularly of the respiratory and alimentary systems. Additionally, lymphoid neoplasia may occur in association with certain autoimmune manifestations. For example, autoimmune hemolytic anemia not infrequently accompanies leukemia. Similarly, myasthenia gravis and thymoma are also associated. These associations are probably indicative of interference with normal immunoregulatory function and in this regard there is evidence that some of the immunosuppressive effects of lymphoid neoplasia arise from misdirected suppressor activity.

Classification of lymphoid tumors

Neoplasms of the lymphoid system have traditionally been grouped into two classes: **leukemias** and **lymphomas**. In leukemia, neoplastic leukocytes predominate in the circulation, whereas lymphomas are confined largely to lymphoid organs. Several forms of these two major divisions are distinguished by the type of cell from which the clone arose, the properties of the neoplastic cells and the clinical signs accompanying the disease (see also Chapter 3).

These various subtypes are identified by their possession of different cell surface and enzyme markers. In particular, the presence of cell surface immunoglobulin is considered to be characteristic of B cells and the presence of a receptor molecule (now called CD2) responsible for the formation of rosettes with sheep red cells is an identifying feature of T cells. Other cell membrane markers which are utilized to identify human lymphoid tumors include: cALL, an antigen common to all acute lymphoblastic leukemias, and DR antigens of human class II MHC (Ia antigens; see p. 20). The presumed differentiation state of various lymphoid tumors designated by these means are depicted in Fig. 2.24. Canine and other animal tumors of lymphoid origin can also be identified presently to a limited extent by using anti-immunoglobulin (B cell) or T cell antisera, but more precise characterization employing antisera to other cell surface markers will undoubtedly soon follow. In contrast to the medical field, many of these animal tumors are now known in the domestic species to be caused by viral infections of particular

Table 2.10. *Lymphoid tumors in domestic animal species*

Tumor	Agent (where known)	Lymphoid cell involved	Immune aberrations
Feline leukemia	Feline leukemia virus (FeLV)	T	Lymphopenia Response to mitogens Infections Allograft rejection
Marek's disease	Marek's disease virus	T (immature)	Response to mitogens CM cytotoxicity IgG Infections
Avian leukosis (ALL)	ALL virus	B (immature)	Infection
Bovine leukosis (BL)	BL virus	B	IgM Infections
Canine lymphosarcoma	—	B	Infections Autoimmunity
Myeloma	—	B (plasma cell)	Infections Myeloma Ig, other Igs

Ig, immunoglobulin; CM,

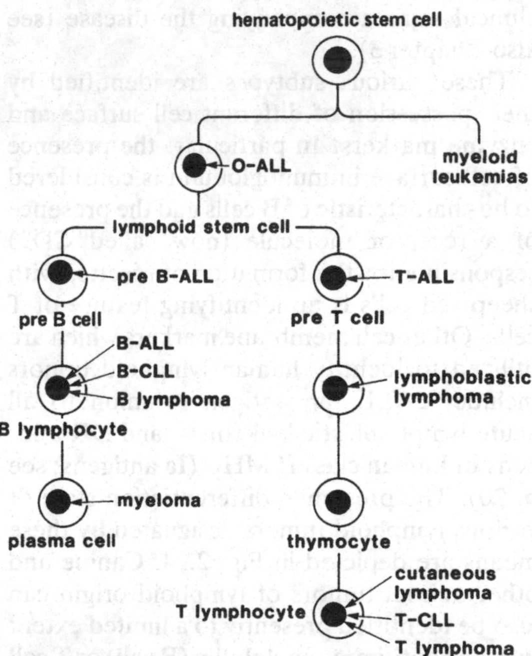

Fig. 2.24. Neoplastic transformation of lymphoid cells may occur at several points along the differentiation pathway and gives rise to different types of disease. 0-ALL, null cell ALL.

target cell stages in lymphoid differentiation. Table 2.10 summarizes some of the more important lymphoid tumors of the domestic species and their consequences.

Myelomas are a group of lymphoid neoplasms resulting from malignant transformation at the plasma cell stage of B cell differentiation. This gives rise to the development of a clone of cells which is actively secreting a single molecular type of immunoglobulin protein. Plasma cell tumors are known as myelomas or plasmacytomas and their product as a myeloma protein or M protein. The latter can belong to any immunoglobulin class and in some cases only particular fragments of the immunoglobulin molecule are secreted, for example light chains (Bence-Jones proteins) or the Fc portion of the heavy chain ('heavy chain disease'). Since such cells are not subject to control, their rapid proliferation and active M protein secretion leads to an increasing concentration of their product in the plasma which can be readily detected by characteristic changes in the serum protein profile in electrophoresis.

These conditions have been reported in dogs, cats, horses, cattle, pigs and rabbits. As well as profound immunosuppression, other clinical manifestations include hyperviscosity (particularly with IgM-producing myelomas – macroglobulinemia), renal failure (due to the toxic effect of Ig light chains on renal tubules) and hemorrhage (as a consequence of the binding of clotting components to M proteins).

Additional reading

Barta, O. (1981). Laboratory techniques of veterinary clinical immunology: review. *Comp. Immun. Microbiol. Infect. Dis.* **4**: 131–60.

Fudenberg, M. M., Stites, D. P., Caldwell, J. L. and Wells, J. V. (eds.) (1980). *Basic and Clinical Immunology*. Los Altos, CA, Lange Medical Publications.

Klein, J. (1982). *Immunology: The Science of Self-Non-self Discrimination*. New York, John Wiley and Sons.

Perryman, L. E. and Magnuson, N. S. (1982). Immunodeficiency disease in animals. In *Animal Models of Inherited Metabolic Diseases*, pp. 271–86. New York, Alan R. Liss Inc.

Ryder, L. P. and Svejgaard, A. (1981). Genetics of HLA disease association. *Ann. Rev. Genet.* **15**: 169–92.

Schultz, R. D. and Adams, L. S. (1978). Immunologic methods for the detection of humoral and cellular immunity. *Vet. Clin. N. Am.* **8**: 721–53.

Tizzard, I. (1981). *An Introduction to Veterinary Immunology*. Philadelphia, W. B. Saunders Co.

Unanue, E. R. and Benacera, B. (1983). *Textbook of Immunology*. Baltimore, Williams and Wilkins.

Jennifer N. Mills and V. E. O. Valli

3 The hematopoietic system

The hematopoietic system is composed of a remarkable variety of cells. Included are those circulating in the blood and their ancestors in marrow and progeny in the tissues. Also included are cells whose function is to remove both senescent cells from the bloodstream and any foreign material, especially microorganisms that may gain entrance to the body. After birth, the major location for hematopoiesis is the bone marrow, and in the newborn animal all medullary cavities of the skeleton are given over to this purpose. As the demands of body growth subside, hematopoiesis normally retreats to the metaphyses of long bones and to the flat bones of the pelvis, ribs, calvarium and vertebrae. From here, it may re-expand if need be, both into the bony cavities and even into extraskeletal sites such as the liver, spleen and lymph nodes.

Hematopoietic stem cells are extravascular colonists in bone marrow and here they proliferate, differentiate and mature, being finally released as appropriately developed progeny into the circulation. These end cells have acquired sufficient membrane plasticity and movement to penetrate sinusoidal endothelium and leave the marrow.

The hematopoietic cell system consists of a hierarchy in which the progenitor stem cells are capable of unlimited self-renewal and multilineal differentiation, giving rise to all blood cell types via committed precursor cells (Fig. 3.1). Committed precursor cells have a limited capacity for self-renewal and differentiation. The control of stem cell differentiation is little understood but is apparently initiated by the interaction of helper and suppressor lymphocytes and modulated by hormones specific for each cell line (Table 3.1). Any increase in the rate of blood cell production is dependent on an *increase* in stem cell input rather than an increased rate of cell division in committed cell lines. The process is analogous to the setting up of additional assembly lines to increase the production of motor cars. This necessarily means that a delay occurs between the sudden need for increased production and an increased output of cells. This problem is partly solved by reserve mechanisms for an immediate response, which are available for the red cell and granulocyte systems.

The final cell lines are diverse in nature and include red cells for gas transport, neutrophils and macrophages for phagocytosis, platelets for hemostasis, lymphocytes for antibody production and eosinophils and basophils for involvement in immune reactions. The production of individual cell lines may be specifically enhanced and, in some cases, the intensity of demand may have such priority that the production of other lines may suffer significantly. In addition, there is a close relationship between the production of red cells and platelets so that demand for one usually stimulates increased production in both. A similar relationship exists between neutrophil and monocyte cell lines.

The foregoing reveals the hematopoietic

tissue as a dynamic cell-renewal system with considerable reserve and adaptive powers. Inadequate marrow function is incompatible with basic health and untreated marrow failure means certain death. Failure of marrow function may be total or it may be selective to one or more cell lines. It may be congenital, or acquired at any time during life and may be reversible or irreversible. Its causes are many and varied, as are its manifestations.

The hematopoietic system is clinically monitored by examination of the circulating blood and the bone marrow. Variation from the normal state may indicate an adaptive response secondary to a disorder in some other system, or a primary disease of the hematopoietic system itself. Hematologic examinations are a routine part of many clinical investigations and this fact is a reminder that the bloodstream permeates the whole body and is inevitably caught up in the trials and tribulations of its every corner. To simplify the discussion, the hematopoietic system is conveniently divisible into three components: the erythron, the leukon and the thrombon (hemostasis).

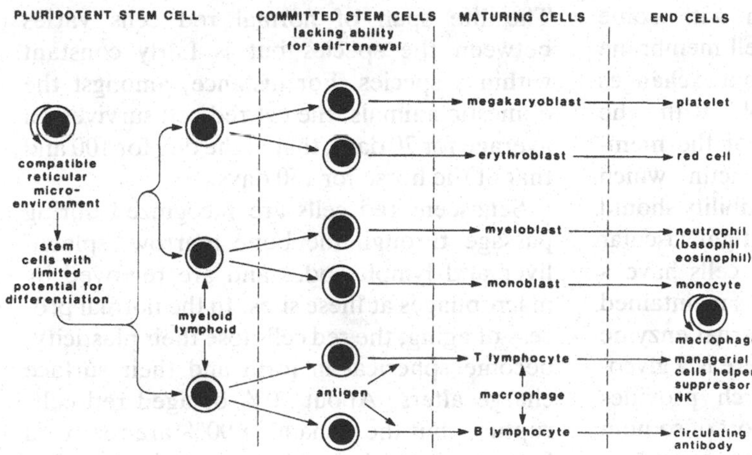

Fig. 3.1. The differentiation of hematopoietic stem cells. NK, natural killer cells.

The erythron

The essential physiologic task of the erythron is to provide a total body transport system for hemoglobin and thus for oxygen. For any species and class of animal there is a consistent concentration of hemoglobin per volume of plasma and it is carried by a consistent number of erythrocytes, but the red cell population is dynamic, with a constant exit of senescent cells balanced by an equal input of new cells from the marrow. For example, an average-sized dog requires the production of one million new red cells per second to maintain homeostasis.

The marrow and the peripheral blood constitute an entity but it is convenient to consider them separately for ease of understanding. Within the marrow, primitive progenitor cells undergo several generations of replication and differentiation, at the end of which an almost mature progeny is released to the circulation. The erythron is evaluated in health and disease by examination of both the ‘end cells’ of the blood and the ‘beginning cells’ of the marrow.

Table 3.1. *Hematopoietic regulators*

Target cells	Stimulator	Suppressor
Committed erythroid stem cells (and erythroid series of cells)	Erythropoietin (mol. wt 40 000)	Erythropoietin inhibition factor (EIF) ≡ $PGF_2\alpha$ (mol. wt 5000)
	Erythropoietin modulators	Tumor necrosis factor
	Androgens	
	Thyroid hormones	Interleukin I
	Adrenal hormones	
	PGE_1 and PGE_2	Estrogen
Committed myeloid stem cell	Serum colony-stimulating factor (CSF) (mol. wt 45 000)	Chalone ≡ PGE (mol. wt 35 000) Lipoprotein
Developing granulocyte	Differentiation factor	
Pro-megakaryoblast	Thrombopoietin (mol. wt 60 000–70 000) Interleukin 3	
Pluripotential stem cells (and possibly all hemopoietic cell lines)	Interleukin 3	
Pro-lymphoid cells (T cells)	Interleukin 2	

PG, prostaglandin.

The circulating red cells

Metabolic features

The highly specialized mammalian red cell, devoid of nucleus, mitochondria and ribosomes, is given over absolutely to its task of gas transport. It is a membrane-bound package of heme and globin with some ancillary enzyme systems. Its cell membrane contains stable protein antigens but exchanges lipids, particularly cholesterol, with the plasma. The inner framework of the membrane consists of spectrin and actin, which allows for deformability and flexibility should the cell need to squeeze through tight vascular spaces. In general, mature red cells have a biconcave disk structure, which is maintained until senescence. The major internal enzyme systems are four in number. There is a glycolytic respiratory pathway, which provides energy for the membrane transport of sodium and potassium and the maintenance of cellular tonicity. A hexose monophosphate shunt system protects globin against oxidant attack and a methemoglobin reductase system maintains heme iron in the ferrous state essential for oxygen transport. Finally a 2,3-diphosphoglycerate system influences the affinity of heme–oxygen binding.

Red cell life span

The life span of normal red cells varies between the species but is fairly constant within a species. For instance, amongst the domestic animals, the cat red cell survives on average for 70 days, that of the dog for 100 and that of the horse for 150 days.

Senescent red cells are recognized during passage through the bone marrow, spleen, liver and lymph nodes and are removed by macrophages at these sites. In the normal process of aging, the red cells lose their plasticity, become spherical in form and their surface charge alters. About 10% of aged red cells rupture, but the remaining 90% are removed from the circulation by macrophages. The ingested cells are degraded, and the heme tetrapyrrole residue is released to the plasma as unconjugated bilirubin. The iron is made available for reutilization.

Evaluation of circulating red cells

In most circumstances the erythron is assessed relatively easily by sampling the peripheral blood and quantifying:

- The hemoglobin concentration.
- The percent of red cells in the blood sample (hematocrit).
- Red cell numbers.
- The number of immature red cells (reticulocytes).

A qualitative appraisal of a thin film blood smear is also carried out, particularly noting the size and shape of red cells.

Quantitative evaluation of red cells

The most consistent value between species of animals is the hemoglobin concentration, which ranges between 100 and 160 grams/liter. If the hemoglobin concentration is below this range, **anemia** is present. Because of a marked variation in red cell size between species, the number of cells transporting the same load of hemoglobin differs markedly. For instance, sheep and goats have small red cells and normally have higher red cell counts than the dog, which has much larger red cells. Within species, red cell counts decline with growth and maturity as red cell size increases. Red cell counts vary between 6×10^{12} and 12×10^{12} cells/liter. The hematocrit values range from 0.30 to 0.50.

Further information can be obtained from calculations derived from the hemoglobin concentration, red cell count and hematocrit. The **mean corpuscular hemoglobin concentration** (MCHC) is derived by dividing the hemoglobin concentration by the hematocrit. This calculation assesses the degree to which the red cells are packed with hemoglobin, which is remarkably consistent across the species (330–350 grams/liter). When the MCHC is in this range there is said to be a normochromic state; when it is below it is a hypochromic state. The MCHC cannot rise above the maximum concentration of red cell hemoglobin. The **mean cell volume** (MCV) may be obtained by dividing the hematocrit by the red cell count or may be measured directly electronically. The MCV is expressed in femtoliters. A typical MCV may be as follows:

$$0.46 \div 6.8 \times 10^{12} = 68 \text{ femtoliters.}$$

When the MCV is in the normal range there is said to be a normocytic state, if below, a microcytic state, and if above, a macrocytic state. The **mean cell hemoglobin** (MCH) is calculated by dividing the hemoglobin concentration by the red cell count and is expressed in picograms. For instance the normal MCH in the dog is:

$$160 \text{ grams/liter} \div 6.8 \times 10^{12} = 23.5 \text{ picograms.}$$

As the MCH depends on red cell numbers there is a great variation in the normal value between species. The relationships between the red cell indices (MCHC, MCH, MCV) and the hematocrit, red cell count and hemoglobin concentration are depicted in Fig. 3.2.

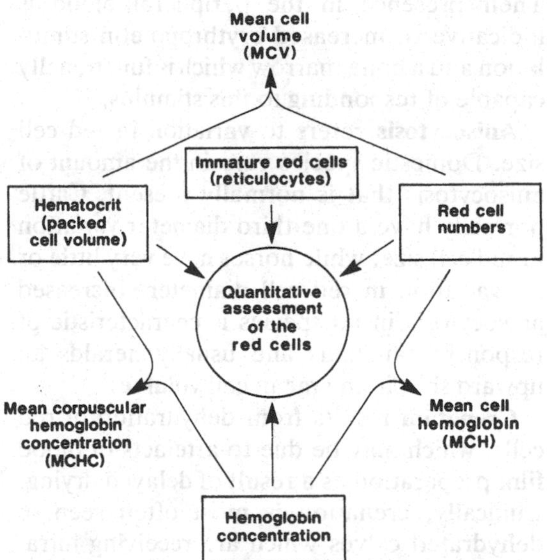

Fig. 3.2. Quantitative evaluation of the erythron revolves around the hematocrit, hemoglobin concentration and red cell numbers, from which red cell indices MCHC, MCH and MCV can be calculated.

Qualitative assessment

In conjunction with quantitative values, various qualitative characteristics of circulating red cells are assessed routinely by the examination of stained blood smears. Red cells normally stain uniformly.

Red cell **polychromasia** is the presence of diffuse or punctate basophilia in peripheral blood red cells stained routinely with a Romanovsky stain, while **reticulocytes** are the same cells demonstrated by supravital staining. In general, the supravital technique is more sensitive and a more accurate measure of marrow erythroid response. Most normal domestic animals have a steady-state level of reticulocytes at 0.1–1% of the total red cells or about 30–60 × 10^9/liter in absolute terms. This level of new cell production which is required to maintain homeostasis can be increased five to six times the normal level or even higher if there is an adequate supply of nutrients. A strong reticulocytosis is usually accompanied by the presence of **shift red blood cells** in the circulation, which are twice the diameter of normal cells and are immature, as demonstrated by the presence of basophilic staining. Their presence in the peripheral blood is indicative of increased erythropoietin stimulation and a bone marrow which is functionally capable of responding to this stimulus.

Anisocytosis refers to variation in red cell size. Domestic species vary in the amount of anisocytosis that is normally present. Cattle normally have a one-third diameter variation in red cell size, while horses have very little or no variation in red cell diameter. Increased anisocytosis in all species is characteristic of responsive anemias and usually heralds an upward shift in the mean cell volume.

Crenation results from dehydration of red cells, which may be due to artefacts of blood film preparation as a result of delayed drying. Clinically, crenation is most often seen in dehydrated calves which are receiving intravenous fluid therapy. While crenation is indicative of loss of cell volume, **poikilocytosis** is indicative of marked red cell shape change with loss of cell membrane. **Keratocytes** and **acanthocytes** are specific types of poikilocytosis. In the former, there is a 'bite-like' loss of a segment of cell periphery, while in the latter there is an asynchrony of cell membrane to volume, with the formation of spicular membrane projections. The acquisition of cholesterol, or altered cholesterol to phospholipid ratios in the red cell membrane may be allied with acanthocyte formation. **Schizocytes** (triangular and star-shaped red cells) may occur when there is intravascular lysis of red cells, for example, when there is endothelial damage, with intraluminal fibrin strands in small arterioles. Red cell membranes are torn as they travel at high velocity and impinge on the damaged endothelium. Poikilocytosis is seen whenever there is inflammation of a large vascular bed such as occurs in enteritis, pneumonitis and dermatitis. The presence of poikilocytosis in horses with a history of colic is highly suggestive of anterior mesenteric parasitic arteritis.

Spherocytes are red cells which have lost portions of the cell membrane and their biconcave shape, to become spheroidal. They appear in blood films as small red cells with increased density and an absence of central pallor. These cells are recognized as abnormal by the body and removed from the circulation by sinusal macrophages of the spleen, liver and bone marrow. The transition from a normal biconcave to a spherical configuration may be accelerated by a specific adherence of antibody to red cell membranes; thus, increased numbers of spherocytes are suggestive of immune-mediated hemolysis.

Target cells (codocytes) result from a reduction in plasticity of red cell membranes and are seen characteristically in animals with hepatic disease and abnormal levels of plasma lipids, which are in dynamic transition from plasma to red cell membrane.

Rouleaux formation (alignment of red cells in rows) is never observed in cattle but is normal in dogs, cats and horses. Rouleaux formation is increased in inflammatory states when there is increased plasma fibrinogen and globulin.

Rubricytes, or nucleated red blood cells, are not found in the peripheral blood of normal animals. Rubricytes may appear in the peripheral blood in response to marrow hyperactivity in what is known as a 'leukoerythroblastic' reaction, which is usually accompanied by neutrophilic immaturity. Rubricytes will appear in the blood in response to acute hypoxia (such as occurs in cattle with traumatic pericarditis and tamponade) and in animals with skeletal fractures. Rubricytosis in the absence of reticulocytosis, or increased numbers of polychromatic red cells, is indicative of stress erythropoiesis with inadequate response. These conditions may occur in poorly responsive anemias or in animals with tumorous invasion of the bone marrow. In occult leukemias, the first indication of serious marrow disease may *not* be the presence of tumor cells in the peripheral blood but a poorly responsive anemia with rubricytosis. In responsive anemias, the peripheral blood rubricytes are late-stage cells, primarily metarubricytes and polychromatic rubricytes, whereas, in cases of marrow invasion, there is asynchronous release with peripheral blood basophilic rubricytes and prorubricytes. It should be remembered that the erythroid response of the bone marrow should be assessed on peripheral blood reticulocyte response and not on the numbers of circulating rubricytes. The qualitative assessment of red cells is depicted in Fig. 3.3.

The erythropoietic marrow cells

Recognizable erythropoietic cells comprise about 50% of the total population of bone marrow cells under normal conditions. They are referred to as the **rubricyte series** and become identifiable once a committed stem cell has produced the first-stage **rubriblast**. There are six morphologically distinct stages of rubricyte maturation before the nucleus is lost and the final red cell appears. The first three, the rubriblast, prorubricyte and basophilic rubricyte stages, comprise dividing cells, while the last three, the polychromatic, normochromic and metarubricyte stages are purely maturation phases. Cell division is limited to two progeny at each stage, although there may be some limited self-renewal by basophilic rubricytes. In general, a single committed stem cell produces eight to ten mature red blood cells. The total production time is

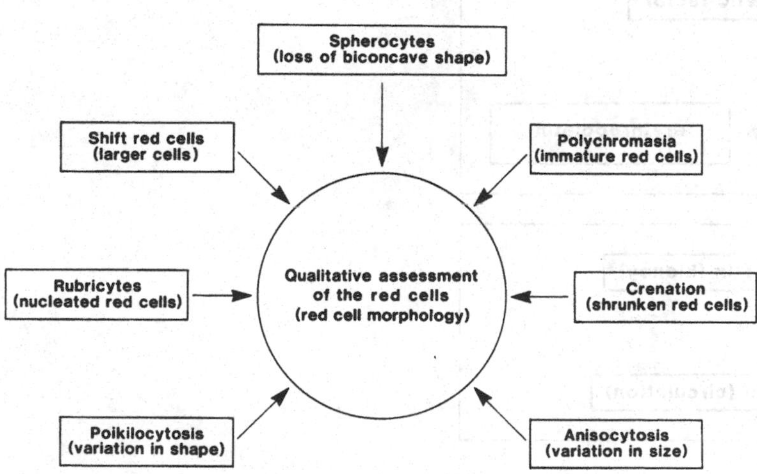

Fig. 3.3. Qualitative assessment of red cells involves particularly changes in shape and size of red cells. Most of these changes give an indication of the disease process and in some cases the etiology.

four days and each stage takes the same proportion of the four-day period.

The central feature of erythroid cytoplasmic maturation is the process of hemoglobinization, which involves the production of heme and globin within the developing red cell. Iron is a crucial component in the heme molecules, and is incorporated into protoporphyrin 9, type III at the heme synthetase step. Iron stored as ferritin (as Fe^{3+}) is ready to use after reduction to the ferrous form within the cell. The copper-containing protein, ceruloplasmin, is active in iron oxidation and reduction and, after oxidation, dietary iron is transported in the plasma by the globulin transferrin. Iron uptake by developing red cells is not regulated, and in conditions of defective porphyrin production, for example in erythropoietic porphyria, iron can accumulate excessively in red cells. However, the production of heme and globin within the immature red cell is precisely balanced, and a decrease in synthesis of one is paralleled by a decrease in manufacture of the other. This situation occurs in iron deficiency anemia.

Control of erythropoiesis

The basal regulator of erythropoiesis is the body wide demand for oxygen, which may change for physiologic or pathologic reasons. For instance, a healthy marrow may compensate for the effects of a failing heart by increasing the oxygen-carrying capacity of the blood, in the same way as it compensates a healthy individual at high altitude or for athletic effort. It will similarly respond to make up any red cell depletion. Direct regulation is applied by the hormone **erythropoietin**, a glycoprotein produced either directly by the kidney or from a circulating precursor activated in the kidney. Erythropoietin acts on the marrow positively to promote red cell production by increasing stem cell input to the rubriblast stage and causing skipped divisions in the proliferative phase (Fig. 3.4). In its absence, red cell production fades as the erythropoietic cells go into a state of 'atrophy'. Conversely, tissue hypoxia will cause erythropoietin production to increase, its plasma concentration to rise, and in turn this will provoke an increased input of stem cells into the rubricyte series.

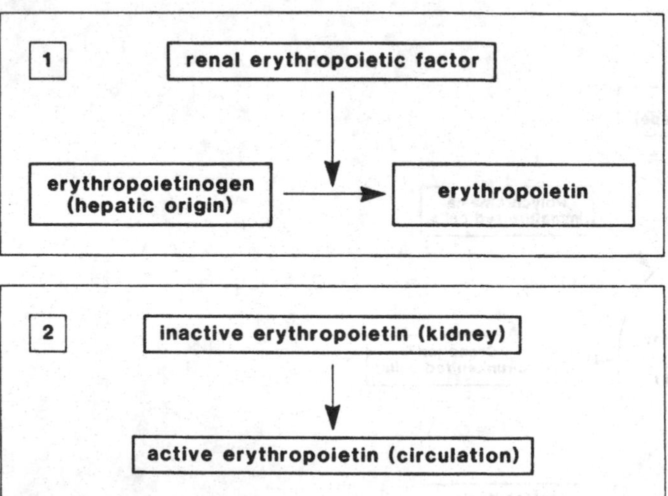

Fig. 3.4. Two mechanisms are proposed for the production of erythropoietin: (1) the production of erythropoietin from a precursor activated by renal erythropoietic factor; (2) the activation of erythropoietin in the kidney.

Features of a normal erythropoietic response

Under the influence of an increased erythropoietin concentration, the normal marrow becomes more heavily populated by erythroid cells and a large number of nearly matured cells are made ready for release. Accelerated production pressure may even cause cell division to be skipped and very large macro reticulocytes to occur, which are twice the diameter of normal red cells. Such reticulocytes in the peripheral blood are called shift reticulocytes and their presence indicates a particularly vigorous marrow response. They may be accompanied by nucleated rubricytes which are not normally found in the circulation. However, the reticulocyte count is always the most sensitive indicator of erythroid responsiveness. For example, in the dog after a severe red cell depletion, reticulocytes may increase in 3–4 days from the baseline level of around 50×10^9/liter to around 1000×10^9/liter. The effect of such a response is an upward shift in the mean cell volume (MCV) as the new larger cells enter the circulation. Although these cells may not have completed hemoglobin synthesis at the time of their release, their total hemoglobin content is unusually high and they will increase the MCH, usually by 2–3 picograms. Because of the large number of polychromatic cells, however, the MCHC may be slightly low. The large new cells are remodeled in the circulation by loss of cell membrane over about 10 days. This means that the MCH and MCV will return to normal values over this period once the production surge has waned.

Anemia

If there is one tissue that epitomizes the principle of balance between cellular renewal and removal it is the erythron. When this principle is contravened, and removal is excessive or renewal is inadequate, the result is anemia, the definition of which is a deficiency of circulating red cell numbers (Fig. 3.5).

An excessive loss of red cells from the peripheral blood may occur in either of two ways.

A loss from the vascular system by hemorrhage or
a loss due to decreased red cell life span (red cell lysis).

Inadequate production of red cells by the marrow also falls into two general categories.

Reduced proliferation of red cell precursors or
defective synthesis of hemoglobin.

Notwithstanding the method by which circulating red cells are either lost or inadequately produced, the clinical signs of anemia are relatively constant and are due to five major factors.

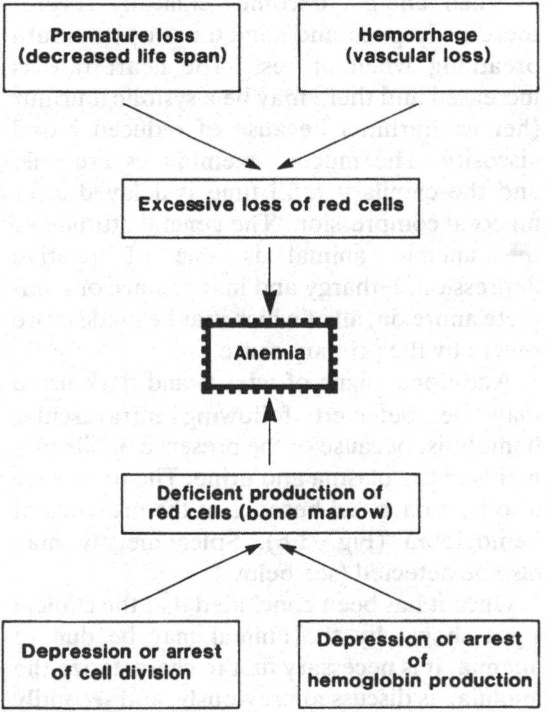

Fig. 3.5. Anemia is basically the result of either an excessive loss, or a deficient production, of red cells, The basic mechanisms are also highlighted.

- Reduction of oxygen transport.
- Reduction of blood volume.
- The rate at which the two above occur.
- The reserve capacity of the splenic, cardiovascular and pulmonary systems.
- The associated effects of the underlying cause of the anemia.

In general, animals with anemia of long standing may show few signs of distress, with hemoglobin levels as low as 80 grams/liter and a hematocrit of 0.25, particularly if they are non-working, non-productive types of animal. Similarly, growing animals may show few signs of distress at levels as low as 60 grams/liter, because the 2,3-diphosphoglyceric acid (2,3-DPG) shift increases the efficiency of the remaining hemoglobin. In contrast, in racing dogs and horses, even a mild reduction in hemoglobin is likely to result in impaired performance, although no signs may be evident when the animal is at rest.

When anemia becomes clinically severe, there is dyspnea and sometimes open-mouth breathing when at rest. The heart rate is increased and there may be a systolic murmur (hemic murmur) because of reduced blood viscosity. The mucous membranes are pale and the capillary refill time is delayed after mucosal compression. The general attitude of the anemic animal is one of relative depression, lethargy and inappetance or complete anorexia, all of which may be made more severe by the primary cause.

Additional signs of icterus and dark urine may be detected following intravascular hemolysis, because of the presence of bile pigments in the plasma and urine. The urine may also be a dark red because of the presence of hemoglobin (Fig. 3.6). Splenomegaly may also be detected (see below).

Once it has been concluded that the clinical signs shown by the animal may be due to anemia, it is necessary first to characterize the anemia, as discussed previously, and secondly to confirm the suspected etiology where possible.

It now remains to discuss anemia under the four general categories mentioned above and to highlight the common causes.

Excessive red cell loss by hemorrhage

The term hemorrhage indicates the escape of whole blood from the vascular compartment. However, the blood may still be confined within the body, in which case the hemorrhage is internal, or it may escape out of the body, in which case the hemorrhage is external. Furthermore, the blood loss may occur suddenly as an acute episode, or it may be drawn out as a chronic process. The character of the resulting anemia will depend upon the foregoing variables.

In **acute** internal or external hemorrhage, a large quantity of blood suddenly escapes into the tissues, a body cavity, or to the exterior, producing an abrupt plunge in circulating blood volume. This will produce hypotension and a state of shock, which requires immediate therapeutic attention. Such events may be induced by trauma, surgery or the spontaneous rupture of vessels within lesions like splenic vascular tumors, abscesses or deep gastric ulcers. Blood clotting defects may also predispose to acute severe hemorrhage. With the correction of the blood volume, a deficit in hemoglobin will become apparent. In most

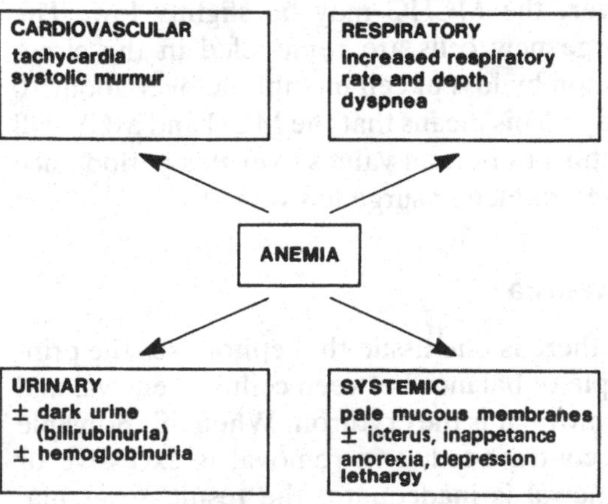

Fig. 3.6. The clinical signs indicative of anemia often involve the cardiovascular, respiratory and urinary systems. Some systemic signs may also be apparent.

instances the bone marrow is competent and responds with a vigorous reticulocytosis and consequent rise in MCV and MCH. Iron uptake and utilization is also increased.

While external blood loss is usually observed, internal bleeding may not be so obvious. When interstitial or cavity bleeding occurs many of the extravasated red cells may re-enter the circulation but will show signs of their ordeal in the form of membrane damage. Thus, in blood smears there may be poikilocytosis and acanthocytosis. In addition, the breakdown of extravasated red cells and release of heme pigment may produce a mild rise in serum bilirubin concentration.

In anemia of **chronic** hemorrhage the situation almost invariably involves unseen (occult) blood loss from the alimentary or urinary tracts. The rate of blood loss is generally not sufficient to precipitate an acute depletion of hemoglobin. The critical factor is a steady depletion of body iron reserves and eventually an iron deficiency anemia. This produces a characteristic hematologic picture which will be discussed in a later section. Lesions producing such blood loss are frequently tumors, chronic inflammatory reactions or parasitic infections.

In any situation in which whole blood, and therefore iron, has been lost from the body, acutely or chronically, there is a need to replenish body iron status. In general it is necessary to supply 3–4 milligrams of elemental iron for each gram of hemoglobin deficit. Standard formulae are available to calculate the requirement in any particular class of animal.

A rough approximation of the extent of blood loss may be determined on clinical examination. Thus, a loss of 15% of blood volume will accelerate the heart by less than 20 beats per minute, while a loss of 30% of blood volume will cause a rise in heart rate exceeding 30 beats per minute. With a 40% loss of blood volume, the animal is unable to stand and may suffer syncope if it attempts to do so. At this level of loss, there may be bradycardia to as low as 20 beats per minute during collapse, which may be accompanied by loss of sphincter control. With loss of 50% or more of blood volume, the animal will be recumbent and in shock.

Red cell loss due to decreased life span (hemolysis)

Many of the anemias encountered have as their central pathogenetic feature the premature lysis of red cells (hemolysis). Such a loss is primarily the result of two mechanisms.

1 An altered stability of the red cell leading to intravascular rupture. (Intravascular hemolysis.)
2 Additions or alterations to the red cell membrane which renders the cell susceptible to removal by the macrophage–monocyte system. (Extravascular hemolysis.)

There are clear examples of (1) or (2) acting alone. However, there are also anemias which have characteristics of both (1) and (2). In both cases the changes may be either congenital or acquired. The major types are depicted in Fig. 3.7.

Hemolysis indicates that red cells are being destroyed prematurely. Additional clinical features of hemolytic disorders are mostly associated with increased hemoglobin catabolism, and diagnostic features may include jaundice, hemoglobinemia, decreased plasma haptoglobin, hemoglobinuria and hemosiderinuria and methemalbuminemia. Intravascular hemolysis is considered more injurious to the body than extravascular hemolysis as free red cell membranes circulate in the peripheral blood, and may trigger disseminated intravascular coagulation. Red cell membrane fragments are also damaging to renal epithelium. Unconjugated bilirubin levels may increase when the rate of production exceeds the rate of conjugation by the liver. Free bilirubin at high levels causes tissue damage, as does free heme. However, protective proteins such as haptoglobin bind to free hemoglobin; hemopexin binds to free heme or oxidized heme, and albumin also has affinity for oxidized heme forming methemalbumin. The presence of methemalbumin is a trade mark of intravascular hemolysis.

Intravascular hemolysis

(Oxidant damage, red cell parasites, microbial and other toxins, immune-mediated anemias, familial disorders)

In this category of anemia the stability of red cells is perturbed and they disintegrate within the plasma, releasing fragments of membrane and hemoglobin. The plasma will be red (hemoglobinemic) and in severe cases hemoglobinuria and icterus will occur.

One mechanism producing this outcome is oxidant damage, when the hemoglobin molecule is denatured and red cell stability is impaired. Oxidant damage occurs when the protective hexose monophosphate shunt pathway is overwhelmed and there is a failure to maintain adequate concentrations of reduced

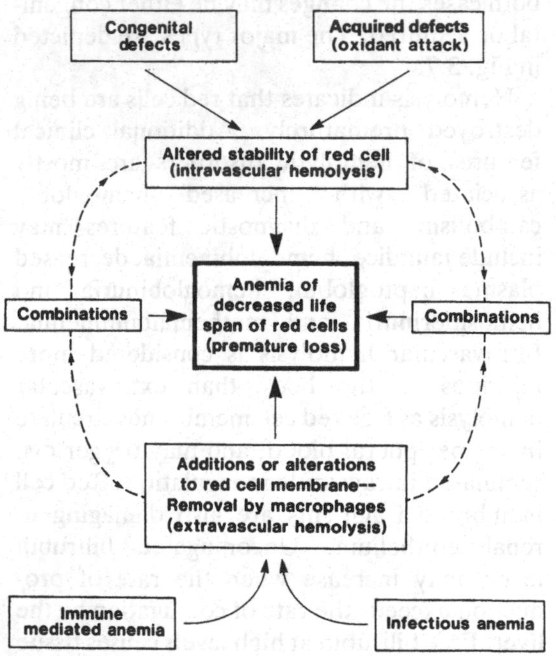

Fig. 3.7. The two broad categories of anemias of decreased life span are intravascular or extravascular hemolysis. However, anemia in a particular case may follow a combination of both intra- and extravascular hemolysis.

glutathione. This results in the formation of disulfide linkages between globin chains and oxidized glutathione. Ferrous iron in hemoglobin may also be oxidized to methemoglobin. Several plant toxins produce the effect, notably from rape, kale and onion; it is also evident in chronic copper poisoning in sheep, in post parturient hemoglobinuria in cattle and in phenothiazine poisoning in horses. In sheep it has been recorded as an inborn deficiency of the enzyme glutathione synthetase.

A feature of this type of anemia is the occurrence of **Heinz bodies** in red cells, which appear as abnormal intracellular inclusions in blood smears. The Heinz bodies are clumps of denatured hemoglobin within intact red cells. Toxic Heinz body anemia usually resolves well if exposure to the toxin ceases. Young red cells are spared because of their high protective enzyme content and the disorder reverses as the marrow makes good the loss with a burst of reticulocyte production. If the hemoglobinuria has led to iron depletion, the new cells may be somewhat deficient in hemoglobin and will appear hypochromic in blood smears. The anemia will thus be characterized by polychromasia, an increased MCV (macrocytic) and a decreased MCHC.

Another mechanism of intravascular hemolysis follows infection of red blood cells by parasites, notably *Babesia bovis* and by some microbial toxins such as those produced by *Leptospira* sp. and *Clostridium hemolyticum*. The venom of some snakes contains potent hemolysins. Disease caused by these agents frequently features hemoglobinuria ('red water'), anemia and icterus.

Sometimes intravascular hemolysis is due to antibody binding to red cells. This is discussed under immune-mediated anemia below (p. 49).

A unique form of intravascular hemolysis is recognized in Basenji dogs as a familial disorder. It is due to an inborn deficiency of the enzyme pyruvate kinase in red cells. The enzyme defect suppresses the activity of the glycolytic pathway and deprives the red cells of ATP. Without this energy source, the membrane ion pumps fail, cation regulation is lost and the cells rupture. Young red cells, still equipped with mitochondria, are able to maintain their integrity via aerobic energy production, but eventually succumb as they reach maturity.

Affected animals reach normal size and body weight, but are lethargic and have pale mucous membranes and often enlarged spleens. Their anemia is typically of moderate severity (hemoglobin 60 grams/liter), an MCV of about 85 femtoliters (macrocytic) and reduced MCHC (hypochromic). An outstanding feature is the reticulocyte count, which may reach 10^{12}/liter in a total red cell count of 2.5×10^{12} to 3.0×10^{12}/liter. The defect is inherited as a homozygous recessive trait and asymptomatic heterozygotes can be identified because they share with homozygotes a detectable surplus of red cell 2,3-DPG. This accumulates because of the enzyme deficiency.

Extravascular hemolysis

(Red cell parasites, immune-mediated anemias, intrinsic red cell defects)

The macrophage–monocyte system that removes red cells from the blood is highly tuned to detect any abnormal feature that will single out individuals in the passing parade of millions. As has been mentioned, red cells that have completed their allotted time span are recognized by virtue of changed shape, increased rigidity and altered surface charge. In addition, normal pre-senile red cells have a remarkable propensity to adsorb foreign materials such as toxins, drugs and viruses. They can also become infected by parasites such as *Hemobartonella*, *Eperythrozoon*, *Anaplasma* and *Babesia* sp. Under these conditions they too become targets for phagocytosis, as a means of efficient removal of the offending agent.

Finally and most importantly, the binding of antibodies to red cells will cause them to assume the spherical shape of senescence and to become targets for phagocytosis. Antibodies, either heterologous or homologous, may be directed against red cell parasites, foreign agents adsorbed to the red cell, or against the cell membrane itself.

The priming of red cells for premature removal will give rise to anemia when the rate of cell removal exceeds the rate of renewal. It is probably accurate to say that the binding of antibody and complement are the factors most likely to provoke excessive removal and hence anemia.

Specific hemolytic anemias

Immune-mediated anemias

Basic to all immune-mediated anemias is the binding of antibody, with or without components of complement, to the red cell surface. This is routinely detected by employing the **Coombs' test**. In this test, red cells of the patient are reacted with an antiserum directed against the globulin and complement of the patient's own species. These antisera are usually raised in rabbits or sheep. When the antiserum encounters the red cells coated with antibody and perhaps complement, the red cells agglutinate, as they are bound together by the ensuing immune combination (Fig. 3.8). Unfortunately the test may yield false negatives as, although hemolysis *in vivo* can occur with relatively few antibody molecules per red cell, at least 500 molecules must be present per cell for the test to be positive *in vitro*.

In immune-mediated anemia associated with the binding of IgA or IgG molecules (see Chapter 2), complement is not usually fixed, and the antibody–red cell reaction is maximal at 37 °C. When the antibody is IgM, complement is usually fixed and the reaction is maximal at 2 °C. For this reason the reactions are said to be 'warm type' or 'cold type', respectively. An animal with a 'cold type' reaction may suffer bouts of intravascular hemolysis after severe exposure to the cold sufficient to chill the blood in the extremities.

This type of anemia is frequently characterized by the presence of many spherocytes (spherical red cells) in the blood, and in most cases the bone marrow precursors are unaffected, so that a regenerative response is present. Not uncommonly, there is simultaneous antibody binding to platelets, so that anemia may be accompanied by thrombocytopenia.

The terms **primary** or **idiopathic** are used when the cause of the immune-mediated anemia is unknown. This is usually the case in animals. It is termed **secondary** or **symptomatic** when the triggering cause is known, such as in drug-induced, infectious or isoimmune disease. True autoimmune anemia, involving antibody generated by the patient against its own unaltered red cell antigens, is extremely rare. In spite of this, idiopathic immune anemia is frequently referred to as 'autoimmune hemolytic anemia'.

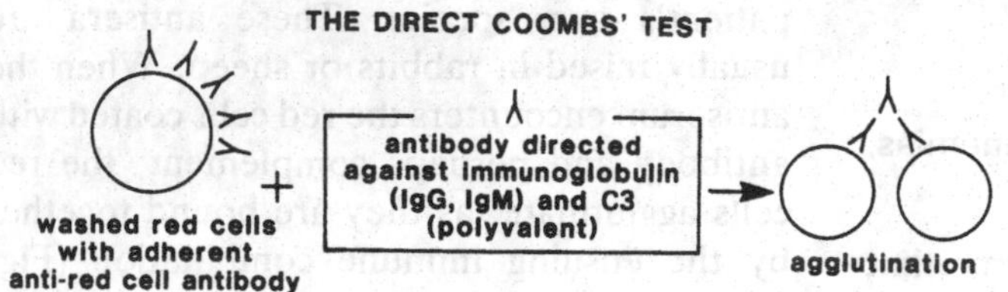

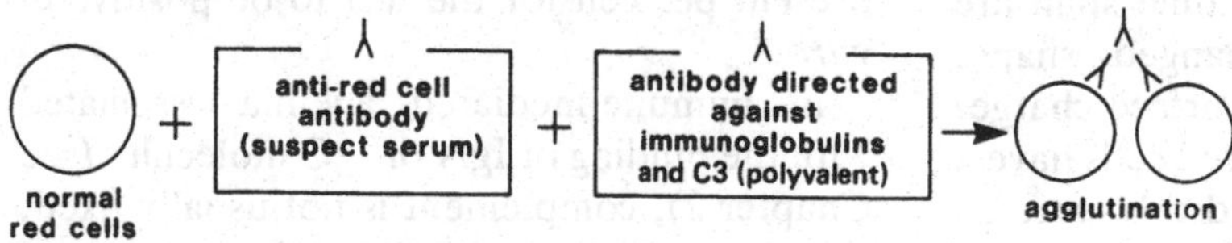

Fig. 3.8. The Coombs' test. There are two variations of this test. In the direct test, red cells from an animal with suspected immune-mediated hemolytic anemia are extensively washed. The anti-red cell antibody remains attached to the red cell. These red cells are reacted with polyvalent antibody and if the test is positive the red cells agglutinate. The indirect test (not often used in dogs) requires the use of normal red cells, which are first reacted with the suspect serum. The remainder of the procedure is as for the direct test. Ig, immunoglobulin; C, complement factor.

Primary or idiopathic immune hemolytic anemia: Immune anemia of unknown cause is most common in dogs and cats and is less common in horses and cattle. It probably occurs in all species. The disease is indistinguishable from drug-induced hemolysis except that the etiology is never identified. Occasionally, a dog will present with a spectrum of diseases including symmetrical facial dermatosis, polyarthritis, glomerulonephritis, immune anemia and thrombocytopenia, which is similar to systemic lupus erythematosis of man. The cause is unknown.

The disease in all species is characterized by a moderate to severe anemia that is hypochromic, macrocytic and usually highly responsive, with reticulocyte counts six to ten times the normal level (400×10^9 to 600×10^9/liter) and occasionally as high as $900 \times$

10^9/liter. There is usually prominent spherocytosis and the formation of red cell aggregates or 'rafts' at the butt of blood films. Not all cases are Coombs positive and the absence of a positive test should not exclude the diagnosis if the other signs are appropriate. In recurrent cases, the anemia is accompanied by icterus and neutrophilic leukocytosis and monocytosis.

Isoimmune hemolytic anemia: This is a disease of the newborn animal due to immunization of the dam by fetal red cell antigens during pregnancy. When the maternal antibodies are ingested in the colostrum, immune-mediated anemia may eventuate with potential for both intravascular and extravascular hemolysis (Fig. 3.9). The disease is best recognized in horses, when mares are immunized by fetal red cells which enter the maternal circulation via spontaneous focal placental hemorrhages. The problem occurs only when the stallion passes on antigens to the foal which are not present in the mare. Several pregnancies are usually required for the accumulation of sufficient antibody. The severity of the disease depends upon the amount and type of antibody absorbed by the foal. Dyspnea and collapse may occur as early as 8 hours or as late as 5 days after birth. There is a strongly positive Coombs' test and the red cells may undergo spontaneous agglutination *in vitro*. The serum or milk of the dam will agglutinate the foal's red cells and usually those of the stallion. The bone marrow responds fairly sluggishly, and, as is usually the case in horses, reticulocytes and polychromasia are not seen in the peripheral blood, but the MCV does rise.

Spontaneous transplacental sensitization occurs in swine and the disease has also occurred by vaccination of dams with a crystal-violet-inactivated hog cholera vaccine. Gestation and parturition are normal and the disease occurs after suckling. The isoantibodies are most often against the Ea red cell antigen, but the clinical disease is predominantly an immune-mediated thrombocytopenia with megakaryocytic hypoplasia in the marrow. Pallor, inactivity, dyspnea and jaundice may occur 1 to 4 days after birth, with

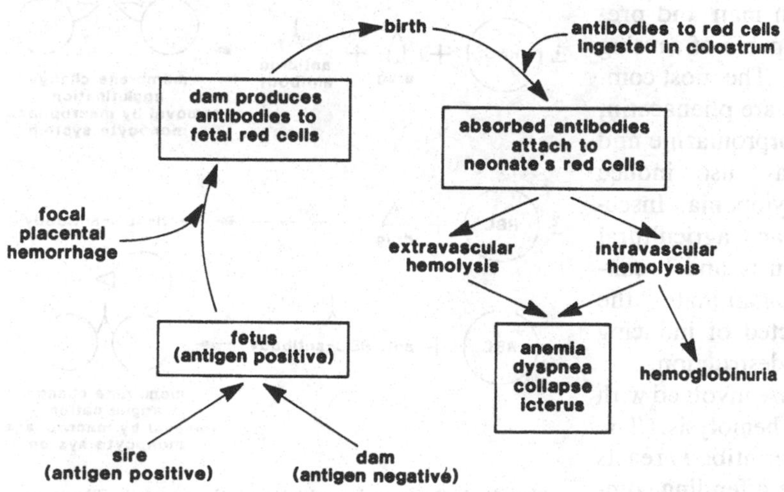

Fig. 3.9. Isoimmune hemolytic anemia is most commonly seen in the foal. The basis is the production of anti-red cell antibodies by the dam. A similar form has been observed in piglets.

some animals dying. Most, however, improve until day 10 or 11 when severe petechial hemorrhages develop and many piglets die at this time. The earlier signs are largely due to hemolysis and the later signs to thrombocytopenia. The red cells are Coombs positive and the dam's serum and milk will agglutinate the piglet's red cells and presumably platelets.

Spontaneous maternal sensitization to calf red cells is not recognized. However, vaccines for the prevention of babesiosis and anaplasmosis which are prepared from blood may contain red cell antigens that immunize the dams. If the bull shared red cell antigens with the vaccine donor then the calves will carry these antigens and may develop isoimmune anemia after suckling. Gestation is normal and the clinical syndrome varies somewhat with peracute illness at 1–2 days of age with hemoglobinuria, but mild cases may be seen with the anemia most severe at 5 days of age, followed by slow recovery. The marrow is hyperplastic but inadequately responsive. The calf's red cells are Coombs positive and the dam's milk agglutinates the calf's red cells.

Drug-induced hemolytic anemia: A number of drugs have been associated with Coombs'-positive hemolytic anemia in man and presumably most of these are capable of causing immune hemolysis in animals. The most commonly involved drugs in man are phenacetin, penicillin, sulfonamides, chlorpromazine and dipyrone. These drugs may also induce immune-mediated thrombocytopenia. Insecticides and other industrial and agricultural chemicals and their by-products and breakdown products which contaminate the environment are also suspected of inducing immune-mediated blood cell destruction.

Two types of mechanisms are involved with drug- or chemical-induced hemolysis. The most obvious case is where the antibody reacts only with cells exposed to the offending compound. Proof of this etiology can be obtained by incubation of non-affected donor cells with the suspected compound and their subsequent agglutination by the serum of the affected animal. The second mechanism involves production of an antibody that will react with normal red cells in the absence of the inciting compound. This latter disease is thus like an 'auto' immune hemolytic anemia, except for the history of specific exposure and exacerbation of the disease with retreatment. Drug-induced hemolysis is usually highly specific, though drugs with similar structure may cause hemolysis in the same individual. The amount of exposure required to cause sensitization is highly variable and penicillin-induced hemolysis is characteristically associated with prolonged high-level therapy. The two mechanisms are depicted in Fig. 3.10.

Infectious hemolytic anemias

Equine infectious anemia: Of the infectious hemolytic diseases in animals equine infectious anemia is perhaps the best characterized. This disease is caused by an arthropod-borne retrovirus and affects horses, mules and

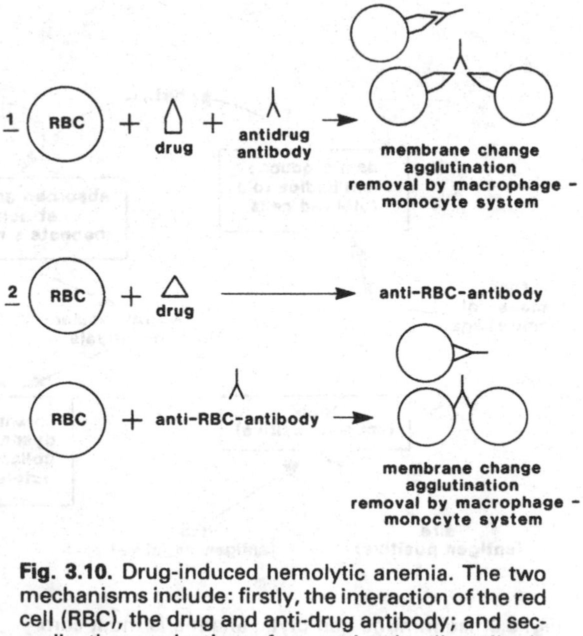

Fig. 3.10. Drug-induced hemolytic anemia. The two mechanisms include: firstly, the interaction of the red cell (RBC), the drug and anti-drug antibody; and secondly the production of an anti-red cell antibody following interaction of the drug and the exposed red cells.

donkeys. Infection is followed by life-long viremia which results in the adherence of virus to red cells and platelets. The cellular destruction is immune mediated. The onset of disease occurs after a 6–10 day period of incubation and coincides with the appearance of complement-fixing anti-viral antibody which is not protective but destroys red cells and platelets in an 'innocent bystander' reaction. The blood cells bearing C3 and immune complexes are selectively removed by macrophages, resulting in intracellular hemolysis. In acute febrile exacerbations, thrombocytopenic-dependent microvascular injury causes red cell fragmentation and intravascular hemolysis. The fragmentation is evidenced by distorted red cells on stained blood films, while the erythrophagocytosis results in the presence of iron-bearing monocytes (sideroleukocytes) in peripheral blood buffy coat films. The red cells are Coombs positive during the febrile disease and the infection is confirmed by agar gel immunodiffusion of patient serum against splenic-derived viral antigen (the Coggin's test).

Leptospirosis: In most domestic and many wild species, leptospirosis causes a multisystem disease characterized by anemia, icterus, hemoglobinuria, nephritis, hepatitis and abortion. The variation in pathogenicity of the many *Leptospira* serovars appears to reside largely in the activity of their respective hemolytic toxins. The early acute phase of the disease, which is often fatal, is due to intravascular red cell destruction by hemolysins, but in those animals which survive, the adherence of the toxins to red cells and the appearance of antibody results in Coombs-positive extravascular hemolysis which extends into the recovery period.

Anemias due to parasites: In a wide range of domestic and wild species, hemolytic anemia may be caused by intra-erythrocytic parasitoses, the most important of which are babesiosis and anaplasmosis. In hemobartonellosis, eperythrozoonosis, bartonellosis and malaria, infection is less often associated with anemia. The intravascular trypanosomes are extracellular organisms that, like equine infectious anemia virus, result in prolonged antigenemia, with an exaggerated but ineffective immune response that results in immune hemolysis.

Babesiosis and anaplasmosis are both tick-borne infections of great importance in cattle, characterized in enzootic areas by passively acquired immunity in calves and clinical disease in yearlings. The spleen is of pivotal importance in such infections of the blood and any event which impairs or overloads splenic function, such as splenectomy, hemorrhage or intercurrent infection, may lead to recrudescence in carrier states. A major distinction between these two diseases is that babesiosis is characterized by intravascular hemolysis and hemoglobinuria, while hemolysis in anaplasmosis is extravascular and the urine is merely bile tinged. Hemobartonellosis is similar to anaplasmosis in pathogenesis and, although the organism is much more widespread than anaplasma, which is limited by vectors, the disease hemobartonellosis is relatively uncommon. Cats with clinical hemobartonellosis frequently have occult lymphoma, which impairs splenic function. The critical diagnostic feature in all these diseases is the recognition of organisms in blood smears.

Hemolytic anemia due to mechanical damage to red cells: Systemic diseases which result in endothelial injury may cause the formation of intraluminal fibrin strands which impede red cell flow. When these conditions arise schizocytes are formed. These appear initially as cells with marked shape variations and some regain spherical form to become microcytes. There is significant hemoglobin loss during red cell fragmentation and hemoglobinuria may result. Vascular injury of this type may occur in disseminated intravascular coagulation and in valvular endocarditis involving the aortic and pulmonary outflow tracts. The somatic migration of parasites particularly *Strongyloides stercorales* and *Strongylus vulgaris* may cause red cell fragmentation where the injury to vessels may be more serious than the anemia.

Reduced proliferation of red cell precursors

Marrow erythropoiesis may be severely slowed or completely halted by a number of chemical intoxications, viral infections, irradiation or immune reactions. The process may involve the disappearance of cells from the marrow following the wholesale destruction of stem cells, or, alternatively, the cellular production line may be blocked at some point and numerous precursors may be present but unable to complete their development (Fig. 3.11). When erythropoiesis is suppressed in this way, a non-regenerative anemia develops progressively as the population of red cells already circulating is reduced by the removal of aged cells. This may take some time in animals with long-lived red cells. Because there is a steady depletion of normal red cells and no marrow input, the anemia is usually normocytic and normochromic.

Although the red cell series may be affected exclusively in this way it is more usual for several or all cell lines to be involved. Total

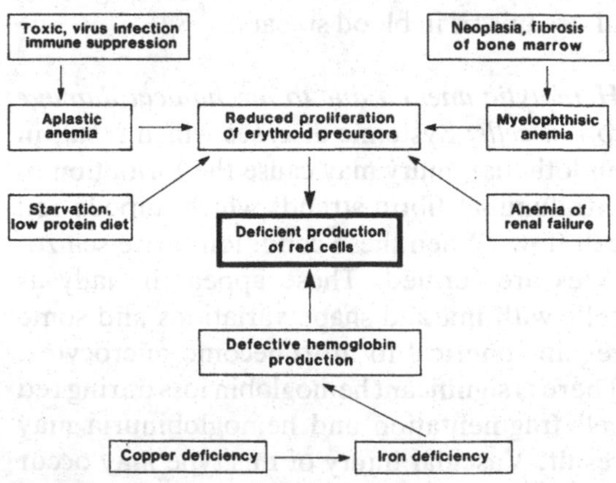

Fig. 3.11. Deficient production of red cells has two broad mechanisms, destruction or mitotic arrest of erythroid precursors and depression or arrest of hemoglobin production. There are a number of causes for each.

marrow aplasia or hypoplasia will give rise to pancytopenia, a deficit in all circulating elements of marrow origin. In pancytopenia, anemia will be the last abnormality to appear because red cells and platelets have the greatest longevity. The **marrow suppression** may be **reversible or irreversible.** Feline parvovirus causes an acute bone marrow depletion which is rapidly reversed if the animal survives. Because of the rapidity of the process, only the short-lived white cells are affected in the circulation and there is a panleukopenia but no anemia. Feline leukemia virus, on the other hand, can cause selective suppression of erythropoiesis and a chronic irreversible anemia, with either a depletion of red cell precursors in the marrow or a blockade of maturation. A classic example of myelointoxication is provided by bracken fern poisoning in cattle, in which there is irreversible depletion of all marrow precursors and severe pancytopenia. However, the clinical picture is dominated by the effects of the thrombocytopenia. Several of the drugs used for cancer chemotherapy are also myelotoxic, as are high levels of estrogens in the dog. The antibiotic chloramphenicol is also well known for potential marrow suppression, especially in the cat. It is now suspected that immune reactions can be directed against erythropoiesis, although the points of focus of such reactions have not yet been clarified.

Myelophthisic anemia results from displacement of normal marrow cells by the connective tissue of inflammation in myelofibrosis and by neoplastic cells in leukemia. Because of direct competition for space, myelophthisis is a constant early development in acute myeloid leukemias. Myelophthisis is seen most often in dogs and cats with lymphoma and in calves with the sporadic form of bovine lymphoma.

Anemia of uremia occurs in all species but is seen most frequently in dogs and cats. Anemia is constantly seen in chronic renal failure, where the mechanisms include sequestration of iron due to inflammation, hemolysis by retained metabolites, loss of erythropoietin and loss of blood from the kidney itself.

Anemia of endocrine dysfunction is mild and never of primary concern. Animals with adrenal and thyroid hypofunction have a mild anemia that is masked by plasma volume contraction. Castration results in a rise in hemoglobin levels in females and a drop in males to a similar median level most likely maintained by adrenal function.

Nutritional anemia occurs in all species and is most severe in neonates. Since there is concurrent hypoproteinemia and plasma proteins are preferentially replaced, an increased plane of nutrition is followed by reticulocytosis and an initial drop in hemoglobin as the rise in plasma proteins results in plasma volume expansion and hemodilution.

Defective hemoglobin production

Although there are occasional examples of anemias associated with defective production of globin, for all practical purposes anemias in this category result from deficient heme production. It is largely due to deficiencies in either iron or copper or to the effects of chronic inflammatory disease elsewhere in the body.

Iron deficiency anemia

Iron deficiency anemia occurs in all species and is usually the result of either inadequate iron in the diet, particularly in young animals, or chronic blood loss such as occurs with intestinal parasitism, or hemorrhage from neoplasms.

Iron is required not only for heme synthesis but also for a number of respiratory enzymes including the cytochromes. Thus, iron deficiency causes lethargy by metabolic impairment as well as via the effects of anemia. The effects of an iron deficiency can be divided into those due to iron depletion, iron deficient erythropoiesis, or fully developed iron deficiency anemia.

Iron depletion is the normal situation in young growing animals where, although the blood picture is normal, iron reserves are absent. Iron deficient erythropoiesis occurs when iron reserves are absent and there is a mild reduction in red cell mass accompanied by mild hypochromia. In iron deficient anemia there is, in addition to depleted reserves, a significant reduction in red cell mass with a non-regenerative, hypochromic, microcytic anemia accompanied by poikilocytosis, thrombocytosis and hypersegmented neutrophils. Rubricyte maturation is delayed and total erythroid marrow is increased with a shift to increased proportions of late-stage cells. The serum iron is reduced to less than 10 micromoles/liter while total iron binding capacity is increased to 75 to 100 micromoles/liter.

Milk is deficient in iron, and the newborn are in a precarious state having no iron reserves, an inadequate dietary intake of iron and a rapidly expanding body weight and plasma volume. The state appears truly 'physiologic' to the extent that, provided it is not too severe, it confers a measure of protection against invading bacteria which also require iron for proliferation. As would be expected, the most effective pathogens require the least iron and the body activates iron-scavenging proteins in inflammatory disease to deny invading bacteria access to iron. Iron supplementation to neonates (especially if parenteral) may temporarily saturate the iron transport proteins and cause iron toxicity (e.g. in vitamin E-deficient animals) and reduce resistance to bacterial infection. Thus, the strategy in iron supplementation to the newborn should be to give only sufficient iron to avoid serious anemia until ingestion of solid food begins.

Clinical iron deficiency anemia in young animals is mainly seen in piglets and provides the necessity for injectible iron to be given routinely. If iron treatment of baby pigs has been delayed, care must be taken in capture and restraint, as excessive exercise may be fatal if the anemia is severe. Subclinical iron deficiency anemia occurs in puppies of larger breeds of dogs and in calves but these are not treated regularly. Iron deficiency anemia in adult animals is almost always due to iron loss and, if the cause is not apparent, the primary diagnosis must be sought. Iron administration will remedy the clinical signs, but will not affect initiating causes such as intestinal parasitism or neoplasia.

Anemia of copper deficiency

Copper is required in a number of essential enzyme systems and its deficiency will affect energy metabolism, iron absorption and utilization, pigmentation of skin and hair, nerve conduction and collagen strength. Copper deficiency occurs where the element is deficient in the soil and the resulting disease syndromes have a defined geographic distribution. Conditioned copper deficiency occurs with excess dietary molybdenum, which is also most commonly associated with soil type in well-defined geographical areas. The effects of copper deficiency on skin (hair and wool), vessels and myelin are dealt with in Chapters 11 and 13 and only the hematologic effects of deficient copper will be discussed here.

Copper-deficient anemia is rarely seen and other signs of impaired copper metabolism are usually more apparent. The anemia, is, like iron deficiency anemia, microcytic and hypochromic in pigs and lambs. In adult cattle and sheep it is normo- or macrocytic and hypochromic. In dogs and rabbits, deficiency results in a normocytic/normochromic blood picture, with little or no anemia, but serious bone disease featuring epiphyseal fractures occurs in growing animals. In copper-deficient states, iron stores are often increased, suggesting increased absorption but impaired utilization of iron. In copper deficiency with anemia, repletion has no effect for several days, during which the copper is synthesized into ceruloplasmin. A rise in its serum concentration is followed by reticulocytosis. While the degree and type of anemia associated with copper deficiency varies with species, depigmentation of hair in all species is a sensitive indicator of deficiency. Depigmentation in association with increased body stores of iron, especially in adult animals, is indicative of a copper deficient state.

Anemia of chronic disease

Anemia of chronic disease is the most important cause of anemia due to defective hemoglobinization. As part of the acute inflammatory reaction, the body produces increased fibrinogen as well as proteins which convert serum iron to ferritin. The process is continuous and as long as the inflammatory reaction persists, serum iron is maintained at a low level, even though body iron stores may be normal or increased. Under these circumstances, iron may be rate limiting for erythropoiesis and the anemia is generally of mild degree and is of secondary significance in comparison to the primary disease.

This course of maintaining low serum iron in the process of inflammatory disease is believed to be a defensive mechanism to deny access to iron by invading microorganisms. It is interesting that non-pathogenic bacteria tend to have a very high obligate iron requirement and are completely inhibited under these circumstances, while the serious pathogens such as *Salmonella* and *Pasteurella* are able to function with very low access to iron.

The anemia of chronic disease is compounded by two other mechanisms: suppression of erythropoiesis, possibly by interleukin I and tumor necrosis factor, and a shorter half life of erythrocytes.

Erythropoietic porphyria

Heme defects resulting in erythropoietic porphyria occur in man, cattle, swine and cats. The disease is transmitted as a recessive and heterozygotes are unaffected. An abnormal protoporphyrin isomer is produced which is not capable of combining with iron to form heme. The enzyme uroporphyrinogen III cosynthetase is low in carriers and deficient in affected homozygotes. As a result there is a deficiency of protoporphyrin isomer III, which is the only one that can combine with iron. This metabolic block allows a build up of the precursor isomer I whose porphyrinogen intermediates photosensitize affected animals and can be detected in red cells by fluorescence. The block is never complete and some normal heme is formed. All body cells share the defect but, since the hematologic effects are most prominent, the disease is called erythropoietic porphyria. The aberrant porphyrinogens oxidize to the brown porphyrins which discolor urine, teeth and bone. Protection from sunlight permits prolonged survival and reproduction of affected animals. Anemia is moderate if there is protection from sunlight and severe if not.

The disease in cattle occurs primarily in Holsteins where dissemination of the disease has occurred by artificial insemination from a carrier bull. Calves fail to grow normally and the unpigmented areas of skin and mucosa are photosensitized. Bones are weak and spontaneous fractures occur. The teeth (pink tooth disease) fluoresce in ultraviolet light. The disease in Siamese cats is similar to that in cattle, with pigmentation of teeth and bones, dark urine, anemia, lack of vigor and photosensitization. In man, the facial scarring with hirsutism, red mouth and a tendency to wander only at night are believed to be the basis for legends of werewolves.

The red cell maturation defect of lead poisoning
Lead poisoning is not a cause of anemia in the general sense, but, in the dog, the blood changes are distinctive and often indicate the diagnosis. Environmental lead is ubiquitous and the element is constantly present at low levels in blood, tissue and feces. Plasma lead levels greater than 0.3 parts per million inhibit heme synthesis at three levels: ring closure, protoporphyrin assembly and iron incorporation. The inhibition of the early stages of pyridoxal and CoA succinate union results in the impaired utilization of δ-amino levulinic acid (δ-ALA). Consequently, δ-ALA concentration rises in plasma and it is excreted in urine where its detection is a sensitive indicator of lead poisoning in all species including man. The asynchrony of rubricyte maturation results in the overproduction or abnormal persistence of RNA, which in man and dog aggregates into punctate basophilic bodies or basophilic stippling in young red cells. In addition, rubricytes appear in the blood in discontinuous showers which, in the non- or mildly anemic dog in the absence of polychromasia, is characteristic of lead poisoning. The effects of lead on the nervous and enteric systems (including a coryza syndrome) are more prominent than the hematologic effects. It must be pointed out that ruminants undergoing stress erythropoiesis of any cause will have punctate basophilia and in these species there is no association with lead intoxication.

Hematopoietic proliferative diseases

There is a bewildering array of proliferative diseases that arise from the cells of the hematopoietic system. These vary greatly in their state of differentiation and in the numbers of cell types involved. Some order has been created by subdividing them into three major groups. The first are those classified as myeloproliferative diseases, a group of neoplastic diseases originating from cells produced by the bone marrow (myelos = marrow). The second group are myelodysplastic syndromes, which exhibit some, but not all, of the characteristics of neoplasia. The third are the lymphoproliferative diseases, which are neoplasms of the lymphoid series. Some in this last group could be considered myeloproliferative in that they probably arise from cells of bone marrow origin.

With neoplasia of the bone marrow there is often an associated anemia, thrombocytopenia and leukopenia. This change was attributed to crowding out of the remaining normal cells by the neoplastic cells, but it appears that in many cases all the cells in the marrow are abnormal to varying degrees. It has also been suggested that the anemia may be due to either a suppression of erythropoiesis or an increased destruction of red cells (shortened survival time).

In the cat and the cow, most hematopoietic neoplasms are induced by retroviruses, but a cause has not been identified in other species.

Most hematopoietic neoplasms are of lymphoid origin. Myeloproliferative diseases are rare in farm animals and constitute about 5% of hematopoietic tumors in the dog. They are most common in the cat. A summary of the types of hematopoietic tumor is depicted in Fig. 3.12.

Myeloproliferative disease

The myeloproliferative diseases are characterized by the medullary and extramedullary proliferation of one or more of the bone marrow cell lines. The acute leukemias are tumors of the more primitive cell lines and are characterized by a short course of usually one to two months from the time of diagnosis. In contrast the chronic leukemias result in the overproduction of more or less mature cells and a long clinical course of one to three years.

Acute myeloid leukemias

Collectively, the acute myeloid leukemias are characterized by marked non-responsive anemia, and, in most cases, by epistaxis and/or bloody diarrhea due to thrombocytopenia. These diseases become clinically apparent when 50% or more of the bone marrow is overtaken by tumor cells. Some aspect of marrow failure, either anemia, neutropenia with sepsis and/or thrombocytopenia with bleeding becomes apparent. There are a number of morphologic types.

Acute myeloblastic leukemia is characterized by a relatively mild increase in total nucleated cells with a preponderance of myeloblasts and very few mature neutrophils, accompanied by anemia and thrombocytopenia.

Promyelocytic leukemia is similar to myeloblastic leukemia, except that the cells have somewhat more cytoplasm with increased granulation and their progeny tend to mature to the myelocyte and the metamyelocyte stages.

Myelomonocytic leukemia is a concurrent tumor of the monocytic and neutrophilic cell lines. Clinically it is similar to acute myeloblastic leukemia and cytologically there are blasts and impaired maturation of both cell lines with phthisic anemia and thrombocytopenia. **Erythroleukemia** is a disease with concurrent neoplasia of the erythroid and myeloid systems which is characterized by an atypical peripheral blood rubricytosis, with severe anemia and absence of reticulocytes, and by myelocytic leukemia, with neutropenia and thrombocytopenia. **Erythremic myelosis** is a tumor of the erythroid system characterized by severe anemia, with marked rubricytosis consisting of rubriblasts to metarubricytes, with the latter predominating. The MCV is usually increased but reticulocytes are rare or absent and there is often neutropenia and thrombocytopenia. **Megakaryoblastic leukemia** is characterized by anemia, neutropenia and thrombocytopenia, with the presence in the peripheral blood of cells similar to myeloblasts with scanty cytoplasm frequently arranged in wispy strands with a fine pink granulation. In all the acute leukemias, the bone marrow aspirates are diagnostic and there is extensive phthisis of normal cells.

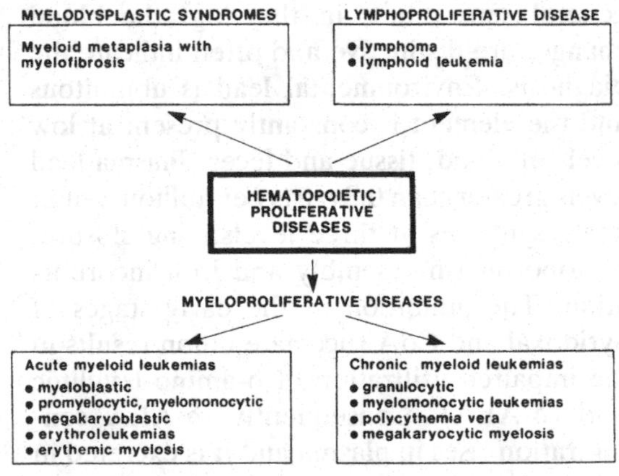

Fig. 3.12. Hematopoietic proliferative diseases include those that arise from cells normally present in the bone marrow (myeloproliferative disease) and those of lymphoid origin. Included are proliferative diseases that are of a severely hyperplastic nature (myelodysplastic syndromes).

Chronic myeloid leukemias

As a group, the chronic myeloid leukemias tend to be slowly progressive diseases which, because of their low mitotic rate, are relatively unresponsive to chemotherapy. In contrast to the acute myeloid leukemias, where the diagnosis is fairly obvious from the examination of peripheral blood, the chronic myeloid leukemias may, in some cases, resemble hyperplastic benign responses. A distinguishing feature of the chronic myeloid leukemias is that there are always blast cells present in the peripheral blood, whereas these would not be present in benign conditions.

Chronic granulocytic leukemia is characterized by a marked increase in white cell counts, usually in excess of 100×10^9/liter. The leukemia is most often of neutrophil origin, although it may arise from eosinophil or basophil cell lines. The anemia is usually of moderate severity and poorly responsive, and thrombocytopenia is of moderate severity. Most cases terminate in an accelerated phase, with the appearance of increasing numbers of immature cells in the peripheral blood. Splenomegaly and hepatomegaly are common in late stages, with irregular involvement of the lymph nodes, which are most often smaller than normal.

Megakaryocytic myelosis is a disease of the platelet system characterized by a massive increase in peripheral blood platelets usually of the order of 1000 to 3000×10^9/liter and a fairly normal neutrophil level and moderate anemia.

Primary polycythemia (polycythemia vera) is the result of autonomous hyperactivity of erythroid marrow, independent of erythropoietin, leading to the massive overproduction of mature erythrocytes but without a reticulocytosis. **Secondary polycythemia** is most commonly the expression of a physiologic response by a normal bone marrow to a powerful stimulus by erythropoietin. It may occur in a healthy individual as a response to a sustained high level of physical activity or to a prolonged sojourn at a high altitude. It may also occur in individuals with poor circulatory or respiratory function, as a compensatory maneuver. In all such cases it is an appropriate response. More rarely, secondary polycythemia of a less appropriate nature occurs when erythroprotein is secreted by a tumor, usually of the kidney, or if its secretion is stimulated by a lesion compromising renal blood flow. In secondary polycythemia, reticulocytes would be expected in the peripheral blood in most cases. In both inappropriate secondary and primary polycythemia, the number of circulating red cells may exceed 15×10^9/liter, the hemoglobin 250 grams/liter and the hematocrit 0.75. This provokes severe hyperviscosity of the blood, with resultant circulatory embarrassment and tissue hypoxia.

Chronic myelomonocytic leukemia is characterized by a low leukocyte count, typically of about 4×10^9/liter with a predominance of monocytoid cells, mild neutropenia and thrombocytopenia and non-responsive anemia. This disease has been termed the preleukemic syndrome and chronic smouldering leukemia. In the chronic myeloid leukemias, as in the acute leukemias, the bone marrow is always hyperplastic and aspirates assist the diagnosis.

Myelodysplastic syndromes

The myelodysplastic diseases include lesions where there is an unusual hyperplasia in the absence of an identifiable cause or target for increased cell production. They are not clearly neoplastic. **Myeloid metaplasia with myelofibrosis** is largely a disease of dogs and cats, characterized by a long course with insidious onset. The leukocyte and platelet counts are usually increased in the early stages but become decreased later, with moderate to severe non-responsive anemia. The marrow is hyperplastic, with replacement of fat and characteristically megakaryocytic and granulocytic hyperplasia with erythroid atrophy. Marrow is at first easily aspirated and, in later stages as the disease progresses and there is increased marrow stroma, aspiration becomes difficult or impossible and a marrow core is required to obtain sufficient tissue to confirm the diagnosis. The disease appears to occur in cats as part of the non-neoplastic diseases caused by feline leukemia virus and in other species by chronic marrow irritation of any cause. The marrow at first has increased reticulin fibers and finally collagen fibers are present with a concurrent reduction in density of hematopoietic cells and increasing signs of marrow failure.

Atypical myeloid reactions

The **leukemoid reaction** is characterized by a neutrophilic leukocytosis, with significant immaturity to the extent that leukemia is suspected. The disease occurs most frequently in the cat and dog, and the marrow is hyperplastic, with relatively synchronous cellular maturation. Most cases are related to an occult inflammatory focus and terminate in recovery. The **leukoerythroblastic reaction** is characterized by the concurrent presence in the peripheral blood of immature rubricytes and granulocytes. Typically, the nucleated erythroid cells are as young as basophilic rubricytes and the granulocytes as young as myelocytes. The syndrome occurs in all species and is most often seen in the dog in association with a reactive marrow typical of the response to immune hemolytic anemia. It should be recognized that this reaction involves cells in their normal maturation sequence without the presence of blasts and these changes should not be mistaken for leukemia.

Lymphoproliferative disease

The lymphoproliferative disorders may present as a spectrum of diseases, including lymphoma and lymphoid leukemia and combinations of the two. **Lymphoma** is a lymphoid proliferation which involves primarily the lymph nodes, spleen and liver with irregular involvement of other organs, while **lymphoid leukemia** affects primarily the bone marrow, spleen and liver. Lymph node involvement is inconsistent and there may even be lymph node atrophy. Most lymphomas in their late stages involve the bone marrow at least focally and tumor cells are present in the blood. In these cases it may be difficult to determine the primary origin of neoplasia unless there is a bulky peripheral tumor.

Neoplastic transformation can occur in both the B and T arms of the lymphoid system and, if these cells retain functional capability, there may be specific clinical signs associated with either type. Tumors may produce immunoglobulin (B cell), causing a hyperviscosity syndrome and some T and B cell malignancies mimick hyperparathyroidism. In general, malignancy in an immature clone of lymphocytes in an early stage of their histogenetic development is likely to occur in the bone marrow and the resulting disease will be a leukemia. In contrast, malignancy arising in a mature clone of lymphocytes which have undergone gene rearrangement and are now in a state of activation is more likely to be a peripheral disease which presents as lymphoma. Further, the immature or developmental clone lymphoid tumors tend to occur in young animals, while the mature or activation clone lymphoid tumors tend to occur in mature or aged animals.

Lymphomas are classified (*a*) by clinical staging to indicate the extent of body involvement with tumor and (*b*) by histologic type to indicate the aggressiveness of the tumor, which dictates clinical prognosis and also response to therapy. The diagnosis of lymphoid tumors is usually carried out by examination of blood and bone marrow for leukemias and by fine needle aspiration of lymph nodes or palpable superficial lesions for lymphomas. Confirmation by surgical biopsy and histopathology is required to substantiate the diagnosis of malignancy and to establish the histologic subtype of tumor.

In order to define the degree of disease progression at the time of diagnosis, and to measure the response to therapy, a staging system has been developed by the World Health Organization which is similar to that used in man.

Involvement

- Stage I – limited to a single node of lymphoid tissue in a single organ (excluding bone marrow).
- Stage II – many lymph nodes in a regional area.
- Stage III – generalized lymph node involvement.
- Stage IV – liver or splenic involvement plus all of stage III.
- Stage V – manifestations in the blood, bone marrow, and/or other organs.

Each stage is sub-classified into those with and without, systemic signs.

Histological classification of lymphomas
The histologic classification of lymphomas is based on nuclear size and shape, chromatin pattern, number, size and positioning of nucleoli, mitotic rate and volume, arrangement and character of cytoplasm and intercellular relationships. The most differentiated lymphomas are composed of small cells which mimick benign lymphocytes cytologically and architecturally, have a low mitotic rate and have the longest natural course. In contrast, lymphomas which are poorly differentiated architecturally and cytologically are composed of larger cells, have a high mitotic rate and tend to have a short clinical course. Paradoxically, these are potentially curable diseases because their high turnover rate renders them more susceptible to chemotherapeutic treatment. For clinical relevance, lymphomas have been divided into low, intermediate and high grade types, according to their clinical behavior in the untreated state.

Hemostasis

The arrest of bleeding is such an everyday occurrence that it is almost taken for granted, but such efficiency on the part of the body belies the complexity underlying it. In a normal animal there is a fine balance between too little, too much and just the right amount of hemostasis, which is governed by a number of separate but interacting mechanisms. Some inhibit blood clotting, while others enhance it.

On occasions, the hemostatic mechanism fails, leading either to prolonged bleeding or to intravascular coagulation. The failure may be temporary or permanent with, in general, those present at birth being permanent and those acquired during post natal life temporary.

There are four major arms of the hemostatic mechanism (Fig. 3.13).

The blood vessel.
The thrombon.
Coagulation factors.
Counter coagulation factors.

Although the factors involved in hemostasis work in unison, for the purpose of clarity, each will be considered separately. Emphasis will be placed on the disorders of the thrombon and coagulation factors, as these two constitute the vast majority of deficiencies in hemostasis.

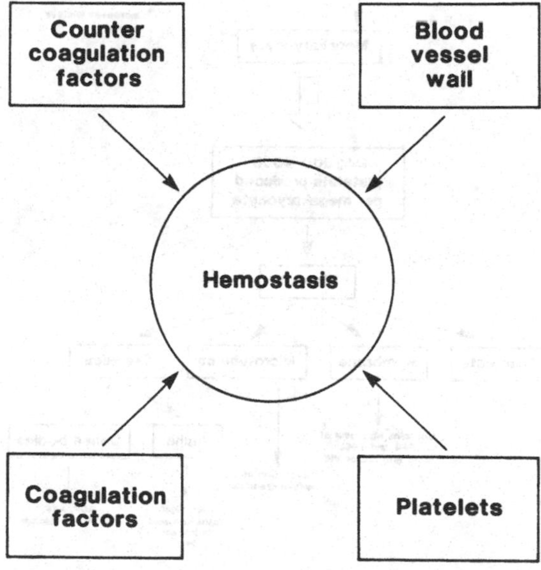

Fig. 3.13. Hemostasis is contributed to by four major areas. Platelets and coagulation factors to repair or stem vascular defects, and the blood vessel walls and counter-coagulation factors working to limit activation of platelets and coagulation.

The thrombon

The thrombon is a relatively new term coined to encompass all the facets of platelet production and function. Platelets are rather nondescript circulating fragments of megakaryocyte cytoplasm, but they play a powerful role in hemostasis and are increasingly recognized as participants in inflammation. The major features of the thrombon are depicted in Fig. 3.14.

Platelet production (thrombopoiesis)

Platelets are produced in bone marrow from the cytoplasm of their large precursor cell the megakaryocyte, which differentiates from a multipotential stem cell. The earliest identifiable precursor of the megakaryocytic series is the megakaryoblast, a binucleate cell with deeply basophilic cytoplasm and prominent nucleoli.

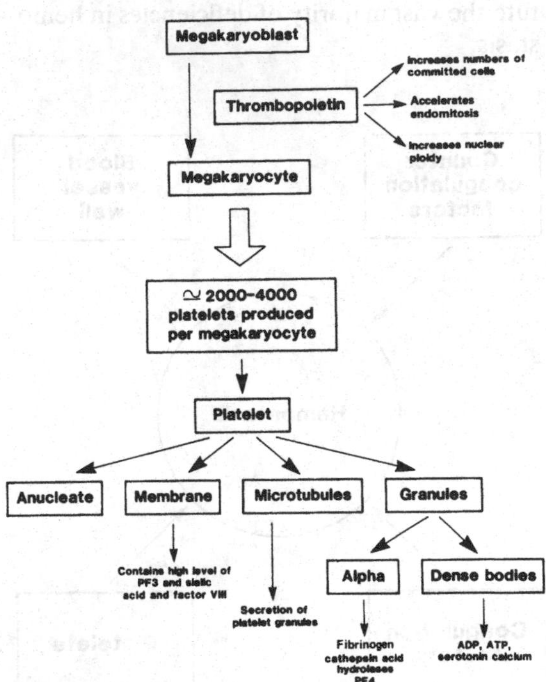

Fig. 3.14. The function of the thrombon is to produce platelets which are discrete anucleate bodies produced by megakaryocytes under the influence of thrombopoietin. Platelets are sensitive to released ADP and aggregate under its influence. Platelets also contribute factors (PF) to the clotting sequence.

In contrast to other cells in the bone marrow, the megakaryocyte nucleus undergoes nuclear division without cytoplasmic division (endomitosis), yielding a polyploid cell with up to 32 times the normal chromatin content of somatic diploid cells. The degree of ploidy depends on stimulation from the hormone thrombopoeitin; the greater the ploidy, the greater the number of platelets produced, with approximately 150–200 platelets being derived from the cytoplasm for each nuclear lobe. Under normal circumstances, nuclear proliferation ceases at the eight-nucleus stage (or 16N) and maturation of the cytoplasm proceeds. The nuclei fuse and the nucleus becomes multilobulated. The cytoplasm then increases in quantity and becomes more eosinophilic and granular.

Tubular channels that will form the demarcation membranes of the platelets then appear in the cytoplasm. These are invaginations of the plasma membrane of the megakaryocyte which split the granular cytoplasm into many separate units, to become the plasma membranes of the future platelets. Filaments of megakaryocyte cytoplasm are extended into the marrow sinusoids where the individual platelets are fragmented, detached and released into sinusoidal blood. Between 2000 and 4000 platelets are released by the average megakaryocyte. The remaining bare nucleus of the megakaryocyte is then processed by marrow macrophages.

The time required to produce platelets from the megakaryoblast stage is about 4 days, and the normal life span in circulation is approximately 10 days. Newly formed platelets differ from their older cohorts: they are larger, denser, metabolically more active and are more functionally effective. Young platelets are preferentially sequestered into the spleen for several days prior to their entry into the circulating blood. Approximately 30% of platelets in the blood are normally concentrated within vessels in the spleen. Platelet destruction and removal appears to be due to both random destruction and to age-dependent removal by macrophages in the spleen and liver.

Regulation of thrombopoiesis

Regulation of platelet production is considered to be hormonal, similar to that controlling the output of erythrocytes and granulocytes. The stimulatory hormone, thrombopoietin is a glycoprotein and is considered to act in several ways:

- Directing multipotential stem cells into the committed thrombopoietic stem cell pool.
- Accelerating megakaryocyte endomitosis and/or maturation of platelets.
- Increasing the nuclear ploidy of megakaryocytes, thus increasing the yield of platelets.

The maximal marrow response to thrombocytopenia (reduced numbers of platelets in circulating blood) is approximately eight times the basal rate. The mechanisms controlling the stimulation and release of thrombopoietin are not understood, nor is its site of production known. However, it is clear that there is a positive relationship between erythrocyte and platelet production, as demonstrated by thrombocytosis (increased numbers of platelets) in certain regenerative anemias (particularly iron deficiency anemia), in renal polycythemia and in chronic hypoxia.

With a marked increase in thrombopoiesis **shift** or **stress platelets** appear in the peripheral blood. These are large platelets one to two times the size of normal red cells.

Platelet structure

In mammals, the platelet is anucleate and discoid, containing a very high surface sialic acid content and some adherent coagulation factors (the fluffy coat). Encircling microtubules provide structural support, and an elaborate system of cytoplasmic canaliculi and vacuoles provide a secretory pathway for the release of platelet granule products. Two types of granules are present in the cytoplasm. The alpha granules are the most numerous and contain substances such as fibrinogen and platelet factor 4. The less numerous electron dense granules, or dense bodies, are storage organelles for ADP, ATP, serotonin and calcium. Storage ADP is to be distinguished from metabolic ADP, which contributes to the energy-requiring mechanisms of platelets. The platelets contain acid hydrolases within lysosomes, glycogen and a contractile protein (thrombasthenin). Fibrinogen and thrombasthenin constitute about 30% of the protein content of platelets.

An important coagulant factor known to be provided by the platelets is platelet factor 3, a complex of phospholipids which is closely associated with platelet and granular membranes.

Platelet function

Hemostatic plug

Platelets assist in the maintenance of vascular integrity, by sealing small endothelial defects. They act in the initial phase of arresting bleeding by the formation of a hemostatic plug. The process can be considered to occur in five stages and is depicted in Fig. 3.15.

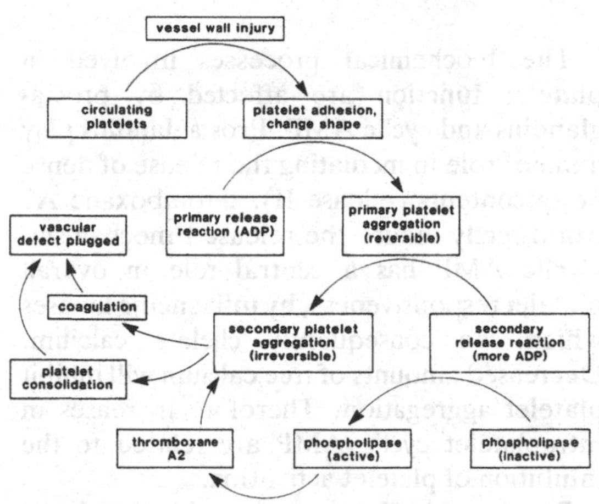

Fig. 3.15. Platelet aggregation. The sequence of events is initiated by vessel wall injury and culminates in the consolidation of platelets and plugging of the vascular defect.

1 Platelet adhesion occurs when they are exposed to subendothelial surface components of injured vessel walls, such as collagen, elastin and basement membranes. This results in a transformation of discoid platelets into spiny forms and the release of some ADP (primary release reaction, release I). Such platelets may be referred to as being activated.
2 Primary platelet aggregation (reversible) is the result of the primary release reaction. If this is extensive, large amounts of ADP are released from the electron dense granules (secondary release reaction, release II) resulting in a secondary wave of aggregation.
3 Secondary platelet aggregation (irreversible) occurs to seal the vascular defect. Platelet factor 3 becomes available for interaction in the coagulation cascade.
4 Consolidation or contraction of the plug by contraction of the protein thrombasthenin, induced by thrombin.
5 Fibrin stabilization, in which the products of the activated coagulation pathway induce the polymerization of fibrin to form a stable fibrin network throughout the clot. Finally, after the vascular defect is sealed, the plasminogen system causes dissolution, or fibrinolysis, of the clot.

The biochemical processes involved in platelet function are affected by prostaglandins and cyclic AMP. Prostaglandins play a major role in mediating the release of dense body contents (release II): thromboxane A2 can directly induce the release I mechanism. Cyclic AMP has a central role in overall platelet responsiveness, by influencing kinases which can consequently chelate calcium. Decreased amounts of free calcium will inhibit platelet aggregation. Therefore increases in intraplatelet cyclic AMP are related to the inhibition of platelet activation.

Platelet function can be assessed by a variety of *in vitro* tests, including platelet aggregation induced by agents such as ADP, collagen, adrenalin and thrombin, and the clot retraction test. An *in vivo* test is measurement of the primary bleeding time.

Coagulation

A number of platelet factors have been identified which have roles in the coagulation process. The most prominent of these is platelet factor 3 (PF-3). It is a complex of phospholipids which is required in the intrinsic pathway of the coagulation cascade, primarily in the activation of factors X and II. PF-3 becomes available for this function only after platelet aggregation is induced by ADP and thrombin; that is, after the secondary or irreversible phase of platelet aggregation. PF-3 is in the form of micelles on which surface-dependent reactions in the coagulation mechanism can take place. Activated platelets (PF-3) can also trigger the contact factors XII and XI of the coagulation cascade. Platelets are also involved in the synthesis and release of major components of the factor VIII complex.

Other functions

Platelets exhibit phagocytic properties, adhering to and engulfing viruses, fat droplets, iron and immune complexes. Platelets may thus provide the mechanism for clearing particulate matter from the blood. Among the proteins synthesized by platelets are factors promoting wound healing (growth factor from alpha granules), vascular permeability and leukocyte chemotaxis. Platelets contain most of the body's supply of serotonin, and act as a transport mechanism for various substances such as epinephrine, potassium and serotonin.

Assessment of the thrombon

The state of the thrombon is evaluated by examining the numbers, morphology and function of platelets in the blood and the numbers of megakaryocytes in the bone marrow.

Platelet numbers are assessed on a fresh peripheral blood sample either manually or by an automated particle counter. Normal platelet numbers in the blood range from 100 $\times 10^9$ to 800 $\times 10^9$/liter. Clinical signs associated with low platelet numbers may occur when numbers drop below 50 × 10^9/liter. However, because platelets may vary in their functional capacity, a small number of large platelets may prevent bleeding when numbers are as low as 10 × 10^9/liter. Platelet volume and diameter may also be measured electronically.

Platelet function can be assessed using a variety of laboratory tests to determine platelet adhesiveness, primary and secondary aggregation and release reactions. Another test of function in the presence of normal numbers is the primary bleeding time, an *in-vivo* method involving the time taken for bleeding to stop after a standardized skin incision has been made. It tests the competence of platelet plug formation and involves platelet and vessel wall factors. A major disadvantage is the difficulty of standardization of this test in animals.

Platelet morphology is evaluated using a Romanowsky stained blood smear. The presence of basophilic and larger 'shift' platelets usually indicates an increased thrombopoietic effort and is seen in conditions of increased platelet destruction and in regenerative anemias. Small platelets are seen in anemias such as iron deficiency anemia. A mixture of small and very large platelets can be seen in bone marrow dysplasias, and giant shift platelets are a relatively constant feature of feline leukemia virus infection.

Megakaryocyte appearance and numbers are assessed from a smear after bone marrow biopsy.

Disorders of the thrombon

Platelet disorders can be considered as quantitative or qualitative, inherited or acquired, and related to altered production in the marrow or altered utilization or destruction in the circulation. Acquired thrombocytopenias dominate as causes of pathologic bleeding in animals. Increases in platelet numbers are often incidental findings. A general outline of thrombon disorders is given in Fig. 3.16.

Clinical signs of platelet disorders

A reduction in platelet numbers (thrombocytopenia) or activity (thrombasthenia) may result in petechial or ecchymotic hemorrhaging. This occurs particularly in areas where capillaries are subject to intense hydrostatic pressure or trauma, for example the tongue, soft palate and gastrointestinal and lower urinary tracts. Clipping the hair coat can readily exacerbate petechial hemorrhages locally. In severe cases, neurologic signs may occur due to cerebral hemorrhage.

Increases in either platelet numbers (thrombocytosis) or aggregation may result in an increased tendency for microthrombus formation. In rare cases this may cause occlusion of large vessels leading to ischemic necrosis of extremities.

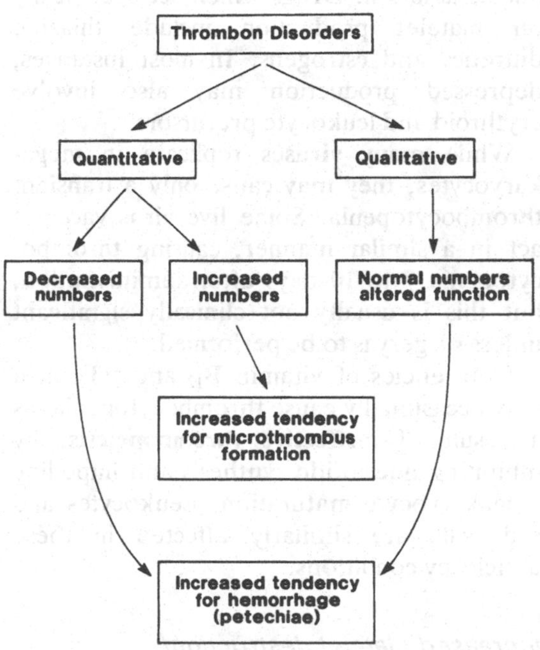

Fig. 3.16. Most thrombon disorders ultimately result in an increased tendency to bleed which is usually seen as petechiae. Platelets may be reduced in either number or function. Occasionally, excessive numbers of platelets lead to microthrombus formation.

Quantitative disorders

Depression of production

There are a host of causes that depress the production of platelets leading to a thrombocytopenia. Megakaryocytes can be destroyed, physically crowded out of the bone marrow, or their rate of division can be depressed.

Thrombocytopenias following reduced production can be divided broadly into those that are present at birth and those that develop later in life.

Congenital causes include thrombopoietin deficiency, hereditary thrombocytopenia, some *in-utero* viral diseases and the maternal ingestion of certain drugs such as thiazide diuretics.

Acquired conditions include idiopathic marrow aplasia, neoplastic marrow infiltration and bone marrow destruction by some viral and rickettsial (e.g. *Ehrlichia canis*) infections, and myelosuppression by drugs and irradiation. Drugs which act specifically on platelet production include thiazide diuretics and estrogens. In most instances, depressed production may also involve erythroid and leukocyte precursors.

While many viruses replicate in megakaryocytes, they may cause only a transient thrombocytopenia. Some live virus vaccines act in a similar manner, causing thrombocytopenia 7 to 10 days after administration, but this is usually not clinically significant unless surgery is to be performed.

Deficiencies of vitamin B_{12} and folic acid may occasionally cause thrombocytopenia, as a result of **ineffective thrombopoiesis**, by inhibiting nucleotide synthesis and impeding megakaryocyte maturation. Leukocytes and red cells are similarly affected in these deficiency conditions.

Increased platelet destruction/consumption

Immune-mediated thrombocytopenia

The premature destruction of platelets may be associated with an antibody directed against either platelets or material attached to platelets. The antibody may be of autoimmune, isoimmune or alloimmune origin. It may also have developed following exposure of the animal to certain drugs, chemicals or viruses (symptomatic).

Autoimmune (idiopathic, primary) thrombocytopenia is said to occur when no identifiable cause can be found for the development of the anti-platelet antibody. Thrombocytopenia may occur in isolation or it may be a part of other immune-mediated diseases such as autoimmune hemolytic anemia or systemic lupus erythematosus.

Secondary or symptomatic immune-mediated thrombocytopenia has been shown to be caused by the development of antibodies against certain drugs, chemicals or viruses. Two mechanisms are involved. The first entails the production of an antibody directed against the drug or chemical. The drug–antibody complex then attaches to the platelet. The second follows the production of antibody directed against the combination of the material and the platelet. A number of drugs have been identified to act in either of these ways including penicillin, phenylbutazone and sulfonamides.

Isoimmune thrombocytopenia follows the acquisition of maternally derived anti-platelet antibodies either via colostrum or transplacentally. This type has been seen most commonly in neonatal pigs. One cause of the development of anti-platelet antibodies in the sow was vaccination, during pregnancy, with hog cholera vaccines containing crystal violet. Antibodies directed against red cells may also occur.

Thrombocytopenia of alloimmune origin has been described following multiple platelet transfusions.

In most cases of immune-mediated thrombocytopenia, the marrow megakaryocytic response is excellent and has all the features of a vigorous thrombopoiesis. As there are many young and efficient platelets in the circulation, clinical signs may not be apparent until thrombocytopenia becomes extreme (usually less than 20×10^9/liter).

Acquired, non-immune-mediated thrombocytopenia

Massive thrombocytopenia may be secondary to a number of disorders, such as septicemia, disseminated intravascular coagulation (DIC), arterial thrombi, burns, reactions to some drugs, and cavernous neoplasia such as hemangiosarcoma. The common factor appears to be widespread injury to the vasculature. In these conditions, widespread microthrombus formation, triggered by infections or toxins, result in consumption of platelets, some coagulation factors and occasionally hemolytic anemia.

Platelet sequestration

In disorders involving splenomegaly, platelet numbers may fluctuate and may be moderately reduced but are not as low as in immune-mediated thrombocytopenia. There may be some premature destruction of platelets in the enlarged spleen, but the condition is caused primarily by passive pooling of a large fraction (up to 90%) of the total platelet mass. This is brought about by the slow passage of platelets through the tortuous channels of the splenic sinuses.

Thrombocytosis

Elevated platelet counts may follow autonomous production, such as in myeloproliferative disorders, from overproduction in response to thrombopoietin increases (secondary thrombocytosis) and transiently after removal of an enlarged spleen.

Myeloproliferative disorders

An extremely rare marrow neoplasm affecting only megakaryocytes is termed megakaryocytic myelosis. In this condition there are extremely large numbers of platelets in circulation (thrombocythemia) and myelophthisis with evidence of aplasia of the erythroid and myeloid series. Polycythemia vera, granulocytic leukemia and myelofibrosis may also be associated with marked thrombocytosis. Platelets in these conditions do not have normal functional capabilities.

Secondary thrombocytosis

Overproduction of platelets occurs in many chronic inflammatory diseases, in recovery from acute inflammatory conditions, in association with regenerative anemias, particularly following acute hemorrhage, and with iron deficiency anemia. In these cases there may be a rebound or 'sympathetic' thrombopoietic stimulatory effect in association with increases in other hematopoietic hormones. Platelets function normally in these conditions.

Qualitative disorders

Disturbances to platelet function can be congenital or acquired. Platelet numbers are normal, but there is a prolonged primary bleeding time in affected patients because of deficiencies in platelet adhesion, aggregation or release.

Congenital disorders of platelet function (thrombopathy)

These rare disorders result from defects in adhesion or aggregation of platelets (thrombasthenias), or in the release reaction of ADP in response to exposure to collagen. The latter may result from insufficient stores of ADP in dense granules (storage pool deficiency) or from abnormalities in the mechanism for its release. A variety of other congenital thrombopathies occur, such as glycogen storage diseases and Von Willebrand's disease. The latter is a combination of a factor VIII deficiency and a platelet function defect.

Acquired disorders of platelet function

The administration of aspirin and other non-steroidal anti-inflammatory drugs will inhibit platelet function. Aspirin acetylates and blocks cyclo-oxygenase activity in the platelet, directly inhibiting the primary steps in the release I reaction of electron dense granules, and consequently inhibiting secondary aggregation. In uremia there is interference with platelet adherence and aggregation. The mechanism is not understood but is believed to be due to the presence of a low molecular weight metabolite of urea, and to decreased platelet factor 3 activity.

Other drugs inhibit platelet activity by increasing intraplatelet cyclic AMP levels. These include prostaglandin E1, glucagon, caffeine and aminophylline. Other substances inhibit primary aggregation by acting on platelet membranes or surface fluffy coats. These include procaine, macroglobulins and paraproteins.

Platelet function *in vivo* also may be altered in metabolic diseases (such as diabetes mellitus), renal failure, in severe liver disease and myeloproliferative disorders.

Coagulation

The second major arm of the hemostatic mechanism is totally directed toward the formation of fibrin, a large insoluble protein formed from the precursor fibrinogen in plasma. Fibrin is a large strand-like protein and has the major attribute of sealing minor or major defects in the endothelium lining blood vessels. Because of its physical arrangement, it has the capacity to trap platelets and other particulate constituents of the blood.

However, to arrive at the final product of fibrin, there are a host of other proteins, material from platelets and finally calcium which must first interact. The sequence of events or cascade leading to the formation of fibrin is an amplification mechanism to mobilize a sufficient amount of fibrin from a relatively small start. It may be thought of as an exercise in body economy, where, to achieve a stable clot, only large amounts of one or two proteins are necessary. There are checks and balances in the system which prevent the untoward production of fibrin, the so-called counter-coagulation factors inhibiting the production of fibrin or enhancing its breakdown once it has been formed.

The coagulation factors so far identified have by convention been designated by roman numerals (Table 3.2). Most are produced by the liver, although major components of one (factor VIII) are produced by platelets and endothelial cells. Another important feature is that some factors (II, VII, IX, X) require vitamin K for their production. The coagulation cascade may be separated broadly into three major pathways, the intrinsic (or intravascular) system, the extrinsic (or tissue) system and the common pathway (Fig. 3.17). The intrinsic system is composed of coagulation factors in the plasma. This portion of the cascade is triggered by contact of its initial protein factor XII (Hageman factor) with a number of substances which come into contact with it, usually after damage to vascular endothelium (Fig. 3.18). The extrinsic system is triggered by tissue injury which releases tissue thromboplastins (Fig. 3.19). Both intrinsic and extrinsic systems come together and continue via the common pathway (Fig. 3.20).

So far, the activation of the coagulation pathway has been highlighted. Probably just as important, but by no means as well defined, are the mechanisms that inhibit the activation of clotting, the so-called counter-coagulation factors.

Disorders of coagulation (coagulopathies)

The absolute requirement for an efficient system of coagulation can be easily seen when there is only a deficiency of one factor, particularly of a factor close to the formation of fibrin.

In general the closer that the deficiency occurs to fibrin, the more serious are the consequences. Indeed, prothrombin and fibrinogen deficiencies are incompatible with life. Probably the best-known example of a congenital defect is that of factor VIII, the disease usually being termed hemophilia A. It is characterized clinically by prolonged bleed-

Table 3.2. *Coagulation factors*

Factor	Name	Vit. K dependent	Special characteristic
I	Fibrinogen	F	
II	Prothrombin	K	
III	Tissue thromboplastin	Lipoprotein	
IV	Calcium	—	Clinically not associated with coagulation disorders
V	Labile factor	F	Very labile to time and temperature
VII	Stable factor proconvertin	K	Stable, present in aged serum
VIII	Anti-hemophilic globulin	F	Labile to time and temperature
IX	Christmas factor, plasma thromboplastin component	K	Stable, present in aged serum
X	Stuart Prower factor	K	
XI	Plasma thromboplastin antecedent	C	
XII	Hageman factor, glass activation factor	C	Deficiency only causes defect *in vitro*
XIII	Fibrin stabilization factor	F	

F, Fibrinogen group – non-enzymatic protein.
K, Vitamin K-dependent enzymes.
C, Contact factors enzymes.

ing from wounds, the appearance of hematomas from blunt trauma, and bleeding, particularly into body cavities and joints. In this particular case, the factor VIII molecule is incomplete.

Coagulopathies originate in three ways. There may be a congenital factor deficiency, an acquired deficiency of production of one or more factors or there may be an excessive consumption of coagulation factors. All result in similar clinical signs of ecchymotic hemorrhages and hematomas. Hemorrhage most commonly occurs into body cavities and joints and there is prolonged bleeding from wounds or venepuncture sites. An outline of the coagulopathies is given in Fig. 3.21.

Congenital deficiencies may result from either reduced production or synthesis of inactive factors. It is rare to have more than one factor involved, and most are inherited and of a recessive character. Some are sex linked. The most commonly observed congenital coagulopathies are listed in Table 3.3.

Acquired deficiencies of production of coagulation factors usually involve deficiencies of several factors. Causes include interference with vitamin K synthesis in Warfarin or Dicoumarin poisoning, and in severe liver disease. Less commonly, individual coagulation factors may be inhibited selectively or inactivated by chemicals (heparin, protamine) or immunoglobulins produced in various disease states, for example amyloid inactivates factor X; specific antibody to coagulants may appear in myeloma.

Increased consumption of coagulation factors is an important secondary disease process; disseminated intravascular coagulopathy (DIC) (or secondary fibrinolysis) can be triggered by other disease processes such as infections, neoplasia and heat stroke. The massive activation of coagulation and platelet aggregation leads to depletion of fibrinogen related factors and platelets. After such depletion the affected animal is unable to react to calls for clotting.

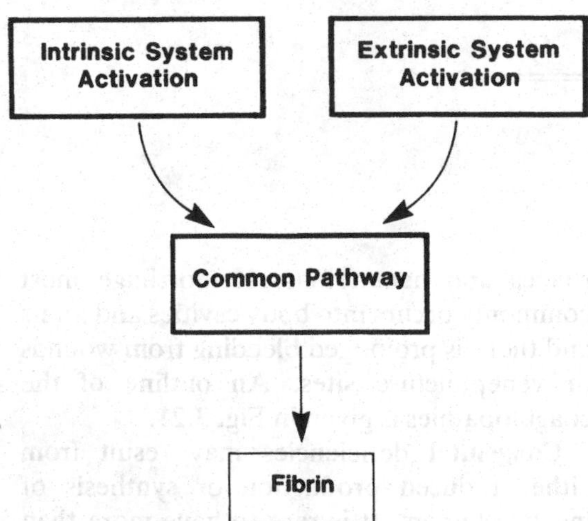

Fig. 3.17. The coagulation cascade is directed toward the production of fibrin via the common pathway. Common pathway activation derives from activation of either the intrinsic or extrinsic system.

The blood vessel

The endothelium lining all blood vessels of the body provides a benign environment for circulating blood. Endothelial cells separate platelets and coagulation factors from the material that excites them to aggregate and coagulate, respectively. They also contain and possibly secrete inhibitors which modulate platelet aggregation, namely prostacyclins (prostaglandin PGI_2). PGI_2 is the most potent inhibitory prostaglandin reported. It inhibits both the expression of fibrinogen receptors on the platelet and the secretion of dense granule constituents.

Primary diseases of blood vessels, particularly those that lead to disruption of the endo-

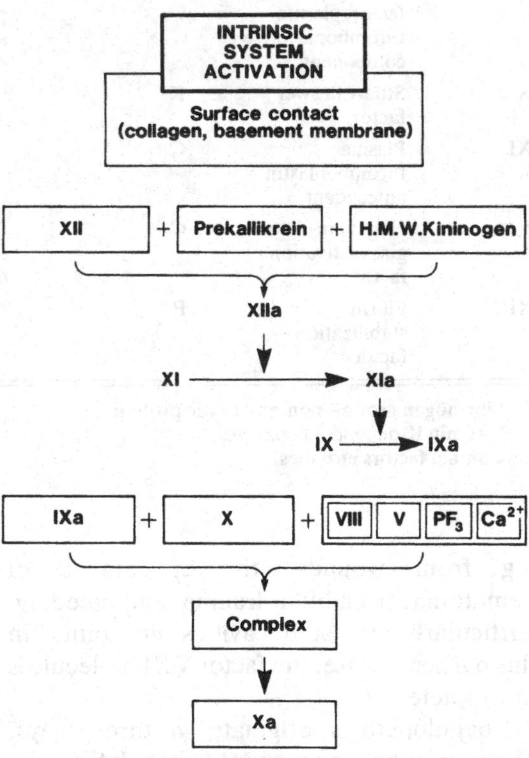

Fig. 3.18. Intrinsic system activation follows contact with substances such as collagen and basement membranes lining blood vessels. This in turn moves through a rather complex sequence leading to the production of activated factor X. H.M.W., high molecular weight; PF, platelet factor; a, activated.

thelium, lead to the formation of thrombi within the vessels. Of these, viral and rickettsial infections are the most important. In a number of viral diseases, including hog cholera and infectious canine hepatitis, there is destruction of vascular endothelial cells. The result is widespread exposure of sub-endothelial collagen and basement membrane, both of which are potent platelet aggregators and initiators of the intrinsic coagulation cascade. Direct endothelial damage by endotoxin is another cause acting by a similar mechanism.

Other diseases of blood vessels include **allergic purpura**, which is a widespread vasculitis characterized by cutaneous ecchymoses, petechiae and edema. The reaction may be triggered by drugs, antibiotics and constituents of food or vaccines which lead to the deposition of immune complexes in capillaries and arterioles. Endothelial damage follows the fixation of complement by the deposited immune complexes. Allergic purpura may be seen in horses

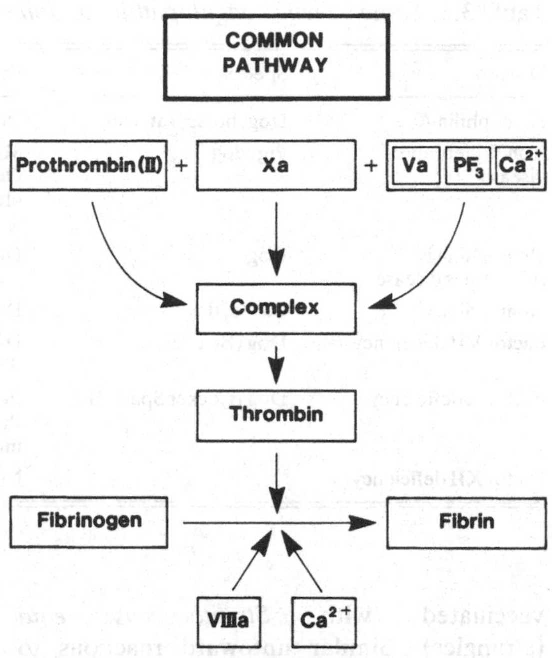

Fig. 3.20. The central feature of the common pathway is the production of fibrin which is initiated by the combination of activated factor X and a number of other factors. PF, platelet factor; a, activated.

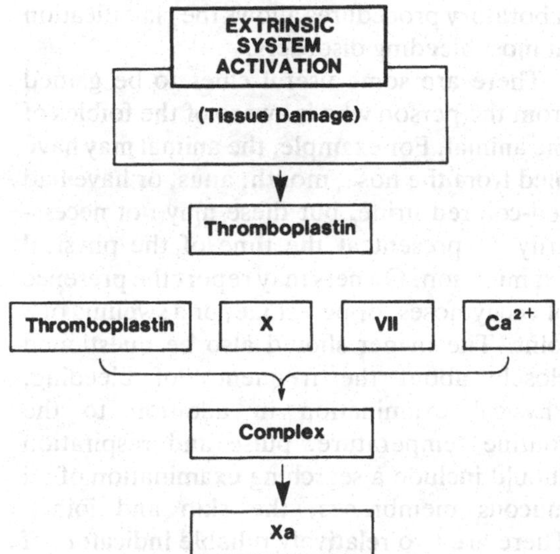

Fig. 3.19. While the extrinsic system also results in the activation of factor X, the trigger is tissue thromboplastin which combines with X, VII and Ca^{2+}. a, activated.

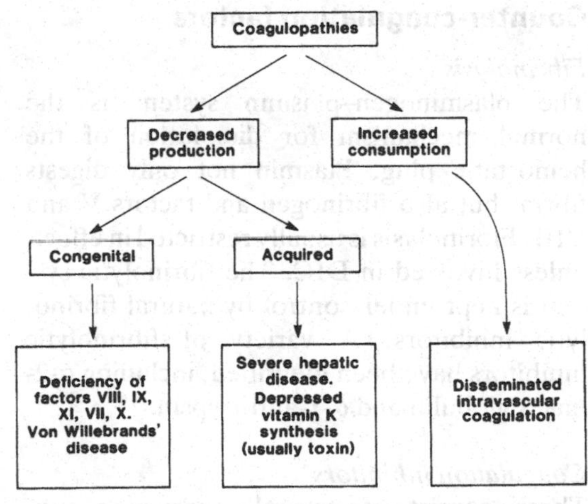

Fig. 3.21. Coagulopathies result from either decreased production or increased consumption. All are manifest by a tendency to bleed with ecchymoses particularly associated with decreased production of coagulation factors.

Table 3.3. *Some genetic coagulopathies of domestic animals*

Disease	Species	Features
Hemophilia A	Dog, horse, cat, cattle	Defect of factor VIII
Von Willebrand's disease	Pig, dog	Mild disorder. Deficiency of factor VIII plus platelet function defect
Hemophilia B (Christmas disease)	Dog	Deficiency of factor IX
Hemophilia C	Cattle, dog	Deficiency of factor XI
Factor VII deficiency	Dog (Beagle)	Homozygotes – mild disorder
Factor X deficiency	Dog (Cocker Spaniel)	Fatal in neonates. Produces fetal mummification
Factor XII deficiency		No clinical disease

vaccinated with *Streptococcus equi* (strangles). Similar untoward reactions to vaccination may be seen occasionally in dogs and pigs. Scurvy (vitamin C deficiency) and some congenital collagen disorders increase vascular fragility and permeability leading to the formation of petechiae and ecchymoses and prolonged bleeding time.

Counter-coagulation factors

Fibrinolysis

The plasminogen–plasmin system is the normal mechanism for dissolution of the hemostatic plug. Plasmin not only digests fibrin, but also fibrinogen and factors V and VIII. Fibrinolysis is usually restricted in effect, unless involved in DIC. The fibrinolytic system is kept under control by natural fibrinolytic inhibitors. A variety of fibrinolytic inhibitors have been identified, including α-2-macroglobulin and α-1-anti-trypsin.

Coagulation inhibitors

These consist of naturally occurring substances including antithrombins, heparin cofactor II, protein C and α-2-macroglobulin. Decreases of antithrombin levels occur in DIC and glomerulopathies (nephrotic syndrome) resulting in a hypercoagulable or thrombotic state.

Clinical evaluation and laboratory diagnosis of bleeding disorders

A combination of the history, physical examination and some relatively simple laboratory procedures allows the classification of most bleeding disorders.

There are some useful clues to be gained from the person who is aware of the foibles of the animal. For example, the animal may have bled from the nose, mouth, anus, or have had red-colored urine, but these may not necessarily be present at the time of the physical examination. Owners may report the presence of ecchymoses, or petechiae, or a swelling of a joint. The owner should also be questioned closely about the frequency of bleeding. Physical examination in addition to the routine temperature, pulse and respiration should include a searching examination of all mucous membranes, the skin and joints. There are two relatively reliable indicators of platelet versus coagulation factor defects. The **presence of petechiae is almost always diagnostic of platelet abnormalities** and the development of **spontaneous hemarthroses is**

usually limited to coagulation factor deficiencies (Fig. 3.22).

One should not be left with the idea that all bleeding disorders are primary defects of hemostasis. Care should be taken to rule out the possibility that the bleeding observed is not secondary to some other disease such as viral or bacterial infection. When other primary disease is present, additional clinical signs are usually present.

Laboratory assessment

Once a disorder of hemostasis is suspected and a complete blood count carried out, a limited number of screening tests may be performed to define the defect. They are a platelet count, a partial thromboplastin time, a prothrombin time, thrombin time and possibly a bleeding time, although in practice the bleeding time is difficult to standardize. These five relatively simple tests allow the clinician to decide whether the problem is basically a platelet disorder, an intrinsic system, an extrinsic system or a common pathway problem (Fig. 3.23).

The evaluation of platelet numbers and morphology has been discussed in detail earlier in the chapter. The reader is referred to p. 64 for a review.

The **partial thromboplastin time** (PTT) measures the activity of both the intrinsic and common pathways. A partial thromboplastin such as cephalin or inosithin (soya phosphatide) is added to a citrated blood sample and clotting time is measured at 37 °C after the addition of calcium. The clotting time is prolonged if there is a depression in coagulation factor activity. The **prothrombin time (PT)** measures the activity of the extrinsic (factor VII) and common pathways. It is initiated by adding a 'complete' thromboplastin such as an extract of rabbit brain to plasma.

The **thrombin time** tests the integrity of the common pathway from fibrinogen to fibrin. In this test, thrombin is added to plasma and the clotting time measured. Table 3.4 shows the usefulness of the three coagulation factor tests to decide if the bleeding is the result of an intrinsic, extrinsic or common pathway defect.

Additional specialized tests

Depending on whether the hemostatic abnormality is platelet or coagulation factor derived, there are a number of specialized tests to assess platelet function and to define the specific coagulation factor defect. For platelets, functions such as adhesion and aggregation, the release reaction and coagulant activity may be assessed.

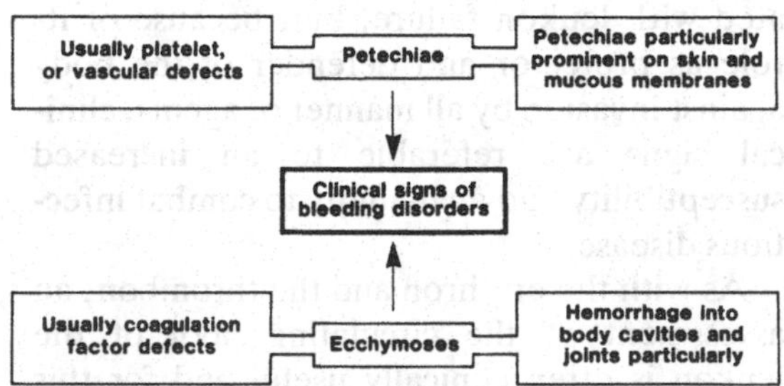

Fig. 3.22. Among the clinical signs of bleeding disorders are petechiae, which are of platelet or vascular origin, and ecchymoses, which are usually of coagulation factor origin.

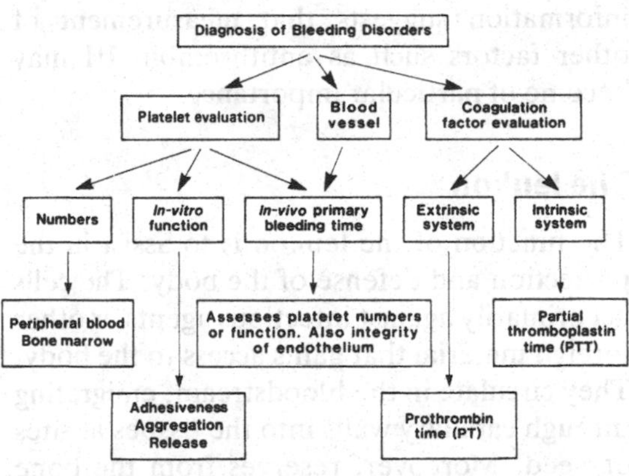

Fig. 3.23. The laboratory assessment of bleeding disorders includes the evaluation of: platelet numbers and function, coagulation factors, and in some cases blood vessels. Initial steps include enumeration of platelets, the PT and PTT. As disseminated intravascular coagulopathy (DIC) is a common disorder, fibrinolytic activity is also assayed (see the text).

Table 3.4. *Evaluation of a hemostatic abnormality with three screening tests*

	Test		
Group	PTT	PT	Thrombin time
Intrinsic pathway abnormality	Prolonged	Normal	Normal
Extrinsic pathway abnormality	Normal	Prolonged	Normal
Common pathway abnormality	Prolonged	Prolonged	Usually normal

PTT, partial thromboplastin time; PT, prothrombin time.

Measurement of fibrinolytic activity

There are numerous times after the screening tests have been performed when both platelet numbers are low and coagulation factor clotting times are prolonged. In this circumstance the most likely cause is a consumption coagulopathy following DIC. Both platelets and coagulation factors have been consumed, usually because of widespread endothelial damage. To confirm this suspicion the concentration of fibrin degradation products (FDP) is assayed. It is an indirect measurement of the activity of the fibrinolytic system, particularly of plasminogen. Relatively recent information suggests that measurement of other factors such as antithrombin III may become of particular importance.

The leukon

The function of the leukon is to assist in the protection and defense of the body. The cells act primarily against infectious agents or other foreign material that gains access to the body. They circulate in the bloodstream, emigrating through capillary walls into the tissues at sites of need. Moreover, reserves from the bone marrow can be mobilized at short notice, increasing the numbers in the bloodstream and ultimately in the tissues. As such the leukon is a **responder**. If the stimulus to respond is sufficiently great, it is reflected in an absolute or relative change in number and type of cell in the bloodstream. Such responses are readily quantified and are used as indicators of disease processes elsewhere in the body. The quantification of the response is also used to predict the outcome of particular disease states.

In contrast to its ability to respond, the leukon or components of it may **fail primarily** in three ways: all or parts of it may be congenitally deficient; its cells may be attacked directly and destroyed by infectious or toxic agents; and finally its function may be compromised by proliferative or neoplastic disease. The leukon may also **fail secondarily** by exhaustion and/or suppression of its reserves, as in an overwhelming bacterial infection. In this circumstance the cells of the leukon are not targeted primarily by the invading organism but die in the line of duty.

There are no specific clinical signs associated with leukon failure, but, because of its role as protector and defender of the body against invasion by all manner of agents, clinical signs are referable to an increased susceptibility and an inability to combat infectious disease.

As with the erythron and the thrombon, an assessment of the circulating pool of the leukon is often clinically useful and for this reason is carried out commonly. However, a single total and differential white cell count carries with it some inherent deficiencies. It gives only an indication of the state of a portion of the leukon at any one time, omitting the state of the marrow reserves and the remaining portion of the leukon in the circulation that is said to be marginated (those cells closely associated with vessel walls) (Fig. 3.24). For these reasons it is desirable to examine a number of blood samples over time and, if possible, the bone marrow also. If considered necessary, there are more sophisticated, but cumbersome, tests of function that may be carried out.

With these introductory sentences in mind, the leukon will be discussed initially as an indicator of disease processes in the body and then in terms of functional leukon failure. Little emphasis will be placed on lymphocyte structure and function, as this is covered in Chapter 2.

The cells of the leukon

The leukon may be divided conveniently into two major arms, the myeloid series and the lymphoid series. Cells of the myeloid series arise from a common stem cell in the bone marrow which divides and differentiates into either the myeloblast or the monoblast. The **myeloblast**, through further division and differentiation to the promyelocyte, myelocyte, metamyelocyte and band forms, ultimately produces the neutrophil, the eosinophil or the basophil (Fig. 3.25). These three cells are known collectively as the granulocytes. The **monoblast**, through a similar process, results in the production of the monocyte, the immediate precursor of the macrophage. Similarly, the lymphoid series arise from a common stem cell leading finally to committed cells of the bursal equivalent-derived group ('B' cells) and the thymus-derived group ('T' cells).

Each fully differentiated cell of the leukon has its own particular set of functions which is reflected in its structure. So, the plasma cell which is concerned with antibody production has abundant rough endoplasmic reticulum and Golgi apparatus, whereas the cytoplasm of the neutrophil is almost entirely filled with membrane-bound granules, bursting with weapons of destruction such as acid hydrolases and proteases.

The production and release of each cell type of the **granulocyte series** from the bone

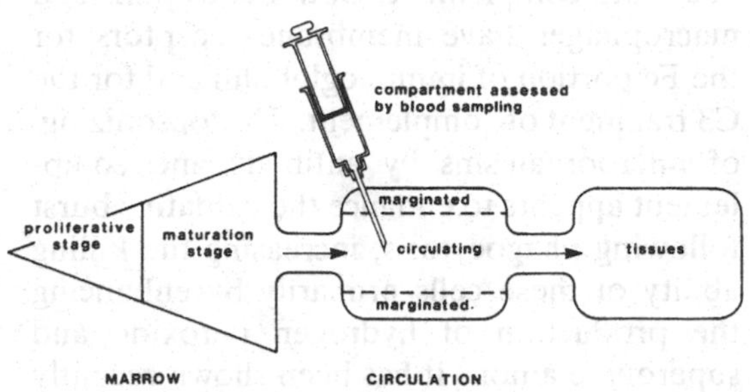

Fig. 3.24. A peripheral blood sample gives only an indication of the state of a portion of the leukon at any one time.

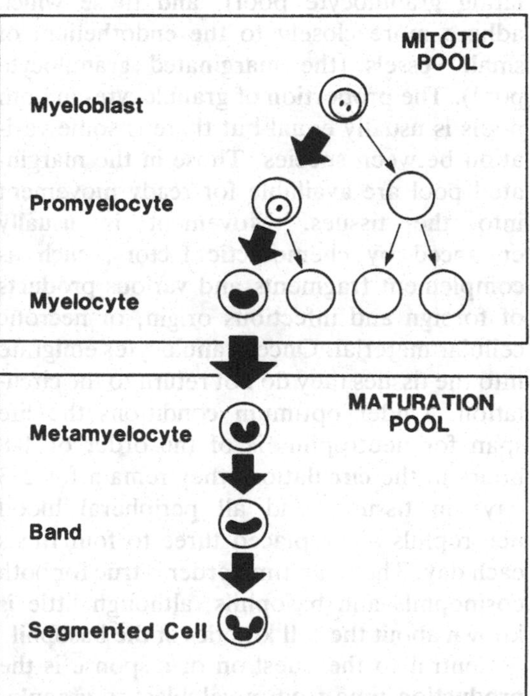

Fig. 3.25. Division and differentiation of the myeloblast, which ultimately produces the neutrophil, eosinophil and the basophil.

marrow is exquisitely regulated by a number of factors including colony-stimulating factor (CSF) and leukocytosis-inducing factor (LIF). CSF acts by increasing both stem cell input and effective proliferation (the percentage of immature cells that reach maturity) and LIF increases the egress of leukocytes from the bone marrow. In some cases the increased leukocyte output is at the expense of red cell precursors.

On increased demand, the maturation stages are released into the circulation in order of cell maturity, the most differentiated being released first. Release, at least for neutrophils, is under the control of LIF. The increased release of mature and immature cells comes from the marrow granulocyte reserve (MGR). The MGR is composed of the metamyelocyte, band and segmented cells. Once released into the circulation the granulocytes compartmentalize into two interchangeable pools: those freely circulating (the circulating granulocyte pool), and those which adhere more closely to the endothelium of small vessels (the marginated granulocyte pool). The proportion of granulocytes in both pools is usually equal but there is some variation between species. Those in the marginated pool are available for ready movement into the tissues. Movement is usually enhanced by chemotactic factors, such as complement fragments and various products of foreign and infectious origin, or necrotic cellular material. Once granulocytes emigrate into the tissues they do not return to the circulation. Under optimum conditions the life span for neutrophils is of the order of 4–8 hours in the circulation; they remain for 2–3 days in tissues, and all peripheral blood neutrophils are replaced three to four times each day. The same time order is true for both eosinophils and basophils, although little is known about the cell kinetics of the basophil.

Central to the question of response is the production time from myeloblast to granulocyte. For neutrophils it is of the order of 6 days. So if most of the precursors are used or destroyed, the granulocytes of the bone marrow may be reconstituted and release resumed within a relatively short time. Indeed, when the need is great, increased CSF reduces myelocyte death, resulting in a quicker response (2–3 days).

Marrow production of monocytes proceeds more rapidly than that of the granulocytes (< 2 days), but unlike the granulocytes there is no monocyte reserve in the bone marrow. Under conditions of increased monocyte demand, the cell cycle is shortened, which is associated with an increase in the numbers of immature monocytes released into the circulation. Their transit time in the circulation is about 20 hours and they do not re-enter the circulation. In the organs and tissues monocytes differentiate into macrophages, which have different roles and properties according to their location. Collectively they constitute the monocyte–macrophage system of the body.

Functions of cells of the leukon

The commitment to defend the body is reflected in the specialized function of each of its members (Fig. 3.26).

Phagocytosis is not limited to neutrophils and macrophages, but is developed to the highest degree in these cells. Eosinophils are not noted for their phagocytic ability except for the ingestion of immune complexes, whereas neutrophils and macrophages will attempt to phagocytose almost anything. In the case of microorganisms, the cell's phagocytic ability is aided by the presence of antibody and complement. Both neutrophils and macrophages have membrane receptors for the Fc portion of immunoglobulin and for the C3 fragment of complement. The 'opsonizing' of microorganisms by antibody and complement appears to enhance the oxidative burst following phagocytosis, increasing the killing ability of these cells primarily by enhancing the production of hydrogen peroxide and superoxide anion. It has been shown recently that macrophages also secrete a number of substances, including lysozyme, hydrogen peroxide and lysosomal enzymes.

Macrophages are also intimately concerned with the immune response, processing antigen for presentation to lymphocytes. Neutrophils are chemotactically attracted to particular areas in response to a number of chemical agents including those generated by the inflammatory response such as C3 and C5 complement fragments. Macrophages move more slowly and are usually attracted into the tissue by bacterial products, complement fragments and lymphokines. Both neutrophils and macrophages contain abundant lysosomes, which fuse with the membrane-enclosed, phagocytosed material. Enzymes active at low pH (3.5–4.0) and the production of peroxides of various types form the major mechanisms of destruction of the phagocytosed material.

Eosinophils also have cytoplasmic receptors for complement. There are also specific chemotactic factors for eosinophils and they are found in tissues often associated with mast cells in hypersensitivity reactions. Eosinophils appear to modulate the IgE immediate hypersensitivity response. They are also usually associated with parasitic infections. One of the components of their lysosomal granules, major basic protein, is a potent cytotoxin for certain parasites. Compared to the other members of the myeloid series comparatively little is known about the basophil. It is characterized by the presence of abundant cytoplasmic granules which contain both histamine and heparin. Basophils are capable of phagocytosis and in most species have membrane receptors for IgE and appear to have a functional relationship with mast cells. Basophils also release heparin in response to a postprandial lipemia. The released heparin activates lipoprotein lipase.

Leukon response and failure

The term leukon response refers to the ability of one or more of the cell lines comprising the leukon to combat effectively or to contain whatever insult is thrust its way, changing to meet the needs of the body and returning to normal again once the insult has been removed. In contrast, leukon failure refers either to the inability of one or more of the major cell lines to respond initially to an insult, or to be unable to respond further following a primary response.

Response or failure are usually separable, but there are times when response merges with failure. It is often a question of degree, a transition phase. For example, the leukon may be able to respond initially to a bacterial infec-

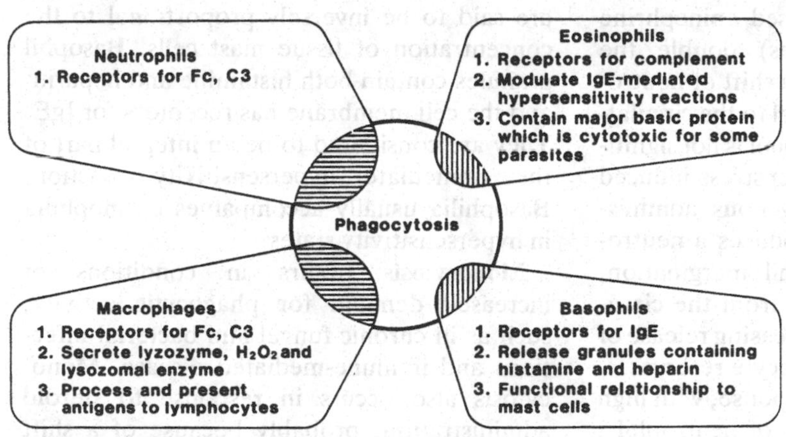

Fig. 3.26. Function of cells of the myeloid series. Note that all are phagocytic, but it is developed to the highest degree in neutrophils and macrophages.

tion, only to lapse into failure and to lead possibly to the death of the animal following the exhaustion of all available reserves.

The leukon response

For the myeloid series, the ability to respond is related to the functional reserve capacity of the bone marrow. It also requires a normal bone marrow structure and function. The response usually reflects disease in other organ systems within the body or it may reflect a response to a physiologic process such as the release of endogenous epinephrine. The response is influenced by two primary factors: **the rate of input from the bone marrow, and the rate of exit from the bloodstream**.

Response is usually manifest by an absolute increase in one or more members of the myeloid series. So there may be a neutrophilia, eosinophilia, basophilia, or monocytosis. As part of the response immature forms not normally released from the bone marrow are released, giving rise to a **left shift** (increase in circulating numbers of immature cells).

The most common causes of a **neutrophilia** include stimuli such as bacterial infections or tissue necrosis. Causes of a neutrophilia also include physiologic neutrophilia (or pseudoneutrophilia) following endogenous epinephrine release in response to fear, excitement or exercise. The released epinephrine will transiently (20 minutes) double the neutrophil count because of a shift of neutrophils from the marginated pool to the circulating pool. However, the total pool is not significantly altered. Similarly, either stress-induced endogenous release or exogenous administration of corticosteroids produces a neutrophilia by reducing neutrophil margination, decreasing neutrophil egress from the circulation into the tissues and increasing release of cells from the marrow granulocyte reserves.

Occasionally, a marked response, with high cell numbers and immaturity of neutrophils, may be found at levels usually seen only with granulocyte neoplasia, and is termed a **leukemoid reaction**. It occurs rarely, in response to a variety of disorders such as chronic infections or neoplasia in other tissues. The mechanism of this response is not known but it probably resides in an effect on the factors concerned with the normal regulation such as CSF.

Neutrophilia with myeloid and erythroid immaturity in the circulation may also be observed in some conditions such as bone marrow neoplasia, myelofibrosis, or immune-mediated anemias. It is termed **leukoerythroblastosis** and is probably the result of altered marrow production and indiscriminate sinusoidal release.

Eosinophilia may be seen with hypersensitivity reactions, particularly in parasitic infections where there is continued antigenic stimulation with extensive larval migration, such as in some helminth infections. Eosinophilia is also seen with particular types of gastrointestinal or respiratory disease characterized by massive infiltration of the submucosa with eosinophils. Although the etiology is unknown, it is likely to be allergic in origin. Corticosteroid levels are also important modulators of peripheral eosinophil numbers. When steroid levels are high, eosinophil numbers are often decreased.

Basophils, although of different origin, appear to share a functional relationship with mast cells, where circulating basophil numbers are said to be inversely proportional to the concentration of tissue mast cells. Basophil granules contain both histamine and heparin, and the cell membrane has receptors for IgE. They are considered to be an integral part of the immediate hypersensitivity reaction. Basophilia usually accompanies eosinophilia in hypersensitivity states.

Monocytosis occurs in conditions of increased demand for phagocytic activity, such as in chronic fungal and bacterial infections and immune-mediated disease. Monocytosis also occurs in response to steroid administration, probably because of a shift from the marginated to the circulating pool.

It is not often that only two states exist, that of either compensation (response) or failure;

there is in many cases a transition phase from response to failure. Nowhere is this more evident than with the dynamics of the leukon. There are numerous instances of this 'in-between' phase. For example, during a systemic bacterial infection the leukon may respond initially, then begin to become exhausted, particularly after the consumption of the marrow granulocyte reserves. Depending on the circumstances, the leukon may finally gain the upper hand and combat the infection or alternatively move into failure. The transition phase is also observed with neoplasia of the leukon, particularly in the early phases when there are sufficient remaining normal cells in the leukon capable of responding. In contrast to the bacterial infection, neoplasia ultimately ends in failure.

At least for neutrophils, there are indications of which phase the leukon is in. When there is an increased demand placed on neutrophil numbers, progressively more immature cells are released, which is termed a **left shift**. When the mature cell numbers exceed immature numbers, it is termed a **regenerative left shift** and it carries with it a relatively good prognosis. However, when immature cell numbers exceed the number of mature neutrophils, it is termed a **degenerative left shift**, which carries with it a poorer prognosis. Another indicator of the state of the leukon is the presence of **toxic changes in neutrophils** (see p. 82).

Species variation in the leukon response

There is marked species variation in the functional capacity of the leukon to respond, particularly to infection. The pig and dog are able to mount the largest acute neutrophilic response, whereas the mature ruminant is more likely to have a neutropenia (a decrease in circulating neutrophil numbers) and a marked left shift in acute infectious states. Cows with chronic infections such as pulmonary or hepatic abscesses will show a marked neutrophilia. This suggests that one of the major differences is in the rate of the response.

The nature of the infective agent is most important in determining the leukocyte response. Generally, localized Gram-positive bacterial infections cause leukocytoses, whereas bacteremic Gram-negative infections, acute viral infections and systemic infections are more likely to cause leukopenia (decreased numbers).

Guidelines for leukocyte interpretation are:

1 the absolute increase in leukocytes indicates individual or species response, whereas
2 the degree of left shift and toxic changes indicate the marrow response.

In all species, a leukopenia with persistent neutrophil immaturity is a poor prognostic sign.

Pseudoneutropenia (sequestration of neutrophils) is seen when cells shift from the circulating pool to the marginated pool, and occurs with acute endotoxic shock, hypersensitivity and hemolysis. Cells may also marginate in a temporarily enlarged spleen following the administration of barbiturates or some tranquilizers such as acetylpromazine.

Leukon failure

When one component or, more rarely, all members of the leukon fail to meet the needs of the body, it not only fails to protect the body against virulent organisms but also against organisms that are normally of low virulence and little consequence. Except for the congenital deficiencies or absences of one or more cell lines, failure is a dynamic state. The animal slides into failure and as such there is a transition phase. Failure also emphasizes the contribution of each cell line to the whole, particularly those cells primarily concerned with phagocytic function, the neutrophils and macrophages, and those concerned with assisting phagocytosis, the B lymphocytes and the numerous T lymphocyte functions.

Leukon failure may be temporary or permanent; whichever it is, it is related to the disease affecting the leukon. Almost without exception, congenital defects and neoplastic transformation of particular cells of the leukon result in an irreversible failure, whereas some viral infections and plant, drug or bacterial toxins are at least potentially reversible if the animal can be nursed through the critical phase of susceptibility until the marrow is able to regenerate. Many of the acquired causes of reversible leukon failure do not affect leukocytes exclusively; other organs and tissues are often compromised, giving rise to a set of clinical signs, one of which is leukon failure (Fig. 3.27).

Congenital leukon failure

When present, which is rare in domestic animals, congenital disorders of phagocytic function wreak havoc on the affected animal. Probably the best recognized, but still little understood, is the granulocytopathy syndrome in Irish Setter dogs. This is an autosomal recessive trait characterized by the reduced ability of neutrophils to kill bacteria. Another is cyclic hematopoiesis of Gray Collie dogs, also inherited and of a recessive character. In this disease it is not only neutrophils that are affected. Both the erythroid elements and megakaryocytes are also affected, but it is the failure to maintain adequate numbers of neutrophils that is critical. The neutropenia is cyclic, occurring at intervals of 10–11 days.

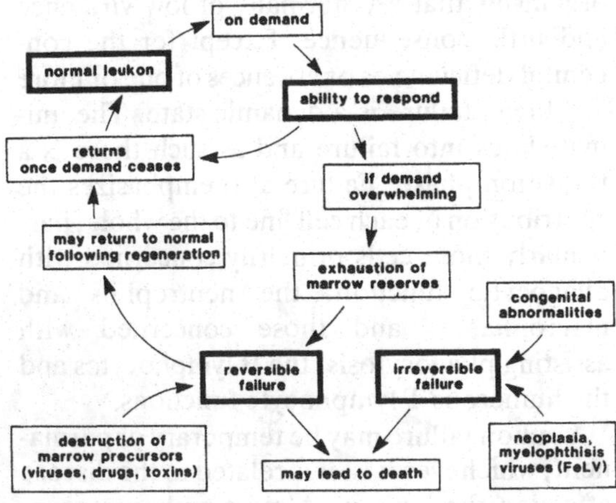

Fig. 3.27. The inter-relationship between the leukon response and reversible or irreversible leukon failure. FeLV, feline leukemia virus.

There is usually about a 3 day period of neutropenia followed by a 6–7 day period of normal to neutrophilic levels. Affected animals die prematurely of bacterially related diseases.

Chediak–Higashi syndrome of Persian cats, mink, Hereford cattle, mice and man is a defect in neutrophil bacteriocidal activity because of defective fusion of phagocytic vacuoles with the lysosomal granules. The basis of this defect appears to reside in neutrophil microfilaments. The syndrome is associated with partial albinism and affects all cells of the granulocytic series. Giant lysosomes are found in circulating granulocytes, lymphocytes and monocytes. The syndrome leads to an increased susceptibility to infection and a bleeding tendency because of defects in platelet aggregation. The classical disease in this category, although it has only been described in man, will be discussed: it is chronic granulomatous disease. In this condition both neutrophils and monocytes lack the oxidases necessary to metabolize oxygen to hydrogen peroxide or superoxide. This defect severely limits the ability to kill bacteria intracellularly. Hence patients have an increased susceptibility to staphylococcal and Gram-negative bacterial infections.

Congenital disorders of lymphocyte numbers and function are of more importance and occur more commonly. Discussion of these is found in Chapter 2.

Acquired leukon failure

In contrast to congenital leukon failure, acquired leukon failure occurs much more frequently. It may be the result of a reduced production or increased destruction and apart from neoplasia it is usually reversible. The latter fact of reversibility is of some importance as, if the animal can survive the 'critical period' with or without medical treatment, reconstitution of the leukon occurs quickly. Whether temporary or not, many cases of leukon failure are characterized by low absolute numbers (leukopenia), the most common of which is neutropenia.

Reduced myelopoiesis

Most conditions in this category are also accompanied by marrow aplasia. Common causes include viral infections such as canine and feline parvoviruses, toxic plants (*Pteridium aquilinum*), cytotoxic drugs and lastly irradiation, which is always mentioned but is seen rarely in domestic animals. All have in common the destruction of the proliferative phases of cells of the leukon. A removal of the proliferating cells of the leukon also occurs in space-occupying lesions in the bone marrow, such as myelophthisis from a hematopoietic neoplasm, or myelofibrosis, which is usually idiopathic but can be due to infection with feline leukemia virus in cats.

Increased destruction or utilization of the maturation phases

The most common causes of excessive utilization include overwhelming bacterial or viral infections where the rate of egress of neutrophils to the tissues exceeds the rate of inflow from the marrow. In many of these toxemic states, there is a combination of depletion of both marrow granulocyte reserves and an ineffective bone marrow response. Another rare cause is immune-mediated destruction of mature neutrophils, which can be either idiopathic or drug induced. In contrast to most others, immune-mediated destruction has associated with it marrow hyperplasia.

Neoplasia of the leukon

Only some general comments will be made here as this is dealt with in detail under the erythron (see p. 57).

All members of the leukon may undergo neoplastic transformation and the effects of such on the animal are due almost entirely to 'crowding out' of the remaining normal cells in the marrow giving rise to anemia and thrombocytopenia. There may also be premature release of neoplastic cells into the circulation, which is termed leukemia. The neoplastic cells, probably because of their immaturity, function poorly. In contrast to most cases of leukon failure, where leukopenia is the rule there may be, in cases of leukemia, moderate to marked leukocytosis.

Laboratory assessment of the leukon

As discussed previously, diagnostic tests of the leukon include analysis of peripheral blood and bone marrow, and *in-vitro* leukocyte function tests. Total leukocyte counts may be performed by automated methods, while differential white cell counts are done manually on a Romanowsky-stained blood smear in most veterinary diagnostic laboratories. In practical terms, the careful evaluation of a blood smear will yield much useful diagnostic information with minimum expense. Leukocyte morphology should be evaluated on a rapidly dried, stained blood smear made within a few hours of collection. Neutrophils, eosinophils, lymphocytes and, to a lesser extent, monocytes, can demonstrate diagnostically useful morphologic changes.

Neutrophils

In domestic species segmented neutrophils from normal animals range from 3×10^9 to 7×10^9/liter with carnivores having a higher normal level of neutrophils than herbivores and thus a higher neutrophil/lymphocyte (N/L) ratio. Band neutrophils vary from 0.02×10^9 to 0.2×10^9/liter in domestic species, with the exception that under normal circumstances most herbivores do not have band cells in the peripheral blood. The interpretation of 'left shift' is largely subjective but should involve a consideration of both the **degree of immaturity** and the **numbers** of immature cells. So, immaturity characterized by the increased presence of band neutrophils is generally interpreted as a **mild left shift**, while reactions characterized by both bands and metamyelocytes constitute a **moderate left shift;** bands, metamyelocytes and myelocytes or younger **comprise a marked left shift**. In inflammatory states, cells younger than myelocytes are not seen circulating, while, in myeloid leukemias, promyelocytes and blast cells will be present. These interpretations based on cell immaturity must be considered in the light of the numbers of cells present. When the number of immature neutrophils exceeds the number of mature neutrophils, the reaction is termed a **degenerative left shift** and the prognosis is usually guarded. Thus, a cow with 5×10^9 mature neutrophils/liter and 10×10^9 band, metamyelocytes and myelocytes/liter would be considered to have a marked degenerative left shift. Septicemia may induce this response. The entire neutrophil picture (numbers and immaturity) depicts the balance between tissue demand and the ability of the marrow to supply functional, mature cells.

Toxic changes

These are seen in severe inflammatory and toxemic states and include Döhle bodies. These are blue, angular cytoplasmic bodies that persist from the normal cytoplasmic basophilia of immature granulocytes. Mature neutrophils with these bodies will most likely contain fewer specific granules than normal. This change is generally considered to be the most severe toxic change. **Cytoplasmic vacuolation** may be seen and, although there is a high correlation between neutrophil vacuolation and septicemia in man, this change can be artefactual in blood stored in EDTA anticoagulant for prolonged periods. Cytoplasmic **toxic granulation** is the persistence of primary azurophilic granules in the mature neutrophil, which appear as dark pink granules with Romanowsky stain. **Cytoplasmic basophilia** is a combination of increased RNA and reduced specific granulation which contributes to an overall blue cytoplasmic stain reaction. Nuclear karyolysis may be present in peripheral blood neutrophils in toxemic states. There may be a left shift and nuclear hyposegmentation, with the appearance of 'snake-like' nuclei. This change is more pronounced in septic fluids in contact with bacterial toxins causing various degrees of nuclear swelling and degeneration.

Nuclear hypersegmentation is seen when neutrophils contain more than five nuclear lobes. This can occur with prolonged steroid administration, vitamin B_{12} and folate deficiencies, and in association with malignancy states. It may also occur as a rare, inherited condition. The hypersegmentation represents an abnormality of maturation, rather than being solely due to increased age of the cell. Macropolycytes (giant granulocytes) with hypersegmentation may be seen with vitamin B_{12} and folate deficiencies. The presence of large numbers of hypersegmented neutrophils is sometimes referred to as a 'shift to the right'.

Asynchronous maturation (pseudo-Pelger–Huet anomaly) is diagnosed if the nuclear development appears to be out of step with cytoplasmic development, or if the degree of nuclear segmentation is not in keeping with the chromatin structure. This occurs with increased stress on granulocyte production such that a maturation step is excluded. For example, a metamyelocyte transforms into a segmented cell with a bilobed nucleus, thus eliminating the band stage. This occurs particularly in ruminants with infections. A similar change may occur in the course of myeloproliferative disorders, and is due to an abnormality of chromatin synthesis. **Pelger–Huet anomaly** is a hereditary anomaly affecting all cells of the granulocytic series, where the mature cells are hypersegmented. Two-lobed nuclei are seen in the majority of neutrophils and eosinophils (pince-nez cells). The condition should be distinguished from the left shift of infection. **Hereditary granulation abnormalities** have been described, including the Chediak–Higashi syndrome in cats, cattle and mink and mucopolysaccharide storage disease of cats. In the Alder–Reilly anomaly in man, the specific granules fail to develop, and coarse azurophilic granules remain.

Eosinophils and basophils

Eosinophils vary from 0.1×10^9 to 0.3×10^9/liter in normal animals, with cows developing a transient eosinophilia of up to 10^9/liter at the time of estrus. A persistent eosinophilia in excess of 10^9/liter should be considered part of an immune response and mediated by specific lymphokine. Basophils are rare in the peripheral blood of domestic animals and under normal circumstances none is found on a routine count of 100 cells. The granulation of basophils is distinct, but, if the blood films are not rapidly prepared and dried, the granules may lyse and the cells may not be recognized as basophils. Basophils tend to increase in the peripheral blood in diseases where there is also an eosinophilia.

Monocytes

Monocyte levels vary between 0.3 and 10 × 10^9/liter in normal domestic species, with the lower numbers characteristic of herbivores and cats, while the dog is characterized by a higher level of monocytes under normal conditions. Both the dog and cow respond with a rapid monocytosis in response to endogenous or exogenous steroid as well as maintaining a monocytosis in chronic inflammatory states. Monocytosis in these species, therefore, is not in itself indicative of chronic disease and the numbers of monocytes must be interpreted in the light of case history and other clinical signs.

Lymphocytes

Reactive or immune-transformed lymphocytes are seen with antigenic stimulation and morphologic changes are intense cytoplasmic basophilia (increased ribosomes) often with a pale Golgi zone, and densely staining chromatin. Cells are usually firmly rounded and small or medium in size.

Prominent chromocentering occurs in ruminants and horses with inflammatory conditions. Lymphocyte nuclei may show an irregular chromatin distribution, forming large chromocenters which may resemble dense nucleoli. **Immature lymphocytes** (lymphoblasts or prolymphocytes) found in the peripheral blood often implies the presence of a lymphoid neoplasm, particularly if there is marked variation in nucleolar size and shape. These cells are recognized by the presence of pale blue staining nucleoli, which should be distinguished from chromocenters. The chromatin of malignant lymphoid cells is darker than their benign counterparts with non-uniform distribution (fewer large chromocenters) and there is usually more nuclear detail to be discerned in malignant nuclei. Often, low numbers of lymphoid cells are seen in immune-mediated states. **Cytoplasmic vacuolation in lymphocytes** is rare, but occurs in some storage diseases such as mucolipidosis II, Pompe's disease (also with large cytoplasmic glycogen granules) and in Swainsona poisoning. **Cytoplasmic granulation** is observed in some lymphocytes that contain fine azurphilic (lysosomal) granules. Many of these cells are said to be T effector or natural killer lymphocytes.

Additional reading

Erythron

Archer, R. K. and Jeffcott, L. K. B. (1977). *Comparative Clinical Haematology*. Oxford, Blackwell Scientific Publ. Ltd.

Bunn, H. F. (1972). Erythrocyte destruction and hemoglobin catabolism. *Sem. Hemat.* **9**: 3–17.

Bunn, H. F. and Kitchen, H. (1973). Hemoglobin function in the horse: the role of 2,3-diphosphoglycerate in modifying the oxygen affinity of maternal and fetal blood. *Blood*, **42**: 471–9.

George, J. W. and Duncan, J. R. (1979). The hematology of lead poisoning in man and animals. *Vet. Clin. Path.* **8**: 23–30.

Giddens, W. E., Jr (1975). Feline congenital erythropoietic porphyria associated with severe anemia and renal disease. Clinical, morphologic, and biochemical studies. *Am. J. Path.* **80**: 367–86.

Jacob, H. S. (1966). A pathogenic classification of the anemias. *Med. Clin. N. Am.* **50**: 1679–87.

Lee, G. R. (1983). The anemia of chronic disease. *Sem. Hemat.* **20**: 61–80.

Lumsden, J. H., Valli, V. E. O. and McSherry, B. J. (1975). The hematological response to hemorrhagic anemia in the standard bred horse. *Proc. First Int. symp. Equ. Hemat.* pp. 356–60.

MacWilliams, P. S., Searcy, G. P. and Bellamy, J. E. C. (1982). Bovine postparturient hemoglobinuria: a review of the literature. *Can. Vet. J.* **23**: 309–12.

Martinovich, D. and Woodhouse, D. A. (1971). Post-parturient haemoglobinuria in cattle: a

Heinz body haemolytic anaemia. *N.Z. Vet. J.* **19**: 259–63.
Miller, D. R. (1972). The hereditary hemolytic anemias. Membrane and enzyme defects. *Ped. Clin. N. Am.* **19**: 865–87.
Nagy, S., Deavers, S. and Huggins, R. A. (1971). Tolerance to blood loss in the growing beagle (35747). *Proc. Soc. Exp. Biol. Med.* **137**: 1163–7.
Petz, L. D. and Garratty, G. (1980). *Acquired Immune Hemolytic Anemias*. New York, Churchill Livingstone.
Priester, W. A. (1980). *The Occurrence of Tumors in Domestic Animals*. NCI Monograph 54. Bethesda, U.S. Dept. Health and Human Services, National Cancer Inst.
Schalm, O. W., Jain, N. C. and Carroll, E. J. (1975). *Veterinary Hematology*. 3rd edn. Philadelphia, Lea and Febiger.
Switzer, J. W. and Jain. N. C. (1981). Autoimmune hemolytic anemia in dogs and cats. *Vet. Clin. N. Am.* **11**: 405–20.
Valli, V. E. O. (1985). The hematopoietic system. In *Pathology of Domestic Animals*, vol. 3, 3rd edn, K. V. F. Jubb, P. C. Kennedy and N. Palmer (eds.), pp. 83–236. Orlando, FL, Academic Press.

Thrombon

Dodds, W. J. (1980). Hemostasis and coagulation. In *Clinical Biochemistry of Domestic Animals*, 3rd edn, J. J. Kaneko (ed.), pp. 671–718. New York, Academic Press.
Duncan, J. R. and Prasse, K. W. (1980). *Veterinary Laboratory Medicine*, 2nd edn. Ames, IA, Iowa State Univ. Press.
Handin, R. I. (1977). Hemorrhagic disorders. II. Platelets and purpura. In *Hematology*, W. S. Beck (ed.), pp. 517–46. Cambridge, MA, MIT Press.
Harker, L. A. (1974). *Hemostasis Manual*, 2nd edn, Philadelphia, F. A. Davis Co.
Hilton, B. P. (1979). Blood platelets. *Sci. Prog. Oxf.* **66**: 59–80.
Meyers, K. M. (1985). Pathobiology of animal platelets. *Adv. Vet. Sci. Comp. Med.* **30**: 131–65.
Ratnoff, O. D. and Forbes, C. D. (1984). *Disorders of Hemostasis*. Orlando, FL, Grune and Stratton Inc.
Triplett, D. A. (1978). *Platelet function. Laboratory Evaluation and Clinical Evaluation*. Chicago, American Society of Clinical Pathologists.
Williams, W. J., Beutler, E., Erslev, A. J. and Rundles, R. W. (1977). *Hematology*, 2nd edn. New York, McGraw-Hill Book Company.

Leukon

Boggs, D. R. and Winkelstein, A. (1975). *White Cell Manual*, 3rd edn. Philadelphia, F. A. Davis Co.
Duncan, J. R. and Prasse, K. W. (1985). *Veterinary Laboratory Medicine: Clinical Pathology*, 2nd edn. Ames, IA, Iowa State Univ. Press.
Schalm, O. W., Jain, N. C. and Carroll, E. J. (1975). *Veterinary Hematology*, 3rd edn. Philadelphia, Lea and Febiger.
Thomson, R. G. (1978). *General Veterinary Pathology*. Philadelphia, W. B. Saunders Co.
Valli, V. E. O. (1985). The hematopoietic system. In *Pathology of Domestic Animals*, vol. 3, 3rd edn, K. V. F. Jubb, P. C. Kennedy and N. C. Palmer (eds.), pp. 83–236. Orlando, FL, Academic Press.
Williams, W. J., Beutler, E., Erslen, A. J. and Lichtman, M. A. (1983). *Hematology*, 3rd edn. New York, McGraw-Hill Book Company.

Leonard K. Cullen

4 Acid–base balance

Optimal cellular function occurs when the pH of the extracellular fluid is 7.4, which in terms of hydrogen ion concentration ($[H^+]$) is 40 nanomoles/liter. There are, however, many instances, especially in disease, when the pH of plasma varies from the norm. Failure to maintain pH within the normal range changes the ionization status of chemical groups, altering the activity of enzymes and the integrity of cell membranes. Such pH changes are potentially life threatening and therefore require immediate correction. A state of abnormally low extracellular pH is referred to as **acidosis** and the opposite situation, a rise in pH, as **alkalosis**. Both reflect a failure in the regulation of extracellular $[H^+]$.

Normal metabolic activity continuously generates an acid load which is regulated within the body by:

- Intracellular and extracellular chemical buffers.
- Respiratory adjustment of CO_2 concentration.
- Excretion of acid and the regeneration of body buffer systems by the kidney.

These regulatory mechanisms are complementary and the renal and respiratory systems have considerable compensatory capacity, the former by modulating the excretion of acid or base and the latter by varying the rate of CO_2 venting.

Acid–base disturbances may be categorized broadly into those of either respiratory or metabolic origin. Whereas respiratory acidosis or alkalosis derives from primary respiratory dysfunction, metabolic acidosis or alkalosis may be of renal and alimentary origin. Metabolic acidosis may also arise from the generation of acid, particularly lactic acid, following severe or unaccustomed exercise.

In some situations the disturbance is complex and, in spite of significant losses or gains in acid or base by the body, the plasma pH stays close to normal. An example is the animal with severe vomiting and diarrhea, where large quantities of both acid and base, respectively, are lost from the body.

Mechanisms of regulation

Buffer systems

At the heart of the question of acid–base regulation is the familiar Henderson–Hasselbalch equation, which is derived from the character of a simple buffer system consisting of a weak acid and its conjugate base, and is represented:

$$HA \rightleftharpoons H^+ + A^-. \qquad (4.1)$$

Within limits, the buffer solution will maintain a constant $[H^+]$ when challenged by additional acid or base. On the addition of H^+ to such a solution, the equilibrium shifts to the left, while on the addition of base it shifts to the right.

When the reaction expressed by this equation is at equilibrium, the concentrations of H^+ and A^- are a constant fraction of HA. This equilibrium can be expressed as:

$$K = \frac{[H^+][A^-]}{[HA]}, \quad (4.2)$$

where K is the dissociation constant.

Rearrangement of equation (4.2), provides an expression of $[H^+]$:

$$[H^+] = \frac{K[HA]}{[A^-]}, \quad (4.3)$$

and further:

$$-\log [H^+] = -\log K + \log \frac{[A^-]}{[HA]}, \quad (4.4)$$

which can then be written as the Henderson–Hasselbalch equation:

$$pH = pK + \log \frac{[A^-]}{[HA]}, \quad (4.5)$$

A buffer is most effective when pH and pK are equal and the ratio $[A^-]/[HA]$ is unity.

There are several major buffer systems in the body, and these are located in the blood plasma, red cells, interstitial fluid, intracellular fluid and the mineral matrix of bone.

Bicarbonate buffer

The bicarbonate buffer system makes up about 75% of total plasma buffering capacity, about 40% of the total red cell capacity, most of the interstitial fluid capacity and a significant proportion of the intracellular fluid capacity.

The system centers on the following equation:

$$CO_2 + H_2O \rightleftharpoons H_2CO_3 \rightleftharpoons H^+ + HCO_3^- \quad (4.6)$$

which, if transposed to the Henderson–Hasselbalch equation gives:

$$pH = pK + \log \frac{[HCO_3^-]}{[H_2CO_3]}, \quad (4.7)$$

At body temperature (38 °C) pK = 6.1 and the solubility coefficient of CO_2 is 0.03 mmole/mmHg.*

Thus, the equation may be written

$$pH = 6.1 \times \log \frac{[HCO_3^-]}{0.03 \times P_{CO_2}}, \quad (4.8)$$

where P_{CO_2} is the partial pressure of CO_2 in the plasma.

When the pH is 7.4, the ratio:

$$[HCO_3^-] : 0.03 \times P_{CO_2} = 20 : 1.$$

The plasma concentration of HCO_3^- is normally 24 mmole/liter and P_{CO_2} is about 40 mmHg, so the conditions for an optimal buffer system do not appear to be met. In fact, the bicarbonate system is very efficient because in the body it functions as an open buffer, with the P_{CO_2} being maintained at a constant value (40 mmHg) by the respiratory system. Under these conditions, additional fixed acid will lower the HCO_3^- concentration but change in pH is minimized because of a constant P_{CO_2}.

The bicarbonate system is unable to buffer H_2CO_3 itself. This is carried out within red cells by hemoglobin, protein and phosphate. The enzyme carbonic anhydrase, catalyzes the formation of H_2CO_3 from CO_2 and H_2O. This acid then dissociates (equation (4.6)) and the H^+ is buffered mainly by hemoglobin, with HCO_3^- leaving the red cell in exchange for chloride ions, and entering the plasma.

The kidneys play a key role in the preservation of body bicarbonate reserves in a recovery process termed regeneration of buffer (see below). The respiratory system controls the concentration of CO_2 in the body and in equation (4.8), the respiratory component is represented by P_{CO_2}. The renal, or 'metabolic' component is represented by HCO_3^-.

Phosphate buffer

Under normal physiologic conditions, the function of the phosphate buffer system is

* 1 mmHg = 133.3 Pa. Units in mmHg are used throughout for convenience.

epitomized in the equation

$$H_2PO_4^- \rightleftharpoons HPO_4^{2-} + H^+$$

and it has a pK of 6.8.

This system is of greatest importance in the intracellular compartment, and also within the renal tubular lumens, where it plays a major role in urinary buffering.

Protein buffers

The basic and acidic groups of protein molecules provide great potential for buffering activity and are capable of buffering acids of non-respiratory origin and carbonic acid. Protein buffers are important in the plasma and intracellular locations, with hemoglobin dominant within red cells.

The major buffering capability of the hemoglobin molecule is provided by the imidazole groups of its histidine residues and the N-terminal regions of its valine residues. Hemoglobin is a more effective buffer in its deoxygenated form due to the effect of oxygen on the pK value.

Regulatory responses

The respiratory response

As indicated earlier, the respiratory system is a vital regulator of acid–base status by virtue of the venting of CO_2. When CO_2 accumulates in the plasma, acid is generated (see equation (4.6)), and conversely when CO_2 is depleted the reverse occurs.

The high lipid solubility of CO_2 allows it to cross the blood–brain barrier easily, so that if its concentration rises in the plasma, it rapidly induces a drop in the pH of cerebrospinal fluid. In response to this pH change, the central chemoreceptors stimulate alveolar ventilation within a few minutes in an effort to drive off surplus CO_2.

An abnormally low plasma concentration of CO_2 will reduce ventilation, but this mechanism will be overridden if there is concurrent severe hypoxia. An increase in plasma [H^+] or a decrease in O_2 stimulates the aortic and carotid body chemoreceptors leading to increased ventilation.

Significance of carbon dioxide and oxygen measurements

In considering this section, the reader should also consult Chapter 5 (pp. 107–12), as it is not intended to present here a complete account of pulmonary function tests. Only details of those disturbances likely to alter the P_{O_2} and P_{CO_2} significantly (see Table 4.1) will be given.

The venting of CO_2 by the lungs is critically influenced by their functional ability and therefore the question of O_2 exchange cannot be ignored. P_{O_2} is a valuable indicator of lung function and although it has no influence on acid–base balance, it often indicates the seat of an acid–base disturbance.

The efficiency of gas exchange by the lungs may be assessed by measuring the arterial CO_2 (P_{a,CO_2}), the arterial O_2 (P_{a,O_2}) and the inspired O_2 concentration (P_{i,O_2}). An estimate of the alveolar O_2 concentration (P_{A,O_2}) can be obtained from the equation:

$$P_{A,O_2} = P_{i,O_2} - (P_{a,CO_2} \times 1.1) \qquad (4.9)$$

The efficiency of gas exchange is reflected as the difference between P_{A,O_2} and P_{a,O_2} and is expressed as:

$$P_{A,O_2} - P_{a,O_2}.$$

The value obtained from equation (4.9) for an animal breathing air (P_{i,O_2} = 20.9%) is about 10 mmHg. Values in excess of 20 mmHg indicate decreased gas exchange efficiency under these conditions. When pure O_2 or a 95% O_2 concentration is inspired (usually under gaseous anesthesia) the value is up to 100 mmHg. In this circumstance, a value in excess of 150 mmHg is considered to indicate impaired efficiency.

Carbon dioxide

Arterial P_{CO_2} is a measure of the CO_2 dissolved in the plasma. The amount of CO_2 in the blood is dictated by the balance between its metabolic production and its removal by pulmonary ventilation. The normal range for

Table 4.1. *Possible combinations for arterial oxygen and carbon dioxide*

Inspired oxygen concentration (%)	Arterial oxygen (mmHg)	Arterial carbon dioxide (mmHg)	Possible causes
20.9	95–100	35–40	Normal range
20.9	70 (Hypoxemia)	55 (Hypercapnia)	Hypoventilation
20.9	60 (Hypoxemia)	25 (Hypocapnia)	Reduced gas exchange and hyperventilation due to lung disease
20.9	115 (Hyperoxia)	25 (Hypocapnia)	Hyperventilation
100	380 (Hyperoxia)	60 (Hypercapnia)	Hypoventilation
100	150 (Relative hypoxemia)	60 (Hypercapnia)	Reduced gas exchange and hypoventilation

P_{a,CO_2} is from 35 to 40 mmHg. Values in excess of this indicate hypercapnia, while values below it indicate hypocapnia.

Hypercapnia is usually the result of hypoventilation, but will occur in anesthetized animals when anesthetic circuits with a large dead space are employed, or when anesthetic machines have depleted soda-lime units. In the latter the concentration of CO_2 is increased. Hypocapnia is invariably the result of hyperventilation.

Oxygen

P_{O_2} is usually determined from arterial blood, but it is a measure only of the O_2 dissolved in the plasma and not the amount present in the whole blood. P_{a,O_2} is a direct reflection of pulmonary gas exchange function, and the normal range is 90–100 mmHg. When it falls below this range, there is hypoxemia and if it falls to 60 mmHg or less, the severe hypoxemia is a powerful respiratory stimulant. P_{a,O_2} in excess of 100 mmHg produces a state of hyperoxia. As hemoglobin is 97% saturated when P_{a,O_2} is 100 mmHg, hyperoxia has little further effect on it. Hyperoxia never occurs as a natural disease state, but is always the result of the administration of O_2-rich gas mixtures, usually during anesthesia.

Some well-known causes of hypoxemia are: low concentrations of inspired O_2, hypoventilation, impaired alveolar gas diffusion, ventilation/perfusion ($\dot{V}/\dot{Q}$) mismatch and pulmonary vascular shunts (see Chapter 5, p. 108). An approximation of the expected P_{a,O_2} can be derived by multiplying the percentage concentration of inspired O_2 by 5. For example, an animal breathing air (20.9% O_2) should have a P_{a,O_2} of 100 mmHg, while an animal breathing 50% O_2 should have a P_{a,O_2} of 250 mmHg. Table 4.1 contains some examples of abnormalities in P_{a,O_2} and P_{a,CO_2} together with their causes.

The renal response

Fixed acid production in the body is predominantly the consequence of protein catabolism, with a small contribution from the breakdown of phospholipids and nucleic acids. The H^+ is buffered initially by the intra- and extracellular buffer systems, but the kidneys generate H^+ and secrete them into the urine. Renal excretion of H^+ is carried out via the bicarbonate resorption and phosphate and ammonia excretion mechanisms (Figs. 4.1, 4.2 and 4.3).

The mechanism utilizing **bicarbonate** has as its focal point, the large load of Na^+ and HCO_3^- in the glomerular filtrate. The filtered

Na^+ is exchanged for H^+ which then combines with the HCO_3^- in the filtrate to form H_2CO_3, which yields CO_2 and water. The CO_2 rapidly diffuses into the tubular cells, where much of it forms H_2CO_3 under the influence of carbonic anhydrase. The dissociation of H_2CO_3 provides the H^+ for the sodium exchange process, and bicarbonate ions, which move into the plasma compartment. Because the HCO_3^- returned to the plasma are not those initially filtered, but are new ones 'regenerated' by the renal epithelium, the process is termed renal regeneration of the bicarbonate buffer system (Fig. 4.1). CO_2 that diffuses from the filtrate to the plasma is expelled by the lungs.

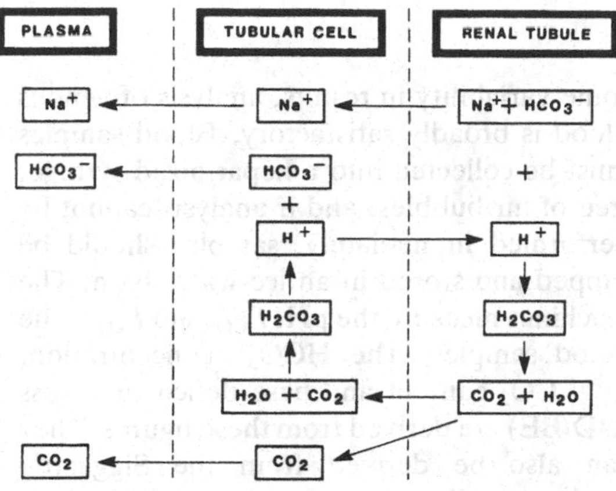

Fig. 4.1. The active retention of Na^+ along with HCO_3^-. Na^+ is exchanged for H^+ and the quantity of HCO_3^- retained by this mechanism is related to the amount of H^+ secreted. (Adapted from Guyton, 1986, by kind permission, see Additional reading.)

The mechanism using **phosphate** is centered on the filtered load of Na_2HPO_4 and its dissociation to yield Na^+ and $NaHPO_4^-$. Once again the exchange of Na^+ for H^+ is a feature, with the H^+ being supplied via the carbonic anhydrase-catalyzed formation of H_2CO_3 within the renal tubular epithelium (Fig. 4.2). The process also regenerates the plasma bicarbonate buffer system.

The mechanism using **ammonia** depends upon the metabolic generation of NH_3 by the renal epithelium, which is a product of the metabolism of glutamine and other amino acids. NH_3 readily diffuses into the tubular lumen, where it combines with H^+ to form the non-diffusible NH_4^+ (Fig. 4.3).

In *both* the phosphate and ammonia mechanisms a large quantity of H^+ can be concentrated in the urine without lowering the pH. In acidosis, when there are excessive quantities of H^+, these renal mechanisms will respond, eliminating more acid in the urine, but may take 4 or 5 days to reach maximum capacity.

Several factors can influence the activity of

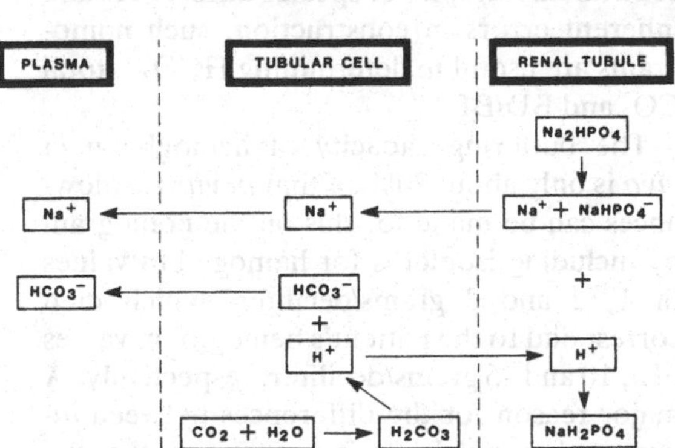

Fig. 4.2. The phosphate buffering mechanism in the kidney. (Adapted from Guyton, 1986, by kind permission, see Additional reading.)

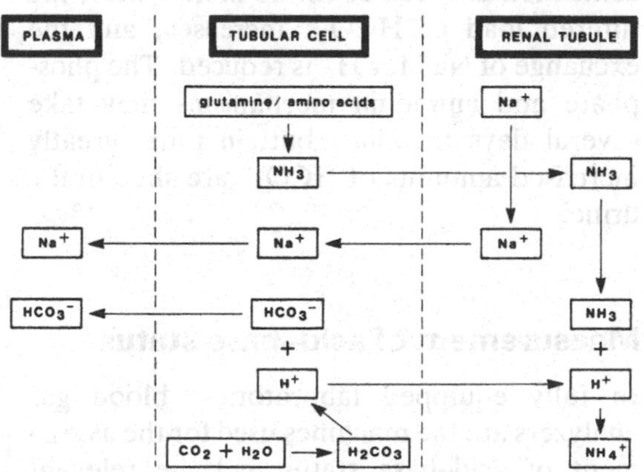

Fig. 4.3. The ammonia buffering system in the kidney. (Adapted from Guyton, 1986, by kind permission, see Additional reading).

Table 4.2. *Measurement of acid–base status*

	Definitions
P_{CO_2}	Partial pressure of CO_2 expressed in mmHg. P_{CO_2} gives an indication of ventilatory efficiency and its relation to metabolic rate
P_{O_2}	Partial pressure of O_2 expressed in mmHg. P_{O_2} gives an indication of lung function
BD/BE	Base deficit (BD) or base excess (BE) seen in acidosis or alkalosis, respectively. BD or BE refer to the quantity of base or acid needed when titrating plasma pH to 7.4 at 37°C while the P_{CO_2} is held at 40 mmHg. Some texts use the term negative base excess (−BE) instead of base deficit
Total CO_2	The sum of the carbon dioxide dissolved in plasma, carbonic acid and bicarbonate. The total CO_2 or CO_2 content is an estimate of the metabolic component of acid–base balance
'Blood gases'	pH, P_{CO_2}, P_{O_2}, HCO_3^- total CO_2 and BD/BE values, usually from an arterial blood sample. The last three values are derived from the pH and P_{CO_2}.

these renal mechanisms. Any factor reducing the resorption of Na^+ may reduce the elimination of H^+ in the urine, thereby predisposing to acidosis. This will occur if there is an expansion of the extracellular fluid volume, causing renal sodium resorption to be reduced. Acid secretion is also suppressed if carbonic anhydrase activity is suppressed by drugs such as acetazolamide.

Conversely, when Na^+ resorption is increased, for example when aldosterone levels are excessive, significant quantities of both H^+ and K^+ can be lost in the urine, and concomitantly HCO_3^- resorption is promoted.

In alkalosis, when the plasma HCO_3^- concentration exceeds about 28 mmole/liter, the filtered load of HCO_3^- increases, and the exchange of Na^+ for H^+ is reduced. The phosphate and ammonia mechanisms may take several days to adjust but, in time, greatly increased amounts of HCO_3^- are shed in the urine.

Measurement of acid–base status

In fully equipped laboratories, blood gas analyzers are the machines used for the assessment of acid–base status and the relevant parameters are shown in Table 4.2. The most reliable data for the assessment are provided by sampling arterial blood, but, in spite of some variability in results, analysis of venous blood is broadly satisfactory. Blood samples must be collected into a heparinized syringe, free of air bubbles, and if analysis cannot be performed immediately, samples should be capped and stored in an ice-water bath. The machines measure the pH, P_{CO_2} and P_{O_2} of the blood samples. The HCO_3^- concentration, total CO_2 content and base deficit or excess (BD/BE) are derived from these figures. They can also be derived from the Siggaard–Anderson alignment nomogram (Fig. 4.4), using the measured pH and P_{CO_2} levels. The nomogram has been constructed from measurements made when known concentrations of acid or alkali were added to samples of human blood of various hemoglobin concentrations. In spite of species differences and inherent errors in construction, such nomograms are useful in determining HCO_3^-, total CO_2 and BD/BE.

The buffering capacity of hemoglobin *in vivo* is only about 20% of that *in vitro*. Allowances can be made for this on the nomogram by including isopleths for hemoglobin values of 1, 2 and 3 grams/deciliter, which then correspond to the patient's hemoglobin values of 5, 10 and 15 grams/deciliter, respectively. A major reason for the differences between *in-vitro* and *in-vivo* buffering capacity is that the bicarbonate buffer system is an open system (as discussed earlier).

Because the water bath of blood-gas

analyzers is kept at a temperature of 37 °C, which differs from the body temperature of animals, adjustments need to be made to the pH, P_{CO_2} and P_{O_2} readings of the blood samples to give the *in vivo* values. pH values decrease and P_{CO_2} and P_{O_2} values increase when the patient's body temperature is higher than that of the water bath of the blood-gas analyzer. Nomograms are available for the correction of blood-gas values when there is a difference between the temperature of the analyzer and the patient's body temperature. Some machines have this correction factor built in to the computer program.

Blood-gas analyzers are too complex and expensive to buy and maintain in most veterinary practices, but information about the metabolic component of acid–base disturbances in animals is available using the Harleco apparatus or Oxford titration. These machines measure total CO_2 and plasma HCO_3^-, respectively, and are very much cheaper to buy. The procedure is rapid and inexpensive. However, there are limitations to the test: all samples must be analyzed at a constant temperature and results need to be interpreted after a complete clinical examination of the patient to ascertain what disease processes, and physiologic compensatory mechanisms, are likely to affect the acid–base status. The technique evaluates the metabolic component only, but accuracy is influenced by respiratory changes.

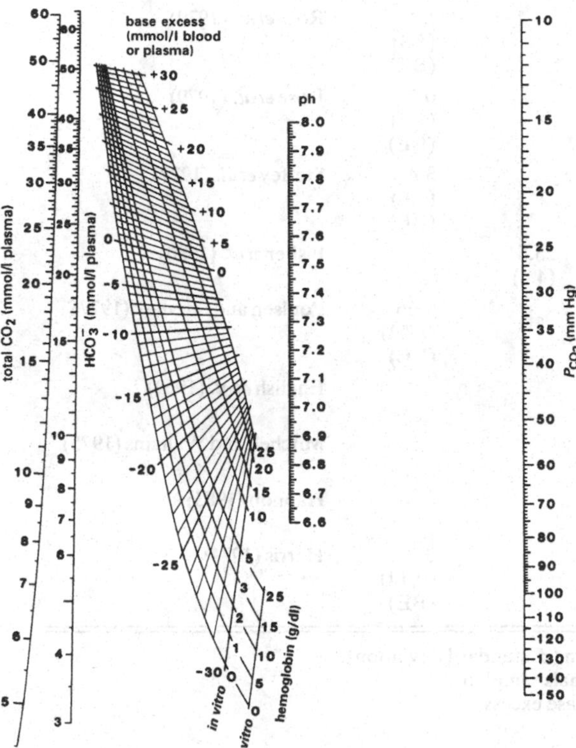

Fig. 4.4. The Siggaard-Anderson alignment nomogram. (Adapted from Siggaard-Anderson, 1963, © Radiometer A/S, by kind permission, see Additional reading.)

Minor differences in the blood–gas values exist between animal species and Table 4.3 shows the normal blood–gas values of different animals. Variation in the BD/BE values between the species may be attributed to a difference in the diet.

Whereas arterial blood samples give results that show less variation than venous samples, it is difficult to get arterial samples in cats, small dogs, pigs and some ruminants. Venous samples are usually analyzed in these animals.

Interpretation of acid–base disturbance

It is important that blood-gas results be interpreted with a knowledge of the patient's history and details from a clinical examination. Although they are not essential, plasma Na^+, K^+ and Cl^- concentrations are very useful when interpreting blood-gas results and allow a much better assessment of the patient to be made.

Blood-gas results can be interpreted as follows. The plasma pH indicates whether the animal is acidotic or alkalotic. The pH value is an overall assessment of all acidotic and alkalotic processes in the body, including primary acid–base disturbances and any physiologic compensatory mechanisms. The primary acid–base disturbance tends to be in line with the plasma pH change, because physiologic compensatory mechanisms are unlikely to overcorrect.

Table 4.3. *Normal acid–base values for the domestic species*

Species (no.)	Blood sample	pH	P_{O_2} (mmHg)	P_{CO_2} (mmHg)	HCO_3^- (mmol/l)	Total CO_2 (mmol/l)	BD/BE (mmol/l)	References
Dog (38)	A	7.45 (0.03)	90.7 (8.8)	31.0 (3.6)	20.9 (2.3)			Cornelius and Rawlings (1981)
Dog (12)	A	7.43 (0.01)		32.6 (0.5)	21.0 (0.4)		1.7 (0.4) (BD)	Carter and Brobst (1969)
Dog (12)	V	7.40 (0.01)		35.0 (0.6)	20.9 (0.4)		2.3 (0.4) (BD)	Carter and Brobst (1969)
Cat (13)	A	7.34 (0.1)	102.9 (15.2)	33.6 (7.1)	17.5 (2.98)	18.4 (3.9)	6.4 (5.04) (BD)	Middleton *et al.* (1981)
Cat (13)	V	7.30 (0.09)	38.6 (11.44)	41.8 (9.12)	19.4 (4.0)	20.1 (4.16)	5.7 (4.6) (BD)	Middleton *et al.* (1981)
Cat (10)	A	7.46 (0.01)	97.2 (2.7)	29.9 (0.6)	21.0 (0.4)			Herbert and Mitchell (1971)
Cat (10)	V	7.39 (0.01)	34.5 (1.1)	37.5 (0.8)	22.4 (0.6)			Herbert and Mitchell (1971)
Horse (8)	A	7.41 (0.03)	96.0 (8)	40.8 (2.6)	25.3 (1.7)		1.1 (1.4) (BE)	Rose *et al.* (1979)
Horse (11)	V	7.39 (0.02)	46.5 (4.7)	43.0 (2.6)	25.4 (1.7)		0.7 (1.7) (BE)	Rose *et al.* (1979)
Horse (12)	A	7.42 (0.01)	96.0 (3)	43.0 (1)			3.4 (0.6) (BE)	Steffey *et al.* (1977)
Cow (10)	A	7.47 (0.03)	103.0 (6.2)	36.8 (7.2)		25.5 (4.4)		Fisher *et al.* (1980)
Cow (2)	V	7.38 (0.03)	33.1 (2.99)	44.1 (2.01)	24.13 (1.42)		0.66 (1.52) (BE)	Poulsen and Surynek (1977)
Sheep (33)	V	7.40 (0.05)		42.1 (4.9)	25.4 (2.6)			English *et al.* (1969)
Sheep (34)	A	7.47– 7.49	85.5– 90.4	31.6– 32.4				Mitchell and Williams (1975)
Pig (40)	A	7.5 (0.002)	83.1 (0.75)	40.0 (0.86)	31.0 (0.52)			Hannon (1983)
Pig (9)	V	7.42 (0.01)	29.4 (1.85)	45.5 (0.87)	28.4 (0.54)		3.8 (0.14) (BE)	Harris (1974)

Blood -gas values of different animal species expressed as mean (± standard deviation).
The figure under each animal species gives the number of animals sampled.
A, arterial sample; V, venous sample; BD, base deficit, BE, base excess.
For references, see Additional reading.

Table 4.4. *Types of primary acid–base disturbances before physiologic compensation occurs*

Simple acid–base disturbances
Respiratory acidosis or alkalosis
Metabolic acidosis or alkalosis
Complex acid–base disturbances
Metabolic
Metabolic acidosis; metabolic alkalosis
Metabolic and respiratory
Metabolic acidosis; respiratory acidosis
Metabolic acidosis; respiratory alkalosis
Metabolic alkalosis; respiratory acidosis
Metabolic alkalosis; respiratory alkalosis
Triple metabolic and respiratory
Metabolic acidosis, alkalosis; respiratory acidosis
Metabolic acidosis, alkalosis; respiratory alkalosis

The P_{CO_2} assesses the respiratory component; that is, the adequacy of ventilatory function and its relation to metabolic rate. The BD/BE assesses the metabolic component or renal function.

Relatively uncomplicated acid–base disturbances occur when there is malfunction of one system and an attempt at correction by the other, as in an animal with a primary metabolic acidosis with respiratory compensation. However, it is often not that simple; in many cases more than one primary acid–base disturbance may be present in a single animal. A list of possible primary uncompensated acid–base disturbances is shown in Table 4.4.

Assistance with the interpretation of blood-gas values from animals with complex acid–base disturbances can be obtained from the pH/HCO_3^- diagram shown in Fig. 4.5. It demonstrates first the relationship between the pH, HCO_3^- and P_{CO_2} values in the normal animal (point A) and the animal with one primary acid–base disturbance. Points B and D represent uncompensated respiratory alkalosis and acidosis, respectively, and points C and E uncompensated metabolic alkalosis and acidosis, respectively, pH/HCO_3^- points that lie between points B, C, D and E usually indicate the presence of either mixed (complex) acid–base disturbances or compensated simple acid–base disturbances. The likely positions for different acid–base disturbances are shown on the figure.

Pathophysiology of simple acid–base disorders

Respiratory acidosis

This disorder occurs when the lungs are unable to eliminate CO_2 at a sufficiently rapid rate, and its fundamental cause is inadequate pulmonary alveolar ventilation (see Chapter 5). The common precipitating conditions are listed in Table 4.5. Under these conditions, the P_{CO_2} rises and the easily diffusible gas enters red cells, where carbonic anhydrase catalyzes the formation of H_2CO_3, which then yields H^+ and HCO_3^-. The latter enter the extracellular compartment in exchange for chloride ions (the process is termed the 'chloride shift') and the H^+ are buffered intracellularly as previously described. H^+ formed in the extracellular compartment may enter tissue cells in exchange for K^+, and the plasma concentration of the latter is expected to rise (hyperkalemia). Although bicarbonate ions are generated by the mass action, the process has an upper limit of about 32 mmoles/liter

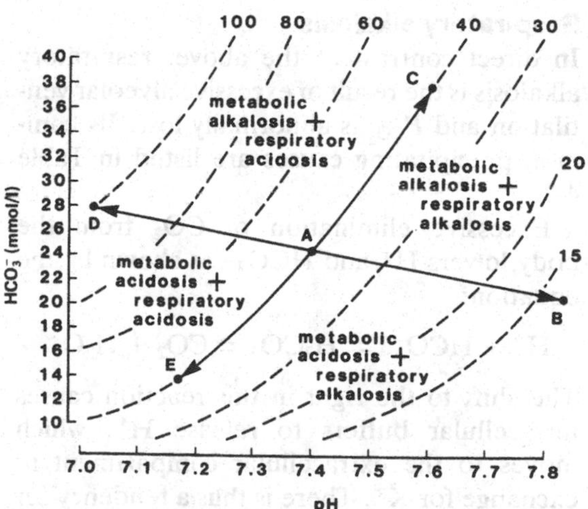

Fig. 4.5. The pH/HCO_3^- diagram showing the relationship between pH, HCO_3^- and P_{CO_2}. The broken lines show isobars for P_{CO_2}.

Table 4.5. *Common causes of respiratory acidosis*

Depression of central respiratory function
Anesthesia
Other drugs
CNS trauma
Increased CSF pressure
Decreased pulmonary function
Airway obstruction
Pneumonia
Pulmonary edema
Pneumothorax
Hemothorax
Hydrothorax
Atelectasis
Abdominal distension
Inadequate mechanical ventilation

CNS, central nervous system; CSF cerebrospinal fluid.

(plasma concentration) when P_{CO_2} increases above about 80 mmHg. Up to this limit, for each 10 mmHg rise in P_{CO_2}, the plasma HCO_3^- will rise by 1 mmole/liter in acute respiratory acidosis, and 3–4 mmoles/liter in chronic disease.

The kidneys respond to respiratory acidosis by excreting both H^+ and HCO_3^-. A case example of respiratory acidosis is shown in Table 4.6.

Respiratory alkalosis

In direct contrast to the above, respiratory alkalosis is the result of excessive alveolar ventilation and P_{CO_2} is abnormally low. Its common precipitating causes are listed in Table 4.7.

Excessive elimination of CO_2 from the body lowers H^+ and HCO_3^- as shown by the equation:

$$H^+ + HCO_3^- \rightleftharpoons H_2CO_3 \rightleftharpoons CO_2 + H_2O$$

The shift to the right in this reaction causes intracellular buffers to release H^+, which moves to the extracellular compartment in exchange for K^+. There is thus a tendency for hypokalemia, which is exacerbated by a renal response to shed K^+ and retain H^+. In addition, renal mechanisms promote the

Table 4.6. *Respiratory acidosis in a horse under general anesthesia (halothane/oxygen)*

pH = 7.055	HCO_3^- = 30.4 mmol/l
P_{CO_2} = 109.4 mmHg	BD = 1.8 mmol/l
P_{O_2} = 121.9 mmHg	

The pH shows a significant acidosis and the markedly elevated P_{CO_2} indicates the respiratory origin of the acidosis. HCO_3^- is consequently elevated. The base deficit (BD) is close to normal. The P_{O_2} is high and the animal is inspiring nearly pure O_2; however, the P_{O_2} should be greater than 121.9 mmHg. The horse is therefore hypoventilating and requires positive pressure ventilation to correct the acidosis.

Table 4.7. *Common causes of respiratory alkalosis*

Hyperventilation due to:
Excessive positive pressure ventilation
Pain
Excitement or fear
Severe arterial hypoxia (primary respiratory disease, anemia)
Hyperthermia
Disease of the central nervous system:
Meningitis
Encephalitis
Cerebral hemorrhage

excretion of HCO_3^- and the retention of Cl^-. Respiratory alkalosis becomes severe when P_{CO_2} falls to about 20 mmHg and pH rises to about 7.6. At this point several protective mechanisms are brought into play. These are peripheral vasoconstriction, a depressed ability of hemoglobin to release oxygen and a stimulus to lactic acid production by red cells via the enzyme phosphofructokinase. In acute respiratory alkalosis, plasma HCO_3^- concentration decreases by approximately 1 to 3 mmoles/liter for each 10 mmHg decrease in P_{CO_2}, while in chronic disease the reduction in HCO_3^- is approximately 5 mmoles/liter. A case example of primary respiratory alkalosis is shown in Table 4.8.

Metabolic acidosis

The disturbance termed metabolic acidosis arises if the kidneys are unable to eliminate H^+

Table 4.8. *Respiratory alkalosis in a dog with pneumonia*

pH = 7.45	HCO_3^- = 18.0 mmol/l
P_{CO_2} = 26.3 mmHg	BD = 5.4 mmol/l
P_{O_2} = 73.4 mmHg	

The minor change in pH suggests alkalosis. P_{CO_2} is low and indicates respiratory alkalosis. The decreased HCO_3^- and base deficit (BD) are consistent with renal compensatory activity. The low P_{O_2} suggests that the primary respiratory alkalosis was caused by hyperventilation from hypoxia. Treatment was directed toward the pneumonia, which included oxygen therapy, thus correcting the acid–base disturbance.

Table 4.9. *Common causes of metabolic acidosis*

Decrease in renal excretion of H^+
- Renal failure
- Carbonic anhydrase inhibitors

Body Loss of HCO_3^-
- Diarrhea
- Proximal renal tubular damage

Failure of renal clearance of excessive H^+
- Diabetic keto-acidosis
- Lactic acidosis
- Intoxications (ethylene glycol, salicylates, methanol)

generated by metabolic activity, or alternatively if there is excessive loss of HCO_3^- from the body. The common causes of metabolic acidosis are listed in Table 4.9.

During metabolic acidosis, the increasing numbers of H^+ are buffered by intracellular and extracellular buffers. In exchange for the intracellular movement of H^+, K^+ enters the extracellular compartment, predisposing to hyperkalemia. This tendency towards hyperkalemia is opposed in those patients with diseases causing the large-scale loss of K^+ from the body, such as acute renal tubular necrosis or severe diarrhea.

If the disturbance is prolonged, some H^+ may exchange with Ca^{2+} in bone. As plasma pH falls and P_{CO_2} rises, chemoreceptors stimulate a compensatory increase in pulmonary alveolar ventilation. An elevated resting respiratory rate is a useful clinical diagnostic sign. This respiratory response is initiated rapidly and may be maximal within a few hours. When respiratory compensation is maximal, the approximate decrease in P_{CO_2} can be expressed as:

$$1.5 \times [HCO_3^-] + 8.$$

Provided they are intact, renal mechanisms adjust to maximize the conservation of HCO_3^- and excretion of H^+ resulting in an increased acidity of the urine. Extensive renal damage will limit the efficacy of the renal response, but in dogs and cats the urinary pH may fall to less than 6 in disorders such as diabetic keto-acidosis.

In metabolic acidosis there is frequently a substantial impact upon the plasma electrolyte profile and this gives rise to the concept of the **anion gap**. All plasma cations are balanced electrically by the major anions, Cl^- and HCO_3^- plus a number of unmeasured anions including PO_4^{3-}, SO_4^{2-}, organic acids and anionic proteins. The contribution of these unmeasured anions can be estimated from the expression:

$$[Na^+] - ([Cl^-] + [HCO_3^-])$$

and is referred to as the anion gap. Its normal value is approximately 8–12 mmoles/liter and values in excess of 15 mmoles/liter are taken to reflect a metabolic acidosis. In this situation unmeasured anions are present in abnormally large concentrations in the plasma.

Anion gap can be used as an aid to the differential diagnosis of the causes of metabolic acidosis; that is, the conditions causing an increase in unmeasured anions compared to those that have no effect on unmeasured anions. To serve as an illustration, Table 4.10 lists some metabolic acidotic states in which the anion gap is increased on the one hand, or normal to decreased on the other.

The therapeutic approach to metabolic acidosis is based upon the rapid removal, if possible, of the basic cause, the correction of imbalances in extracellular fluid volume and electrolyte concentration and the maintenance of effective renal function. If all of

Table 4.10. *The use of the anion gap to determine causes of metabolic acidosis*

Increased anion gap	Normal or decreased anion gap
Endogenous increase in H^+	Decreased renal excretion of H^+
Diabetic keto-acidosis	Distal renal tubular disease
Lactic acidosis	Carbonic anhydrase inhibitors
Exogenous increase in H^+	HCO_3^- loss
Ethylene glycol intoxication	Diarrhea
	Proximal renal tubular disease
Salicylate intoxication	Excessive saliva loss
Renal failure	Chloride increase
Phosphate excretion reduced	Dilutional acidosis from 0.9% (w/v) NaCl infusion
Sulfate excretion reduced	

these can be achieved, normal body mechanisms will correct the acidosis. If, however, the basic cause is difficult to remove and the acidosis is severe, the administration of bicarbonate solution is indicated. This must be done cautiously as serious complications can result from excessively rapid administration. Metabolic precursors of HCO_3^-, such as lactate, acetate, gluconate or citrate may be used in some instances, but also have their drawbacks. Lactate, for instance, requires hepatic metabolism to be converted and is therefore contraindicated in animals with liver disease. An example of primary metabolic acidosis is shown in Table 4.11.

Metabolic alkalosis

Metabolic alkalosis may result from excessive loss of H^+ from the body, a shift of H^+ from the extracellular to the intracellular compartments, or excessive infusion of alkalinizing solutions, such as sodium bicarbonate. Table 4.12 lists some of the common causes.

Pathophysiologic consideration of metabolic alkalosis needs to focus on several points. The relationship between renal excretion of H^+ and regeneration of HCO_3^- has already been highlighted, but equally important at this point is the gastric secretion of H^+, which also generates HCO_3^- which enters the plasma. Normally the H^+ excreted into the gastric lumen are recovered in the small intestine, but should this be prevented by vomiting or gastric sequestration, alkalosis will occur. Furthermore, severe vomiting will cause hypovolemia, to which the renal response is retention of Na^+, excretion of H^+ and regeneration of HCO_3^-. This, of course, will exacerbate alkalosis induced primarily by vomiting. A prominent cause of sequestration is torsion or displacement of the abomasum (see Chapter 7) and it leads rapidly to metabolic alkalosis and hypochloremia.

Hypokalemia predisposes to metabolic alkalosis by promoting the intracellular movement of H^+, in exchange for the K^+ which move out into the plasma in an effort to maintain plasma concentrations. It is in this way that hyperaldosteronism may lead to metabolic alkalosis.

The respiratory response to metabolic alkalosis is triggered by the effect of elevated blood pH on peripheral chemoreceptors. Alveolar ventilation is reduced and P_{CO_2} increases by between 0.5 and 1 mmHg for each 1 mmole/liter rise in HCO_3^- concentration.

The renal response is to excrete an alkaline urine and urinary pH is usually greater than 8. On occasions, however, there is a **paradoxical aciduria** when, in the distal nephrons, Na^+ are exchanged for H^+ rather than K^+, and so the pH of the urine is kept low.

The therapeutic principles involved in treating metabolic alkalosis are the correction of fluid volume and Na^+, Cl^- and K^+ deficits. Replacement of Na^+ and Cl^- alone may

Table 4.11. *Metabolic acidosis in a dog with renal failure*

pH = 7.310	HCO_3^- = 14.6 mmol/l
P_{CO_2} = 29.5 mmHg	BD = 8.8 mmol/l
P_{O_2} = 92.6 mmHg	Na^+ = 124 mmol/l
	Blood urea nitrogen = 37.2 mmol/l

The pH shows acidosis. Both the low HCO_3^- and the base deficit (BD) value indicate metabolic acidosis. The P_{CO_2} is low, indicating a compensatory respiratory response. Both the metabolic acidosis and the electrolyte changes require correction. In this case, $NaHCO_3$ is the appropriate treatment. Calculation from the formula:

$$P_{CO_2} = 1.5 \times [HCO_3^-] + 8 = (1.5 \times 14.6) + 8 = 29.9 \text{ mmHg}.$$

demonstrates the similarity between the measured P_{CO_2} (29.5 mmHg) and the calculated value (29.9 mmHg) and is evidence that the respiratory system has compensated.

Table 4.12. *Common causes of metabolic alkalosis*

- Vomiting
- Excessive diuretic therapy
- Hyperadrenocorticism (Cushing's-like syndrome)
- Pyloric or duodenal obstruction in ruminants
- Excessive sweating in horses

Table 4.13. *Metabolic alkalosis in a cow with left displacement of the abomasum*

pH = 7.55	HCO_3^- = 40.0 mmol/l
P_{CO_2} = 45.9 mmHg	BE = 15.9 mmol/l
P_{O_2} = 86.0 mmHg	Na^+ = 144 mmol/l
	K^+ = 3.5 mmol/l
	Cl^- = 99 mmol/l

The pH shows alkalosis. The high HCO_3^- and the base excess (BE) are consistent with metabolic alkalosis. P_{CO_2} is high, indicating a respiratory compensation. The underlying problem of abomasal displacement was corrected surgically and a mixture of 0.9% (w/v) NaCl with added K^+ was infused. Correction of the fluid volume and electrolyte imbalances allowed the metabolic alkalosis to be corrected by renal mechanisms.

exacerbate the problem in the presence of severe hypokalemia, by further promoting the renal exchange of Na^+ for H^+ and therefore the regeneration of HCO_3^-. Replacement of K^+ is particularly important in hyperaldosteronism.

If regulatory mechanisms are unable to correct the problem, acidifying solutions may be infused, but their administration must be carefully monitored. Isotonic solutions of arginine or lysine hydrochloride, ammonium hydrochloride or hydrochloric acid may be infused. All the compounds, with the exception of hydrochloric acid, require initial hepatic metabolism and are contraindicated when there is hepatic or renal disease. An example of primary metabolic alkalosis is shown in Table 4.13.

The effects of acid–base imbalance on other organs

In addition to the stimulation of respiratory and renal function in an attempt to correct the H^+ imbalance, the function of other body systems, including the nervous and cardiovascular systems, is altered by acid–base changes. An increase or decrease in CO_2 levels and an increase in $[H^+]$ have narcotizing effects on the brain. Consequently, animals with these changes may appear to be depressed.

Increases in CO_2 levels and $[H^+]$ have different effects on the cardiovascular system. When changes are small, circulatory function is usually maintained through stimulation of the sympathoadrenal system. This response prevails over the direct depressant actions of CO_2 and H^+ on the myocardium and vasculature. If acidosis is severe, marked myocardial depression and peripheral vascular dilation ensues, drastically reducing the efficiency of the circulatory system. Respiratory and metabolic alkalosis, on the other hand, have a mild stimulatory effect on the myocardium. Severe metabolic acidosis of alkalosis may induce dangerous cardiac arrhythmias.

Variations in plasma $[H^+]$ influences the actions of most drugs by altering the ratio ionized to unionized and the extent of binding to plasma protein. The unionized form is lipid soluble and can readily diffuse across cell membranes.

Additional reading

Carter, J. M. and Brobst, D. F. (1969). A comparison of the pH of canine capillary, arterial and venous blood. *J. Comp. Lab. Med.* **3**: 19–24.

Cornelius, L. M. and Rawlings, C. A. (1981). Arterial blood-gas and acid–base values in dogs with various diseases and signs of disease. *J. Am. Vet. Med. Assoc.* **178**: 992–5.

English, P. B., Hardy, L. N. and Holmes, E. M. (1969). Values for plasma electrolytes, osmolality and creatinine and venous PCO_2 in normal sheep. *Am. J. Vet. Res.* **30**: 1967–73.

Fisher, E. W., Sibartie, D. and Grimshaw, W. T. R. (1980). A comparison of the pH, PCO_2, PO_2 and total CO_2 in blood from the brachial and caudal auricular arteries in normal cattle. *Brit. Vet. J.* **136**: 496–506.

Guyton, A. C. (1986). *Textbook of Medical Physiology*, 7th edn. Philadelphia, W. B. Saunders Co.

Hannon, J. P. (1983). Blood acid–base curve nomogram for immature domestic pigs. *Am. J. Vet. Res.* **44**: 2385–90.

Harris, W. H. (1974). Haemoglobin, blood-gas and serum electrolyte values in swine. *Can. Vet. J.* **15**: 282–5.

Herbert, D. A. and Mitchell, R. A. (1971). Blood-gas tensions and acid–base balance in awake cats. *J. Appl. Physiol.* **30**: 434–6.

Kelman, G. R. and Nunn, J. F. (1966). Nomograms for correction of blood PO_2, PCO_2, pH and base excess for time and temperature. *J. Appl. Physiol.* **21**: 1484–90.

Middleton, D. J., Ilkiw, J. E. and Watson, A. D. J. (1981). Arterial and venous blood-gas tensions in clinically healthy cats. *Am. J. Vet. Res.* **42**: 1609–11.

Mitchell, B. and Williams J. T. (1975) Normal blood-gas values in lambs during neonatal development and in adult sheep. *Res. Vet. Sc.* **19**: 335–6.

Poulsen, J. S. D. and Surynek, J. (1977). Acid–base status of cattle blood. *Nord. Vet. Med.* **29**: 271–83.

Rose, R. J., Ilkiw, J. E. and Martin, I. C. A. (1979). Blood-gas, acid–base and haematological values in horses during an endurance ride. *Eq. Vet. J.* **11**: 56–9.

Rossing, R. G. and Cain, S. M. (1966). A nomogram relating PO_2, pH, temperature and hemoglobin saturation in the dog. *J. Appl. Physiol.* **21**: 195–201.

Siggaard-Anderson, O. (1963). Blood acid–base alignment nomogram. *Scand. J. Clin. Lab. Invest.* **15**: 211–20.

Steffey, E. P., Wheat, J. D., Meagher, D. M., Norrie, R. D., McKee, J., Brown, J. and Arnold, J. (1977). Body position and mode of ventilation influences arterial pH, oxygen and carbon dioxide tensions in halothane anesthetized horses. *Am. J. Vet. Res.* **38**: 379–82.

David A. Pass and John R. Bolton

5 The respiratory system

The major function of the respiratory tract is to facilitate the exchange of O_2 and CO_2 between the blood and the atmosphere. The tract has two major functional divisions: a gas-transport system comprising the nasal cavity, larynx, trachea, bronchi and bronchioles and a gaseous exchange system comprising alveolar ducts and alveoli. The transport system not only carries gases but also warms, humidifies and filters them. Gaseous exchange in alveoli is maximized by a large surface area and a thin gas-exchange barrier (Fig. 5.1).

Clinical signs of respiratory tract disease depend upon the level (or levels) of the tract involved as well as upon the nature, severity and duration of the insult. In the gas-transport system, involvement of a small area may produce major clinical signs; for example, a local foreign body or area of inflammation in the nasal cavity, larynx, trachea or bronchi can produce violent sneezing, coughing or dyspnea. By contrast, large areas in the lungs may be diseased with little functional impairment. For instance, when there is pulmonary neoplasia, signs of respiratory failure may not appear until two-thirds of the lung tissue are involved.

Not all diseases of the respiratory tract produce clinical signs that might be expected. For instance, in farm animals, chronic bronchopneumonia is common and is usually manifest clinically by illthrift, rather than by coughing and dyspnea. The latter are not evident until the animal is forcibly exercised, which seldom occurs under most conditions of management.

Respiratory function may be secondarily altered by dysfunction of another organ

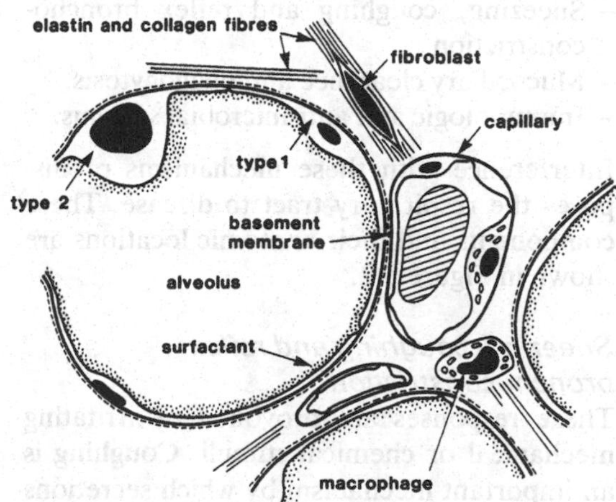

Fig. 5.1. Structure of the normal aveolar wall and interstitium. The alveolus is lined predominantly by thin type 1 epithelial cells and occasional cuboidal type 2 epithelial cells. Surfactant, produced by type 2 cells, coats the surface. In the interstitium capillaries abut the alveolar wall. The thin part of the capillary wall is separated from the alveolar lumen by surfactant, type 1 epithelial cytoplasm and fused basement membranes of epithelium and endothelium. Gases pass across this area into the bloodstream. Collagen and elastin fibers that provide strength and contribute greatly to elastic recoil are specifically arranged spatially between alveoli. Macrophages are also found in the interstitium. Note that there are no lymphatic vessels at this level. Edema fluid is moved through interstitial tissue by a 'pumping action' of alveoli until it reaches lymphatics adjacent to the terminal airways.

system, particularly the cardiovascular system. For example, when the left atrioventricular valve becomes insufficient, the development of pulmonary congestion and edema is reflected by dyspnea. Also, in conditions causing metabolic acidosis, polypnea is a prominent sign, as the respiratory rate is stimulated in an effort to compensate for the acidosis.

Defense mechanisms of the respiratory tract

The respiratory tract is continuously exposed to animate and inanimate particulate matter and therefore requires an efficient mechanism for disposing of such insults. Under normal circumstances, particles are removed by a defense system comprising three basic components:

- Sneezing, coughing and reflex bronchoconstriction.
- Mucociliary clearance and phagocytosis.
- Immunologic and anti-microbial systems.

Interference with these mechanisms predisposes the respiratory tract to disease. These components and their anatomic locations are shown in Figure 5.2.

Sneezing, coughing and reflex bronchoconstriction

These responses are provoked by irritating mechanical or chemical stimuli. Coughing is an important mechanism by which secretions and foreign bodies are removed from the trachea and major bronchi following stimulation of afferent nerve endings in the laryngeal, tracheal and bronchial mucosa. The cough reflex results in both increased velocity of air flow in the major airways and decreased diameter of the airways. During coughing, retained secretions on the mucosa occupy a larger proportion of the airway and the high air-flow velocity forces them towards the pharynx. Although peripheral airways are not directly affected by the cough reflex because of low velocity of air flow, the high intrathoracic pressure developed during coughing probably aids in the movement of material within them towards the larger airways.

Reflex bronchoconstriction is mediated by the vagus nerve and is characterized by an increase in airway resistance. It helps to prevent further penetration of mechanical or chemical irritants.

Clearance by mucociliary activity and phagocytosis

In general, the respiratory epithelium is pseudostratified, ciliated, columnar and contains goblet cells. The surface of the cells is covered by a two-phase fluid layer known as the mucociliary blanket or escalator. At the cell surface, the cilia beat within a thin aqueous layer and their tips drive the thicker overlying mucus layer. The cilia beat synchronously and move the mucus in definite directions. Thus, in the nasal cavity most of the mucus is moved back to the pharynx, and in the trachea it is moved up to the pharynx. The mucus is then either swallowed or in some species expectorated.

When particles enter the nasal cavity, the greater the size the greater the chance of entrapment. Almost all particles greater than 10 micrometers in diameter are retained on

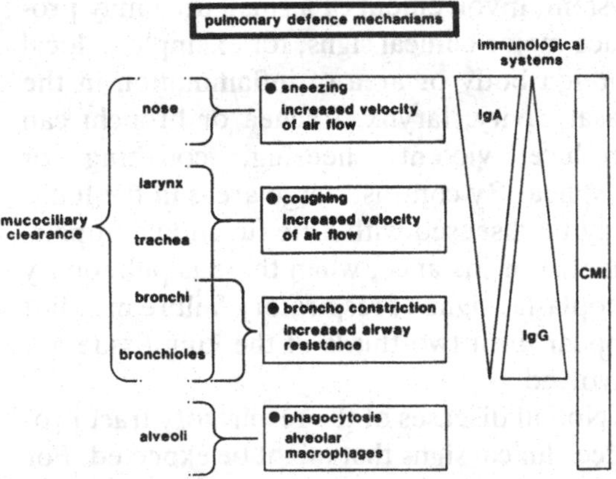

Fig. 5.2. Pulmonary defense mechanisms. A single anatomic location has at least two defense mechanisms to prevent or to curtail the deleterious effects of physical, chemical or infectious agents. CMI, cell-mediated immunity; Ig, immunoglobulin.

the mucus whereas only 5% of those that are 1–2 micrometers in diameter are trapped. As infectious agents are often transmitted in the form of droplet nuclei which are less than 2 micrometers in diameter, they should easily bypass the nose and enter the lower respiratory tract. However, humidification of such droplet nuclei increases their size, enhancing the probability of deposition.

It should come as no surprise that respiratory defense can be compromised by factors that change the composition and rate of production of mucus, the activity of cilia, or both. Anything that changes the rate of ciliary beat changes the rate of mucus flow. For example, rising body temperature affects ciliary activity, increasing it up to 39 °C and then decreasing it at higher temperatures. Hence, fever is likely to result in decreased clearance. Ciliary activity is also decreased by a lowering of body and air temperature, by the administration of certain drugs and by the invasion of epithelial cells by some viruses.

Changes in the amount and composition of mucus have pronounced effects on its flow rate. A hot, dry atmosphere results in evaporation of water and increased viscosity of mucus and if the adverse conditions persist, a crust forms and cilia degenerate. Conversely, excessive production of serous fluid in, for example, acute inflammation, increases the depth of the aqueous layer and raises the mucus layer above the tips of the cilia thereby reducing movement. Decreased viscosity also may result in tearing of the mucus and exposure of the underlying mucosa. This enables infectious agents to gain direct access to the surface of epithelial cells.

The fate of deeply inhaled particles is dependent upon the site of deposition. Those particles deposited above the respiratory bronchiole are eliminated by mucus transport, which may be aided by cough. Those deposited below the respiratory bronchiole are subject to phagocytosis by alveolar macrophages or transported by movement of fluid from the alveolus to the bronchiole. Particles are transported within the phagocytes either on to the mucus or through the alveolar wall and into lymphatics.

Alveolar macrophages are the most important line of defense against infectious agents and arise from a resident population of tissue histiocytes that originate from bone marrow precursors or from blood monocytes. The activity of alveolar macrophages can be markedly depressed by various drugs, hypoxia, acute starvation, emotional stress, acidosis and, most importantly, by some viral infections.

The areas of the respiratory tract that are least well protected are the terminal bronchioles, which are lined by cuboidal epithelium interspersed with groups of ciliated cells, and the respiratory bronchioles, which have no ciliated epithelium. These areas are transition zones where there is a mixture of mucociliary transport and macrophage activity. The walls of these airways are poorly supported by connective tissue, and are subject to collapse in those animals with long respiratory bronchioles, such as the dog and cat. However, these zones are able to undergo rapid repair following injury, and this in itself constitutes a defense mechanism.

From the above discussion it can be seen that the factors influencing mucociliary activity and phagocytosis are those well recognized as being associated with the onset of respiratory disease, namely wet and cold or hot and dry environments, poor nutrition, stress and concurrent infections.

Immunologic defense

The importance of immunologic mechanisms in the defense of the respiratory tract is exemplified by the well-known occurrence of respiratory tract disease in immunosuppressed individuals. All classes of immunoglobulins (Ig) occur in respiratory tract secretions, and in the upper respiratory tract IgA predominates. It is produced by plasma cells in the mucosa and is secreted across the mucosa in its dimeric form. IgA acts principally to block attachment of antigen to the surface of epithelial cells. It is present in normal secretions, but its concentration is increased in inflammatory conditions. Although the concentration of IgA decreases in the lower tract, the concentration of IgG increases.

Cell-mediated immunity is also of great importance. Both T and B cells exist in lymphoid aggregates in the bronchial submucosa (bronchial-associated lymphoid tissue – BALT), which are similar to the Peyer's patches in the gut. The amount of lymphoid tissue increases in response to infection.

The significance of microbial colonization of the respiratory tract

The nasal cavity harbors a microbial flora characteristic of the species and the environment. In cattle for example, there is a basal bacterial flora which is species dependent, including *Pasteurella multocida*, *P. haemolytica*, *Neisseria catarrhalis*, a supplementary flora (members of which are often but not always present) such as *Moraxella* spp., *Streptococcus* spp., and a transient flora which arises from the environment.

The ability of some organisms to colonize the mucosa in spite of the defense mechanisms is in general poorly understood, and many such colonists do not produce disease. Those that are pathogenic may either cause disease in their own right or predispose to other infections by suppressing clearance mechanisms. This is usually achieved by depression of ciliary activity, destruction of cilia, or wholesale destruction of ciliated cells. Some respiratory pathogens attach to the cell surface or to cilia by specific receptors.

The normal lung is bacteriologically sterile below the major bronchi and, when organisms are present, there is usually obvious inflammation and bacteria are present in large numbers. On most occasions, clearance mechanisms are highly efficient, and following aerosolization of the lung with bacteria there is a rapid loss of viability of organisms. The virulence of an organism probably plays a role in its ability to persist in the lung as, under experimental conditions, some virulent organisms are not removed as quickly as non-virulent organisms.

There is no doubt that many respiratory infections involve a combination of microorganisms and, in particular, bacterial infections often occur 5–7 days after viral infections. In most cases it is accepted that this is due to the initial viral infection causing suppression of defense mechanisms.

Although certain viral infections compromise mucociliary clearance mechanisms by damaging mucosal epithelial cells, it appears that the critical factor is damage to the alveolar macrophages. In general, virus titers peak 3–5 days following a respiratory viral infection, and decline by 7–9 days. Histologic changes are maximal at 7–9 days and alveolar macrophage biocidal activity is maximally suppressed at this time. Functional deficits in alveolar macrophages result from a decrease in the number of receptor binding sites, decreased uptake of microorganisms and reduced intracellular destruction of microorganisms.

In some species, respiratory disease occurs as a result of synergism between certain bacteria and viruses. The mechanisms are complex and depend on the species and strain of bacteria and viruses, and on host factors. Examples of true viral–bacterial synergism are few but include swine influenza, caused by swine influenza virus type A and *Haemophilus influenza suis*, some cases of fibrinous pneumonia in cattle caused by bovid herpesvirus I and *P. haemolytica*, and fibrinous pneumonia is sheep infected with parainfluenza 3 virus and *P. haemolytica*.

The upper respiratory tract

General reactions of the mucosa of the nose and tracheobronchial tree

Exposure of the mucous membrane of the upper respiratory tract to irritants, whether infectious or non-infectious, produces similar changes although the time scale can vary.

The causes of such irritation are many and include physical agents (dust), chemicals (sulfur dioxide, ammonia), viruses (herpesviruses, adenoviruses, calciviruses), bacteria (*Bordetella bronchiseptica*), mycoplasma, fungi and parasites (*Dictyocaulus* sp.). In some environments, physical and chemical agents are the major factors predisposing to microbial infections by depressing clearance mechanisms. For example, hot, dry, dusty conditions are major predisposing factors in the development of bacterial bronchitis and bronchopneumonia in horses in some parts of the world. Similarly, ammonia levels in the environment of intensively raised pigs, poultry and laboratory rodents influence the severity of infection with mycoplasmas.

Allergic reactions are well recognized in man and almost certainly occur in some species of animals. In veterinary medicine, the best-recognized disease of this type is chronic obstructive pulmonary disease of horses, referred to under emphysema (see below).

The earliest changes induced by injury are a decrease in the number of cilia and an increase in the number and activity of goblet cells. These changes occur within several days in respiratory virus infections, but may take weeks to months to develop when damage is caused by irritant gases such as sulfur dioxide or smoke. They are accompanied by acute inflammation in the lamina propria resulting in hyperemia, edema, and exudation of neutrophils.

The functional consequences include disturbance of ciliary coordination causing abnormal mucus-flow patterns, increased adhesiveness of mucus, and ciliostasis. With continued irritation there is gradual and progressive metaplasia. Columnar cells become more cuboidal and less ciliated, basal cells proliferate and stratify and the basal lamina becomes multilayered and thicker. With time, the mucosa becomes increasingly stratified and squamous in character.

Experimental production of chronic bronchitis with sulfur dioxide has shown that the metaplastic change and its severity vary with anatomical site. Lesions increase in severity from the trachea distally. In the trachea and major bronchi there are loss of cilia, loss of goblet cells and squamous metaplasia. In bronchioles, epithelial hyperplasia occurs rather than metaplasia and it is accompanied by an inflammatory response. Goblet cells are increased in number along with non-secretory cells. Hyperplasia distorts the cells making them more elongated and tends to force the mucosa into folds and polyps. The bronchiolitis which accompanies the epithelial changes may be due to the resulting failure to clear microorganisms.

Submucosal glands also become hyperplastic and dilated and this, together with hyperplasia of goblet cells results in excessive mucus secretion. Qualitative changes (high viscosity, low elasticity) in mucus also occur particularly when infection is present. The excessive quantity of abnormal mucus is cleared with difficulty.

Hyperplasia of bronchial and bronchioalveolar muscle which occurs in chronic bronchitis probably follows hyperactivity associated with prolonged irritation to the mucosa. This is believed to produce hypertonicity of the bronchi and increasing resistance to air flow. Thus, the overall effects of irritation to the mucosa are increased mucus secretion, decreased clearance and increased airway resistance due to excessive mucus, inflammation and perhaps hypertonicity of bronchial smooth muscle (Fig. 5.3).

Nasal cavity and sinuses

The clinical signs of nasal disease may include sneezing, noisy breathing, increased inspiratory effort, discharges, foul odor and facial deformity. Involvement may be unilateral or bilateral. The character of discharges and odor depends on the type of lesion present. Apart from occasional foreign bodies that obstruct the nasal cavity, the major causes of clinical disease are developmental deformities, inflammation and neoplasia.

Developmental deformities

The most common developmental 'abnormality' is the shortened, cramped nasal cavity of brachycephalic breeds of dog, which may result in some degree of obstruction to inspiration. Many dogs have no impediment other than noisy inspiration, but in others the obstruction may be severe enough to cause cyanosis and collapse on exercise. Stenosis of the nares by the lateral alar cartilages may complicate the problem.

Developmental abnormalities of the hard and soft palate often result in upper respiratory tract disease. Cleft palate, which occurs in all species, is characterized clinically by discharge of milk and other food from the nostrils, sneezing and nasal discharge. Elongation of the soft palate, again seen most commonly in brachycephalic dogs, can cause obstruction of the glottis, with resultant increased inspiratory effort.

Rhinitis and sinusitis

Inflammation of the nasal cavity (rhinitis) can occur independently of, or together with, inflammation of the sinuses (sinusitis). The clinical signs of rhinitis are dependent upon the type and duration of the inflammatory reaction. Rhinitis may be serous, mucoid, mucropurulent, purulent, fibrinous or granulomatous. The dominant clinical signs are nasal discharges and noisy inspiration and expiration, the result of airway obstruction.

The major sign of **acute rhinitis** is a seromucoid nasal discharge which is the result of hyperemia and edema of the mucosa, discharge of mucus from goblet cells and in some cases mild epithelial cell degeneration. These changes occur most commonly in acute viral infections such as infectious bovine rhinotracheitis (IBR), feline viral rhinotracheitis (FVR) and equine viral rhinopneumonitis. They also occur in acute allergic rhinitis. In the viral diseases, the lesions are induced by virus invasion of epithelial cells with subsequent degeneration and accompanying acute inflammation. In allergic rhinitis, antigens such as pollens stimulate a type I hypersensitivity reaction.

Mucopurulent and purulent nasal discharges are usually indicative of bacterial infections, which are mostly of a secondary nature. There are few instances of primary bacterial rhinitis in animals. The classic example is strangles in the horse, caused by *Streptococcus equi*. Usually, bacteria invade the mucosa when clearance mechanisms are hampered by a viral infection and this is a common occurrence in canine distemper.

Some invasive bacteria can produce fibrinous rhinitis but the most common causes of this reaction are IBR in cattle and FVR in cats, which induce a severe fibrinous inflammation by causing necrosis of nasal epithelial cells.

In the majority of cases of acute rhinitis, the inflammatory reaction is limited to the superficial mucosa but, in some infections, deeper structures may be involved. In some cases of FVR infection for example, there is also necrosis of osteocytes in the underlying turbinate bones, with consequent removal of necrotic bone and remodeling of the turbinates.

While prolongation of an acute rhinitis of infectious or allergic cause results in chronic inflammation, some agents produce chronic rhinitis from the outset. This is seen typically

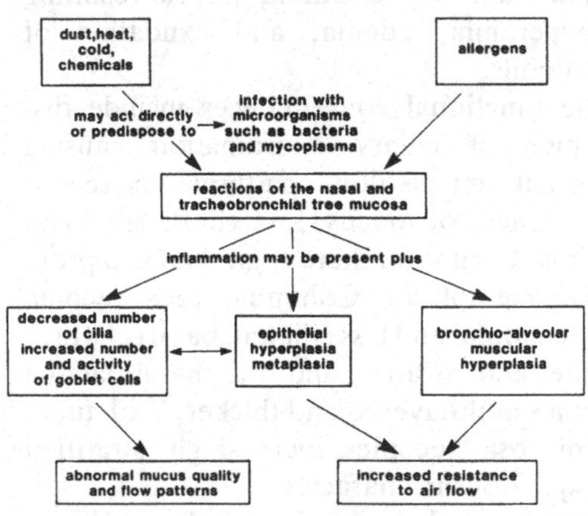

Fig. 5.3. Reactions of the nasal and tracheobronchial tree mucosa. Physical, chemical or infectious agents and allergens induce reactive changes in the mucosa including inflammation and alterations in the epithelium.

in fungal infections such as cryptococcosis and aspergillosis.

In **chronic rhinitis**, in addition to nasal discharges there is usually some degree of obstruction which may be associated with deformity of the nasal cavity and/or face. Obstruction is due to thickening of the mucosa by inflammatory cell infiltrates, hyperplasia and metaplasia of the surface epithelium and hyperplasia of the glands. In some cases polyps formed by these elements project into the nasal cavity. Deformity is produced by expanding aggregates of inflammatory tissue and fibrosis within the confined space of the nasal cavity. Pressure of expanding tissue on bone may result in atrophy of bone such that turbinates and nasal bones become thin, distorted by remodeling, or they may disappear completely. Atrophy of turbinates may also occur if turbinate bone undergoes degeneration or necrosis. This is believed to occur in the disease of pigs known as atrophic rhinitis.

In cattle, repeated exposure of the nasal mucosa to certain allergens produces a chronic pruritic rhinitis, the disease being known as *nasal granuloma* because of the 'granular' appearance of the mucosa. The allergens are believed to be pollens. The 'granules' on the mucosa form when chronic inflammation and fibrosis occur in the lamina propria. The chronic inflammatory tissue pushes the mucosa up between the necks of the glands. The thickening is increased by coexistent hyperplasia of glandular mucosa and squamous metaplasia of surface epithelium. Affected cattle show signs of nasal obstruction which is believed to be due to congestion of the deep veins of the mucosa rather than to mucosal thickening. An outline of acute and chronic rhinitis is shown in Fig. 5.4.

Neoplasia of the nasal cavity and sinuses

Primary neoplastic lesions of the nasal cavity are seen most commonly in the dog, and are generally adenocarcinomas, although fibrous, chondroid and osseous neoplasms also occur. Neoplasia is uncommon in other species. Although connective tissue tumors in the sinuses of young horses are well recognized, it is not certain whether these lesions are neoplastic or dysplastic in nature.

Neoplasms produce clinical signs by occupying space, indicated chiefly by obstruction to air flow, deformity of nasal bones, discharge, irritation and pain.

Epistaxis, or bleeding from the nostrils, may also occur with nasal and sinus tumors in small animals. In horses, epistaxis is more commonly associated with exercise-induced pulmonary hemorrhage or guttural pouch mycosis, with erosion of the internal carotid artery. Erosion into a pulmonary artery by a lung abscess may be associated with acute epistaxis in cattle.

Larynx and trachea

The reactions of the laryngeal and tracheal mucosa to injury are the same as those of the nasal mucosa, and the agents that cause inflammation are essentially the same as those that cause rhinitis.

When the laryngeal mucosa is inflammed, the resultant swelling produces changes in the

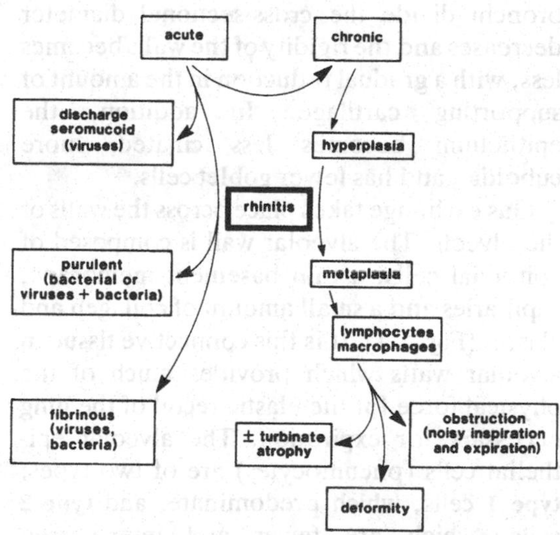

Fig. 5.4. Rhinitis may be acute or chronic in nature. Acute rhinitis is generally exudative and chronic rhinitis proliferative. Acute rhinitis may resolve or become chronic. In some types of rhinitis, only the chronic phase is clinically evident.

voice, pain, and signs of obstruction. The last may be manifest by increased noise during respiration or by dyspnea. Similar signs are produced by tumors of the larynx, or other space-occupying lesions around the larynx.

The major sign of tracheitis is cough, the nature of which depends on the type of inflammation present. Mild acute inflammation associated with increased mucus production is characterized by a moist cough, but as inflammation progresses and the mucosa becomes hyperplastic and metaplastic, any cough is likely to become dry or hacking.

In mammals, the diameter of the trachea and main-stem bronchi is wide, so that signs of obstruction are not usually a feature of tracheitis.

Form, function and dysfunction of the lung

Gases are conducted through the lungs via the bronchi, bronchioles, terminal and respiratory bronchioles, and alveolar ducts. The bronchi are wide semi-rigid tubes of large diameter, lined by pseudostratified, ciliated columnar epithelium containing goblet cells. As the bronchi divide the cross-sectional diameter decreases and the rigidity of the walls becomes less, with a gradual reduction in the amount of supporting cartilage. In addition, the epithelium becomes less ciliated, more cuboidal, and has fewer goblet cells.

Gas exchange takes place across the walls of the alveoli. The alveolar wall is composed of epithelial cells, a thin basement membrane, capillaries and a small amount of collagen and elastin (Fig. 5.1). It is this connective tissue in alveolar walls which provides much of the physical force for the elastic recoil of the lung necessary for expiration. The alveolar epithelial cells (pneumocytes) are of two types: **type 1** cells, which predominate, and **type 2** cells, which are fewer and interspersed between type 1 cells. The nomenclature is unfortunate because the type 2 cell is the precursor of the type 1 cell. Type 1 cells form a continuous layer over each alveolar surface. The cytoplasm is thinly stretched to facilitate diffusion of gases, and the intercellular junctions are tight and relatively impermeable to small water soluble molecules. The type 2 cell is also the major producer of surfactant, a phospholipid-rich material that coats the alveolar surface. It functions to reduce surface tension within the alveoli and in so doing reduces the work required to expand the alveoli (increases compliance). It also promotes the stability of alveoli and helps keep them dry by reducing forces that tend to suck fluid into them.

Replacement of type 1 by type 2 cells is known as alveolar hyperplasia or alveolar epithelialization. This process is a common and a non-specific response to many stimuli, although it is a major lesion in some specific diseases. The pathogenesis of alveolar epithelialization in many cases is obscure. While some agents have a direct effect on epithelial cells, others have an indirect effect. Agents having a direct effect do so either by stimulating hyperplasia *per se* or by causing cell degeneration and death, which stimulates hyperplasia of the surviving cells. Epithelialization is stimulated directly by toxic gases such as nitrogen dioxide, pure oxygen and ozone, by toxic chemicals such as phenols, ethylene dibromide and pyrrols of some pyrrolizidine alkaloids, and by viruses such as parainfluenza-3 virus in calves and sheep, distemper in dogs, caliciviruses in cats and ovine retrovirus.

Pulmonary edema and fibrosis stimulate epithelialization indirectly and the lesion therefore often accompanies inflammation, particularly when it is chronic. No matter what the cause, the effects of epithelialization on respiratory function are similar. Hyperplasia of the epithelium thickens the alveolar wall and increases the distance over which gases must diffuse. It also presumably decreases the 'stretchability' of the wall thereby restricting expansion of the lung during inspiration. Both of these features contribute to the development of restrictive respiratory failure (see below).

The capillaries within the alveolar wall, like all other capillaries, are composed of endothelial cells, but these cells differ in appearance depending on whether they are adjacent to the epithelial cells or to the connective tissue. Endothelial cells that abut the epithelium are thin, featureless and devoid of cell junctions. Those that lie against the interstitial connective tissue are thicker, contain numerous pinocytotic vesicles and commonly form intercellular junctions. They also appear to be much more metabolically active.

It can be seen therefore that oxygen must diffuse across a surfactant layer, a thin epithelial cytoplasm, a thin basement membrane and a thin endothelial cytoplasm, before being taken up by hemoglobin in a red cell.

Lymphatic capillaries are not present in the alveolar wall, but are present within the walls of the respiratory bronchioles, and it is estimated that all blood capillaries are within 1 millimeter of a lymphatic capillary.

Pulmonary lymph arises from fluid which passes through the walls of the thicker alveolar capillary endothelial cells adjacent to the septal connective tissue. Pulmonary lymph is higher in protein than that from other tissues (3.5 grams compared to 2 grams/100 milliliters), is formed continuously and is rapidly removed from the interstitium by drag induced by a pumping action of the lymphatics. The alveoli are kept dry by the efficient removal of lymph and the relative impermeability of alveolar epithelial cells. The lymph drains to collecting vessels around bronchi and the branches of the pulmonary artery. These collecting vessels anastomose with subpleural lymphatics and all the lymph drains to the lymph nodes in the hilus of the lung.

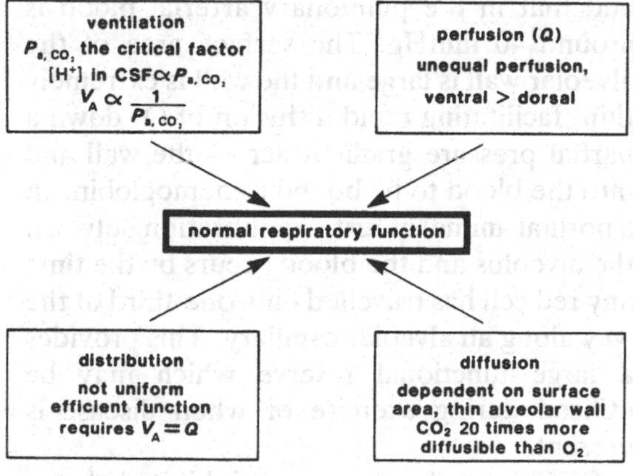

Fig. 5.5. Normal respiratory function has four major aspects, ventilation, perfusion, distribution and diffusion. V_A, alveolar ventilation; CSF, cerebrospinal fluid.

Normal respiratory function

The chief function of the lung is to exchange O_2 and CO_2 between blood and air within alveoli and this is achieved by a combination of four processes: ventilation, perfusion, distribution and diffusion (Fig. 5.5).

Ventilation

Ventilation is the process by which air is moved into the alveoli. Approximately two-thirds of the inspired air reaches the alveoli; the remainder occupies the dead space in the trachea, bronchi and bronchioles. The volume of air entering the alveoli each minute is known as the alveolar ventilation (V_A). V_A is inversely proportional to the partial pressure of arterial carbon dioxide (P_{a,CO_2}) providing there is not excessive metabolic production of CO_2 in the body. If V_A is reduced P_{a,CO_2} increases, and if V_A is increased P_{a,CO_2} decreases. By contrast, increasing the V_A of normal lung does not effectively increase the P_{a,O_2} because the blood leaving well-ventilated alveoli is already almost fully oxygenated.

Under physiologic conditions, ventilation is precisely controlled to keep arterial partial pressures of O_2 and CO_2 within close limits, despite widely varying demands for O_2 uptake and CO_2 elimination. Ventilation is controlled centrally from the medullary respiratory center and the apneustic and pneumotaxic centers in the pons. These centers are themselves sensitive to stimuli arising from receptors located both in the brain and in peripheral structures. Minute to minute changes in respiration are dictated by chemoreceptors situated in the brain, adjacent to the nuclei of the ninth and tenth cranial nerves and responsive to changes in H^+ concentration in the surrounding extracellular fluid. The $[H^+]$ in the extracellular fluid is largely determined by that in the cerebrospinal fluid (CSF), which in turn is determined by the P_{a,CO_2} in the cerebral vessels. As the P_{a,CO_2} rises, CO_2 in the form of carbonic acid diffuses across the blood–brain barrier and dissociates to liberate H^+. A decrease in the pH of CSF increases ventilation and vice versa.

Peripheral control is mediated mainly by chemoreceptors in the aortic and carotid bodies, although regulation is also contributed to by stretch receptors in the lung, baroreceptors in the carotid and aortic sinuses and other receptors in the lung, upper respiratory tract and muscles. The peripheral chemoreceptors respond largely to arterial hypoxemia but they also respond to a fall in blood pH and an increase in arterial P_{a,CO_2}.

Perfusion of the lungs with blood

The lungs are largely perfused by the output of the right ventricle, but not uniformly. Perfusion is affected by pulmonary arterial pressure, pulmonary venous pressure and alveolar pressure.

Gravitational force is also of consequence as it increases hydrostatic pressure in the ventral portions of the lung. Interaction of these factors results in a gradient of blood flow between the dorsal and ventral aspects of the lung, with the ventral portions of the lung having a much greater blood flow than the dorsal regions. This is of little consequence in small animals but is important in large deep-chested dogs and large animals. Pathologic overperfusion occurs in the ventral regions of the lung when venous pressure is increased in left-sided heart failure. Similarly, in laterally recumbent animals, the 'downside' lung suffers overperfusion.

Distribution

The distribution of air throughout the lungs is not uniform but the differences are not as marked as perfusion differences. For an alvelous to function efficiently, the blood flow (Q) to it must be closely matched to V_A. The ideal situation is for an alveolus to have a ventilation/perfusion ratio ($\dot{V}/\dot{Q}$) of 1, but this does not occur evenly throughout the normal lung. Many alveoli are perfused but have little or no ventilation so that $\dot{V}/\dot{Q}$ is < 1 to zero. In this case the particular area of lung is acting as a right to left shunt, since the venous blood perfusing it cannot be oxygenated. The other extreme is for alveoli to be ventilated but not perfused so that $\dot{V}/\dot{Q}$ is > 1. These alveoli act as dead space. The overall value for the lungs is a summation of these conditions, and the oxygen status of the blood entering the left atrium from the pulmonary veins is the net result of the mixing of blood from alveoli in these differing functional states.

Diffusion

The diffusion of gas across the alveolar wall is dependent on the surface area, the differences in partial pressures of the gas in the liquid and gaseous phases, and on the thickness of the tissue through which the gas has to pass. In addition, the rate of transfer is proportional to the solubility or diffusibility of the gas, CO_2 being 20 times more diffusible than O_2.

The P_{O_2} in inspired air is around 100 mmHg and that in the pulmonary arterial blood is around 40 mmHg. The surface area of the alveolar wall is large and the wall is extremely thin, facilitating rapid diffusion of O_2 down a partial pressure gradient across the wall and into the blood to be bound to hemoglobin. In a normal animal at rest, equilibration between the alveolus and the blood occurs by the time any red cell has travelled only one-third of the way along an alveolar capillary. This provides a large functional reserve which may be utilized during exercise or when disease is present.

Diffusion of oxygen is inhibited by a decrease in the alveolar P_{O_2} or an increase in the thickness of the alveolar wall. As CO_2 is much more diffusible than O_2, an increase in the thickness of the alveolar wall does not generally result in a rise in P_{a,CO_2}.

Respiratory failure

The lung, like many other organs, has a large functional reserve and it is not until this reserve is exhausted that respiratory failure occurs. The critical feature of respiratory failure is the inability of the lungs to maintain the arterial levels of O_2 and CO_2 within defined limits. When arterial oxygen levels fall from 90–100 to 60–70 mmHg and CO_2 rises from 40 to 50 mmHg, an animal is in respiratory failure.

The major clinical signs associated with respiratory failure are changes in the rate and depth of respiration, and cyanosis. Cyanosis is a bluish discoloration of mucous membranes caused by an increase in the amount of reduced hemoglobin in the blood. Increased respiratory rate is designated as **polypnea**, decreased rate as **oligopnea** and cessation of respiration as **apnea**. Moderate increase in the depth (amplitude) of respiratory movement is referred to as **hyperpnea** and difficult or labored respiration as **dyspnea**.

Apart from the major clinical signs outlined above, pulmonary auscultation and percussion are often the only other means available to the clinician for assessing an animal with respiratory failure. Even when this assessment is thorough it may often be difficult to relate the findings to the pathologic changes that are present in the lungs.

Lung sounds are produced by the movement of air through airways down to 1–2 mm in diameter. These sounds are then transmitted to the stethoscope head through the lung parenchyma, pleural space and chest wall. This transmission may be affected by disease processes involving both the lung and the pleural cavity. For example, sounds are transmitted more readily through lung parenchyma containing fluid than through normal air-filled lung. Therefore, abnormally loud sounds may be heard over a severely compromised lung that is receiving very little ventilation. Fluid in the pleural cavity, however, tends to reflect sounds so that their transmission to a stethoscope on the chest wall is absent or markedly reduced. Fluid in the lung parenchyma and in the pleural cavity both yield dull sounds with thoracic percussion. When used in conjunction with auscultation, percussion therefore gives the clinician an important clue as to the location of the fluid accumulation within the thoracic cavity.

Lung sounds may be classified into breath sounds and adventitious sounds. **Breath sounds** are produced by the movement of air through the tracheobronchial tree. Their intensity is related to the velocity of airflow. In many species, the attenuation of these sounds is such that they are inaudible near the end of expiration. Normal animals will have increased breath sounds with increased ventilation. Animals with respiratory failure will have increased breath sounds for the same reason. In addition, breath sounds will be increased if the respiratory failure is associated with the accumulation of fluid in the lung parenchyma or with increased airway resistance. Breath sounds of reduced intensity tend to be associated with conditions in which there is an increase in the amount of air within the lung, or fluid within the pleural cavity.

Previously, breath sounds have been referred to as either bronchial sounds or vesicular sounds. The recognition that no sounds are produced in the 'vesicular' or alveolar area or the lung has led to the suggested abandonment of this terminology.

Adventitious sounds are extrinsic to the normal sound-production mechanism of the respiratory system and are superimposed on breath sounds. They are classified as crackles and wheezes. Crackles are defined as discontinuous, non-musical sounds. They may be further characterized by the timing of their occurrence during the respiratory cycle. Wheezes are defined as continuous, musical sounds. They are further characterized as monophonic (single tone) or polyphonic (multiple tones), as well as by the timing of their occurrence during the respiratory cycle.

Previously, crackles have been referred to

as rales and wheezes have been referred to as rhonchi. This suggested change in terminology is also based on new information on the mechanism of sound production. The term rales was used as it was thought that these sounds were produced by air bubbling through fluid in the airways. This appears to be incorrect. More likely, the sounds are produced by the rapid equalization of pressures following the reopening of airways running into diseased areas of lung. Rhonchi were thought to be associated with the resonation of sounds within the airways, similar to the way in which a musical note can be produced by blowing into a pipe. This mechanism also appears to be incorrect. Current evidence suggests that the sounds are produced by the vibration of airway walls in close contact. This close contact may be produced by either intrathoracic airway obstruction, such as occurs in chronic obstructive pulmonary diesase, or extrathoracic airway obstruction, such as with a collapsing trachea.

The character of the changes in rate and depth of respiration is determined by the type of failure. Other signs may be present which will depend on the nature of the lesions in the lung.

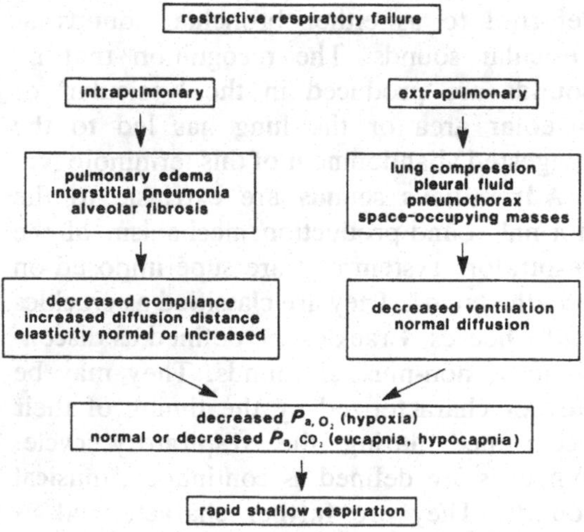

Fig. 5.6. Restrictive respiratory failure may be intrapulmonary or extrapulmonary. Both lead to hypoxia with the dominant clinical feature of rapid, shallow respiration.

There are two basic forms of respiratory failure; **restrictive** (type 1) and **obstructive** (type 2). Classification into type is determined by the character of clinical signs, the physiologic parameters and the lesions present in the lung or thorax. Examination of physiologic parameters such as tidal volume, functional residual capacity, total lung capacity, vital capacity and residual volume are all important for classification. However, in most situations in veterinary medicine it is usually neither practical nor possible to obtain these measurements. It is therefore often not possible accurately to categorize the type of respiratory failure present in animals. However, to comprehend respiratory disease it is essential to understand the basis and features of this classification.

Restrictive respiratory failure

Restrictive respiratory failure is caused by pathologic processes within the lungs or within the thoracic cavity, that restrict inflation of the lungs (Fig. 5.6).

Intrapulmonary lesions producing this type of failure are those that predominantly affect the interstitial tissues (the alveolar and interlobular septa). Common examples are pulmonary edema, interstitial pneumonia (pneumonitis) and alveolar fibrosis. The septa are thickened by the accumulation of fluid, inflammatory cells or fibrous tissue. In interstitial pneumonia and alveolar fibrosis, the thickening is often contributed to by hyperplasia of pneumocytes lining alveoli (Fig. 5.7).

The effect of the thickening of the alveolar walls is two-fold. Firstly, the walls do not stretch as easily or as much as usual and they are therefore stiffer or less compliant. Although they do not stretch easily, elasticity is not decreased, in fact it may be increased, so the alveoli will return to their normal size on expiration. Secondly, the distance across which oxygen has to diffuse is increased.

Another major intrapulmonary lesion that produces restrictive failure is atelectasis of the lung, a pathologic state due to collapse of the alveoli (see below). The alveolar walls are not thickened but because they cannot expand on inspiration, the overall expansion of the lung is restricted.

Extrapulmonary lesions, in the pleural cavity, mediastinum or thoracic wall can also produce restrictive failure. Fluid in the pleural cavities (blood, edema fluid, lymph, inflammatory exudate) causes external pressure restricting movement. Deformities of the thoracic wall and large tumors of the thorax or mediastinum have a similar effect. The effects are the same when air is introduced into the pleural cavities (pneumothorax). The pleural pressure then approaches atmospheric, eliminating the normal negative transpulmonary pressure necessary for lung expansion on inspiration.

Restriction of lung expansion, while decreasing ventilation of the affected portion does not decrease the perfusion of capillaries. Therefore blood leaving these areas is less well oxygenated as the vessels act as right to left shunts. The underoxygenated blood dilutes the fully oxygenated blood flowing from normally ventilated alveoli and thereby reduces the total O_2 concentration in the blood flowing from the lungs. The O_2 concentration is reduced further if diffusion across alveolar walls is restricted. The overall effect therefore is subnormal P_{a,O_2} (hypoxia). P_{a,CO_2} does not increase presumably because of its greater diffusibility. Hypoxia stimulates increased respiratory rate and depth, but as expansion of the lung is restricted, considerable muscular effort is necessary for inspiration. To minimize the work required, the animal develops a rapid, shallow respiration. The increased respiratory rate may eliminate CO_2 excessively and result in lowering the P_{a,CO_2} (hypocapnia).

In summary, restrictive respiratory failure is characterized pathophysiologically by hypoxia and either eucapnia or hypocapnia and clinically by rapid, shallow respiration.

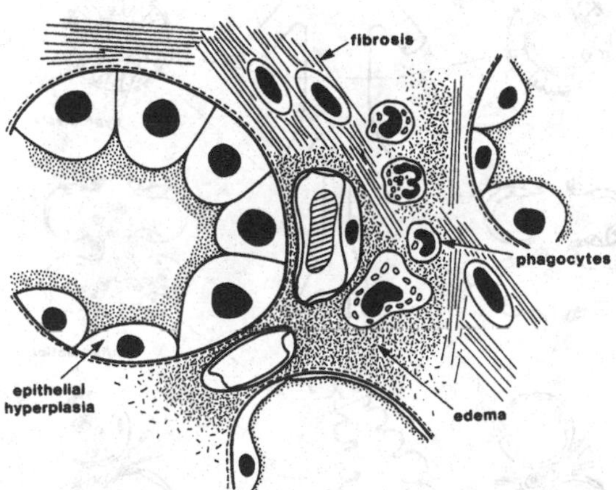

Fig. 5.7. Intrapulmonary restrictive respiratory failure is centered on the alveolus. Edema fluid, inflammatory exudate and excess fibrous tissue thicken the alveolar wall and increase the stiffness of the wall. It takes more muscular effort to expand the wall, but the elasticity within the wall may not be decreased so the alveoli collapse. Hyperplasia of alveolar epithelial cells contributes to the restrictive mechanisms within the alveolar interstitium and also impedes diffusion of O_2 across the wall.

Obstructive respiratory failure

Obstructive respiratory failure is caused by processes within the lungs that reduce ventilation either by obstructing the movement of air in conducting airways or by reducing the elasticity of the connective tissue of the lung parenchyma (Fig. 5.8). The most common example of the former is bronchitis and bronchiolitis where inflammatory exudates, excessive mucus and hyperplastic epithelium partially or completely obstruct the bronchi and bronchioles (Fig. 5.9). Reduction in the elasticity of lung parenchyma occurs in chronic emphysema. Emphysema refers to overexpansion of alveoli, which causes the connective tissues of the alveolar septa to become overstretched and weak. A loss of elasticity results in reduced elastic recoil of inflated alveoli and hence reduced expiratory movement. In essence, obstructive failure occurs when the amount of air drawn into and expelled out of the lungs is insufficient to maintain the arterial partial pressures of O_2 and CO_2 necessary for normal life. It is characterized by reduced P_{a,O_2} (hypoxia) and increased P_{a,CO_2} (hypercapnia). Animals with obstructive failure have both inspiratory and expiratory dyspnea. To overcome the resistance to movement of air they develop an increased rate and depth of respiration.

Pathologic mechanisms and clinical expression of lung dysfunction

Pulmonary edema

Pulmonary edema occurs when the rate of formation of interstitial fluid exceeds the capacity of the peribronchial lymphatics to remove it. The fluid accumulates first in the interstitium around the bronchi and larger vessels and then extends into the septa adjacent to the thick sectors of the capillary walls. This is likely to restrict lung expansion as the compliance of the alveolar walls is reduced, but there is little, if any, restriction to gaseous diffusion. Edema of the interstitial tissues alone is seldom recognized clinically in animals. Fluid does not accumulate in the thin part of the wall until late in the process and only after alveolar filling is advanced. **Fluid accumulation in alveoli is a late, rapid and abrupt event**.

When edema fluid is present in alveoli, diffusion of O_2 and CO_2 is greatly impaired and signs of obstructive respiratory failure appear. In some species, notably cats, pleural effusion may occur concurrently due to edema fluid moving from subpleural interstitial tissue through gaps in the pleural mesothelium. Fluid in the pleural cavity will restrict expansion of the lungs and add to the effects of fluid within the lungs.

The clinical signs of pulmonary edema include an increase in the rate and depth of respiration and a soft, moist cough. Auscultation of the chest typically reveals abnormally

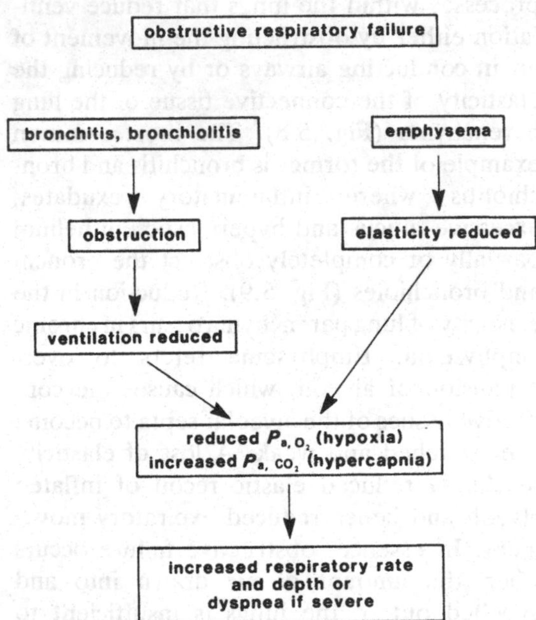

Fig. 5.8. Obstructive respiratory failure is characterized by hypoxia and hypercapnia. Respiratory rate and depth are increased mostly because of the hypercapnia. In the most severe cases, animals are dyspneic.

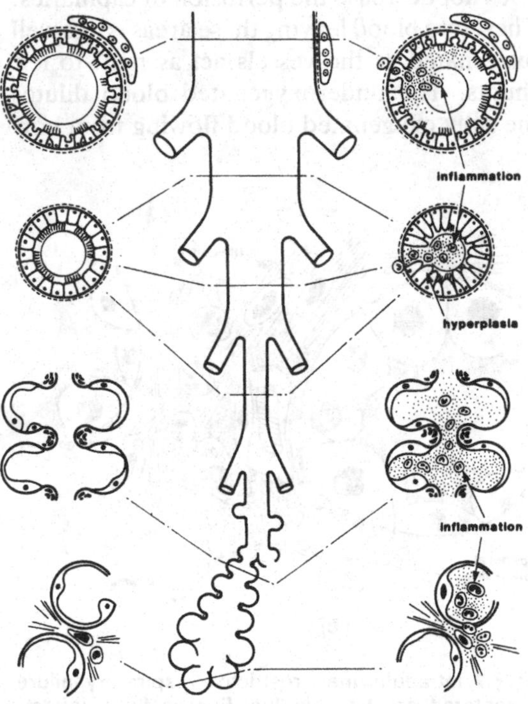

Fig. 5.9. Obstructive respiratory disease. The normal structure of airways and alveoli on the left is compared to the effects of inflammation on the right. Obstruction to airflow is due to excessive mucus production, epithelial hyperplasia and the products of inflammation that block the lumen and obstruct passage of inspired and expired air.

loud breath sounds and late inspiratory crackles, particularly over the ventral lung fields. In severe cases, respiratory distress is extreme, and foam which may be blood tinged, exudes from the nostrils.

At necropsy, the gross appearance of edematous lung is conditioned by the presence or absence of congestion. Edema and congestion usually coexist, as described below for hydrostatic pulmonary edema. Acutely congested lungs are deep red in color and heavy and they fail to collapse fully when handled. Blood and frothy fluid run freely from the cut surface and the airways are filled by a stable foam. This foam is formed by air mixing with the protein-rich mixture of edema fluid and surfactant. If the congestion has been of long duration, the lungs may be mottled by the accumulation of hemosiderin pigment, which imparts a rusty, yellow-brown tinge.

In those areas where edema is not accompanied by congestion, the lungs are pale, wet and heavy and contain stable foam in the airways.

There are three basic pathophysiologic situations in which pulmonary edema arises and these may occur singly or in combination (Fig. 5.10). According to the basic mechanism

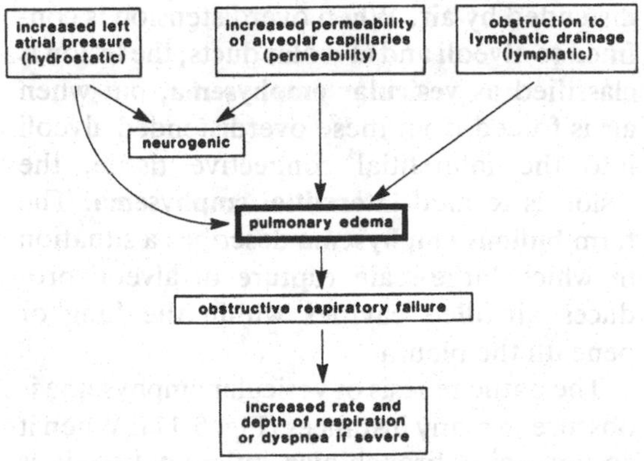

Fig. 5.10. Pulmonary edema may occur following increased hydrostatic pressure, increased capillary permeability, obstruction to lymphatic drainage, or a combination of these. Severe pulmonary edema gives rise to obstructive respiratory failure.

operating, there may be hydrostatic edema, permeability edema, or lymphatic edema.

Hydrostatic edema occurs most commonly when pulmonary venous pressure is raised and the most common cause of this is increased left atrial pressure associated with left-sided or bilateral heart failure (this form of edema may also be called cardiogenic edema; see Chapter 6). Hydrostatic edema may also be caused by hypervolemia resulting from the excessive infusion of intravenous fluids. The edema fluid has low protein concentration (approximately half that of plasma).

Permeability edema occurs when there is increased permeability of capillaries and venules to water and solutes including plasma proteins. This type of edema is caused by inflammation, bacterial endotoxins and exotoxins, the inhalation of pure oxygen, or by toxins such as the pyrroles of some pyrrolizidine alkaloids, and a wide variety of drugs. In all instances, morphologic changes occur in the walls of small blood vessels. There may be separation of endothelial cell junctions (inflammation), gaps in the endothelial cells (pyrroles), or destruction of endothelial cells (inflammation, bacterial toxins and pure oxygen). In all cases the edema fluid has a high protein concentration which may approach that of plasma.

Combinations of hydrostatic and permeability edema occur. **Neurogenic pulmonary edema** occurs in association with acute injury to the central nervous system, particularly if the hypothalamus is involved. Initially there is increased pulmonary hydrostatic pressure associated with a massive transfer of blood from the peripheral to systemic circulation. This process is mediated by adrenalin as it can be prevented or relieved by adrenergic blocking agents. Following treatment, although the hydrostatic pressure drops rapidly, the edema persists due to increased vascular permeability. This is because capillaries are damaged by the rapid, massive increase in pulmonary hydrostatic pressure.

Lymphatic edema, the third basic mechanism of pulmonary edema is due to obstruction of lymphatic drainage by infiltrating neoplastic cells. This is a rare cause of pulmonary edema in animals.

Atelectasis (collapse or incomplete expansion)

In strict definition, the term atelectasis refers to fetal lung that does not fully expand at birth. However, by common usage it also refers to collapse of previously aerated lung, so that it is regarded as being congenital or acquired. Acquired atelectasis is further categorized as being of compression type or obstructive type.

Congenital atelectasis is presumably due to failure of the critical alveolar opening pressure being achieved in the affected areas. The cause of this is not usually discerned. This lesion is quite common but in most cases is of little clinical significance in animals that survive.

Acquired compression atelectasis is due to pressure being exerted on the lung by material such as air, fluid or abdominal viscera which have intruded into the pleural cavity. Deformities of the thoracic wall or expanding masses within the lung may have a similar effect. If the pressure is released reasonably quickly, the process is reversible. However, with time, fibrosis of the pleura and interalveolar septa occurs and complete re-expansion is prevented.

Acquired obstructive atelectasis is due to obstruction of bronchi or bronchioles and this occurs most commonly in animals with bronchitis and bronchiolitis. The lung distal to such obstruction initially fills with fluid and then collapses as the fluid is slowly resorbed. The fluid arises from epithelial secretions and from edema, but the cause of the latter is not clear. Fibrosis of the alveolar interstitium is eventually induced by edema fluid, and secondary inflammation is common. If the obstruction is not removed promptly, full reinflation of the alveoli cannot be achieved.

When atelectasis occurs, development of clinical signs is dependent upon the amount of lung involved. Any clinical signs are due to reduced ventilation accompanied by normal perfusion of capillaries. The atelectatic areas act as right to left shunts so that blood leaving them is poorly oxygenated. When this blood mixes with fully oxygenated blood from normal areas it effectively dilutes the amount of oxygenated blood leaving the lung, thereby causing systemic hypoxia. Small areas of atelectasis can be compensated for by extra dilation of surrounding alveoli, and by constriction of vessels in collapsed areas, which will shunt blood elsewhere. These mechanisms, however, cannot compensate for large areas of atelectasis.

Areas of lung that have undergone recent collapse appear deep red, are depressed below the surface of adjacent normal lung, and have a firm liver-like consistency. With prolonged collapse they become paler and firmer due to increasing fibrosis.

The lung sounds produced by pulmonary atelectasis are variable, depending on the amount of lung involved. Small areas of atelectasis may produce no abnormal sounds. When larger areas are involved, the sounds may be similar to those that occur with pulmonary edema. These are abnormally loud breath sounds and late inspiratory crackles.

Emphysema

The term emphysema refers basically to a state in which part or all of the lung tissue is overdistended by air. When overdistension is confined to alveoli and alveolar ducts, the lesion is classified as **vesicular emphysema**, but when air is forced from these overdistended alveoli into the interstitial connective tissue, the lesion is termed **interstitial emphysema**. The term **bullous emphysema** describes a situation in which large-scale rupture of alveoli produces air-filled cavities within the lung or beneath the pleura.

The pathogenesis of vesicular emphysema is obscure in many instances (Fig. 5.11). When it accompanies bronchiolar inflammation it is thought to result from the enzymatic lysis of elastin fibers in alveolar walls by proteases released from neutrophils and macrophages. When it accompanies acute obstruction of airways, as in acute allergic bronchospasm, it is postulated to be the result of trapping of air. The air is able to enter alveoli on inspiration, but does not exit on expiration, so that the volume of air behind the obstruction increases, dilating and eventually rupturing alveoli.

In vesicular emphysema, although the internal volume of the affected lung is increased, the gas-exchange surface is reduced. Also, stretching and destruction of the connective tissue in alveolar walls results in a loss of elastic recoil, thus reducing the forces necessary for effective expiration. If large areas of lung are involved expiration becomes forced, the volume of the lung increases, the inspiratory excursion lessens and the thorax gradually becomes barrel shaped. Affected animals become hypoxic and hypercapnic, the hypoxia being largely due to mismatching of ventilation and perfusion. Blood leaving the

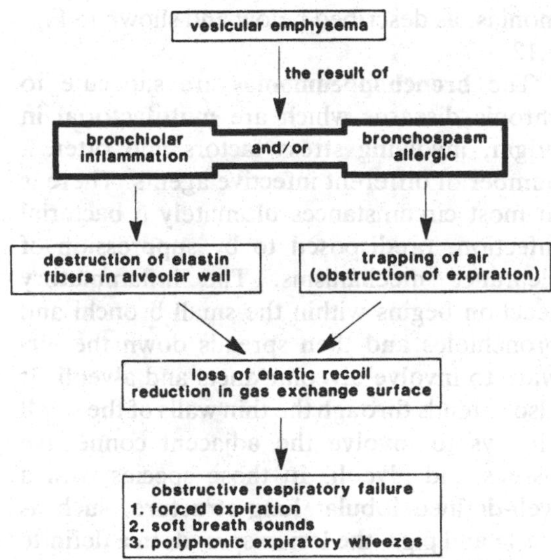

Fig. 5.11. Vesicular emphysema is considered to be a consequence of two major mechanisms, either bronchiolar inflammation or bronchospasm. The latter is usually allergic in origin. Both result in extensive destruction of alveolar walls. The loss of elastic recoil via ventilation/perfusion mismatch leads in severe cases to obstructive respiratory failure.

emphysematous tissue will not be saturated with oxygen and will dilute that leaving normally ventilated parenchyma. Destruction of the gas-exchange surface adds to the effects of ventilation/perfusion mismatching and the P_{a,CO_2} increases due to reduced alveolar ventilation.

The clinical signs associated with emphysema are dominated by expiratory dyspnea. In this condition, the retention of air in the lung decreases the transmission of sounds to the chest wall, producing abnormally soft breath sounds. Polyphonic expiratory wheezes are also characteristic of emphysema. These sounds are produced by the vibrations associated with the collapse and partial closure of multiple airways during expiration.

In practice, emphysema usually coexists with bronchitis/bronchiolitis. This airway involvement tends to produce regularly occurring crackles during the early inspiratory and early expiratory phases of the respiratory cycle. These crackling sounds are thought to be produced by the repeated opening of central airways and by the passage of gas boluses.

There are few diseases in veterinary medicine in which emphysema is the principal determinant of clinical signs. One such is equine chronic obstructive pulmonary disease (COPD) or 'heaves'. This disease is almost certainly caused by hypersensitivity to fungal allergens in the feed, but the pathogenesis is not yet fully understood. The clinical disease results from emphysema, or bronchiolitis, or both. In another disease, known as bovine acute respiratory distress syndrome (also known as 'fog fever' or 'acute bovine pulmonary emphysema', and 'atypical interstitial pneumonia') acute vesicular and interstitial emphysema is the major lesion. This disease has been shown to be a pneumointoxication with 3-methylindole, a rumen metabolite of the amino acid L-tryptophan. The disease occurs when pastures are lush and rich in the amino acid. Chemical toxins from a number of plants including *Zierra arborescens* (stinkwood), *Brassica napus* (rape), *Perilla frutescens* (mintweed) produce a similar lesion and clinical syndrome. A fungus, *Fusarium solani*, which grows on sweet potatoes, is also implicated.

Pulmonary emphysema is also of interest as a pathologic accompaniment to chronic bronchopneumonia of cattle and pigs (see below).

At necropsy, emphysematous areas of lung are pale and voluminous and fail to collapse on manipulation. The tissue is generally dry and may be crepitant. If the lesion is long standing, there may be accompanying bands of fibrosis in the interstitium.

Pulmonary inflammation

The general term pneumonia encompasses all types of inflammatory processes within the lungs. It occurs commonly, as the lungs are exposed to injurious agents directly from the environment via the upper tract, and also via the bloodstream. In the latter case the lungs are unique in being the only organs to receive the entire cardiac output, which furthermore perfuses a vast low pressure capillary bed. It is not surprising therefore that the causes of pneumonia are many and varied and include bacteria, mycoplasmas, chlamydia, viruses, fungi, algae, irritant gases and vapors and particles.

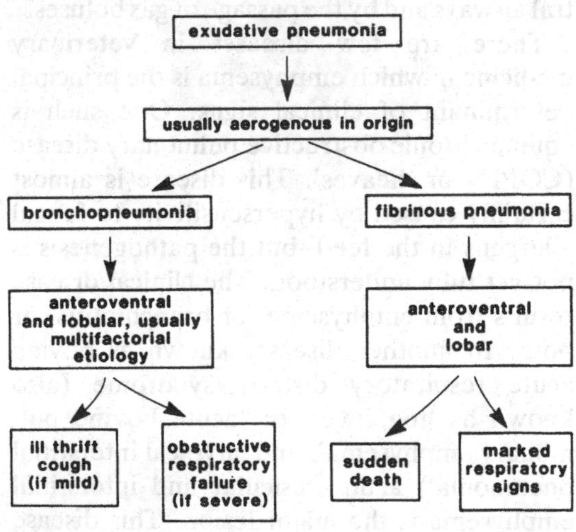

Fig. 5.12. Exudative pneumonias may be subdivided into bronchopneumonia and fibrinous pneumonia. Although both are usually aerogenous in origin, there are differences in pathologic pattern. In general, fibrinous pneumonias are acute to peracute and there is extensive tissue destruction. Bronchopneumonias are often subacute to chronic and spread from lobule to lobule.

The most satisfactory way to classify pneumonia is on a pathologic basis according firstly to the type of inflammation and, secondly, to the anatomical site of its initiation and the pattern of spread. The two major categories are the **exudative pneumonias**, in which the alveolar lumens are filled with fluid and inflammatory cells, and the **proliferative pneumonias**, where the alveolar lumens remain largely clear but the alveolar walls are thickened by hyperplastic epithelial cells and inflammatory cells. This classification is not absolute and considerable overlap can occur.

Exudative pneumonia

The exudative pneumonias are usually aerogenous in origin, and are further classified into bronchopneumonias or fibrinous pneumonias, as described below and shown in Fig. 5.12.

The **bronchopneumonias** are subacute to chronic diseases which are multifactorial in origin, involving stress factors and often a number of different infective agents. There is in most circumstances ultimately a bacterial infection, predisposed to by suppression of clearance mechanisms. The inflammatory reaction begins within the small bronchi and bronchioles and then spreads down the airways to involve alveolar ducts and alveoli. It also extends through the thin walls of the small airways to involve the adjacent connective tissues and alveoli. In those species with a well-defined lobular lung structure, such as cattle and pigs, the lesion spreads in a definite lobule-by-lobule pattern.

The lesions of bronchopneumonia are generally bilateral and fairly symmetrical and tend to involve the cranial and ventral regions of each lung. Recently affected areas of lung become swollen and consolidated by inflammatory exudate and, depending upon the acuteness and severity of the reaction, are deep red or grey-pink in color. As the inflammation resolves, the pneumonic lung collapses due to resorption of exudates. The collapsed lung may undergo fibrosis, or it may reinflate if healing is successful in the airways supplying it. Suppuration and abscess formation sometimes occur and adjacent areas of lung are often emphysematous.

The classical bronchopneumonias are the 'enzootic pneumonias' of pigs and calves. In the pig, infection with *Mycoplasma hyopneumoniae* is primarily often followed by a secondary infection with *Pasteurella multocida*. In calves, events are initiated by stress factors and a primary infection with either parainfluenza 3 virus, *Mycoplasma* sp. or *Ureaplasma* sp. Secondary bacterial infection particularly involves *P. multocida* or *Actinomyces pyogenes*.

In the dog, bronchopneumonia is almost always the result of a primary infection with canine distemper virus, followed by secondary infection with *Bordetella bronchiseptica* or *Escherichia coli*.

In the foal, suppurative bronchopneumonias caused by *Rhodococcus equi* or *Streptococcus zooepidemicus* are well-recognized entities.

Unless large areas of lung are involved, the effect of bronchopneumonia on respiratory function is often not appreciable in an animal at rest. In many instances the clinical expression is illthrift and persistent moist cough. In terms of respiratory function, bronchopneumonia is primarily an obstructive disease, due to impairment of air flow in small airways by exudates and mucosal hyperplasia. The clinical features are as described for obstructive respiratory failure with associated pain, fever and anorexia. Very acute and severe cases will appear clinically as described below for fibrinous pneumonia.

The **fibrinous pneumonias** are acute to peracute exudative pneumonias in which inflammation spreads rapidly from its initial site in small airways to involve large areas of pulmonary lobes in a short time. For this reason, they are also referred to as lobar pneumonias and they frequently terminate fatally. The acuteness and severity of the reaction give rise to a large quantity of fibrin in the exudate, and to extensive vascular damage. This in turn produces hemorrhage, thrombosis and focal necrosis of lung tissue. In the early stages the affected lung is swollen and firm, with fibrin deposited on its pleural surface. The lesions, as in bronchopneumonia tend to be bilateral and to involve the cranioventral regions, but single focal lesions may sometimes occur anywhere in the lung mass. After the initial phase of the disease, necrotic foci within the lung may be walled off to form sequestrums; fibribous exudate in the lung and on the pleura undergoes fibrous organization. This dictates that, even if the animal survives, areas of lung are permanently lost to fibrosis and many pleural adhesions will form. Necrotic sequestrums can act as reservoirs for organisms. Specific fibrinous pneumonias are caused in cattle by *Pasteurella hemolytica* and *Mycoplasma mycoides* (infectious bovine pleuropneumonia), in sheep by *P. hemolytica*, and in pigs by *Hemophilus pleuropneumoniae*.

Non-specific fibrinous pneumonia can also result from aspiration of irritant substances or vomitus.

In contrast to bronchopneumonia, relatively small lesions of fibrinous pneumonia may have dramatic impact. It is not uncommon for animals to be found unexpectedly dead, and in such cases it seems that the fibrinous pneumonia rapidly leads on to fatal toxemia or septicemia. In cases less immediately fatal, respiration is painful, labored and increased in rate and depth. Open-mouthed breathing, an expiratory grunt and cyanosis may be apparent, as may mucopurulent nasal discharge, foul odor on the breath and a persistent deep moist cough. Pyrexia is frequently present.

The abnormal lung sounds produced by exudative pneumonias depend upon the degree of consolidation of the pulmonary parenchyma and the patency of the airways. In the majority of cases of bronchopneumonia, the abnormal sounds are heard predominantly over the ventral and cranial regions of the lung. In fibrinous pneumonias, there is often involvement of the caudal lobes.

Breath sounds are usually louder than normal, especially if consolidation is present. The paradoxical finding of louder than normal sounds over an area of lung that is receiving little or no ventilation is related to the enhanced transmission of sounds through the fluid-filled parenchyma.

In bronchopneumonia, crackles and wheezes are also common, reflecting the bronchial involvement. The crackles tend to occur early in both the inspiratory and expiratory phases of the respiratory cycle. They vary in character from fine to coarse, depending on the quantity of fluid in the airways. Polyphonic wheezes, produced by vibration of the narrowed airways, tend to occur during the expiratory phase of respiration.

Proliferative pneumonia

In the proliferative pneumonias, it will be remembered that the inflammatory process is centered on the alveolar septa, and the terms **interstitial pneumonia**, **pneumonitis** and **alveolitis** are synonymous. Such diseases tend for the most part to be subacute to chronic in nature. The lesions are usually bilateral and diffuse, or multifocal and extensive, and may involve predominantly the dorsocaudal aspects of the lungs.

In the early stages, alveolar septa often become crowded with lymphocytes and macrophages; alveolar pneumocyte hyperplasia may become conspicuous and, with time, septal fibrosis develops. In addition, exudate and surfactant may accumulate to a significant extent within alveoli.

There are several well recognized specific proliferative pneumonias of animals. Those of infectious cause include canine distemper virus pneumonia, ovine and caprine progressive pneumonias, and also in sheep *Mycoplasma ovinpneumoniae* infection. In addition there are hypersensitivity pneumonias, the best recognized of which occurs in cattle sensitized to an antigen of the actinomycete *Micropolyspora faeni*, present in moldy hay.

The clinical signs associated with proliferative pneumonias, as in many other types of pneumonias, depend on the amount of lung involved. Unfortunately, most proliferative pneumonias are characterized by a high degree of irreversible lung damage and the signs reflect a **restrictive respiratory failure** of steadily worsening character. In many cases exercise intolerance may be one of the first signs to become apparent. The increased respiratory effort produces increased breath sounds. Late inspiratory crackles are also common. These are associated with the decrease in the respiratory system compliance that is typical of restrictive pulmonary disease. This predisposes to airway closure.

At this point it should be mentioned that a number of toxins can produce interstitial reactions and resultant restrictive respiratory failure in animals. They have been traditionally classified as interstitial pneumonias but should more correctly be termed toxic **pneumonopathies** because inflammation is not characteristic of the lesion. This group includes the condition known as acute respiratory distress syndrome in cattle (discussed previously) and intoxication by some pyrrolizidine alkaloids (*Crotalaria* sp., *Senecio* sp.), and the bipyridyl herbicide, paraquat. The pertinent features of proliferative pneumonias are shown in Fig. 5.13.

Pulmonary neoplasia

Primary pulmonary neoplasia in domestic animals is not as common as metastasis to the lungs from tumors elsewhere. The major primary tumors are bronchiolar carcinoma and bronchogenic carcinoma and they are most common in the dog. The effect on respiratory function is variable and depends on the number, size and site of the tumors and on associated changes such as necrosis and inflammation within or around tumors, or pleural effusion. Signs are associated with occupation of space within the lung or thoracic cavity, encroachment of the tumor on major airways, and the development of secondary bronchiolitis and bronchopneumonia.

In some cases, clinical signs are not apparent, despite extensive pulmonary involvement. Non-specific signs of weight loss, lethargy, fever and anorexia may occur with or without respiratory signs such as chronic cough, polypnea or dyspnea.

Lameness, swelling and pain in one or more limbs may also be seen due to either metastasis to bone or development of hypertrophic pulmonary osteopathy.

Form, function and dysfunction of the pleural cavity

Disease of the pleural cavity is due primarily either to inflammation of the serous membranes or to the accumulation of air or fluids within the cavities. In the former case, the clinical signs are due mainly to pain, in the latter, to restriction of lung expansion.

The lungs are surrounded by the visceral pleura, which is continuous with the parietal pleura lining the thorax. The pleural cavity is a potential cavity only, because under normal circumstances the apposed surfaces are separated by a very small amount of serous fluid, which holds them together. The forces provided by the connective tissues of the lung that pull inwards and those of the thoracic wall that pull outwards produce a subatmospheric pressure in the pleural cavity. In the normal steady state, the outward forces balance the inward forces so that the lung is partially inflated and the thoracic wall is pulled in. Introduction of air into the pleural cavity causes the lung to collapse and the thoracic wall to spring outwards.

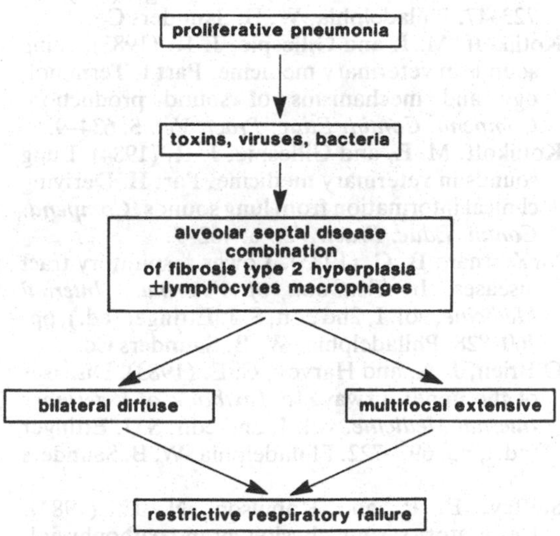

Fig. 5.13. Proliferative pneumonias are centered on alveolar septa. The etiology is diverse and encompasses toxins, viruses and bacteria. All induce alveolar interstitial fibrosis, with the infectious agents inciting a marked infiltration of the alveolar septa with lymphocytes and macrophages. A common finding is alveolar epithelialization.

The mediastinal space is lined by the parietal pleura. It communicates cranially with the deep fascial planes of the neck and caudally with the retroperitoneum via the aortic and esophageal hiatuses. Dorsally and ventrally it has a potential connection with the subcutaneous tissues via the fascial planes. Air introduced into the mediastinal space from the lungs or from a perforated esophagus not only produces pneumomediastinum but also can produce subcutaneous emphysema.

In all young animals, and in cows, sheep and pigs, the mediastinal pleura forms a barrier between the pleural cavities, but in the adult dog, cat and horse the barrier is incomplete. Therefore, accumulation of fluids and air are more likely to occur unilaterally in ruminants and pigs, than in cats, dogs and horses. However, in the latter three species, the mediastinal communications may on occasion be closed by inflammation and unilateral accumulation of exudate may then occur.

The excessive accumulation of fluid in the pleural cavity is prevented by a balance between fluid-forming and fluid-absorbing forces in the pleura. Fluid is drawn into the cavity by the hydrostatic pressure in the capillaries of the parietal pleura in combination with the subatmospheric intrapleural pressure. Fluid moves from the cavity because the pressure in the visceral pleural capillaries, which are supplied by the pulmonary artery, is lower than that of systemic capillaries. Fluid is also removed by abundant lymphatics in the pleura.

Normal functioning of the pleura depends on the pleural surfaces being apposed. If substances are introduced into the cavity and cause separation of the surfaces, the pleural pressure is raised and the expansion of the lungs is compromised. Lung expansion is restricted and the animal develops restrictive respiratory failure.

The most common substances introduced are air, serous fluid, inflammatory exudates, blood and lymph. Fluid or gas in the pleural cavity tend to reflect sounds produced in the lungs. This results in the typical finding of soft, muffled breath sounds.

The presence of air in the pleural cavity is classified as **pneumothorax**. It may be associated with any disease of the lung in which alveoli are ruptured so that air leaks directly into the pleural cavity. Trauma to the lung or penetration of the thoracic wall also leads to pneumothorax.

Hydrothorax is accumulation of excess serous fluid in the pleural cavity. It is generally bilateral and can occur without satisfactory explanation of the pathogenesis. It occurs in congestive heart failure in some species, in those diseases where vascular damage is part of the pathogenesis, such as in mulberry heart disease in pigs and clostridial diseases in ruminants, and in neoplasia of the pleura or the mediastinal lymph nodes in which lymphatic drainage is compromised.

Chylothorax is accumulation of lymph in the pleural cavity. In most cases it is believed to arise from rupture of the thoracic duct. In many cases, however, the origin of the fluid is not evident.

Hemothorax is the presence of blood in the pleural cavity and is generally due to rupture of a vessel. It may be due to rupture of vessels in hyperplastic papillae on the pleura in cases of chronic hydrothorax.

Acute **pleuritis** accompanies some forms of pneumonia, occurs independently, or is part of a more generalized serositis. In the acute phase, respiratory signs are related to pleural pain. Animals have a rapid shallow respiration to minimize respiratory excursion. Pain decreases as the inflammation becomes subacute to chronic. Respiratory signs are then due to accumulation of exudate within the pleural cavity that causes compression of the ventral portions of the lung. The signs of respiratory disease will, therefore, depend on the amount of exudate present. Systemic signs of infection often accompany the lesion.

Additional reading

Blood, D. C., Radostits, O. M. and Henderson, J. A. (1983). Diseases of the respiratory system. In *Veterinary Medicine*, 6th edn, pp. 317–47. London, Bailliere Tindall.

Dungworth, D. L. (1985). The respiratory system. In *Pathology of the Domestic Animals*, vol. 2, 3rd edn, K. V. F. Jubb, P. C. Kennedy and N. Palmer (eds.), pp. 413–556. New York, Academic Press.

Ettinger, S. J. and Ticer, J. W. (1983). Diseases of the trachea. In *Textbook of Veterinary Internal Medicine*, vol. I, 2nd edn, S. J. Ettinger (ed.), pp. 723–47. Philadelphia, W. B. Saunders Co.

Kotlikoff, M. I. and Gillespie, J. R. (1983). Lung sounds in veterinary medicine. Part I. Terminology and mechanisms of sound production. *Compend. Contin. Educ. Pract. Vet.* **5**: 634–9.

Kotlikoff, M. E. and Gillespie, J. R. (1984). Lung sounds in veterinary medicine. Part II. Deriving clinical information from lung sounds. *Compend. Contin. Educ. Pract. Vet.* **6**: 462–7.

McKiernan, B. C. (1983). Lower respiratory tract diseases. In *Textbook of Veterinary Internal Medicine*, vol. I, 2nd edn, S. J. Ettinger (ed.), pp. 760–828. Philadelphia, W. B. Saunders Co.

O'Brien, J. A. and Harvey, C. E. (1983). Diseases of the upper airway. In *Textbook of Veterinary Internal Medicine*, vol. I, 2nd edn, S. J. Ettinger (ed.), pp. 692–722. Philadelphia, W. B. Saunders Co.

Steffey, E. P. and Robinson, N. E. (1983). Respiratory system physiology and pathophysiology. In *Textbook of Veterinary Internal Medicine*, vol. I, 2nd edn, S. J. Ettinger (ed.), pp. 673–91. Philadelphia, W. B. Saunders Co.

Suter, P. F. and Ettinger, S. J. (1983). Pulmonary edema. In *Textbook of Veterinary Internal Medicine*, vol. I, 2nd edn, S. J. Ettinger (ed.), pp. 747–60. Philadelphia, W. B. Saunders Co.

Suter, P. F. and Zinkl, J. G. (1983). Medastinal, pleural, extrapleural, and miscellaneous thoracic diseases. In *Textbook of Veterinary Internal Medicine*, vol. I, 2nd edn, S. J. Ettinger (ed.), pp. 840–99. Philadelphia, W. B. Saunders Co.

West, J. B. (1977). *Respiratory Physiology: The Essentials*, 2nd edn. Baltimore, Williams and Wilkins.

Wayne F. Robinson

6 Cardiovascular system

The normal heart

In the cold light of engineering terminology the heart is a rate-variable, one-way pump, which provides sufficient force to propel a given volume of fluid into a distributing system. These are the fundamental properties of the heart and should be kept uppermost in mind whenever the effects of heart disease are being considered. Also, because of the immense amount of information available on the heart, it is sensible to refer back to these properties when considering cardiac diseases.

Translating from engineering terms into the jargon of the biologist, the heart has the ability to depolarize regularly but variably (heart rate and rhythm), contract forcefully (contractile force) and maintain one way flow (hemodynamics). These themes pervade this chapter when normal and abnormal states are discussed.

The intrinsic heart rate (automaticity)

Few myocytes have an ability to undergo spontaneous depolarization. Mostly, they have stable electrical potentials, and are concerned with the business of contraction.

The specialized myocytes that exhibit automaticity are part of the **myocardial conduction system**, which comprises the sinoatrial (SA) node, the interatrial conduction fibers, the atrioventricular (AV) node, the AV trunk, the left and right crura and the cardiac conducting fibers (Fig. 6.1). Each of these anatomically distinct units depolarize spontaneously, but there is a hierarchy of automaticity, with the most excitable, the SA node, being the **dominant pacemaker**. It governs the intrinsic rate of the heart and all the other units are subservient to its dominance. The remaining portions of the conduction system behave as **latent pacemakers**.

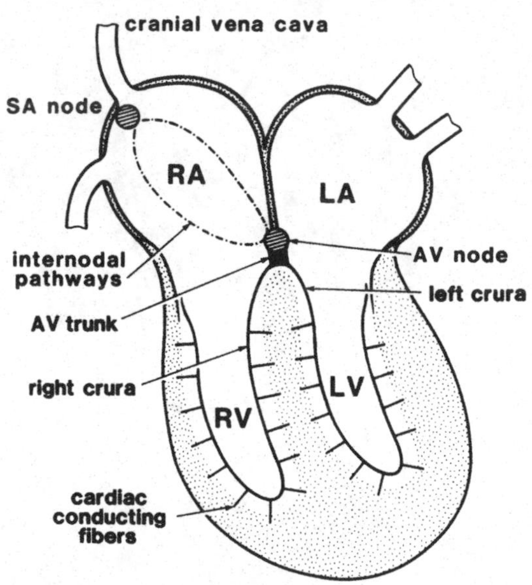

Fig. 6.1. The myocardial conduction system: major species differences include ramification of the cardiac conducting fibers which can reach the subepicardium in some species (not shown). RA, LA, right and left atria; RV, LV, right and left ventricles; AV, atrioventricular; SA, sinoatrial.

Graded automaticity remains as a protection or fail-safe system, for, if the SA node fails to depolarize, the next pacemaker down the hierarchy assumes the dominant role.

The SA node is composed of a small group of specialized myocardial fibers located in the right atrium at the junction of the cranial vena cava and the right auricular appendage. The fibers of the SA node have relatively leaky membranes and depolarize spontaneously after a critical threshold is reached. A rapid phase of depolarization ensues (Fig. 6.2). This encourages the depolarization of adjacent atrial fibers, which gradually spreads throughout both atria, and also through specialized conduction pathways to the AV node.

The AV node is another specialized group of myocytes situated in the interatrial septum, just above the right AV ring. Conduction through the AV node is comparatively slow and has to be of sufficient strength to penetrate the AV trunk. From here, the depolarization wave spreads through the right and left crura, ramifying through the network of cardiac conducting fibers. It finally spreads through the ventricular myocardium. There is quite a species difference in the location of cardiac conducting fibers. In the dog, cat, and man, the fibers terminate in the subendocardium, but, in ruminants, pigs and horses, the fibers reach the subepicardial surface. The configuration of the electrocardiogram (ECG) is greatly influenced by the disposition of cardiac conducting fibers within the myocardium and the subsequent ventricular depolarization.

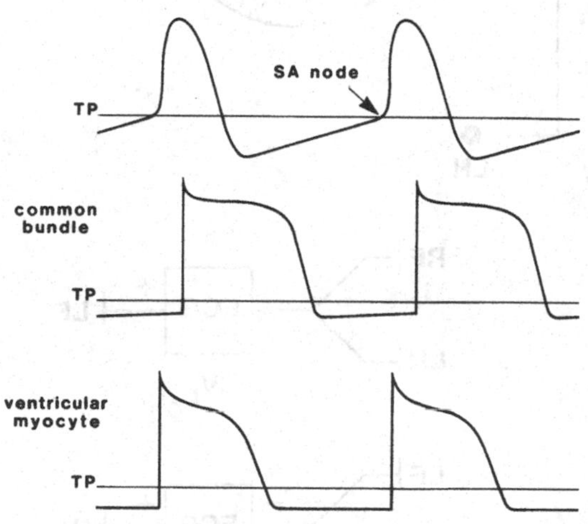

Fig. 6.2. The automaticity of the heart depends on a pacemaker depolarizing spontaneously. The rate of a pacemaker is in turn dependent on the slope of diastolic depolarization. The sinoatrial (SA) node has the greatest slope and so is the dominant pacemaker, whereas the common bundle, for example, has a much slower rate of diastolic depolarization and does not normally reach the threshold potential (TP). It is therefore a latent pacemaker. Ventricular myocytes have stable diastolic potentials and only depolarize when stimulated by the activation process.

Modification of the heart rate

Parasympathetic influence

Parasympathetic innervation is via the left and right vagus nerves, with the right mostly innervating the SA nodal area and the left mostly the AV nodal area. Stimulation of the vagus markedly slows heart rate, but has little, if any, effect on the force of contraction. The traffic of impulses down the cardiac branches of the vagus is governed by aortic and carotid baroreceptors operating via the vasomotor center. The phase of the respiratory cycle also affects vagal tone. On inspiration, vagal tone decreases and allows the heart rate to increase. A decreased heart rate coincides with expiration. The change in heart rate concordant with the respiratory cycle is termed sinus arrhythmia and is commonly observed in most domestic animals.

Sympathetic effects

One of the first things one comes across as a student of physiology is the profound effect of catecholamines on the body, and one of the easiest to document is the increase in heart rate and force of contraction. The heart contains both alpha and beta receptors for catecholamines. The alpha receptors when stimulated appropriately, induce an increase in heart rate. Stimulation of beta receptors increases both rate *and* force of contraction.

The electrocardiogram

It is not the aim of this chapter for readers to become instant experts on the genesis and interpretation of the ECG but, because it is a most important adjunct to aid in the diagnosis of cardiac disease, the salient features of the ECG will be given.

The ECG is a record of the electrical events which occur in the heart and result in an electrical field distributed throughout the body. This is a rather dry, standard definition of the ECG, but it is effectively telling us that the regular depolarization and repolarization of the heart can be detected on the surface of the body using a suitable instrument. The ECG is used to record the potential difference between any two sites on the body but, for historical reasons and for the sake of standardization, specified sites on the body are used. The most common are leads attached to the four limbs and to specified areas on the thorax. There are two limb lead systems used. The first is the **standard lead system**, which is composed of three leads designated I, II and III. Each is bipolar. Lead I measures the potential differences between both forelimbs, lead II between the left hindlimb and the right forelimb and lead III between the left hindlimb and the left forelimb (Fig. 6.3). The second system is the **unipolar**, or **augmented limb leads**. Lead aV_R measures the potential difference between the right forelimb and the combination of the left forelimb and left hindlimb; lead aV_L is between the left forelimb and the

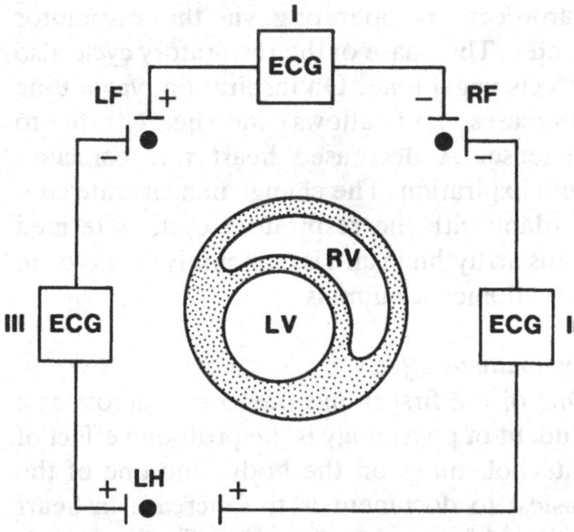

Fig. 6.3. The standard electrocardiograph (ECG) lead system: lead I detects the potential difference between the left and right forelimbs, lead II between the right forelimb (RH) and left hindlimb (LH), and lead III between the left forelimb (LF) and left hindlimb. LV, RV, left and right ventricles.

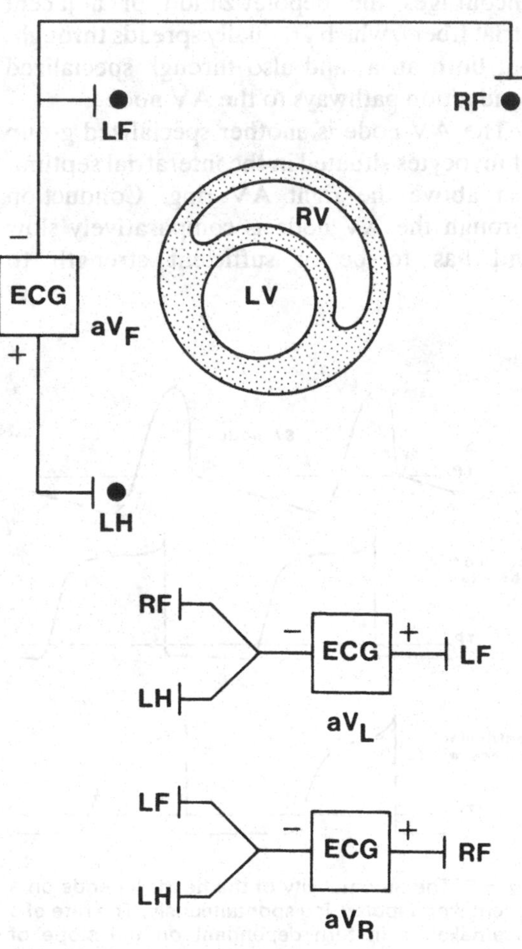

Fig. 6.4. The augmented ECG limb lead system: lead aV_R detects the potential difference between the right forelimb and a combination of the left forelimb and left hindlimb; lead aV_L between the left forelimb and a combination of the right forelimb and left hindlimb; and lead aV_F between the left hindlimb and the left and right forelimbs. For abbreviations, see Fig. 6.3.

combination of right forelimb and the left hindlimb; and lead aV_F is between the left hindlimb and the left and right forelimbs (Fig. 6.4). A chest lead system, the **unipolar chest leads** (the V leads), measures a potential difference between a lead placed on the chest and a central terminal to which the right and left forelimbs and the left hindlimb are attached. V leads are designated V_1 to V_{10}, depending on the position of the leads on the chest (Fig. 6.5).

The ECG detects potential difference (voltage) over time. For this to occur there must be different areas of electrical states within the heart, and indeed this occurs with both depolarization and repolarization. In its resting state, there are no regional electrical differences within the heart and an isoelectric reference line is established. When depolarization is progressing, one or more boundaries exist between the depolarized muscle and that in the resting state. The sequence of depolarization has been discussed, the P wave being the result of atrial depolarization, the QRS complex of ventricular depolarization and the T wave of ventricular repolarization (Fig. 6.6). The form of the P wave and QRS complex is substantially influenced by: (1) the distance of the recording lead from the area of electrical activity; (2) the magnitude of the potential difference; and (3) the geometry of the advancing depolarizing front.

It is important to realize that depolarization, or more accurately the deflections seen on the ECG, are the result of the net changes and are not in most instances proportional to the total amount of depolarization.

The genesis of the electrocardiogram (activation process)

The final deflections seen on the ECG are relatively simple and are the summation of a number of more complex depolarization and repolarization fronts. These sequential changes are often referred to as the activation process. For the P wave, the first half is con-

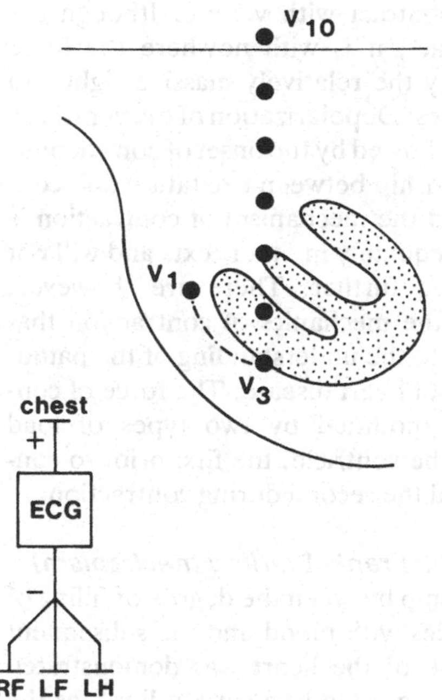

Fig. 6.5. The unipolar chest leads detect the potential difference between a lead placed on a number of designated areas on the chest and the combination of the right forelimb, left forelimb and the left hindlimb. For abbreviations, see Fig. 6.3.

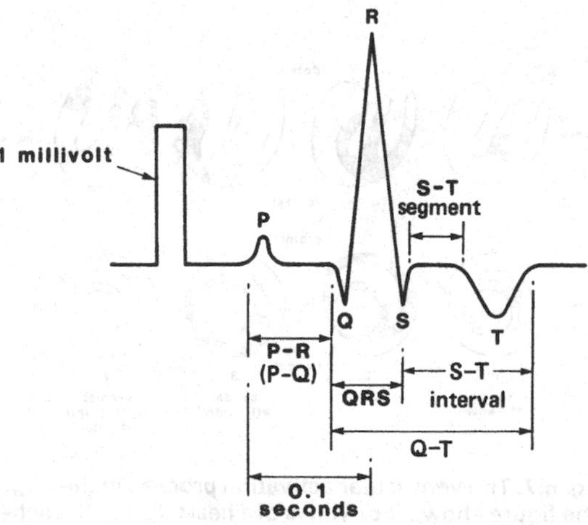

Fig. 6.6. The normal ECG consisting of a P wave (atrial depolarization) and QRS complex (ventricular depolarization) and a T wave (ventricular repolarization). A number of parameters including P–R interval (P–Q interval), QRS complex height and width and S–T segment interval are usually measured.

tributed to by right atrial depolarization and the second half by left atrial depolarization. The form of the P wave varies greatly between the species.

Ventricular depolarization is manifest as the QRS complex, which commences some time after the atria have repolarized. The delay between the P wave and the onset of the QRS complex is due mostly to the depolarization wave traveling through the AV node, the common bundles and the cardiac conducting fibers. The configuration or waveform of the QRS varies greatly, depending on which lead is examined. Once again by convention, the first negative deflection is the Q wave, the first positive deflection the R wave, and the S wave is a negative deflection occurring after a negative (Q) and/or positive (R) wave. The positive and negative deflections in the QRS complex are generated by the wave of depolarization that flows through the heart. The QRS configuration varies markedly between species, but is characteristic for a species. The dog will serve as an example of the genesis of the QRS complex. Ventricular depolarization (the activation process) begins in the interventricular septum (IVS), the boundary being directed toward the right, ventrally and cranially. It continues to involve the IVS, most of the right ventricular wall and some of the left ventricular free wall. In this portion of the activation process there is no net boundary and thus it is electrically silent. The remainder of the left ventricular free wall is depolarized, giving rise to a boundary that is directed to the left, caudally and ventrally. The final area to be depolarized is the base of the ventricles and the IVS, whose boundary is directed cranially, dorsally and to the left in most cases. The activation process is depicted in Fig. 6.7.

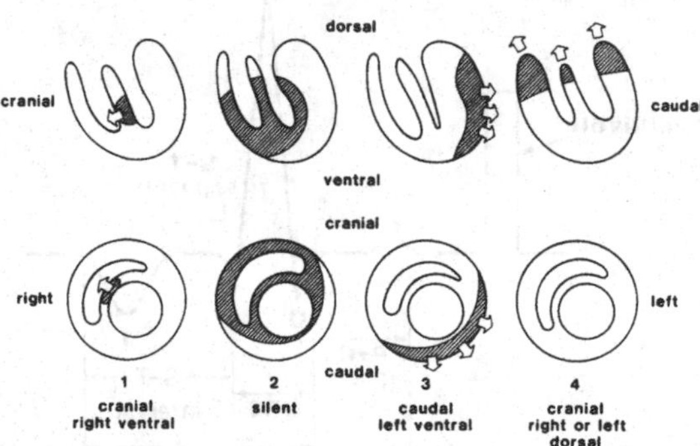

Fig. 6.7. The ventricular activation process in the dog. The figure shows four views of a heart during the activation, the top row is a lateral projection and the bottom row a dorsal projection. The left of the figure shows the initial phase (1) of activation followed to the right by the second (2) third (3) and fourth (4) phases of activation. The designations at the bottom of the diagram (e.g. cranial, right, ventral) indicate the mean direction of depolarization front.

The S–T segment and T wave represent the slow and fast phases, respectively, of ventricular repolarization. It appears that repolarization is not simply the reverse or mirror image of depolarization. Different boundaries apply with repolarization.

Myocardial mechanics (performance)

The second major property of the heart is its ability to contract with vigor. Although the atria contract, it is with nowhere the force exhibited by the relatively massive right and left ventricles. Depolarization of the ventricles is quickly followed by the onset of contraction. The relationship between excitation and contraction and the mechanism of contraction is covered adequately in other texts and will not be discussed further. There are, however, aspects of the mechanics of contraction that are central to the understanding of the pathophysiology of heart disease. The force of contraction is modified by two types of load placed on the ventricle, the first prior to contraction and the second during contraction.

Preload (the Frank–Starling mechanism)

A relationship between the degree of filling of the ventricles with blood and the subsequent stroke work of the heart was demonstrated many years ago. Up to a certain limit, as the volume load on the ventricle is progressively increased, individual fibers are increasingly stretched and the ensuing contraction is faster. This is an intrinsic property of the myocardium, sometimes termed heterometric autoregulation. The tension on the ventricular wall at end diastole is equivalent to the stretch exerted on the fibers and is termed the preload (Fig. 6.8).

Afterload

Whereas preload refers to the initial load placed on the ventricle prior to the commencement of contraction, another determinant of the final mechanics of contraction is the afterload placed on the ventricle during contraction. Afterload is related to the aortic–pulmonic resistance/diastolic pressure, which is in turn related to the impedance of blood flow.

Contractility

Contractility is a measure of the amount of tension the contracting myocardium can develop or, more strictly, the rate of development of tension. A number of substances, including calcium, norepinephrine and digitalis glycosides, increase cardiac contractility. They all cause the myocardial muscle to shorten faster and contract more vigorously.

Ejection fraction

Not all the blood present in the ventricle is expelled with each contraction. The fraction

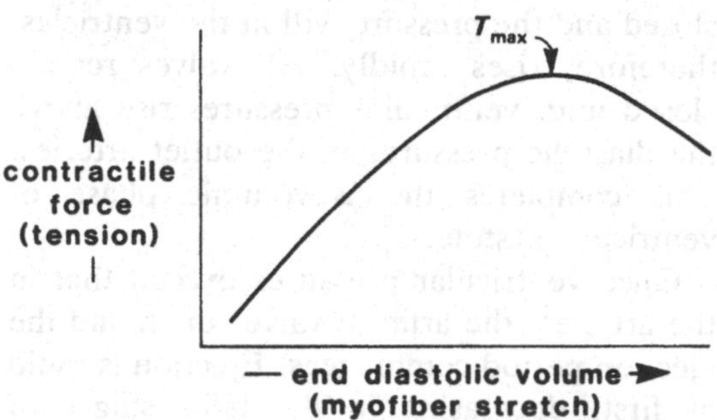

Fig. 6.8. The effect of preload (end diastolic volume/myofiber stretch) on contractility as the myofiber is stretched. There is a corresponding increase in the force of contraction up to a maximum (T_{max}). Stretching beyond this point results in a progressive decrease in contractility.

of the total volume actually ejected is termed the ejection fraction. It is the amount ejected (the stroke volume) divided by end diastolic volume. The normal ejection fraction varies from 60 to 75% of end diastolic volume. Ejection fraction is increased by factors that increase contractility or decrease afterload. It is decreased by factors that decrease contractility or increase afterload.

Unidirectional (one-way) flow

The final property of major importance is the maintenance of a unidirectional flow of blood through the ventricle. It is also important to emphasize that one-way flow should be relatively unimpeded.

Relatively unimpeded one-way flow is a consequence of a system of valves within the heart allowing **entrance but not exit** from the atria to the ventricles (the atrioventricular valves) and **exit but not re-entry** from the ventricles (the semilunar valves). The flow of blood from the atria through the ventricles to the great vessels depends firstly on the valves being pliable and yet resilient and, secondly, on the resistance to filling or ejection being within normal limits. The resistance to inflow is negligible, but the resistance to ventricular outflow is considerable, especially in the case of the left ventricle. To assess adequately the two features of one-way flow and relatively unimpeded flow, the flow of blood through the heart should be examined by contrast radiography and probably pressure measurements within heart chambers and great vessels, but a good first step is to assess the heart sounds. This at least provides information on the structural integrity of the valves of the heart.

Heart sounds

Two major heart sounds, the first and second, are detectable with each heart beat in all domestic species. Two minor sounds, the third and fourth, may also be detected, especially in horses and, to some extent, in cattle.

The **first heart sound** (S_1) which is louder, longer and lower than the second heart sound coincides with the onset of ventricular systole.

It is a group of irregular vibrations with a frequency of 30 to 45 per second and is composed of preliminary, main and final vibrations. Although there is some debate about it, the sound probably originates from three sources.

1 Vibrations of the AV valves during and after their closure.
2 Vibrations originating from eddies in the blood ejected through the aortic orifice into the broader sinuses of Valsalva.
3 Vibration of the ventricular muscle.

The **second heart sound** (S_2) coincides with the closure of the aortic and pulmonic valves at the onset of the diastolic period. It results from vibrations of slightly higher frequency, ranging from 50 to 70 per second.

A **third heart sound** (S_3) is caused by a rapid inflow of blood from the atrium to the ventricles. This sound is not normally heard on auscultation of most domestic species. However, it is heard in a significant proportion of normal horses.

The **fourth heart sound** (S_4), associated with atrial systole and the ejection of blood from the atrium into the ventricle, is not normally detectable by auscultation in many species. However, it is heard commonly in the horse.

From the foregoing, it is apparent that a number of combinations of heart sounds may be heard in the normal horse. They are S_1 S_2; S_1 S_2 S_3; S_4 S_1 S_2; and S_4 S_1 S_2 S_3.

Cardiac output

Cardiac output constantly changes with the changing metabolic needs of the body. The changes in cardiac output are achieved predominantly through variations of heart rate and, to a lesser extent, of stroke volume.

Cardiac output is the volume of blood pumped per minute; it can be increased fivefold from resting to maximal effort. As the total volume of blood within the vascular system does not change, it is the **rate of flow** that is modified with exercise. In other words, an increased cardiac output reflects almost entirely an increased velocity of flow. The blood that is moved is moved more quickly, with the heart acting as an accelerator of the rate of flow of blood.

As heart rate increases, an adequate volume for each contraction is maintained by decreasing the duration of the cardiac cycle, with the stroke volume remaining relatively constant. The maintenance of stroke volume is achieved by the ventricles filling more rapidly and emptying more rapidly. The rate of systolic events is increased mostly by sympathetic stimulation.

The cardiac cycle

The two ventricles function as intermittent pumps, coordinated to empty and fill together. The period of emptying is the systolic phase of the cycle, and the period of filling the diastolic phase. Each phase is further subdivided into an isometric (or isovolumic) period, and an isotonic period. The cardiac cycle, which is depicted in Fig. 6.9, may be followed by monitoring the electrical events associated with it.

The cycle commences with the P wave on the ECG reflecting atrial depolarization, this is followed by the 'a' wave which reflects a small rise in atrial pressure. The atrial pressure then returns to its diastolic level. As the atrium relaxes, the ventricle depolarizes, which is manifest as the QRS complex, signalling the first phase of ventricular systole. As the pressure within the ventricles rises, the AV valves close. At this time the arterial valves are closed and the pressure within the ventricles, therefore, rises rapidly. All valves remain closed until ventricular pressures rise above the diastolic pressures in the outlet arteries. This completes the isovolumic phase of ventricular systole.

Once ventricular pressures exceed that in the arteries, the arterial valves open and the ejection period commences. Ejection is rapid at first, decreasing in the later stages of systole. Following this isotonic phase of ventricular contraction, the pressure in the ventricles falls without a change in volume. The arterial valves close when the pressures exceed that in the ventricles. Ventricular pressures continue to fall without a change in volume and, when the pressure falls below that in the atria, the AV valves then open. During diastole, the blood stored in both the atria and the veins enters the ventricle. Initially the pressure in the atria is relatively high and inflow of blood is rapid.

Heart disease and heart failure

Heart disease is a general term which includes all abnormalities of heart structure or function, be they congenital or acquired. Heart disease ranges from cardiac abnormalities discovered incidentally during physical examination or necropsy, to those that are a central feature of the physical examination or necropsy. For example, a dog may have an abnormal heart sound associated with disease of the left AV valve, but no associated clinical signs of cardiac failure. Alternatively, another dog with left AV valve disease may have signs of cardiac failure. Both animals have heart disease, but only the second animal has heart failure.

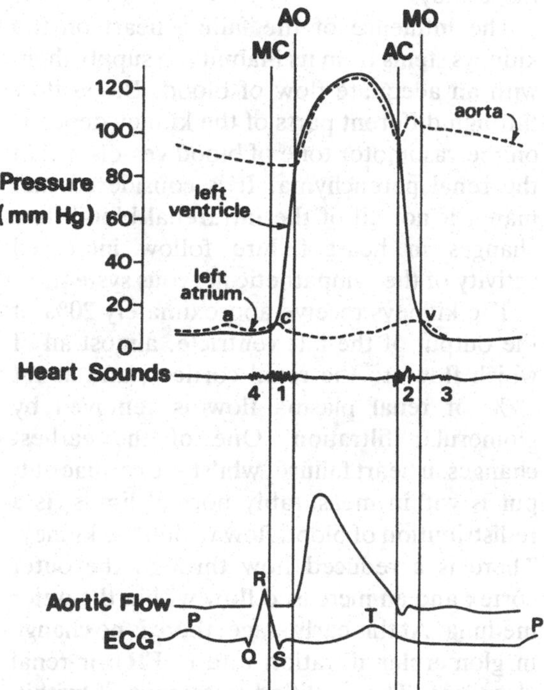

Fig. 6.9. The cardiac cycle, only the events for the left side of the heart are shown. The essential features are the relationships between the electrical events (ECG); aortic flow; heart sounds; pressure changes within the atrium, ventricle and aorta; and the opening (aortic valve opening (AO), mitral valve opening (MO)) and closing (mitral valve closure (MC), aortic valve closure (AC)) of the heart valves. For details, see the text.

The term **heart failure** (or cardiac failure) denotes a situation in which the heart is diseased, all compensatory mechanisms are exhausted, and characteristic clinical and pathologic signs are present. In this pathophysiologic state, the heart is unable to meet the metabolic requirements of the animal. However, heart failure is a rather general term. The addition of **congestive** to heart failure gives a little more specificity to the clinical state. **Acute congestive heart failure** refers to a rapid onset of the clinical signs of congestive heart failure. Congestive heart failure is characterized by vascular congestion and edema fluid within the interstitium and in body cavities. The congestion and fluid accumulation may be associated predominantly with either the **systemic** or the **pulmonary** circulation, and is termed **right-sided** or **left-sided** congestive heart failure, respectively. Right-sided congestive heart failure is characterized by dependent peripheral edema, fluid in body cavities and heptomegaly. Left-sided congestive heart failure is characterized by pulmonary edema, leading to polypnea (rapid breathing), cough and sometimes, in severe cases, dyspnea (difficult breathing). It should be remembered that congestive heart failure is not an etiologic diagnosis, it is the end product of numerous causes.

One of the most common causes of congestive heart failure in dogs is left AV valve disease, where, in most cases, there is no defect in the force of ventricular contraction. When the myocardial muscle fails to contract vigorously it is termed **myocardial failure**. Congestive cardiomyopathy of large-breed dogs is a common disease that causes such failure.

Not all cases of heart failure are of the congestive variety. Whereas in congestive heart failure the clinical manifestations are more or less constant, there is a set of clinical signs associated with heart failure of an intermittent type. The most prominent of these are weakness and syncope. The affected animal may periodically stagger, fall and lose consciousness. These are ephemeral episodes and the animal appears to be normal in between. The cause of such episodes is usually a substantial intermittent change in heart rate and rhythm, resulting in a precipitous drop in cardiac output. Although it is not entirely satisfactory to do so, this type of failure is termed **acute heart failure**. The relationship between heart disease and heart failure is depicted in Fig. 6.10.

Circulatory failure is another broad term used to describe a state which may or may not be the result of heart failure. It is characterized by a drop in effective circulating blood volume. Common causes are acute internal or external hemorrhage, dehydration or endotoxic shock.

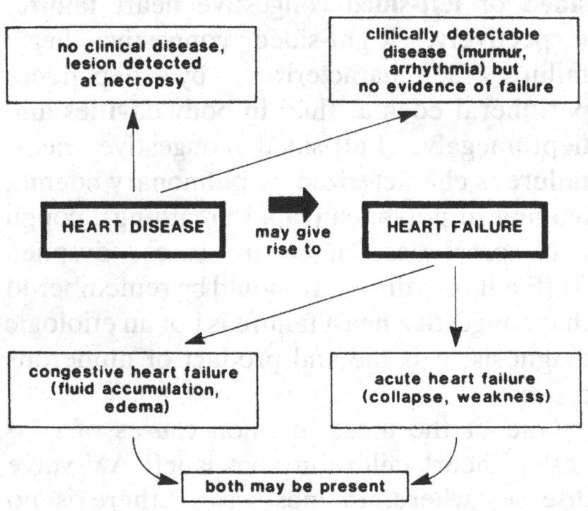

Fig. 6.10. Relationship between heart disease and heart failure. Heart disease may be present without attendant clinical signs or may give rise to either acute or congestive heart failure.

The pathogenesis of heart failure

The clinical signs of heart failure are the end result of two basic pathophysiologic changes. The first is accumulation of fluid, the second, tissue or organ ischemia. Depending on the cause of the heart failure, both effects may be present or, as is more usual, one may predominate. The pathogenesis of heart failure has occupied the time and minds of numerous investigators for a long period, yet still there seems to be no simple unifying hypothesis. Maybe there are no simple answers.

The first undisputed fact is that the expansion of the body's fluid volume in congestive heart failure results from the retention of sodium and water. Secondly, the sodium and water accumulation must necessarily involve the kidneys.

The influence of the failing heart on the kidneys stems from its inability to supply them with an adequate flow of blood. Blood flow through different parts of the kidney depends on the vasomotor tone of blood vessels within the renal parenchyma. It is considered that many, if not all of the intrarenal blood flow changes in heart failure follow increased activity of the sympathetic nervous system.

The kidneys receive approximately 20% of the output of the left ventricle, almost all of which flows to the renal cortices, and about 20% of renal plasma flow is removed by glomerular filtration. One of the earliest changes in heart failure, whilst the cardiac output is within measurably normal limits, is a redistribution of blood flow within the kidney. There is a reduced flow through the outer cortex and an increased flow within the outer medulla. At this early stage, there is no change in glomerular filtration rate (GFR) or renal clearance. The qualified statement of 'within measurably normal limits' implies that minor changes in cardiac output which may have effects on regional kidney flow are not easily quantified.

As cardiac output becomes progressively depressed, renal cortical flow is further limited and outer medullary flow remains elevated.

However, there is a *less* than proportionate drop in GFR compared with renal blood flow, resulting in an increased filtration fraction. Proportionally more sodium moves through the glomerular filter and is delivered to the proximal convoluted tubular epithelium. A greater number of sodium ions are resorbed as the rate of tubular resorption of sodium remains constant. Also, because of the increased filtration fraction, local plasma osmotic pressure in the efferent arteriole increases, causing greater resorption of sodium and water (Fig. 6.11).

The alterations in renal blood flow in heart failure also increase the activity of the **renin–angiotensin–aldosterone** system, producing more sodium resorption from the distal convoluted tubule. There is also evidence of increased activity for the water-retaining effects of anti-diuretic hormone (Fig. 6.11).

It should be noted that none of the hormones previously mentioned (angiotensin, aldosterone, anti-diuretic hormone, or the catecholamines) can produce the edema of congestive heart failure if administered singly. In addition, once a new steady state has been reached, the hormonal state returns to relatively normal limits. Lastly, the mechanisms that are brought into play are not exclusive to the syndrome of congestive heart failure. Any situation that leads to a drop in effective circulating blood volume will activate the renal sodium- and water-retaining mechanism. The fundamental difference between these states and congestive heart failure is that the blood volume in congestive heart failure is already more than adequate for the task at hand but, because of the drop in cardiac output, less blood is reaching the kidneys. The volume changes in congestive heart failure should therefore be viewed as the result of an integrated response by the body to compensate for the inability of the heart to respond to the normal needs of the body.

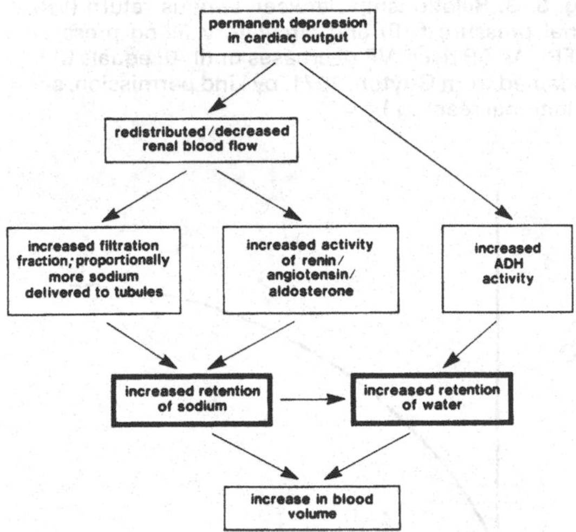

Fig. 6.11. Pathogenesis of heart failure involves sodium and water retention by the kidney, following redistribution of blood flow to the kidney. Increased anti-diuretic hormone activity also contributes to the retention of water. All lead to an increase in blood volume.

Effects of an increase in blood volume

In a sense, the expansion of the blood volume is a trade off. By increasing blood volume, venous return is enhanced, and, in turn, cardiac output is enhanced. However, the trade off is that the balance between capillary hydrostatic pressure and plasma osmotic pressure is disturbed. This leads to an increase in the amount of fluid in the interstitial spaces and body cavities (Fig. 6.12).

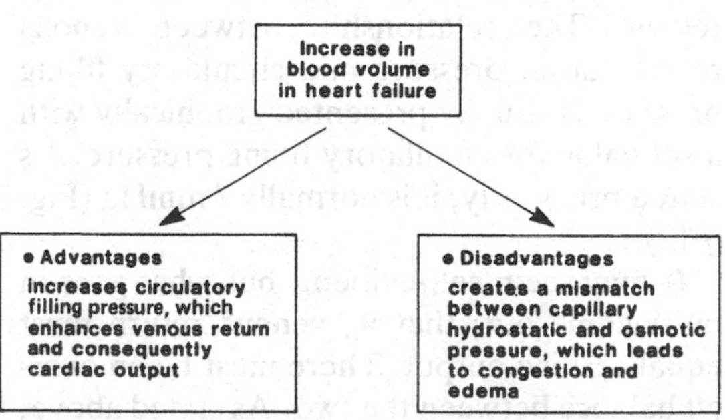

Fig. 6.12. Increased blood volume in heart failure has both advantages and disadvantages. The latter contributing to the clinical signs of congestive heart failure.

The benefit of augmenting venous return

The continuing normal function of the heart depends primarily on three interdependent factors:

Cardiac output,
venous return and
circulatory filling pressure.

Cardiac output and venous return are two terms with which most students are familiar and they are well understood, but circulatory filling pressure is a rather novel term.

Circulatory filling pressure, or mean systemic pressure, is the 'tightness' with which the circulation is filled with blood when the circulation is static. It is a concept central to an explanation of the events that follow the onset of heart failure. Circulatory filling pressure has been measured experimentally, and in the normal animal it is approximately 7 mmHg.*

The circulatory filling pressure is dependent on a number of factors including blood volume, peripheral resistance, venous tone and vascular compliance. All of these factors combine to produce the relative 'tightness with which the circulation is filled'.

Circulatory filling pressure influences venous return and is given by the relationship:

Venous return $\propto$ (circulatory filling pressure $-$ atrial pressure)

Thus, the lower the atrial pressure the higher is the venous return, and the higher the circulatory filling pressure the greater is the venous return. The relationship between venous return, atrial pressure and circulatory filling pressure is usually presented graphically with a set value for circulatory filling pressure. As stated previously, it is normally 7 mmHg (Fig. 6.13).

It may seem self-evident, but what goes in must come out; that is, venous return must equal cardiac output. There must be an overall balance between the two. As stated above, venous return is directly proportional to the circulatory filling pressure minus the atrial pressure, but cardiac output is directly proportional to atrial pressure (Fig. 6.14). There appears to be a conflict between the two, venous return requiring as low an atrial pressure as possible, whereas cardiac output requires as high an atrial pressure as possible. An equilibrium must, and is, reached between the two, because if atrial pressure is too high the atria will not fill adequately and if atrial pressure is too low the ventricle will not fill.

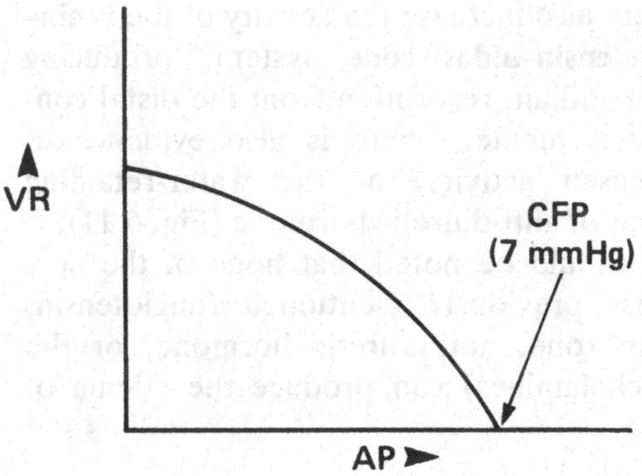

Fig. 6.13. Relationship between venous return (VR), atrial pressure (AP) and circulatory filling pressure (CFP). As AP rises, VR decreases until AP equals CFP. (Adapted from Guyton, 1971, by kind permission, see Additional reading.)

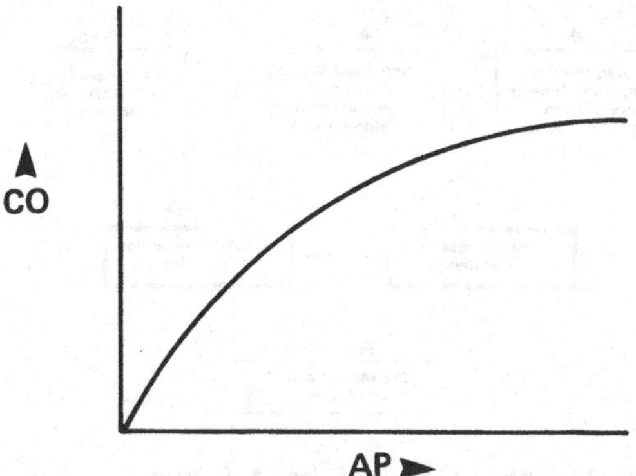

Fig. 6.14. Relationship between cardiac output (CO) and atrial pressure (AP). As AP increases, CO increases. (Adapted from Guyton, 1971, by kind permission, see Additional reading.)

* See note, p. 86.

What occurs is that atrial pressure reaches a pressure at which venous return equals cardiac output (Fig. 6.15). When the heart begins to fail, the cardiac output per unit of atrial pressure falls and a second set of curves can be generated, with a second equilibrium point (B, Fig. 6.16).

As already discussed, one compensatory mechanism in heart failure is the retention of sodium and water, which serves to expand the fluid present in the extracellular space, including the blood. Venous tone is another; it too is increased, mediated by the action of the sympathetic nervous system. These two mechanisms combine to increase circulatory filling pressure, thereby increasing venous return until a new steady state is reached. A new relationship now exists between venous return and cardiac output (C, Fig. 6.17).

As indicated earlier, all three parameters – venous return, atrial pressure and circulatory filling pressure – are interdependent, and a new equilibrium is reached between all three to attempt to satisfy the needs of the animal.

In summary, if the graphs of the normal animal, the animal in initial heart failure and the animal in compensated failure are combined, it can be seen that the initial drop in cardiac output following heart failure is compensated for by the renal retention of sodium and water. The blood volume is expanded and circulatory filling pressure is increased. The venous return and cardiac output are raised to point C in Fig. 6.17.

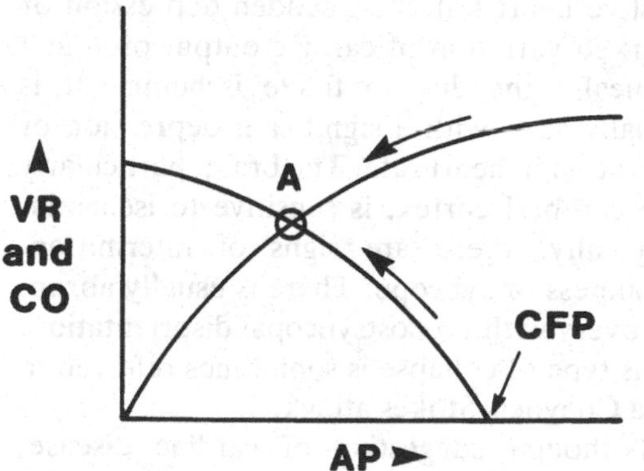

Fig. 6.15. The equilibrium reached at point A between venous return (VR), cardiac output (CO) and atrial pressure (AP). (Adapted from Guyton, 1971, by kind permission, see Additional reading.)

The detrimental effects of loss of balance between hydrostatic pressure and osmotic pressure

Extracellular fluid volume is divided into two compartments, the interstitium and plasma, which, in health, are maintained within relatively narrow limits. There is constant, but

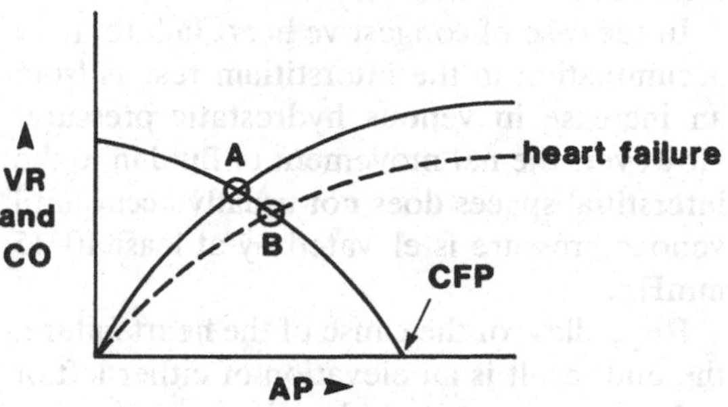

Fig. 6.16. When the heart fails, a new equilibrium is reached between venous return (VR), cardiac output (CO) and atrial pressure (AP) at point B. AP is higher than normal and VR and CO are depressed. CFP, circulatory filling pressure. (Adapted from Guyton, 1971, by kind permission, see Additional reading.)

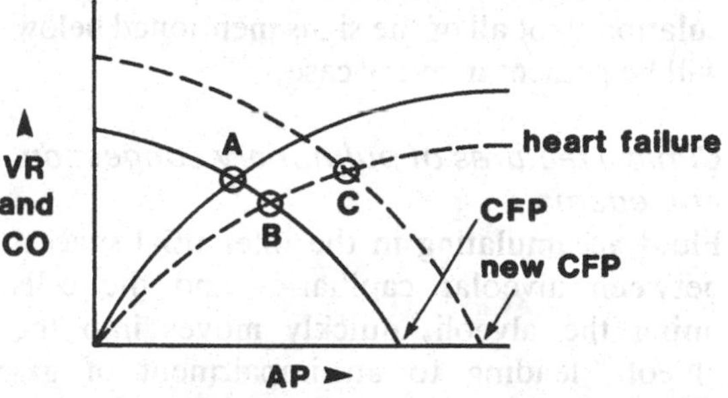

Fig. 6.17. Following renal compensation in heart failure, circulatory filling pressure (CFP) is increased, enhancing venous return (VR) and cardiac output (CO) to near normal levels, but the price paid is an elevated atrial pressure. The new equilibrium is at point C. (Adapted from Guyton, 1971, by kind permission, see Additional reading.)

equal, interchange between the two which is dependent on the relative difference between capillary hydrostatic pressure and plasma osmotic pressure.

In a normal animal, plasma osmotic pressure remains constant at 25 mmHg, whereas capillary hydrostatic pressure falls from approximately 33 mmHg at the arterial end of the capillary to 15 mmHg at the venous end. Fluid moves from the plasma to the interstitium at the arterial end of the capillary and from the interstitium to the plasma at the venous end. When an alteration in either plasma osmotic pressure or capillary hydrostatic pressure occurs, the balance between fluid outflow and inflow is lost.

In the case of congestive heart failure, fluid accumulation in the interstitium results from an increase in venous hydrostatic pressure. However, the net movement of fluid in to the interstitial spaces does not usually occur until venous pressure is elevated by at least 10–15 mmHg.

Regardless of the cause of the heart failure, the end result is an elevation of either left or right atrial pressure and an increased venous hydrostatic pressure in either the pulmonary or the systemic systems, respectively. This leads to venous congestion and the development of edema. Depending on the pressure within the pulmonary or systemic circuits, signs predominate either in the lungs, or in those areas drained by the systemic venous circulation. Not all of the signs mentioned below will be present in every case.

Clinical features of pulmonary congestion and edema

Fluid accumulating in the interstitial spaces, between alveolar capillaries and the cells lining the alveoli, quickly moves into the alveoli, leading to an impairment of gas diffusion across the alveolar septal wall.

Polypnea and dyspnea

In its mildest degree little, if any, alteration in respiratory rate or depth will be evident at rest, but the defect will become apparent with exercise. The affected animal will refuse to continue to exercise normally, or will be polypneic and/or dyspneic. As the extent of the alveolar edema increases, dyspnea will be observed at rest. The alteration in respiratory rate and depth is due to increased blood CO_2 levels, as the medullary respiratory center is exquisitely sensitive to minor increases in blood CO_2 levels.

Quality of cough

Although present in left-sided heart failure because of fluid in the larger airways, cough is present in many other conditions. While heart failure should be suspected, the upper and lower respiratory tract should be examined for the presence of primary respiratory disease.

Clinical features of increased systemic venous pressure

For reasons that are unclear, the site of fluid accumulation varies markedly between the species. Dependent subcutaneous edema is most evident in cattle and horses, but is most unusual in the dog. In the last of these species, ascites predominates and is manifest usually as a distended abdomen.

Tissue or organ ischemia

In contrast to the signs observed with congestive heart failure, a sudden depression or marked variation of cardiac output produces clinical signs due to tissue ischemia. It is usually seen with a significant depression or elevation in heart rate. The brain, particularly the cerebral cortex, is sensitive to ischemia. Typically, there are signs of intermittent weakness or syncope. There is usually abrupt recovery with no postsyncopal disorientation. This type of collapse is sometimes referred to as a Cheynes–Stokes attack.

Although suggestive of cardiac disease, syncope is also produced by primary neurologic or endocrine abnormalities. It is wise to check these systems also.

Pathophysiologic patterns of heart disease

The initial clinical differentiation of an abnormality of cardiac function depends on the direct or indirect assessment of the heart's three fundamental properties. They are the ability regularly to depolarize (rate and rhythm), the normality of blood flow (hemodynamics), and the vigor of contraction (contractility). Of the three, the rate and rhythm of the heart is easiest to assess clinically: the normality of flow is more difficult and requires, at least initially, an assessment of the heart sounds and the characteristics of the pulse. Lastly, the vigor of contraction is most difficult to assess without the use of special techniques such as echocardiography. In everyday practice, the presence or absence of a depression of ventricular contractility is based on the other features being within normal limits.

In order to confirm which of the heart's prime functions is abnormal, it is usually necessary to resort to the use of diagnostic aids following the physical examination. In this regard an ECG, plain and contrast radiography and echocardiography will help clinicians to confirm or to deny their initial clinical impression.

Arrhythmias

As discussed previously, under normal conditions the rate is under strict control. The **intrinsic** rate of the heart is governed by the automaticity of the SA node and **extrinsically** modified by the autonomic nervous system.

An arrhythmia is a disturbance of heart rate, rhythm or conduction and almost always arises from aberrations of the intrinsic system. There are two major patterns. Either a previously subservient focus arises and becomes the dominant pacemaker, or the normal orderly pattern of conduction is disturbed.

The emergence of a newly dominant or ectopic pacemaker gives rise to disorder of **impulse formation**, and an interruption of the normal pattern of conduction is designated a disorder of **impulse conduction**. In general, disorders of impulse formation produce increases in heart rate, whereas disorders of impulse conduction produce decreases (Fig. 6.18). Some of these disorders produce no clinical signs, whilst others produce marked clinical signs and may be life threatening. Although some idea of the type of arrhythmia may be gained by physical examination, an ECG is essential for its classification.

Disorders of impulse formation

The appearance of a focus that depolarizes more rapidly than the SA node allows the ectopic focus to become the dominant pacemaker and to govern the heart rate and rhythm. These ectopic pacemakers are renegades, because they do not usually respond to calls from the autonomic nervous system to modify their activity.

Disorders of impulse formation can therefore be subdivided into two categories. Firstly, there may be **enhanced automaticity** of the conduction system, where the atrial conduction fibers, portions of the AV node, AV trunk, left and right crura, or the cardiac conducting fibers, increase their rate of depolarization. Secondly, there may be **abnormal**

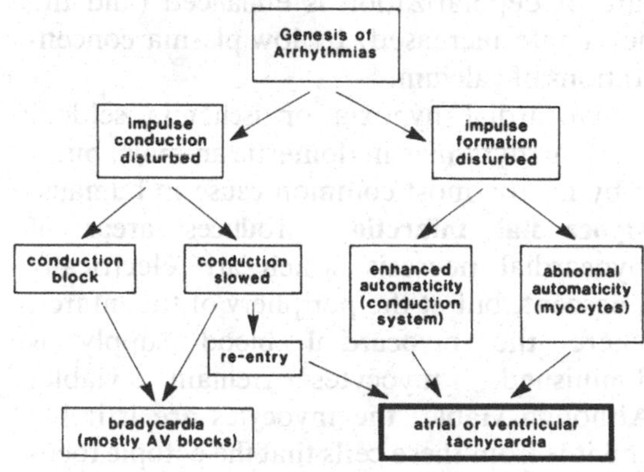

Fig. 6.18. Arrhythmias originate from two broad mechanisms and arise from either disorders of impulse formation or conduction.

automaticity, where impulse formation arises somewhere in the normally quiescent atrial or ventricular myocyte mass. For this to occur the myocyte membrane must be destabilized so that the affected myocyte is partially depolarized. In this situation, the myocyte is teetering on the brink and will fully depolarize given half a chance.

The clinical significance of the abnormality depends to a large extent on the **location**, and the **frequency**, of depolarization of the ectopic focus. As a general rule, ectopic foci within the ventricles are more serious than those that arise in the atria, and also, the greater the frequency of occurrence, the more serious the arrhythmia (Fig. 6.18).

Disorders of impulse formation can arise following many different types of structural injury to the myocardium. Prominent causes of myocardial cell injury include encephalomyocarditis infection in pigs, parvovirus infection in puppies, vitamin E and selenium deficiency in production animals, digitalis intoxication, or the extension of a valvular bacterial endocarditis into the myocardium. The reader should not be left with the notion that all ectopic pacemakers have a morphologic base. Abnormalities of the plasma concentrations of potassium, calcium or magnesium alter the automaticity of the heart without any morphologic change. Thus the rate of depolarization is enhanced (and the heart rate increased) by low plasma concentrations of calcium.

Myocardial hypoxia or ischemia seldom cause arrhythmias in domestic animals, but it is by far the most common cause in humans. **Myocardial infarction** produces areas of myocardial necrosis which are electrically quiescent, but at the periphery of the infarct, where the myocardial blood supply is diminished, myocytes remain viable. Although viable, the myocytes are irritable and it is from these cells that the ectopic focus arises. A similar pattern of injured, but viable, myocytes occurs in a number of the infectious diseases and nutritional deficiencies in domestic animals.

Disorders of impulse conduction

The primary determinants of electrical conductivity within the heart are the effectiveness of the impulses produced by the depolarization of higher pacemakers, and the excitability of downstream fibers. There are a number of factors that may modify the two previous statements, but the principle still holds. In conduction disorders, the depolarization pattern may be facilitated or impaired, not just through the specialized conduction system, but through the myocardium as well.

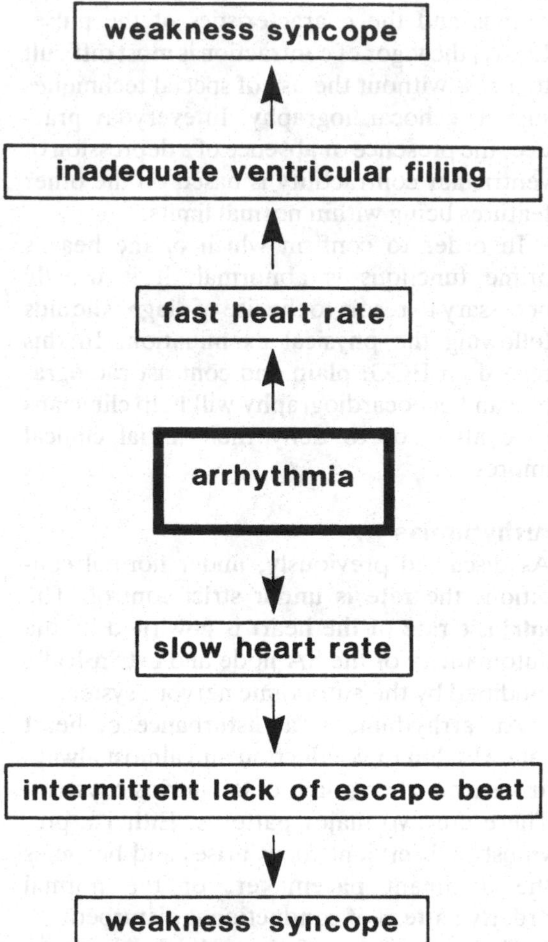

Fig. 6.19. An arrhythmia may produce either a fast heart rate or a slow heart rate, both of which lead to weakness and syncopal episodes, but through different mechanisms.

The excitability of downstream fibers will be dealt with first. The most common is a depression or a block in conduction. This can be at any level, from the SA node to the cardiac conducting fibers. If there is a block, then the fibers with the next most pronounced automaticity assume the role of pacemaker. The block may be either partial or complete. Heart blocks are designated according to both the level at which they occur and by the degree of block. Thus, conduction block at the level of the AV node is first degree if there is prolongation of conduction time through the AV node, second degree if there is transference of some impulses, and third degree if there is no impulse transmission through the AV node at all (Figs. 6.18, 6.22).

If there is slowed conduction through injured atrial or ventricular fibers, a phenomenon termed **re-entry** may occur. As the impulse reaches an injured area, the wave of depolarization moves slowly through this area and, by the time it has traversed this zone and emerges, adjacent normal fibers are capable of being depolarized again. The original wave of depolarization has re-entered the myocardium. This state is one of the mechanisms used to explain some types of increase in heart rate (tachycardia).

Re-entry may result from the same set of causes listed under disorders of impulse formation. Conduction blocks are often focal and have varied causes.

Pathophysiology of arrhythmias

Once some or all control of the heart rate has been lost, the affected animal is prone to great variation in cardiac output. Because most arrhythmias are intermittent, the clinical signs are usually the result of tissue or organ ischemia. Although the end result is the same whether there is a tachycardia or a bradycardia, the mechanisms differ.

In a **tachycardia** (fast heart rate) whether regular or irregular, there is insufficient time for the ventricle to fill during diastole. The premature beat fires before sufficient blood has entered the ventricle. There is also the effect of a weaker than normal ventricular contraction with premature beats. This is probably the result of inadequate myocardial fiber stretching. On contraction, little or no blood enters the great vessels. Consequently, the premature beat is often characterized by only the first heart sound. The AV valves close as the ventricle contracts, but insufficient pressure is generated to overcome the aortic diastolic pressure and fully open the semilunar valves. If the valves are opened, only a small proportion of the blood in the ventricle is ejected. If the semilunar valves do not open, there is no pulse, but if the valves do open, there is a weak pulse.

In clinically significant **bradycardias** (slow heart rate), weakness and syncope probably follow the temporary lack of an escape beat. For example, with a third-degree AV block, only the ability of the ventricular conduction system to depolarize spontaneously keeps the heart beating. It is in a sense, the depolarization of the last resort. If it fails, then asystole ensues and the affected animal collapses and becomes unconscious. If the ventricular conduction fibers manage to depolarize, the animal regains consciousness. If not, sudden death occurs. The pathophysiology of arrhythmias is depicted in Fig. 6.19.

The reader should not be left with the impression that all arrhythmias invariably and constantly give rise to clinical signs. An animal may have a history of weakness or syncope, but may not have these clinical signs at the time of physical examination, even though an arrhythmia may be detected. However, the arrhythmia predicates the possibility of future clinical signs or sudden death.

The electrocardiographic differentiation of arrhythmias

The major use of electrocardiography across the species is for the classification of arrhythmias. Its importance is not only for confirmation of a presumptive diagnosis but also as an aid for prognosis and ultimate treatment. The foregoing discussion emphasized the causes of arrhythmias and the two broad

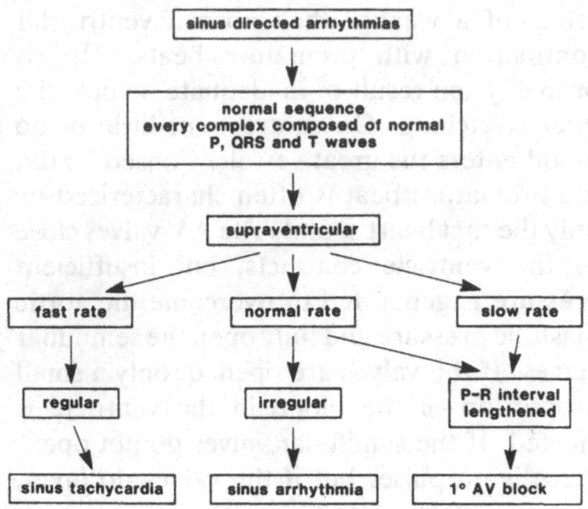

Fig. 6.20. Sinus directed arrhythmias are characterized by normal P, QRS and T waves, but the rate may be fast and regular; normal and irregular, or, slow and regular or irregular. Most are vagally mediated.

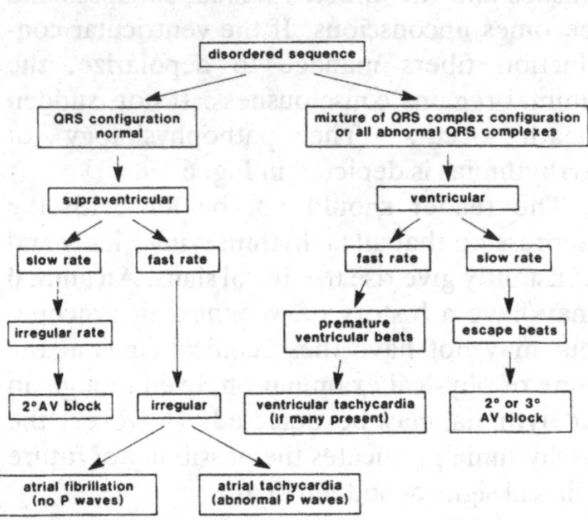

Fig. 6.21. When an ECG trace has some or all P, QRS, and T sequences that vary from normal (disordered sequence), then the configuration of the QRS complex becomes of importance in differentiating the arrhythmia. If the QRS complexes are of normal configuration, then it may be assumed that the arrhythmia is arising from above the common bundle (supraventricular). If the QRS complex is abnormal (usually wide and bizarre in conformation), it usually indicates that the arrhythmia has arisen from within the ventricle. Note that not all possibilities are shown.

mechanisms leading to arrhythmias and, as was said at the time, much can be gained from history and physical examination. As with most things, there is no substitute for experience and everyday familiarity with the ECG trace. There is, however, a checklist that allows a classification of arrhythmias into a number of categories. Once within these categories, subtle variations may be defined by consulting the appropriate textbooks.

The questions to ask after obtaining what is hopefully an acceptable trace are:

1 Is the sequence of complexes composed of normal P, QRS and T waves or are there additions to, or deletions from, the sequence?

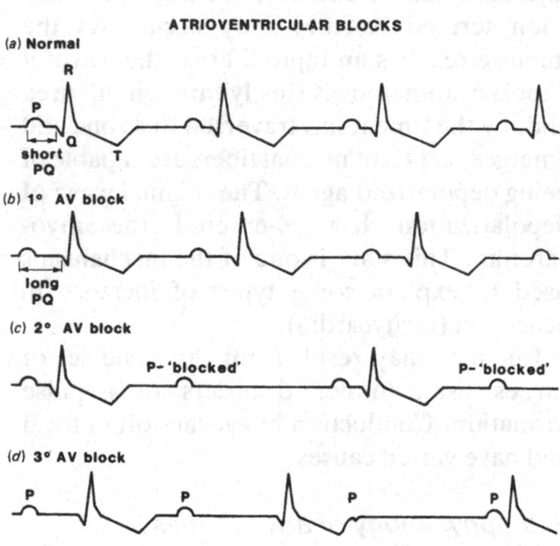

Fig. 6.22. AV blocks are the most common of the conduction blocks and may be incomplete (1° and 2° AV block) or complete (3° AV block). The delaying, intermittent or complete blocking of impulses through the AV node and trunk are almost always associated with a slow heart rate. *(a)* The normal relationship between the P waves QRS complexes and T waves. *(b)* A prolonged PQ interval of 1° AV block. However, all impulses move through the AV nodal area. *(c)* 2° AV block with the regular occurrence of P waves and an intermittent loss of QRS complexes (P – 'blocked'). *(d)* 3° AV block where P waves and QRS complexes occur independently. With 3° AV block the ventricular rate is slow and the QRS complexes may be of normal configuration (shown) or may be wide and bizarre (not shown). (By kind permission of Dr R. L. Hamlin.)

2 Are all the P waves, QRS complexes and T waves of the same form or do they vary?
3 Is the heart rate fast, normal or slow?
4 Is the heart rhythm regular or irregular?
5 Are the time intervals between P, QRS and T waves of normal, shortened or lengthened duration?

The application of these questions is shown in Figs. 6.20 and 6.21 and examples of the use of this approach are contained in Figs. 6.22, 6.23 and 6.24.

This short introduction to the classification of arrhythmias should not be considered as other than a whetting of the appetite. Only the common arrhythmias have been mentioned. The complexities and exceptions to the general rules await.

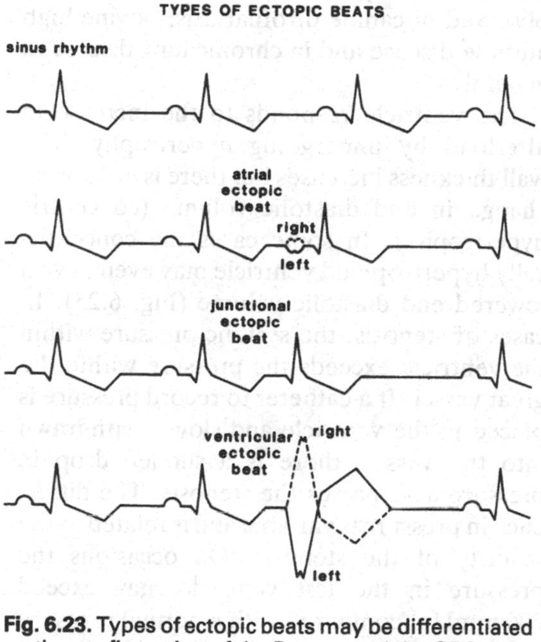

TYPES OF ECTOPIC BEATS

sinus rhythm

atrial ectopic beat

right

left

junctional ectopic beat

ventricular ectopic beat

right

left

Fig. 6.23. Types of ectopic beats may be differentiated on the configuration of the P wave and the QRS complex. An ectopia that arises in an area of the atrium other than the SA node has an appearance different from that of the normal P wave, but is followed by a QRS complex of normal configuration. An ectopia that arises in the AV nodal area (junctional ectopic beat) is not preceded by a P wave, but is of normal configuration. In contrast, an ectopia that arises within the ventricular myocardium has a bizarre appearance. There is also no related P wave. (By kind permission of Dr R. L. Hamlin.)

Hemodynamic disturbances

The term hemodynamic, although not entirely satisfactory, will be used to designate the group of diseases characterized by either **an abnormal pattern of blood flow through the heart and great vessels** or **an impedence to chamber inflow or outflow**.

There are a large number of congenital and acquired diseases which are associated with a hemodynamic disturbance, and a number of systems have been used to classify them. The existence of more than one system of classification reflects the difficulties encountered in providing a framework that encompasses the features of this type of heart disease. Hemodynamic disturbances reflect alterations in either **pressure (afterload)** or **volume (preload) loads**, affecting one or both ventricles.

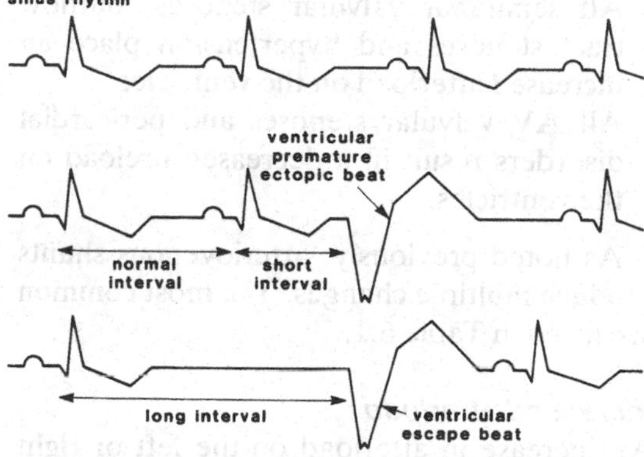

Fig. 6.24. Ventricular ectopic rhythms. Although both premature ventricular ectopic beats and ventricular escape beats can be similar in configuration (wide and bizarre), they differ in timing and genesis. Ventricular premature beats occur after a short interval (prematurely) pre-empting the usually dominant pacemaker (the SA node). By contrast ventricular escape beats arise from a previously latent pacemaker because previously dominant pacemakers (usually the SA node) fail to fire. Ventricular escape beats occur after a long interval reflecting their inherently slow rate of diastolic depolarization. (By kind permission of Dr R. L. Hamlin.)

Table 6.1. *Hemodynamic changes in arteriovenous shunts*

Defect	Load changes			
	LA[a]	LV[a]	RA[a]	RV[a]
Atrial septal defect	↑ Preload	—	↑ Preload	↑ Preload
Ventricular septal defect	↑ Preload	↑ Preload	—	↑ Preload ↑ Afterload
Patent ductus arteriosus	↑ Preload	↑ Preload	—	↑ Afterload
Tetralogy of Fallot	↓ Preload	↓ Preload	—	↑ Afterload

[a]LA, left atrium; LV, left ventricle; RA, right atrium; RV, right ventricle.

Most disorders only have a single pre- or after-load change imposed on the ventricles. This encompasses all of the valvular disorders, which are usually either insufficiencies (failure to close) or stenoses (narrowing of the orifice). Multiple pre- and afterload effects occur in the arteriovenous shunts such as patent ductus arteriosus and ventricular septal defects. The general rules are as follows:

- All valvular insufficiencies place an increased preload on the ventricle.
- All semilunar valvular stenoses, outflow tract stenoses and hypertension place an increased afterload on the ventricles.
- All AV valvular stenoses and pericardial disorders result in a decreased preload on the ventricles.

As noted previously, arteriovenous shunts produce multiple changes. The most common are listed in Table 6.1.

Increased afterload

An increase in afterload on the left or right ventricle occurs when there is increased resistance to ventricular ejection. The resistance usually takes the form of a **stenosis** in the region of the left or right ventricular outflow tract. In the case of the right ventricle this is usually because of a thickened pulmonic valve. For the left ventricle it is usually a ring of fibrous tissue just below the aortic valve, although valvular or supravalvular stenoses occur. If the lesion is of an acquired nature, the cause is commonly a bacterial endocarditis. A special cause of increased afterload on the right ventricle comes in the form of increased pulmonary arterial pressure, alternatively called pulmonary arterial hypertension or cor pulmonale. This is commonly observed in canine dirofilariasis, bovine high altitude disease and in chronic lung disease in general.

The ventricle responds to the increase in afterload by undergoing hypertrophy. The wall thickness increases but there is little or no change in end diastolic volume (concentric hypertrophy). In some cases the concentrically hypertrophied ventricle may even have a lowered end diastolic volume (Fig. 6.25). In cases of stenosis, the systolic pressure within the ventricle exceeds the pressure within the great vessel. If a catheter to record pressure is placed in the ventricle and slowly withdrawn into the vessel, there is a sudden drop in pressure as it passes the stenosis. The difference in pressure is variable and is related to the severity of the stenosis. On occasions the pressure in the left ventricle may exceed 200 mmHg, compared with a normal pressure of 120 mmHg. In pulmonary arterial hypertension, both the arterial and ventricular pressures are raised.

Increased preload

An increased volume load placed on either ventricle can be the result of a number of congenital or acquired diseases. It may follow insufficiency of the AV or semilunar valves, or a redirection of flow between ventricles, atria or great vessels. In the normal animal, blood flow is unidirectional, whereas in volume overloads, a proportion of blood flow during each contraction is bidirectional. To maintain cardiac output, the reduction in unidirectional blood flow must be compensated for. Heart rate is increased and there is an increase in ventricular end diastolic volume. By increasing end diastolic volume, effective stroke volume is maintained (Fig. 6.26).

For example, with AV valvular insufficiency, a proportion of end diastolic blood volume is ejected back into the atrium, and effective stroke volume is depressed. By increasing left ventricular end diastolic volume, the same volume enters the atrium retrogradely, but the volume of blood entering the aorta is increased. It may not reach the normal expected value, but it is increased. There is also the added feature of an increased force of ventricular contraction because of increased stretch on the myofibers. Similar mechanisms come into play for abnormalities such as patent ductus arteriosus (PDA), ventricular septal defect (VSD) and semilunar valvular insufficiency. A by-product of increased preload is myocardial hypertrophy of the eccentric type, where myocardial mass is increased and end diastolic volume is increased, but wall thickness is close to normal.

Decreased preload

The capacity of the ventricles to fill during diastole is restricted by AV valve stenoses and diseases of the epicardium and pericardium, such as epicardial fibrosis, pericarditis, hemopericardium and hydropericardium. There is a restriction to inflow or a diastolic underload, resulting in a depression in diastolic volume and a reduced cardiac output (Fig. 6.27). Apart from an increase in heart rate, there are few cardiac mechanisms that can be brought into play to overcome such problems. However, retention of fluid by the kidneys, secondary to reduction of renal plasma flow, increases capillary filling pressure, thus enhancing venous return. This provides to some extent the pressures required to fill the ventricles during diastole.

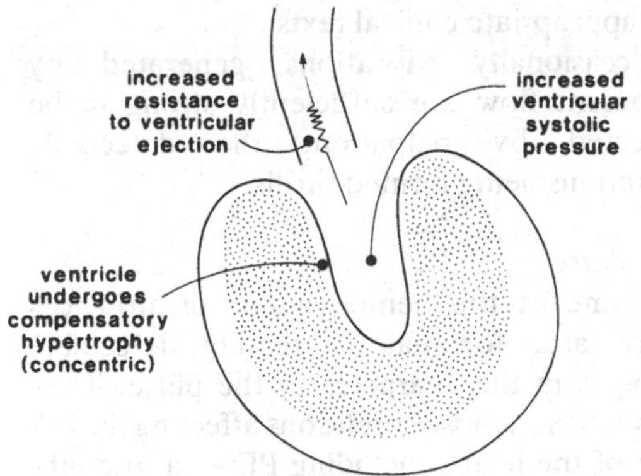

Fig. 6.25. Increased ventricular afterload. The involved ventricle undergoes concentric hypertrophy following an increased resistance to ventricular ejection. End diastolic volume may be reduced. The shaded area shows the outline of a normal ventricle.

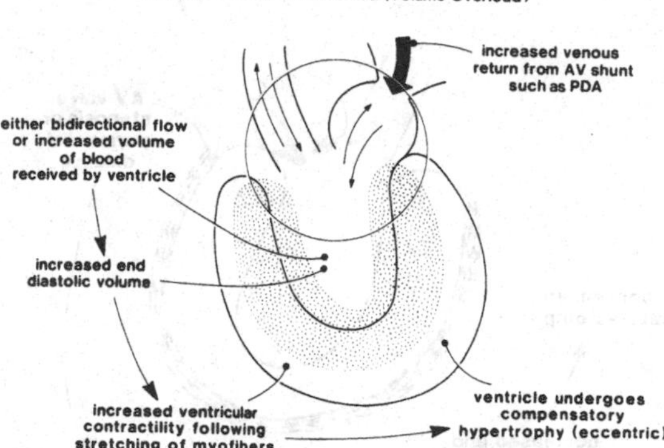

Fig. 6.26. Increased ventricular preload may result from (1) bidirectional flow from insufficient AV or semilunar valves, or (2) increased volume of blood from intracardiac or extracardiac arteriovenous shunt. The ventricle dilates and undergoes eccentric hypertrophy. The shaded area shows the outline or a normal ventricle. PDA, patent ductus arteriosus.

Clinical features of hemodynamic disturbances

Abnormal heart sounds

A prominent feature of most hemodynamic disturbances is the presence of abnormal sounds on auscultation. Exceptions are pulmonary and systemic hypertension in which abnormal sounds are not detectable. There may be a depression in the intensity of normal heart sounds, because of fluid accumulation in the pericardial sac, as in hemopericardium or hydropericardium. Friction rubs may also be detected, for example in pericarditis. In forms of pericarditis where both fluid and gas are present, splashing sounds may be heard.

The most common of the abnormal heart sounds are called **murmurs** and they are generated by turbulent blood flow. Normally blood flow is 'streamlined' but this streamlined flow is broken up once a critical velocity (V_c) is reached. The critical velocity is directly proportional to the viscosity of the blood (η), and, inversely proportional to its density (ϱ) and the radius of the vessel (R). The relationship is given as follows:

$$V_c = \frac{\mathrm{Re}\,\eta}{\varrho R},$$

where Re is the Reynold's number. Transposing the equation:

$$\mathrm{Re} = \frac{V_c\,\varrho R}{\eta}.$$

When Re exceeds a critical level, laminar flow becomes turbulent and murmurs arise. The most common generators of turbulent flow are a change in the radius of a vessel and an increase in the velocity of blood flow.

Murmurs are categorized by their:

timing,
location,
intensity (loudness), and
pitch (frequency).

Cardiac murmurs may occur during systole or diastole, may be heard over the left or right side of the chest, may be soft or loud over the base or apex of the heart, and may vary in frequency. The timing and location of the murmur are the two most valuable indicators of the type of abnormality producing the murmur. The characteristics of murmurs associated with particular hemodynamic disturbances are shown in Table 6.2. A complete characterization of murmurs may be found in the appropriate clinical texts.

Occasionally vibrations generated by turbulent flow are sufficiently strong to be detected by palpation, the detectable vibrations being termed thrills.

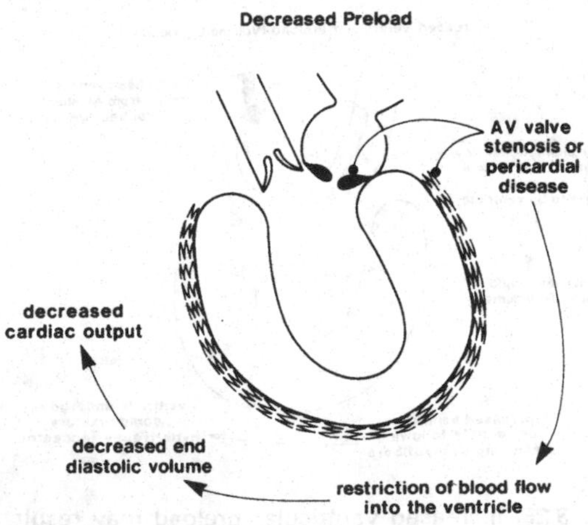

Fig. 6.27. Decreased ventricular preload follows AV valvular stenosis or myocardial restriction such as pericardial fibrosis. End diastolic volume decreases because of restricted inflow.

The pulse

In some of the hemodynamic disturbances there are, fortunately, readily detectable changes in the character of the pulse. They only occur in those conditions affecting the left side of the heart, including PDA, aortic/subaortic stenosis and aortic insufficiency. The reason for the readily detectable alteration in the pulse is the variation in pulse pressure. In most cases of aortic stenosis, systolic pressure is usually lower than normal, producing a

Table 6.2. *Murmurs and hemodynamic disturbance*

Abnormality	Murmur Timing	Murmur Location
Insufficiencies		
Left AV valve	Systolic	Left apex
Right AV valve	Systolic	Right apex
Aortic valve	Diastolic	Left base
Pulmonic valve	Diastolic	Left base
Stenoses		
Left AV valve	Diastolic	Left apex
Right AV valve	Diastolic	Right apex
Aortic valve	Systolic	Left base
Pulmonic valve	Systolic	Left base
Arteriovenous shunts		
Atrial septal defect	Systolic	Left base
Ventricular septal defect	Systolic	Right apex/base
Patent ductus arteriosus	Continuous	Left precordium (cranial to heart)

weak pulse. A strong pulse which rises then falls quickly may be seen with both PDA and aortic insufficiency because of the lowering of diastolic pressure and a widening of the pulse pressure. In the case of a PDA, the blood flows into the lower pressure zone of the pulmonary artery. In aortic insufficiency, blood flows back into the ventricle, whose diastolic pressures approach zero.

The presence of cyanosis

Cyanosis at rest may be observed in a number of the hemodynamic arteriovenous shunts. For this to occur, blood flow must be from right to left, that is from the venous to the arterial circulation. In most cases the flow is left to right, but when venous pressure exceeds arterial pressure the shunt will reverse. This may occur occasionally with atrial septal, or ventricular septal, defects, and PDA.

Myocardial failure (deficit in contractility)

Myocardial failure means that the myocardium itself is unable to generate sufficient force to maintain output commensurate with the demands of the body. Weakness of contraction is really a disorder in systolic function. As such, the heart is unable to expel the required amount of blood per contraction, stroke volume decreases, ejection fraction decreases and cardiac output falls. End diastolic pressure and volume increase.

There are two major categories of myocardial failure. The first is when there are **insufficient numbers of ventricular myocytes** for effective contraction to occur. The second is when there are adequate numbers of myocytes, but they **contract ineffectively** (Fig. 6.28).

The loss of ventricular myocytes usually follows infectious or nutritional disorders; two examples are chronic myocardial fibrosis in dogs following infection with canine parvovirus, and fibrosis in domestic ruminants following vitamin E/selenium deficiency. Even though both start as myocyte necrosis, the pattern of fibrosis is quite different. Parvoviral myocardial fibrosis is relatively diffuse throughout the heart, whereas the fibrosis of vitamin E/selenium deficiency is of larger focal blocks of myocyte loss with replacement fibrosis.

Generalized ineffective contraction of myocytes is encompassed by the term **congestive (dilated) cardiomyopathy**. This rather broad term, literally meaning 'disease of cardiac muscle' often has idiopathic as a prefix, which adds to the mystery of these diseases. What is known, is that these diseases have as their central feature a failure of the ventricular myocardium to contract effectively. They are best known in the dog, especially the giant breeds, such as Great Danes and Saint Bernards. They present in both left- and right-sided congestive heart failure, and the clinical picture is often complicated by the presence of arrhythmias such as atrial fibrillation. It was not until the complicating factor of atrial fibrillation was recognized as such, that it was realized that the underlying cause of the cardiac failure resided in the ventricle. Since that time, cardiomyopathies have been investigated intensively and thought about constantly, but the pathogenesis and the etiology are still obscure.

Combinations

The presentation of cardiac disease as either an arrhythmia, hemodynamic disturbance or myocardial failure is convenient, but it is not the whole truth. The situation often arises where an animal may present with more than one pathophysiologic problem. For example, the underlying abnormality in dogs with dilated cardiomyopathy is a deficit in myocardial contractility, but it is often complicated by the presence of an arrhythmia such as atrial fibrillation. There are also some cases in which cardiomyopathy has a mild hemodynamic disturbance superimposed. Because of the myocardial failure, the AV rings dilate, leading to AV valvular insufficiency, and a systolic murmur is often heard. The preceding is a most extreme example, but the lesson taken from it is that it is up to the clinician to decide what is the primary abnormality; it can be difficult, but there are many precedents to fall back on. For example, a large-breed dog with atrial fibrillation and signs of congestive heart failure is most likely to have cardiomyopathy. Similarly, old small-breed dogs, with loud left apical systolic murmurs and in left-sided heart failure, most probably have a primary hemodynamic abnormality, even though supraventricular or ventricular arrhythmias may be present.

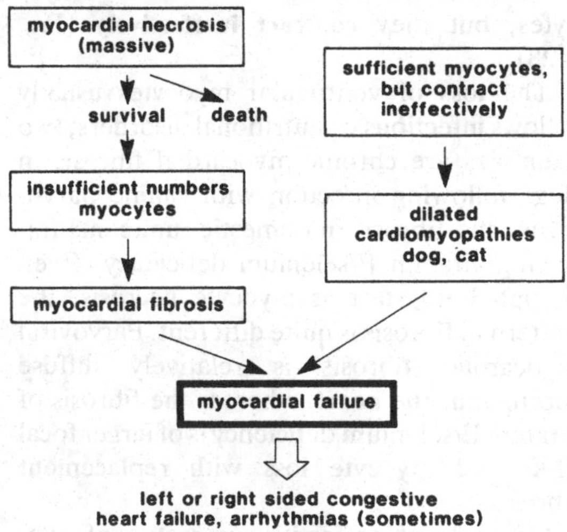

Fig. 6.28. Myocardial failure (failure to pump) arises from severe myocardial fibrosis or ineffective contraction of ventricular myocytes.

Morphologic patterns in heart disease

The three broad anatomical divisions of the heart (the myocardium, the mural and valvular endocardium, and the pericardium), because of their special nature, react differently to insult. At this juncture, only acquired disease will be discussed. Congenital heart disease will be discussed separately.

The myocardium

There are few specific diseases that affect primarily the myocardium, although it is often caught up in the hurly burly of a systemic disease such as bacterial septicemia, or malignant neoplasia. Such effects are usually multifocal and there is usually evidence of remote disease. Those diseases that direct all their force primarily toward the myocardium are usually viral, nutritional or, to some extent, parasitic. There is also the group of diseases known as the cardiomyopathies.

The diseases affecting the myocardium

emphasize some of the characteristics of it. The myocardium, because of its high energy requirements for contraction, is critically dependent on an adequate blood supply. The coronary vessels that supply the myocardium are end arteries and, as such, render the myocardium susceptible to alterations in regional blood flow. A reduction in coronary blood flow compromises the activity of myocytes in regional areas, particularly subendocardial areas. Flow to these areas is influenced by the systolic and diastolic pressure within the aorta, and the systolic and diastolic pressures within the ventricle. Also the end arterial nature of the myocardial blood supply makes it particularly susceptible to ischemia following embolism, the most frequently observed example being bacterial embolism, either as part of a systemic infection or as a consequence of bacterial endocarditis. Disease of the coronary vessels, be they the larger coronaries or smaller intramyocardial arterioles, create conditions suitable for regional ischemia.

The myocardium in concert with skeletal muscle also exhibits a sensitivity to the peroxidation of membranes with either vitamin E or selenium deficiency. It regularly results in multifocal myocardial necrosis in those species particularly sensitive to such deficiency, such as domestic ruminants and pigs. Arrhythmias usually of ventricular origin often lead to sudden death.

In contrast to skeletal muscle, the myocardium has little capacity for repair. Once a cardiac myocyte is lost it cannot be replaced. There is some hyperplasia of myocytes in the first few weeks of life, but following this period no mitotic activity occurs. Myocytes lost are replaced by fibrous tissue. Fortunately, there is a moderate excess of myocytes and comparatively large numbers may be lost without any observable effect on the performance of the animal. The remaining myocytes compensate for the loss by undergoing hypertrophy. The major features of myocarditis, myocardial necrosis and the cardiomyopathies are depicted in Fig. 6.29.

Myocardial hypertrophy

The myocardium has a remarkable capacity to adapt to an increase in workload, be it due either to increased physiologic demands, such as sustained exercise, or to some pathologic alteration within the heart, such as stenosis of the aortic valve. It does so by each myocyte accumulating extra machinery required for contraction. Upon the imposition of an increased workload, each myocyte increases in size, predominantly due to increases in the number of mitochondria and the contractile proteins actin and myosin.

A beautiful example of physiologic hypertrophy follows birth. *In utero*, because of high vascular resistance in the pulmonary circuit and the presence of the ductus arteriosus, the mass of the right ventricle approximates to that of the left ventricle. Following birth, pulmonary vascular resistance drops precipitously, the ductus arteriosus closes and systemic vascular resistance increases following closure of the umbilical arteries. As a result, the mass of right ventricle increases much more slowly than the newly afterloaded left ventricle. In the adult, the ratio left to right ventricular mass is about 3:1.

The ventricular muscle also undergoes hypertrophy following the imposition of a pathologic increase in workload, be it an

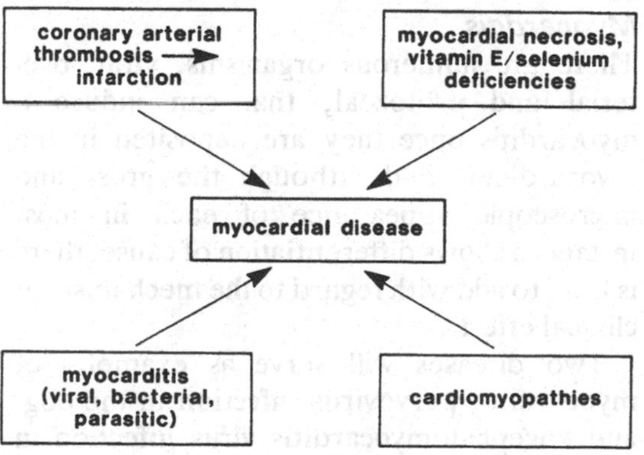

Fig. 6.29. Myocardial disease may be primarily inflammatory, thrombotic, necrotic or idiopathic (cardiomyopathic) in origin.

increase in preload or afterload. The common denominator seems to be an increase in ventricular wall tension. The mechanism for the translation of a mechanical load into a biochemical response has yet to be elucidated. Many parameters have been measured but all appear to be responses rather than the initiator. There has been discussion about mechanical and biochemical differences in physiologic versus pathologic hypertrophy, particularly with respect to mechanical performance. However, at this stage, it is enough to realize that ventricular mass changes with a change in workload and whether pathologic or physiologic it is an adaptation to stress.

The appearance of a hypertrophied ventricle varies with the load placed on it. In cases of an increase in preload, end diastolic volume is increased as well as mass. However, the wall thickness remains about the same. Such a response is termed **eccentric hypertrophy**. With an increase in afterload, end diastolic volume does not change, but myocardial mass does. The resulting ventricular wall is thicker and is **concentrically hypertrophied**.

Hypertrophy takes time to develop. The time lapse from the imposition of the increased workload to the attainment of the required increase in mass is of the order of a month. It is also worth noting that what can be done can be undone. It is a reversible reaction.

Myocarditis

There are numerous organisms, viral, bacterial and protozoal, that can induce a myocarditis once they are deposited in the myocardium, and although the gross and microscopic appearance of each in most instances allows differentiation of cause, there is little to add with regard to the mechanism of clinical effect.

Two diseases will serve as examples of myocarditis, parvovirus infection in the dog, and encephalomyocarditis virus infection in the young pig.

These viruses find the myocyte a pleasant place in which to reside and multiply. The result is multifocal myocyte necrosis with an associated inflammatory response. The myocyte degeneration and necrosis set the stage for the common clinical finding of sudden death, presumably following ventricular tachycardia and fibrillation. There is evidence derived from Coxackie virus myocarditis in mice that most of the myocyte damage is immunologically mediated.

The cardiomyopathies

The term cardiomyopathy was originally used in man to classify a group of primary diseases of the myocardium that were, and in most cases still are, of unknown cause or association. Since that time, the cardiomyopathies have become recognized as some of the more important heart diseases in domestic animals, especially in the dog and cat.

Cardiomyopathies have been subdivided into three categories, hypertrophic, dilated (congestive) and restrictive forms. In the **hypertrophic** form there is symmetric or asymmetric hypertrophy of the left ventricle. This **inappropriate hypertrophy** of the myocardium may either obstruct the ventricular outflow tract, or impair ventricular filling during diastole. In man, hypertrophic cardiomyopathy is inherited as an autosomal dominant character.

Dilated (congestive) cardiomyopathies are characterized morphologically by a dilated heart, but there is little microscopic evidence for the dilation. There is a lowered force of contraction during systole.

The **restrictive** form of cardiomyopathy exhibits marked endomyocardial fibrosis reducing ventricular compliance and therefore diastolic filling.

Canine cardiomyopathies

The most commonly recognized canine cardiomyopathy is the **dilated or congestive** form occurring in young to middle-aged large-breed dogs such as Saint Bernards, Irish Wolfhounds, Great Danes and German Shepherds. This type of cardiomyopathy is characterized clinically by the sudden onset of varying degrees of left- and right-sided heart failure, which is often complicated by atrial fibrillation. There is cardiomegaly with increased end diastolic volume and poor contractile function. Soft systolic murmurs, indicative of left and right AV valvular insufficiency may be heard on auscultation. The prognosis is poor, with mean survival times of six to twelve months after the onset of treatment.

Necropsy findings in typically affected cases are those of congestive heart failure. Both ventricular chambers, particularly the left are markedly dilated and may be hypertrophied. The AV rings are dilated, the endocardium may be opaque due to subendocardial fibrosis, and there may be atrial thrombosis. There may also be some evidence of multifocal myocardial degeneration, necrosis or fibrosis.

There are two further cardiomyopathies in dogs that are variations of those observed in large-breed dogs. The first is in Doberman Pinschers. The necropsy findings are those described for large-breed dogs, with the possible addition of the presence of scattered lymphocyte infiltration of the ventricles. Clinically they are different, in that the dogs are almost always in sinus rhythm and commonly exhibit ventricular arrhythmias. Atrial fibrillation is uncommon. The second is seen in English Cocker Spaniels. It appears to be of familial origin, with a high incidence of subclinical disease. There is ECG evidence of left or biventricular hypertrophy and some of these dogs either may be found dead or may develop acute left ventricular failure. Some of the older dogs may have mild endocardosis as an incidental finding.

Hypertrophic cardiomyopathy is much less common than the dilated form. Associated clinical syndromes include sudden death, death during anesthesia, and congestive heart failure. Disproportionate thickening of the ventricular septum may cause the ratio interventricular wall thickness to left ventricular free wall thickness to exceed 1.3:1, and there may be histologic evidence of myofiber disarray in the ventricular septum.

The etiology and pathogenesis of these conditions is unknown but they have, with the exception of the hypertrophic form, the common denominator of lowered contractility and ventricular dilation. There appears to be a familial tendency in the Doberman and Cocker Spaniel. Methods demonstrating some immunologic involvement or catecholamine dysfunction in the disease in man have yet to be applied to dogs.

Feline cardiomyopathy

This is probably the most common and well-described cardiomyopathy in domestic animals. The condition was only recognized as an entity in the comparatively recent past, but iliac thromboembolism, a common consequence of cardiomyopathy, was originally described more than 50 years ago. There is a wide range in the age of onset of clinical signs from seven months to 24 years. Presenting clinical signs include lethargy, anorexia, dyspnea, tachypnea and occasionally abdominal distension. Murmurs, most often associated with left or right AV valvular insufficiency and arrhythmias of various types are frequently encountered. Approximately one-third of cats present with unilateral or bilateral thromboembolic hindlimb ischemia.

The cause or causes are at present unknown. There has been a suggestion that the condition is infectious in origin but there is little evidence to support this.

Because of the wide spectrum of the findings at necropsy, the condition has been divided into subgroups: (1) endomyocarditis; (2) congestive; (3) symmetrical hypertrophic; (4) asymmetrical hypertrophic and (5) restrictive cardiomyopathies. The **endomyocardial form** is observed most frequently in young cats in which there are often subendocardial petechiae and ecchymoses with plaques of fibrin attached to the endocardium. The subendocardium and myocardium contain a focal or diffuse infiltration of inflammatory cells with neutrophils predominating. There is often accompanying focal myofiber necrosis. A number of the cases may be of a more chronic nature. There is also a variable presence of atrial thrombi.

In **congestive (dilated) cardiomyopathy**, there is generalized cardiomegaly, due to extreme dilation of the atria and ventricles with atrophy of the papillary muscles and trabeculae carnae. Atrial thrombi are uncommon. Histopathologic examination shows only mild interstitial edema and fibrosis of the myocardium.

Cats with **symmetrical hypertrophy** constitute group 3. The left ventricular free wall, papillary muscles and ventricular septum are hypertrophied with a decrease in left ventricular volume.

The fourth group exhibit **asymmetrical hypertrophy** of the ventricular septum, particularly where the hypertrophied septal wall encroaches into the left ventricular outflow tract. Aortic thromboembolism is present in approximately 20% of cases. Once again, there is a possibility of the finding of left atrial thrombosis. In **restrictive cardiomyopathy** there is severe endocardial thickening and fibrosis and there may be mural thrombosis. Left atrial enlargement is marked. Microscopically, the endocardium is thickened by hyaline-like material and fibrosis. Often, there is a mixed inflammatory response in the myocardium.

The endocardium

The endothelial lining of mural and valvular endocardium is in intimate contact with flowing blood on one side and its basement membrane on the other. This endothelium differs little from that lining every blood vessel in the body, but, because of its location, damage to it may have rather serious consequences, particularly so in the case of the valvular endothelium. For reasons that are not at all clear, the valvular endothelium is a favored place for bacteria to lodge. Once the bacteria are in place, the endothelium is quickly lost, leading to a train of events culminating in the formation of large thombi, which impede blood flow through the chambers of the heart.

Endocarditis

Inflammation of the endocardium is almost always valvular and is usually bacterial in origin. There are occasional instances of endocarditis caused by fungi or wandering parasites.

The development of bacterial endocarditis requires a sustained or recurrent bacteremia. This is most commonly produced by a neonatal septicemia, mastitis, metritis or a focal abscess elsewhere in the body. Bacterial endocarditis begins on the free margins of the valves that oppose each other at the time of valve closure and although there is no evidence in animals of pre-existing valvular disease predisposing to the development of endocarditis, the frequency of endocarditis can be increased by increasing the workload on the heart. There is also evidence that some bacteria selectively adhere to the valvular endothelium in much the same way as other pathogens attach to epithelial cells in mastitis and enteritis.

Once endothelial damage has been initiated by the adhering bacteria, platelets and fibrin quickly attach and a large friable mass becomes evident. Such a mass is termed a vegetation, leading to the common term **vegetative endocarditis**. The offending bacteria persist for prolonged periods in the vegetations. With time, the primarily valvular reaction may encroach on to the mural endocardium. Without adequate chemotherapy, endocarditis seldom resolves. It does, however, move into a chronic phase, with a healing reaction beginning at the base of the valve.

Depending on the size of the vegetation and the extent of valvular destruction, the affected valve may become functionally insufficient or stenotic. There is also the specter of fragments of the vegetation breaking off and lodging in remote organs. As such, emboli may or may not contain bacteria; embolic abscessation or just embolic infarction are common sequelae to endocarditis. Endocarditis of valves on the right side of the heart frequently leads to pulmonary abscessation but rarely infarction.

Conversely, emboli arising from the left side of the heart may produce infarction or abscess formation in a number of organs, but most notably in the renal cortex. The major features of endocarditis are shown in Fig. 6.30.

Degenerative endocardial disease

Significant non-inflammatory endocardial disease occurs only in the dog and is in fact the most common cardiovascular disease of the dog. The lesion is valvular and is termed **endocardiosis**. Many dogs have endocardiosis but relatively few develop clinical signs associated with it. The prevalence of the disease increases with age. Of dogs reaching 16 years of age, 75% have the lesions typical of endocardiosis. The cause of endocardiosis is not known, but it occurs with greater frequency in the chondrodystrophic breeds.

Endocardiosis is recognized grossly by shrunken, distorted and thickened AV valves. The left AV valves are most commonly involved, the right AV valve less so and the semilunar valves rarely. The lesion is the result of the proliferation of loose fibroblastic tissue in the spongiosa of the valve accompanied by the deposition of acid mucopolysaccharides. There is also collagen degeneration of the valvular fibrosa.

As with most diseases, there is a spectrum of change in endocardiosis. In its mildest form, only a few small discrete nodules are present on the margins of the valve. At its most severe, there is gross distortion of the valve by gray-white nodules and plaques. The cusps are thickened, contracted and irregular.

As the disease progresses, the affected valves become increasingly insufficient leading to atrial dilation and eccentric ventricular hypertrophy. The left atrial endocardium may be irregularly thickened by fibrosis (jet lesions), following prolonged regurgitation of blood from the ventricle to the atrium.

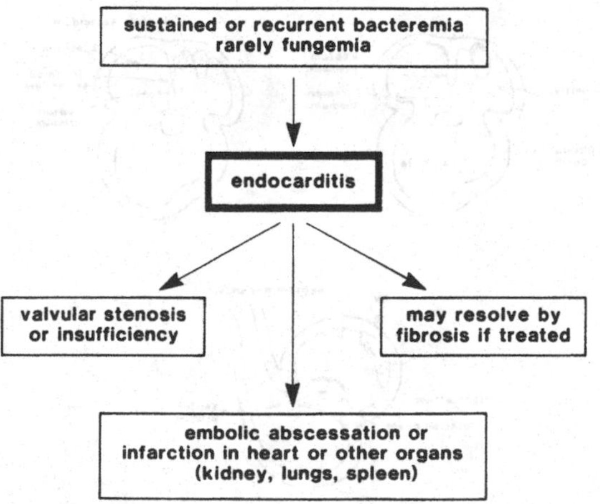

Fig. 6.30. Endocarditis is usually bacterial or rarely fungal in origin and leads to valvular stenosis or insufficiency and commonly emboli in other organs.

The pericardium

The pericardium is a thin fibroelastic sac that encloses the heart. A small amount of fluid is present within the pericardium to facilitate movement (reduce friction) between the epicardium and pericardium.

Clinically significant pericardial disease revolves almost entirely around its inability to stretch quickly. Any rapid accumulation of a fluid, be it a transudate, exudate or blood, severely compromises the ability of the ventricles to fill with blood during diastole. There is also the problem of chronic pericardial or epicardial fibrosis following pericarditis producing similar effects on ventricular filling.

Hemopericardium follows rupture of a large vessel or of a chamber of the heart. It is seen with rupture of the intrapericardial aorta in horses, in atrial rupture in dogs with endocardiosis and in rupture of the aorta or of a coronary artery in pigs. Recently hemopericardium of unknown cause has been described in a number of large-breed dogs. The injudicious use of cardiac puncture to obtain blood may also result in hemopericardium.

Pericarditis, especially fibrinous pericarditis, often accompanies systemic infectious disease in many species. As such, fibrinous pericarditis does not usually have a significant effect on heart function. In contrast purulent pericarditis does. It is seen classically in the cow, following traumatic perforation of the pericardium by a foreign body originating in the reticulum. Purulent pericarditis is sometimes seen in cats and horses in association with a purulent pleuritis. Severe congestive heart failure, usually right sided, may be seen in the subacute and chronic stages of purulent pericarditis.

Congenital heart disease*

Some time in the early weeks of pregnancy the fetal heart begins to beat. At first the contractions are uncoordinated and the flow of the newly formed blood is irregular. Soon, however, the lining of the primitive heart tube becomes organized into valves and an endocardiac cushion which forms between the sacculations of the ventricles and primitive atrium. The heart's four primitive chambers are arranged in linear sequence and named caudally to rostrally, **sinus venosus**, **atrium**, **ventricle** and **bulbus cordis**, the last of these continuing into the **truncus arteriosus**.

The longitudinal heart tube undergoes a three-dimensional S-shaped bending to form a more recognizable heart form, with the primitive atrium and sinus venosus sitting on top of the primitive ventricle and bulbus cordis. The primitive atrium is incompletely separated from the ventricle by a median endocardial cushion formed from dorsal and ventral endocardiac swellings coming together. It is flanked on either side by two rings of endocardiac jelly, which will later become organized into the AV valves.

The partitioning of the atrium: this begins with a sheet of mesodermal tissue (septum primum) growing vertically downwards from the dorsal atrial wall towards the endocardial cushion. The space between the leading edge of this septum and the endocardiac cushion is the **foramen primum**. As the septum primum approaches the endocardial cushion and the foramen primum becomes smaller, a second foramen (**foramen secundum**) begins to form in septum primum. A second sheet of mesodermal tissue (septum secundum) begins to grow towards the endocardiac cushion parallel to, and to the right of, the septum primum. This septum is much thicker than the septum primum and never quite reaches the endocardiac cushion so that a foramen (**foramen ovale**) is formed between the leading edge of septum secundum and the endocardiac cushion. The thin, lower part of septum primum acts as a valve flap, which, if the pressure rises in the left atrium, will be pushed to the right to close the foramen ovale (Fig. 6.31).

Once partitioning of the primitive atrium has taken place, further development is merely growth and modification of the atrial walls. Both atria enlarge; the right expands to incorporate the sinus venosus and the termination of the vena cavae, and the left engulfs the pulmonary veins.

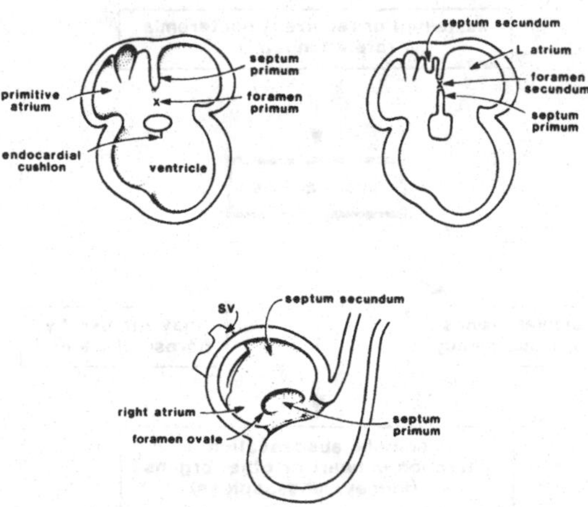

Fig. 6.31. Development of the interatrial septa. Downgrowth of the septum primum is followed by the septum secundum and foramen secundum. The final foramen ovale allows right to left atrial blood flow **in utero**. L, left; SV, sinus venosus.

* In association with Dr Sheila S. White.

Partitioning of the ventricles and bulbus cordis: This occurs simultaneously with partitioning of the atrium. It does not involve the upward growth of a septum towards the endocardial cushion as one might expect, but rather the outward and downward growth of the ventricle and bulbus walls. The heart at this early stage is made of extremely maleable mesenchymes so that, when blood is forced through the left and right AV valves, the lateral tendency of its flow pushes the ventricular walls outward, leaving behind a septum in the midline (Fig. 6.32).

An interventricular foramen is left between the free edge of the interventricular septum and the endocardial cushion. This hole is closed by tissue from the interventricular septum, the left and right bulbar ridges (whose development is explained below), and the endocardiac cushion. The major contribution to closure is from the bulbar ridges.

Partitioning of the truncus arteriosus and bulbus cordis: This is brought about by the development of two opposing ridges of endocardiac jelly which form within. As these ridges grow towards each other they spiral round the inside, dividing the truncus arteriosus and bulbus cordis into two spiraling tubes. The cranial part is associated with the developing branchial arch arteries and the caudal or cardiac ends open into the ventricles over the interventricular septum. The bulbar ridges descend to close over the interatrial septum in such a way that the upper half of the bulbus cordis, which is associated rostrally with the fourth branchial arch artery (the future aorta), opens into the left ventricle and the lower division of the truncus arteriosus bulbus cordis opens into the right ventricle (Fig. 6.33).

At the cardiac end, the right bulbar ridge, in moving across to fill the interventricular foramen, contributes to the right AV valve, which subsequently becomes tricuspid. At the rostral end, the bulbar ridges contribute to

Fig. 6.32. The left and right ventricles form as a result of the volume and pressure effects on the free walls leaving the interventricular septum. The dashed lines demonstrate this progressive development from a single chamber to two chambers.

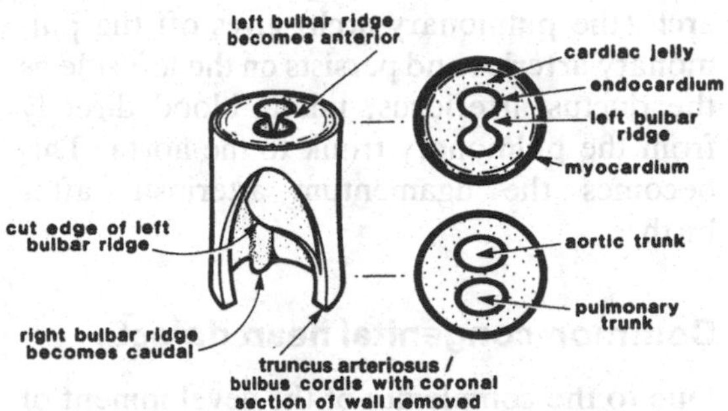

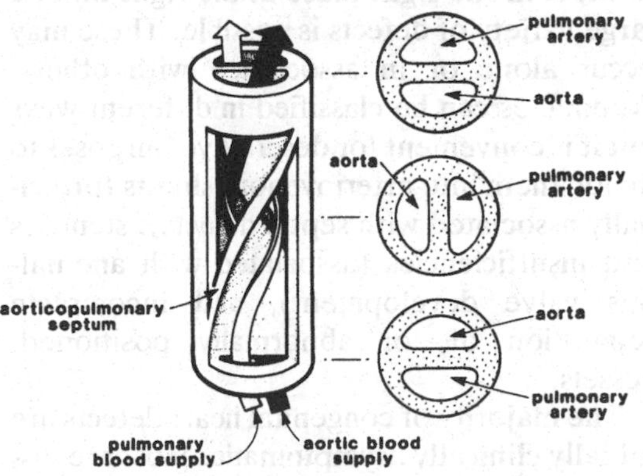

Fig. 6.33. Partitioning of the truncus arteriosus/bulbus cordis into the aorta and pulmonary artery. The bulbar ridges grow toward the center and spiral to produce the two vessels.

splitting off the fourth and sixth branchial arch arteries.

Formation of the semilunar valves: This is associated with the formation of the conobulbar ridges and endocardiac jelly. When the truncus arteriosus has divided almost completely, a pair of endocardiac swellings develop at right angles to the conobulbar ridges. These swellings, together with contributions from each of the bulbar ridges, become hollowed out on their upper aspect and form the semilunar valves.

Development of the branchial arch arteries: Of the six branchial arch arteries, only the fourth and sixth are of potential pathologic significance. The fourth arch artery on the left side forms the arch of the aorta and on the right side the subclavian artery. The sixth aortic arch (the pulmonary arch) gives off the pulmonary arteries and persists on the left side as the ductus arteriosus, taking blood directly from the pulmonary trunk to the aorta. This becomes the ligamentum arteriosus after birth.

Common congenital heart defects

Due to the complexity of the development of the heart and the number of parts which have to meet in the right place at the right time, a large variety of defects is possible. These may occur alone or in association with others. Anomalies can be classified in different ways but it is convenient for descriptive purposes to divide them into arteriovenous shunts (principally associated with septal defects), stenoses and insufficiencies (associated with anomalous valve development), and incomplete separation of, or abnormally positioned, vessels.

The majority of congenital heart defects are initially clinically asymptomatic. Most are discovered incidentally during routine physical examination. That is not to say that the defect will not produce heart failure in the ensuing months, but it must be recognized that the presence of a murmur indicative of congenital heart disease will not necessarily be followed by heart failure.

The development of heart failure is a function primarily of the size of the defect and to some extent the type of defect. Complex defects such as the tetralogy of Fallot invariably produce clinical signs of failure, whereas other more simple ones such as pulmonic stenosis may not. Of those that result in clinical signs, almost all are of a hemodynamic character.

One of the prime reasons for differentiating defects is from a genetic point of view. Many congenital heart defects, particularly in the dog, are inherited. Those proved to be so include patent ductus arteriosis in Poodles, pulmonic stenosis in Beagles and subaortic stenosis in Newfoundlands. Others, whilst not proved to be inherited, are breed predisposed. There are numerous examples of these.

Arteriovenous shunts/septal defects

Atrial septal defect

As described previously, the foramina or ostia in the atrial septal wall are primum and secondum together with ovale. Failure of the foramen primum to close is really failure of the septum primum to meet with the endocardiac cushion. Foramen secundum may expand excessively so that the septum secundum is unable to cover the hole. Another circumstance which may result in a septal defect occurs if there is a failure of fusion of the dorsal and ventral endocardial swellings which form the endocardial cushion. This defect results in the formation of a common AV canal. The amount of blood flowing across the defect following birth is dependent on the size of the atrial septal defect, the inflow resistance into the ventricles, and most likely the resistance to ventricular outflow. With most defects, the flow is from left to right, placing a volume overload on the right atrium and ventricle, increasing blood flow through the lungs and increasing the volume load on the left atrium (Fig. 6.34). A systolic murmur, best heard over the left base of the heart, is the result of relative pulmonic stenosis, as a much greater volume of blood has to flow through the normal pulmonary outflow tract. Flow through the atrial septal defect itself does not produce a murmur.

In some instances the flow through the septal defect may reverse, becoming right to left and leading to cyanosis. This usually only occurs following an increase in pulmonary vascular resistance.

Ventricular septal defect

This is generally failure of the interventricular foramen to close and so is usually found high in the interventricular septum. As mentioned above, the closure of this foramen is complex, with tissue being contributed from a number of sources. Consequently this defect is one of the more common with the clinical manifestations dependent on its size. In the fetus, intraventricular pressures approximate each other, so the presence of an interventricular foramen is of little significance.

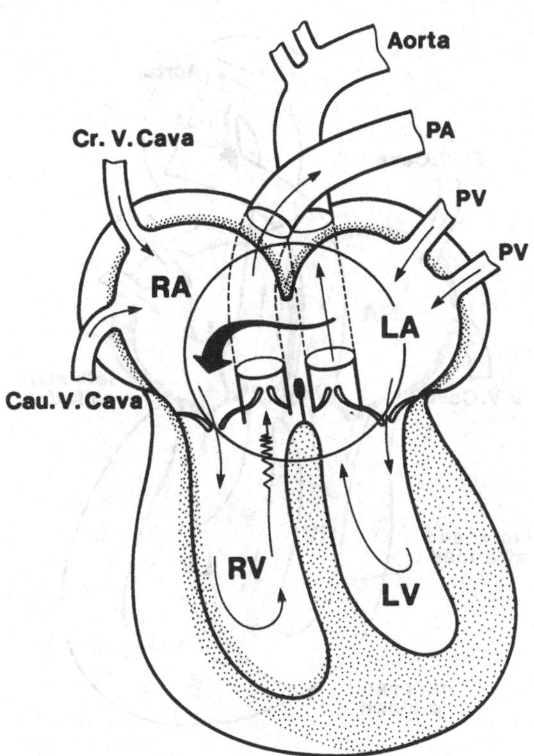

Fig. 6.34. An interatrial septal defect usually results in increased blood flow from the left (LA) to the right (RA) atrium. The right ventricle (RV) dilates under an increased preload. A relative pulmonic stenosis is induced because of the increased volume in the right ventricle. A right base systolic murmur is characteristic. LV, left ventricle; PV, pulmonary vein; PA, pulmonary artery; Cr. V. Cava, cranial vena cava; Cau. V. Cava, caudal vena cava.

However, after birth, following the reduction of pulmonary vascular resistance and the increase in systemic vascular resistance, there is a tendency for blood to flow from the left to the right ventricle through the ventricular septal defect. When the defect is large, ventricular systolic pressures are approximately equal. Under these circumstances the left to right flow is largely determined by differences in the pulmonary and systemic vascular resistances.

There is consequently a large flow of blood through the pulmonary vasculature, left atrium and ventricle. To accommodate the defect and to maintain systemic arterial output, the left ventricle dilates. There is thus an increased preload on the left ventricle which, if it exceeds the capacity of the left ventricle, results in left-sided failure. There is an increased afterload on the right ventricle (Fig. 6.35). Once again a systolic murmur is associated with the defect, mostly located on the right side of the chest, midway between heart base and apex.

Small ventricular septal defects are often found incidentally on physical examination or at necropsy.

Patent ductus arteriosus (PDA)

Under normal circumstances, the ductus arteriosus closes functionally within a few hours of birth, but it may remain patent for up to five days in some normal animals. Patency beyond this time is considered pathologic. The severity of the clinical signs seen with a PDA is related to the extent to which the lumen remains open. As the connection between the pulmonary trunk and the aorta remains open there is a retrograde flow from the aorta to the pulmonary arteries. This occurs during both systole and diastole. Also, because of the open communication, aortic diastolic pressure is lower, resulting in a widened pulse pressure. A PDA is characterized clinically by a continuous or 'machinery' murmur. If the patency is gross, the resultant increased venous return from the pulmonary circulation causes a volume overload on the left atrium and ventricle with compensatory hypertrophy and, many times, failure. The open communication between great vessels equalizes the pressure between them leading to an increased afterload on the right ventricle with consequent right ventricular hypertrophy (Fig. 6.36). There is occasionally a reversal of the left to right shunt following the development of pulmonary hypertension. In such cases the flow is from right to left leading to cyanosis of the posterior half of the body, because the ductus arises after the brachiocephalic trunk, which supplies the cranial half of the body.

Conotruncal anomalies

The final septum associated with the heart is that formed by the bulbar ridges. Malformation of the cardiac end, which is associated with the closure of the interventricular septum, results in three defects: a ventricular septal defect, pulmonic stenosis and an overriding aorta. Together with compensatory right ventricular hypertrophy this condition is called the tetralogy of Fallot. The severity of the defect is influenced by the extent of the pulmonic stenosis. The reduced diameter of

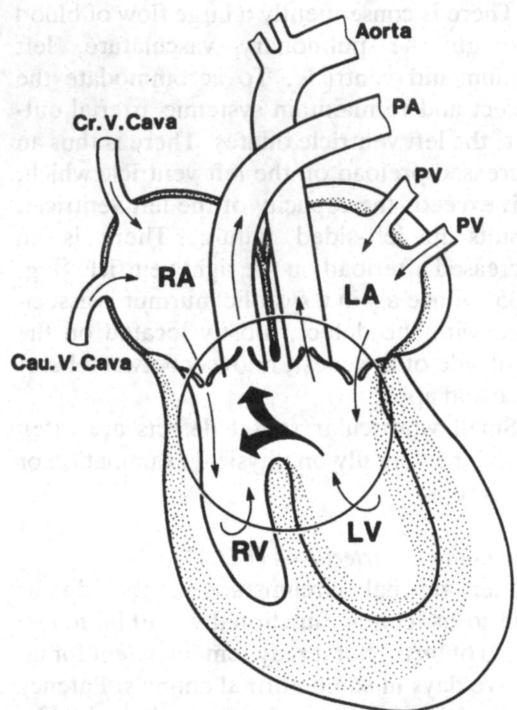

Fig. 6.35. Blood flow in a ventricular septal defect is usually left to right creating both an increased pre- and afterload on the right ventricle. The increased volume of blood returning through the pulmonary circuit places an increased preload on the left atrium and ventricle. For abbreviations, see Fig. 6.34.

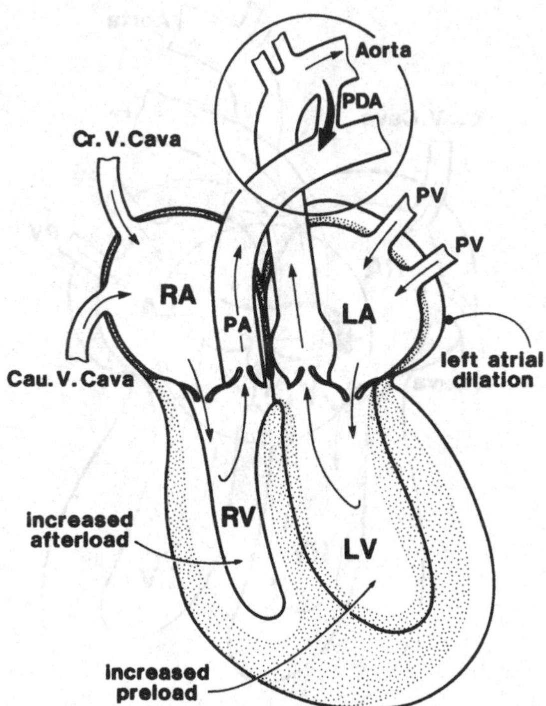

Fig. 6.36. In most cases, blood flow through a patent ductus arteriosus (PDA) occurs during both systole and diastole from the aorta into the pulmonary artery. The right ventricle hypertrophies concentrically because of the increased afterload (pulmonary arterial hypertension). The left atrium and ventricle hypertrophy eccentrically because of the increased preload (increased volume of blood returning from the pulmonary circuit). For abbreviations, see Fig. 6.34.

the inlet to the pulmonary trunk leads to flow from right ventricle through the ventricular septal defect into the aorta. The animal attempts to compensate for the chronic hypoxemia by producing more red cells (Fig. 6.37). The animal is therefore usually stunted, cyanotic and polycythemic with a reduced exercise tolerance. This is a not uncommon condition in dogs, of which Keeshunds are overrepresented.

Stenoses and insufficiencies

(Associated with anomalous valvular development)

Semilunar valvular stenosis is due to anomalous development of the endocardiac jelly components associated with forming the semilunar valve.

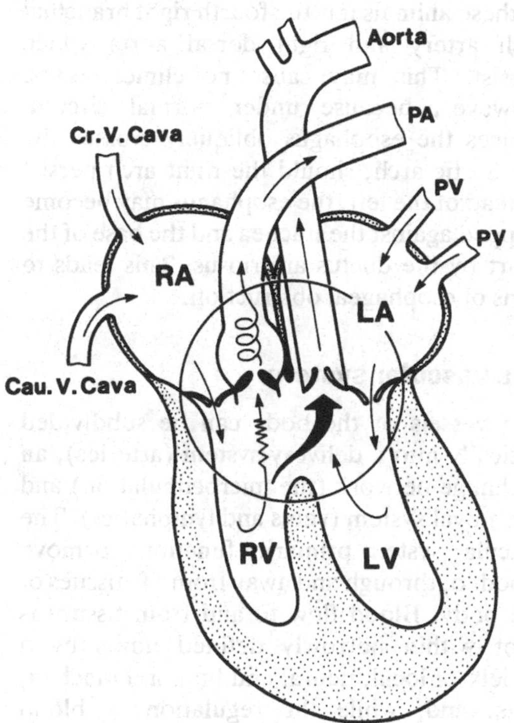

Fig. 6.37. Tetralogy of Fallot. The severity of the pulmonic stenosis determines the predominant direction of flow. If severe pulmonic stenosis is present, flow is into the aorta via the ventricular septal defect. The right ventricle hypertrophies concentrically because of the increased afterload placed on it. For abbreviations, see Fig. 6.34.

Pulmonic stenosis

This is a not unusual occurrence in dogs (particularly Bulldogs, Beagles, Poodles and Fox Terriers). The reduced diameter of the pulmonary valve results in an increased afterload on the right ventricular wall, which concentrically hypertrophies. There is post-valvular dilation of the main pulmonary artery, due mainly to an increased velocity of flow into that great vessel. Turbulent flow may also contribute to the dilation (Fig. 6.38). The murmur of pulmonic stenosis is usually systolic, high pitched and located over the left base of the heart.

Aortic stenosis or subaortic stenosis

This is a less common lesion but is the left-sided equivalent of pulmonic stenosis.

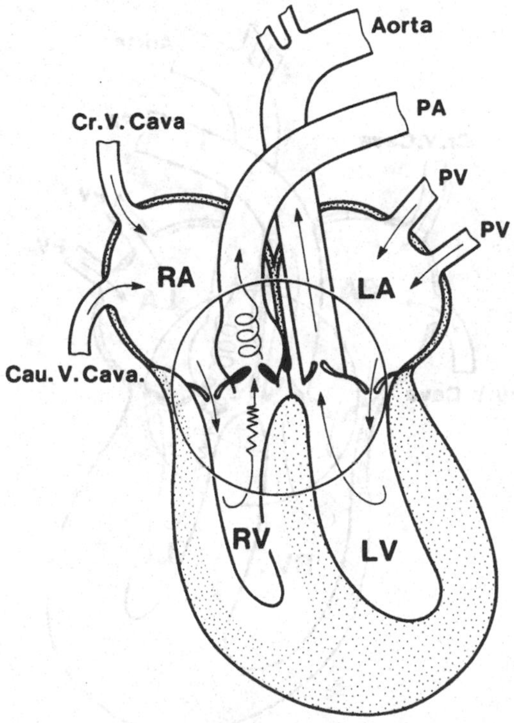

Fig. 6.38. Pulmonic stenosis is usually valvular and places an increased afterload on the right ventricle, which undergoes concentric hypertrophy. There is post-stenotic dilation of the pulmonary artery. For abbreviations, see Fig. 6.34.

Because of the resistance to left ventricular outflow, an increased afterload is placed on the left ventricle. Poststenotic dilation of the ascending aorta is also a prominent feature (Fig. 6.39). Because of the increased left ventricular afterload and the relatively poor supply of blood to the coronary arteries, subendocardial hypoxia occurs. This often leads to sudden death following ventricular arrhythmias. The murmur associated with aortic stenosis is systolic, often late systolic and heard over the left base of the heart. Aortic stenosis is also characterized by a weak pulse strength following low systolic pressure because of restricted aortic outflow.

Valvular insufficiency
Malformation of the left AV valve complex, whilst unusual in most species, is a common cardiac anomaly in the cat, as is dysplasia of the right AV valve. Both result in mild to massively dilated atria and an increased preload on their respective ventricles and are characterized by mixed frequency holosystolic murmurs.

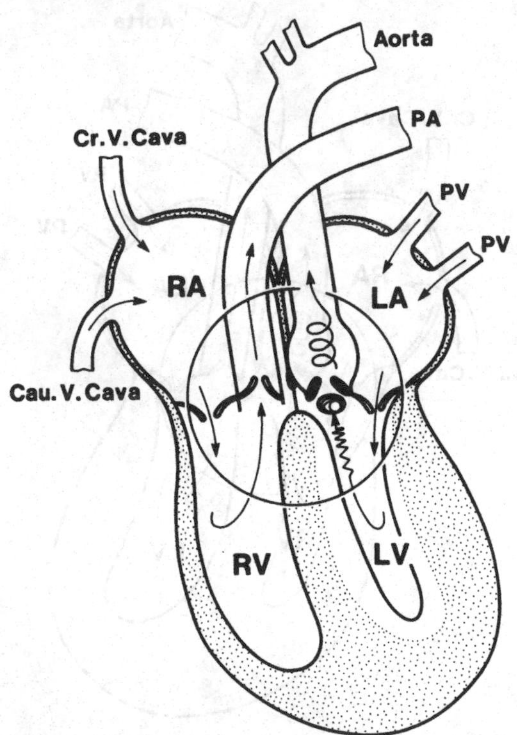

Fig. 6.39. Aortic and subaortic stenosis places an increased afterload on the left ventricle, which concentrically hypertrophies. There is poststenotic dilation of the aorta. For abbreviations, see Fig. 6.34.

Endocardial cushion defects (common atrioventricular canal)
Failure of adequate development of the endocardial cushions, which contribute to the formation of the AV valves and the atrial and interventricular septums, results in what is termed a common AV canal. Defects of this nature are among the most common in the cat and the pig.

Vascular ring anomalies

Persistent right aortic arch
In these animals it is the fourth right branchial arch artery and right dorsal aorta which persist. This may cause no clinical signs. However, because under normal circumstances the esophagus obliquely crosses the left aortic arch, should the right arch persist instead of the left, the esophagus may become trapped against the trachea and the base of the heart by the ductus arteriosus. This leads to signs of esophageal obstruction.

The vascular system

The vessels of the body can be subdivided basically into a delivery system (arteries), an exchange network (the microcirculation) and a removal system (veins and lymphatics). The vascular system primarily functions to move blood to, through and away from the tissues of the body. Blood flow to and from tissues is kept within narrowly defined limits by a variety of local, neural and humoral mechanisms, and, while the regulation of blood pressure and flow is of importance in many instances, it appears to be of little consequence in primary vascular diseases affecting domestic animals. The major sequelae of vascular disease are **obstruction to flow**, the **development of edema** and **hemorrhage**.

Ischemia

A diminution in, or a cessation of, blood flow to one or many tissues is termed **ischemia**, in other words, a failure to deliver adequate amounts of blood to meet a particular organ's metabolic needs. Ischemia is the province of arteries and arterioles and may be localized or generalized. Ischemia may result from intense vasoconstriction following a profound loss of blood, the ingestion of vasoconstricting agents, and widespread vascular microthrombosis. It may also, and more usually, follow physical obstruction of an artery supplying particular tissues. In this case the cause is organic vascular disease.

Whether clinical signs of localized ischemia develop depends on the relationship between (1) the caliber of the vessel involved and (2) the presence or absence of vascular anastomoses.

In general terms, clinical signs associated with ischemia are more likely to develop as the caliber of the obstructed vessel increases and the degree of vascular anastomoses decreases. The outcome of this relationship varies from organ to organ. In those with a highly anastomotic vascular network, the caliber of the vessel obstructed needs to be comparatively large, whereas in those with a poorly developed network the vessel obstructed need only be small. The organ affected is also of some importance. For example, a small area of ischemia in the heart or brain may produce severe clinical signs or death, but in contrast large areas of ischemia of the renal cortex may pass unnoticed.

The causes of localized ischemia are usually either thrombi or emboli. A thrombus that develops at a site of endothelial damage may be non-occluding, partially occluding or occluding, whilst an embolus almost by definition is occluding. An embolus travels in the bloodstream until it reaches a vessel too small to traverse. In fact, almost all emboli are thromboemboli; that is, fragments of a thrombus. A common example of a thromboembolus occurs in cats with cardiomyopathy. In this disease thrombi are often present in the left atrium, fragments of which break off, become emboli and commonly lodge in the abdominal aorta or an iliac artery.

Fortunately, there are few agents that cause ischemia and not all result in the development of clinical signs. Further, the most common are localized and produce well-defined clinical syndromes.

In general, the clinical signs produced reflect the organ involved. There are, for example, two syndromes that produce unilateral or bilateral hindlimb deficits and they are aortic/iliac thrombosis in the horse and the previously mentioned in aortic iliac thromboembolus in the cat. Both result in lameness and often paralysis of one or both hindlimbs. The affected limbs are cold and often painful, with the absence of a palpable pulse.

Thrombi or thromboemboli arising from an arteritis of the cranial mesenteric artery of horses may occasionally occlude the arterial supply to the intestine. The cause is larvae of the nematode *Strongylus vulgaris*. Clinical signs observed in these cases vary from an intermittent colic to a severe unremitting colic, which usually results in the death of the animal. Among the parasitic causes of ischemia probably none is more common than *Dirofilaria immitis* in the dog. In this infection, adult worms often obstruct major pulmonary arteries giving rise to areas of compromised but usually not infarcted lungs. Clinical signs related to pulmonary arterial blockage include a cough and occasionally the coughing up of blood (hemoptysis).

A rare, somewhat multifocal ischemia is observed in dogs with polyarteritis nodosa. The most common signs are neurologic, following thrombosis of meningeal arteries.

Peripheral ischemia may occur in cattle with the ingestion of vasoactive agents such as ergotamine from fungally infected pastures. Affected cattle often exhibit ischemic necrosis of the extremities.

Edema

Edema is the excessive accumulation of fluid in extravascular spaces. Once again, as with ischemia, edema may be localized or generalized and, whilst edema cannot usually be ascribed to just one arm of the vascular system as is ischemia, some indication of the type of vessel involved can be gained from the distribution of the edema.

Edema may be classified broadly into two groups. One that arises following alterations in capillary or lymphatic pressure (non-inflammatory) and the other following an increase in capillary permeability (inflammatory edema).

Non-inflammatory edema

The flux of fluid across the vessels of the capillary bed is a balance of osmotic, capillary and interstitial hydrostatic pressure. Changes in osmotic pressure are of importance from a differential diagnostic point of view. A decreased plasma osmotic pressure is primarily the result of a decrease in plasma albumin concentration, which in turn may follow either a decreased hepatic production or an excessive loss of albumin from a number of tissue sites. Excessive albumin loss includes protein losing nephropathies and enteropathies. Decreased production occurs with chronic hepatic failure or from a prolonged negative dietary protein balance. The resulting edema is generalized, but is not due to a primary vascular problem.

Alterations in capillary hydrostatic pressure come in a variety of guises with a number of inciting causes, most commonly an increase in venous hydrostatic pressure. Among the most common is edema due to congestive heart failure or physical obstruction of venous outflow by for instance a thrombus in a large vein. The edema observed in congestive heart failure is in a special category, as the edema affects either the systemic or pulmonary circulation. Of course, bilateral failure may be present, giving rise to both pulmonary and systemic edema. The pathogenesis of congestive heart failure has been discussed in detail earlier in the chapter.

Physical obstruction to venous outflow in general produces localized edema, the location of the edema being dependent on the vein or veins obstructed. However, because of the abundant collaterals within the venous system, there needs to be an obstruction of a large draining vein for edema to occur. Physical venous obstruction is usually due to venous thrombosis following inflammation of veins (phlebitis) or a stagnation of venous flow. Phlebitis has a number of causes including focal bacterial infections. The other major cause of physical obstruction to venous outflow is veno-occlusive disease, which is seen in some cases of severe hepatic fibrosis. Edema in this case is centered on the peritoneal cavity with the accumulation of ascitic fluid.

The final disturbance of the balance of fluid flow across capillaries is an increase in interstitial pressure following obstruction of the flow of lymph. Again, the resultant edema is usually of a localized nature. Probably the most spectacular form of this type of edema is seen in the rare cases of congenital absence of lymphatics, where edema of massive proportions may be reached. There are also specific, but relatively rare, examples of inflammation of lymphatics (lymphangitis) giving rise to edema, particularly of the limbs. A specific entity in the dog of obstruction of intestinal lymphatics gives rise to a protein-losing enteropathy.

Inflammatory edema

The consequences of the acute inflammatory reaction are well known to all students of pathology. It remains to emphasize that edema is an important facet of that response, but, in contrast to non-inflammatory edema, inflammatory edema is usually accompanied by the other cardinal signs of inflammation, namely heat, redness, pain and loss of function.

Hemorrhage

The clinical manifestations of hemorrhage associated with primary vascular disease depend on the type of vessel affected. Sudden massive hemorrhage occurs after rupture of a large artery or vein and the site and cause are usually readily ascertained, either while the animal is alive or more frequently on necropsy. For instance, horses on rare occasions rupture the aorta at its root, giving rise to massive intrapericardial and sometimes intrathoracic hemorrhage.

Much more common is primary damage to capillary endothelium from a variety of causes, but particularly systemic viral and bacterial infections. While there are usually accompanying systemic signs of these diseases, petechiae may be prominent. The development of petechiae derives from primary endothelial damage and the widespread activation of platelet aggregation and the clotting sequence. This widespread consumption of platelets and clotting factors is termed disseminated intravascular coagulation (DIC). Prominent causes of direct endothelial damage leading to DIC are canine adenovirus I infection, hog cholera virus infection and Gram-negative septicemias. As is well known, DIC due to other causes unaccompanied by endothelial damage can also lead to the appearance of petechiae.

Pathologic and etiologic patterns of vascular disease

The vessels of the body, as with other systems, exhibit the spectrum of changes common to other organs. There are prominent examples of degenerative, inflammatory and neoplastic change, but the pathologic pattern observed depends to a large extent on the anatomic structure of the vessel and for this reason reactions will be discussed under the three broad headings of arteries, veins and lymphatics.

Arteries

Degenerative changes in arteries, although of profound significance in man, are of little clinical consequence in domestic animals. There are two terms applied to the most common of degenerative changes.

The term **arteriosclerosis** (hardening of the arteries) is applied when an artery is hardened, has lost its elasticity and the lumen is narrowed from a combination of proliferative and degenerative changes affecting the media and intima. **Atherosclerosis** is the above combined with degenerative fatty changes. The major consequence of atherosclerosis is breaching of the endothelium leading to thrombus formation, which, as indicated above, is rare in domestic animals. The disease in man is multifactorial but includes diet, genetic predisposition, and such additions as smoking and possibly stress.

Hyaline degeneration of arteries is a common microscopic finding in a number of diseases with differing pathogeneses. The term hyaline refers to the amorphous appearance of the intima and media of a muscular artery or arteriole, with loss of the normal cellular structure. It is also sometimes seen with arteritis. The hyaline material is in some cases deposited plasma protein and in other cases a necrosis of vascular smooth muscle. In either case, the diagnostic usefulness of hyaline degeneration lies in the diseases with which it is regularly associated. This naturally varies with the species, but some common diseases with widespread hyaline changes include edema disease, hepatosis dietetica, mulberry heart disease and cerebrospinal angiopathy in pigs, and uremia in dogs.

Hypertrophy of the smooth muscle of muscular arteries and arterioles in the lungs is an identifiable antecedent to right-sided heart failure in cattle exposed to high altitudes and is also part of congenital cardiac diseases where pulmonary overperfusion occurs. Examples are the left to right shunts of PDA, atrial and ventricular septal defects. Medial hypertrophy of pulmonary arteries is also sometimes seen in cats, where there are no attendant clinical signs and the etiology is unknown.

Inflammation of arteries (arteritis) results from a variety of causes, with the clinical significance of this change relating to the development of thrombosis and ischemia. The inflammatory change may affect predominantly the endothelium, as in bacterial septicemias such as salmonellosis and erysipelas and in viral diseases such as hog cholera and

African swine fever. A predominantly medial and adventitial change is observed in a number of viral diseases such as equine viral arteritis and malignant catarrhal fever. Some fungi also have a propensity to invade arteries, producing arteritis with thrombosis.

Notwithstanding these causes, parasitic (verminous) arteritis is of considerable significance, with prominent examples being dirofilariasis in dogs and *Strongylus vulgaris* infection in horses. *Dirofilaria immitis* is a filarid parasite that lives in the main pulmonary arteries and the right ventricle of the most commonly affected species, the dog. The life cycle is well defined, with circulating microfilariae ingested by mosquitoes of various genera. After development in the intermediate hose, microfilariae in the mouthparts of the mosquito are injected into a susceptible host when the mosquito feeds. Some three to four months later, following development in the subcutis and muscle, the immature parasites reach the right ventricle. After a further two months of maturation, the female parasite is capable of producing microfilariae.

The clinical signs of dirofilariasis are related to the parasite's effect on the pulmonary arteries. They are two-fold. The first is progressive pulmonary arterial narrowing and obliteration, with accompanying hypertension. This leads to the development of right ventricular hypertrophy and, in severe cases, right-sided heart failure. The affected pulmonary arteries exhibit florid myointimal proliferation, giving a characteristic 'shaggy' or villous appearance to the arterial lining. The second major effect is sometimes massive pulmonary arterial thrombosis. It is produced when either live or dead worms become entrapped in the smaller arteries. This especially occurs after adulticide therapy. This particular event is heralded clinically by bouts of coughing and, in severe cases, hemoptysis.

The verminous arteritis associated with *S. vulgaris* infection is but one aspect of the clinicopathologic pattern seen in this disease. However, arterial lesions seen serve as an example of verminous arteritis. As part of its life cycle, the fourth stage larvae of *S. vulgaris* migrate retrogradely along or in the intestinal arteries until the cranial mesenteric artery is reached. Lesions within this artery vary from small intimal tracks to large occluding thrombi, and in many cases larvae are still present in the lesions. Arterial lesions are not confined to the cranial mesenteric artery and may be found in the aorta, renal and other arteries.

Although this verminous arteritis is quite common in horses, clinical signs resulting from it are comparatively rare. When present, colic of varying severity results from the effects of large intestinal ischemia and rarely infarction.

There is finally a group of conditions that have an immunologically mediated arteritis as one of their features; the pathogenesis is discussed in the chapter on the immune system.

Veins

As previously discussed, venous disorders of clinical significance are those relating to thrombosis or other obstruction to venous flow leading to congestion and edema.

Thrombosis of veins (phlebothrombosis) and inflammation of veins (phlebitis) often occur together and may be due to direct effects on the endothelial lining of veins by agents such as bacteria or by extension of an inflammatory lesion in adjacent tissues. Probably the most important specific causes of thrombophlebitis in certain areas of the world are parasitic in nature. Schistosomiasis occurs commonly in central Africa and Asia and to a lesser extent in other countries, with the species involved being *Schistosoma* and *Ornithobilharzia*. The adult trematodes are found in the veins in various areas of the body, depending on the species involved. Whilst lesions of chronic phlebitis are caused by the adult parasites, the more important lesion results from the deposition of parasite eggs in small venules. An intense granulomatous inflammatory response ensues.

Lymphatics

Because of their nature and function, lymphatics are not only involved with primary disease, but are also inevitably caught up in a wide range of disease processes such as inflammation and neoplasia. As is well known, many tumors metastasize via lymphatic channels. In such circumstances, it is only when the larger lymphatics become obstructed that lymphedema is observed.

There are, by contrast, a miscellany of congenital and acquired diseases of lymphatics where the outstanding lesions are limited to lymphatic rupture, obstruction or inflammation.

Congenital lymphedema, either localized or generalized, occurs rarely in most species. Lymphatics may be absent (aplastic), hypoplastic or hyperplastic and dilated. Lymphedema in the latter follows valvular incompetence. In all cases, peripheral and sometimes central lymph nodes may be hypoplastic or absent.

Lymphangitis occurs in the hindlimbs of horses under the title sporadic lymphangitis. While the cause appears to be bacterial and a variety of pyogenic organisms have been isolated from affected cases, there is doubt about their primary involvement. The typical clinical pattern is of acute lameness, hindlimb edema and prominent, irregularly hard and swollen lymphatics. Draining lymph nodes are enlarged, with larger lymphatics containing inflammatory exudate. **Ulcerative lymphangitis** is also a disease of the horse, and occasionally of cattle, primarily by infection with *Corynebacterium ovis*. It is characterized by progressive inflammation of subcutaneous lymphatics with periodic focal abscessation and ulceration of the overlying skin. **Epizootic lymphangitis** of horses is caused by *Histoplasma farcinimosum* and has a clinical pattern similar to that of ulcerative lymphangitis. However, it may also involve the nasal mucosa and conjunctivae. This disease is only of regional significance.

Rupture of lymphatics is really only of significance when the thoracic duct is involved. The result is chylothorax, an accumulation of lymph within the thoracic cavity. This must be distinguished from a number of other causes of pleural effusion. The cause or causes of chylothorax include trauma or malignancy, but in many cases it is unknown.

Additional reading

Antoni, H. (1983). Function of the heart. In *Human Physiology*, R. I. Schmidt and G. Thews (eds.), pp. 358–96. Berlin, Springer-Verlag.

Braunwald, E. (1980). *Heart disease. A Textbook of Cardiovascular Medicine*, vols 1 and 2. Philadelphia, W. B. Saunders Co.

Else, R. W. (1980). Clinico-pathology of some heart diseases in domestic animals. In *Scientific Foundations of Veterinary Medicine*, A. T. Phillipson, L. W. Hall and D. R. Pritchard (eds.), pp. 328–49. London, William Heinemann Medical Books Ltd.

Ettinger, S. J. (1983). *Textbook of Internal Veterinary Medicine. Diseases of the Dog and Cat*, 2nd edn. Philadelphia, W. B. Saunders Co.

Ettinger, S. J. and Suter, P. F. (1970). *Canine Cardiology*. Philadelphia, W. B. Saunders Co.

Guyton, A. C. (1971). *Textbook of Medical Physiology*, 4th edn. Philadelphia, W. B. Saunders Co.

Hurst, J. W. (1978). *The Heart, Arteries and Veins*, vols 1 and 2, 4th edn. New York, McGraw-Hill Book Company.

Kunze, R. S. and Wingfield, W. E. (1981). Acquired heart disease. In *Pathophysiology of Small Animal Surgery*, M. J. Bojrab (ed.), pp. 178–84. Philadelphia, Lea and Febiger.

Patterson, D. F. (1976). Congenital defects of the cardiovascular system of dogs: studies in comparative cardiology. *Adv. Vet. Sci. Comp. Med.* **21**: 1–35.

Robinson, W. F. and Maxie, M. G. (1985). The cardiovascular system. In *Pathology of Domestic Animals*, vol. 3, 3rd edn, K. V. F. Jubb, P. C. Kennedy and N. Palmer (eds.), pp. 1–81. Orlando, FL, Academic Press.

Rudolph, A. M. (1974). *Congenital Diseases of the Heart*. Chicago, Year Book Medical Publishers Inc.

Schramroth, L. (1982). *An Introduction to Electrocardiography*, 6th edn. Oxford, Blackwell Scientific Publications.

Sodeman, W. A. and Sodeman, T. M. (1979). *Pathologic Physiology – Mechanisms of Disease*, 6th edn. Philadelphia, W. B. Saunders Co.
Van der Werf, T. (1980). *Cardiovascular Pathophysiology*. Oxford, Oxford University Press.
Vick, R. L. (1984). *Contemporary Medical Physiology*. Menlo Park, CA, Addison-Wesley Publishing Company, Medical Division.

John R. Bolton and David A. Pass

7 The alimentary tract

Compared with the other organ systems of the domestic animals, the alimentary system is remarkable for its degree of diversity between the species. Despite marked differences in diet, anatomy and digestion, the alimentary tract of each species manages to extract from the material ingested the basic nutrients necessary for maintenance, growth, work, pregnancy and lactation. The alimentary tract also eliminates the indigestible dietary components as well as some of the animal's waste by-products. Again, there is a remarkable diversity in the shape and consistency of this final product.

The volume of information concerning the normal and diseased alimentary system of the domestic species is increasing daily. An in-depth coverage of all this information is obviously beyond the scope of this chapter and beyond the level necessary for preclinical veterinary students. Although it was considered necessary to pay particular attention to some specific species problems such as dysfunction of the ruminant stomachs, an overall attempt has been made to present the general principles of alimentary pathophysiology.

The alimentary tract is usually separated, for the sake of discussion, into upper and lower sections, the upper tract comprising the mouth and esophagus and the lower tract comprising the stomach and small and large intestine. Discussion in this chapter will be directed primarily toward dysfunction of the lower tract. Whilst there are numerous diseases of the upper tract that are of great importance, such as the vesicular diseases, most are species specific and are adequately covered elsewhere.

The oropharynx and esophagus

Disease of the oropharynx is usually manifested clinically by inappetance, excessive salivation, difficulty in swallowing, retching and halitosis. Fortunately, the basis of these clinical signs is usually revealed by a thorough physical examination. Lesions may be present on the oral mucosa, gingiva, periodontal tissues or may involve the teeth. Most of these abnormalities are acquired, although congenital lesions such as cleft palate do occur.

Excessive salivation, anorexia and pain on eating are the major clinical signs associated with inflammation of the oral cavity (stomatitis), which may be caused by a variety of chemical, physical or infectious agents. There are numerous examples of infectious diseases in which stomatitis is a prominent lesion, including feline rhinotracheitis virus and calicivirus infections, the vesicular stomatitides (foot and mouth disease, vesicular exanthema and vesicular stomatitis) and the erosive stomatitides (bluetongue, virus diarrhea, rinderpest and malignant catarrhal fever). Primary oral bacterial infections are uncommon, with most bacterial stomatitides resulting from secondary infection of traumatized or inflammed mucosa, particularly that of the gingiva. Probably the most specific bacterial stomatitides are those due to *Actinobacillus lignieresi* in cattle ('wooden tongue') and *Fusobacterium necrophorum* in calves, deer and kangaroos.

Clinical signs from dental disease are not uncommon and irregularities of wear are of significance in grazing animals in some areas of the world. Dental disease in companion animals includes the accumulation of masses of bacteria which firmly adhere to the teeth (dental plaque) and may ultimately mineralize (dental calculus, tartar). There are also examples of tumors arising from various tissue components of the mouth such as fibrosarcoma and those of dental origin including epulis and ameloblastic odontomas.

The basis of normal deglutition

While the act of swallowing appears outwardly to be a relatively simple affair, it is in fact a quite complex reflex that can be divided into the five stages outlined below:

In the **oral stage**, food that has been taken into the mouth and masticated is formed into a bolus at the base of the tongue. This stage of swallowing is voluntary and may be affected by abnormalities of the oral cavity, tongue and trigeminal (nucleus and cranial nerve V), facial (nucleus and cranial nerve VII) or hypoglossal (nucleus and cranial nerve XII) nerves.

The **pharyngeal stage** is involuntary and is initiated when a bolus of food is propelled into the pharynx by the base of the tongue. During this stage, the internal nares and larynx are sealed and food is moved towards the esophagus. In addition to a normal tongue, this stage is dependent on a normal pharyngeal and laryngeal conformation, and on normal glossopharyngeal (nucleus and cranial nerve IX) and vagus (nucleus and cranial nerve X) nerves.

During the **cricopharyngeal stage**, the bolus of food moves through the pharyngesophageal junction, also referred to as the upper esophageal sphincter. The activity of this sphincter is controlled by the vagus nerve. It is normally actively closed. When the animal is swallowing, inhibition of the spontaneous vagal discharges allows the sphincter to open to a diameter that is proportional to the size and consistency of the entering bolus.

The passage of the bolus into the esophagus starts the **esophageal stage**. The bolus is propelled down the esophagus by an uninterrupted peristaltic wave during this stage. This depends on normal esophageal musculature and on intact vagal and sympathetic innervation. There are marked species differences in the relative proportion of esophageal striated and smooth muscle. This has significant clinical implications in motor end plate diseases such as myasthenia gravis and with some snake envenomations.

In the final **gastroesophageal stage**, the lower esophageal sphincter opens to allow food to pass into the stomach. This stage is dependent on the same factors as the preceding stage.

Clinical features and pathologic mechanisms of abnormal deglutition

Any inability effectively to swallow food is termed **dysphagia** and it may have its source at any of the five stages described above. The clinical evaluation of dysphagia must take this into account. The oropharynx may be examined fairly easily for the presence of structural lesions or foreign bodies. If none can be detected, a neurologic evaluation of the relevant cranial nerves should be performed, as described in Chapter 13.

Animals with dysphagia should be examined for signs of nasal regurgitation of feed or aspiration pneumonia. Both findings may be an indication of partial or complete failure of the pharyngeal stage of deglutition.

Impairment of the cricopharyngeal, or third, stage of deglutition is brought about by failure of the upper esophageal sphincter to open fully. This failure to open is termed **achalasia**, and usually has a neurogenic basis.

Failure of the esophageal stage may result from primary degeneration of the esophageal muscle, as in canine polymyositis, from dys-innervation of the muscle, or from achalasia of the lower esophageal sphincter. The net result is atony and flaccidity of the esophagus, which becomes increasingly impacted and dilated. This condition is referred to as **megaesophagus**. In most instances, megaesophagus is an idiopathic disease and is well recognized in Great Dane, German Shepherd and Irish Setter dogs and in Siamese cats. Affected animals are in poor condition and characteristically regurgitate food at varying times after eating. As with the failure of the pharyngeal stage, food may pass through the internal nares or into the trachea, which may give rise to secondary rhinitis and pneumonia. Should the lower sphincter be incompetent, secondary esophagitis may be induced by bacterial fermentation of food retained in the esophagus and by the reflux of gastric acid and bile. Megaesophagus can be usefully demonstrated by the use of contrast radiography, a major aid in diagnosis.

Esophageal function may also be impaired by internal and external obstructions. Internal obstruction or 'choke' is usually due to swallowed foreign bodies or improperly chewed food, such as potatoes, corn-cobs and turnips. The sites at which foreign bodies lodge are usually where the esophagus deviates or is normally restricted, such as over the larynx, at the thoracic inlet, at base of the heart, and cranial to the diaphragmatic esophageal hiatus. External compression may be caused by space-occupying lesions in the neck or mediastinum, but in the dog are most commonly associated with developmental 'vascular ring' anomalies.

The monogastric stomach

Functional anatomy

In the monogastric animal, ingesta is held in the stomach for a time to be mixed and partially digested under acid conditions before being passed on to the small intestine. The factors most critical to normal gastric function relate to motility, the patency of the pyloric outlet and the structural and functional integrity of the mucosal lining.

The mucosa, it will be remembered, has a non-glandular esophageal region and a tri-partite glandular zone comprising the cardiac, fundic and pyloric regions. The relative dimensions of these regions vary between the species, and in the ruminants the esophageal region is immensely developed to form the forestomachs.

In basic terms, there are two distinct regions of gastric motility. The intrinsic motility of the smooth muscle in both regions is modulated primarily by neurotransmitters and gastrointestinal hormones. The pyloric region of the stomach acts like a pump regulated by a pacemaker located on the greater curvature of the gastric corpus. The regular rhythmic contractions of this pump triturate digesta and move small particles into the duodenum. Larger particles are 'sieved' by the pump, being retained in the stomach for further size reduction and digestion. The fundic region of the stomach serves as a reservoir for ingested food. Slow, phasic contractions of this region, under the influence of the vagus nerves and the gastrointestinal hormone gastrin, move digesta into the pyloric region. Although not proven for all domestic species, the emptying of liquids from the stomach is regulated by contractions of the fundic region, occurring independently of the pyloric contractions.

Gastric secretions are provided by the glandular portions of the mucosa, particularly the fundic region. The parietal cells of fundic glands secrete hydrochloric acid and the chief cells secrete pepsinogen, which autocatalyzes to pepsin in the ambient acid conditions. Pepsin performs the major digestive function of the stomach by hydrolyzing protein, preparing the way for proteolytic digestion in the intestine.

The secretion of acid is regulated by three agents: histamine, acetylcholine and the hormone gastrin. Histamine is released from mast cells, located in the lamina propria adjacent to parietal cells, and acetylcholine from postganglionic parasympathetic axons in this region. **Gastrin** is produced by neuroendocrine C cells in the pyloric and duodenal mucosa, and is delivered to parietal cells via the bloodstream. The hormone **secretin** tends to oppose the action of gastrin, inhibiting gastric acid secretion. Secretory mechanisms are activated by cephalic, gastric and intestinal stimuli, such as the sight, smell and taste of food, the presence of digested proteins in the pylorus and duodenum, and by distension of the pylorus.

The acid/pepsin mixture poses a potential threat to the gastric mucosa itself, which necessitates a protective mechanism. This mechanism is multifaceted, its major component consisting of the tight junctions between the gastric mucosal epithelial cells, known as the **gastric mucosal barrier**. In addition, the surface epithelial cells secrete bicarbonate, which provides a chemical buffer system within the medium of the overlying layer of mucus. It appears that the secretion of bicarbonate is stimulated by certain prostaglandins. Gastric mucus is secreted by the surface epithelium of the entire glandular stomach and by the glands of the cardiac and pyloric regions.

In addition to bicarbonate, the gastric mucosa also secretes sodium, potassium and chloride ions, whose secretion may become highly significant in gastric disease.

Clinical expression of gastric dysfunction

There is a collection of signs which alerts the clinician to the possibility of gastric disease.

A monogastric animal with significant gastric disease will usually exhibit abdominal pain of varying degree, loss of weight, a depressed appetite and vomiting (in the species able to do so). Not infrequently, there is anemia, caused by blood loss or occasionally because of impaired iron absorption. Severe intragastric bleeding will result in vomiting of blood (hematemesis) or the passage of partially digested blood in the feces (melena).

Abdominal pain of gastric origin is due to the release of chemical mediators in inflammatory lesions, particularly areas of ulceration. Characteristically, it occurs shortly after eating, in association with an increase in gastric acid secretion. **Loss of body weight** results from impaired digestion, a reduction of food intake and from the exudation of plasma protein, which may be considerable in some diseases. **Intragastric leakage of blood** may occur insidiously from neoplasms or chronic inflammatory lesions. At the other end of the scale, quite large vessels may be eroded by ulcerative lesions and the resultant hemorrhage may be severe enough to cause rapid death. Blood loss may occur at any level across this spectrum of intensity. In some gastric diseases, abdominal distension may occur, particularly in the case of acute tympany or torsion. **Severe dilation of the stomach** can be associated with acute shock and circulatory failure and is discussed later under gastrointestinal obstruction (see p. 189).

Vomiting is a reflex act, under the control of the medullary vomiting center, and may therefore be unrelated to primary gastric disease. It is common in the dog, cat and pig, but is rarely seen in the horse. The vomiting center may be stimulated directly via autonomic nerve impulses, or indirectly via a closely associated chemoreceptor trigger zone. An automatic stimulus may be provided by fear, pain or excitement, or irritation of the peritoneum or abdominal organs. The chemoreceptor zone may be stimulated by toxins and drugs, or abnormal motion (as in motion sickness). **Vomiting on its own therefore does not confirm a primary gastric lesion**. It must also be distinguished from regurgitation, which is a relatively passive process resulting from overloading of the stomach, or obstruction of the lower esophageal sphincter. Vomiting by contrast is a vigorous, or even violently active, process and the gastric contents may be expeled with considerable force and much exertion. The frequency and force of vomiting and the character of the vomitus are features which may be of value in determining the ultimate cause.

Profuse and persistent vomiting can have serious metabolic consequences, primarily due to the loss of water, hydrogen and chloride ions from the body. This results in dehydration, hypovolemia, hypochloremia and metabolic alkalosis. There may also be hypokalemia due to both direct loss of potassium in vomitus and loss via the urine. While the urine generally reflects the alkalotic state in being alkaline itself, there sometimes occurs a paradoxical aciduria in dogs and cattle. The detail of these electrolyte disturbances is discussed in Chapter 4.

Pathologic mechanisms of gastric disease

Clinical disease due primarily to gastric dysfunction is not particularly common in monogastric domestic animals, but common enough to warrant this discussion. The disorders will be discussed according to the underlying pathologic processes (Fig. 7.1).

Gastric ulceration

Clinical disease due to gastric ulceration is most common in the foal and the pig, and is occasionally seen in the dog. It should also be pointed out that gastric ulcers are commonly found at necropsy but have usually been without significant clinical effect. These lesions are usually regarded as 'stress' ulcers, related in some way to the effects of glucocorticoids, which have a potentiating effect on other ulcerogenic agents.

Various prostaglandins (PG) have been found to play a critical role in ulcerogenesis, particularly prostacyclin (PGI_2). Such agents stimulate secretion of bicarbonate into the mucus layer and also seem to have a general protective effect on the gastric mucosa, but the mechanisms are poorly understood. In this context, drugs that inhibit prostaglandin synthesis, such as salicylates and other non-steroidal anti-inflammatory agents, are potent ulcerogens. Agents such as detergents and lipophilic substances also promote ulceration by direct cytotoxicity to gastric epithelium. Once the gastric mucosal barrier is breached, acid and pepsin continue the attack and the lesion proceeds to expand into the depths of the mucosa and beyond. Ulceration and inflammation, with resulting pain and all the other consequences may ensue, as previously discussed. The most serious outcomes of gastric ulceration are erosion of a large blood vessel, or perforation of the wall and the initiation of peritonitis. Although most heal rapidly if appropriate therapy and management are instituted, the healing of very deep ulcers may leave some residual distortion due to cicatrization.

In the pig, massive fatal hemorrhage from deep ulcers of the cardiac zone is a common occurrence. In this species, gastric ulceration has been related to dietary and genetic factors.

In the dog, gastric ulcers are produced by aspirin administration and may occur in association with mast cell tumors, which secrete histamine into the bloodstream. Bleeding from gastric ulcers often causes hematemesis and melena in the dog.

In foals, vomiting cannot occur, and gastric ulceration is manifested as illthrift and post-prandial pain, which is expressed as grinding of the teeth (bruxism) and salivation. The mechanism of ulcer production in foals is not understood but some are associated with the administration of non-steroidal anti-inflammatory drugs.

Gastritis

Primary gastritis, with diffuse inflammation of the gastric mucosa is a rare disorder. Acute bouts of vomiting mostly reflect the ingestion of contact irritant or allergenic substances which are rapidly rejected and ejected before significant inflammation can develop. In dogs, there have been a few reports of a chronic hyperplastic gastritis, analogous to Menetrier's disease in humans. The cause is unknown.

Gastric retention
This term refers to a chronic condition resulting from obstruction of the pylorus. Eating is followed by dilation of the stomach and then, in some species, vomiting or regurgitation. The vomitus consists of scarcely digested food and contains no bile. The pylorus may be blocked by material in the lumen (obturation), because of lesions within its wall, or from lesions compressing it externally. Obturation results from foreign bodies or mucosal polyps, while intramural lesions include chronic ulceration and fibrosis, neoplasms, or muscular hypertrophy. Congenital pyloric stenosis is also a well-recognized disorder in puppies.

Gastric neoplasia
Mucosal, stromal and lymphoid neoplasms may arise within the stomach and produce clinical signs of chronic gastric dysfunction. The lesions are frequently complicated by secondary ulceration and will clinically mimick a chronic gastritis. As mentioned above, obturation of the pylorus may occur. Clinical diagnosis is difficult and relies upon radiography, endoscopy and exploratory surgery.

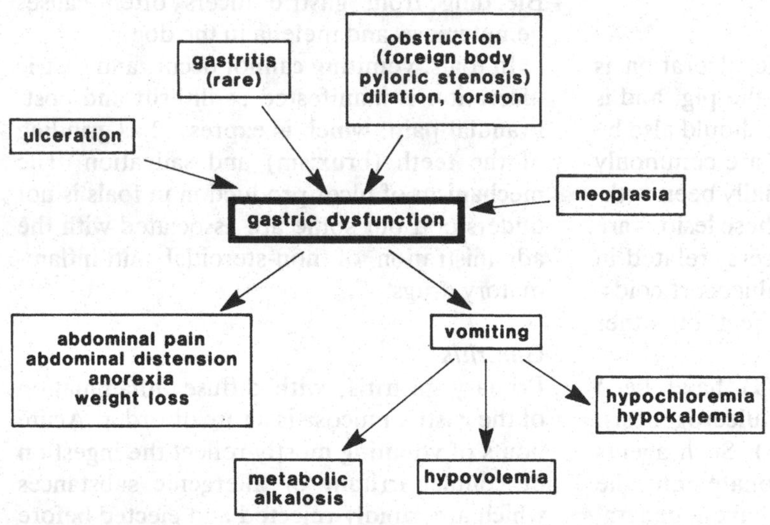

Fig. 7.1. Primary gastric dysfunction may arise from a number of causes. A common set of clinical signs is produced, the most notable being vomiting. However, it should be remembered that the act of vomiting is not only due to primary gastric disease and may be due to other systemic diseases such as uremia or to stimuli such as fear.

The ruminant stomachs

The normal function of the ruminant forestomachs depends upon complex interactions between the motility of these organs and the microbial population of the luminal contents. Although the ruminant digestive system is capable of utilizing a wide range of comparatively low-grade nutrients, dysfunctions of fermentation and motility occur commonly in practice, often having a severe adverse effect on the animal.

Whilst it is convenient to consider disorders of fermentation and motility separately for the purpose of discussion, it is usually impossible to do this clinically. For example, neurogenic reticuloruminal stasis is associated with reduced mixing and emptying of ruminal contents and hence a reduction in the ruminal fermentation rate. Conversely, in animals that have not eaten for 2–3 days, the marked reduction in the numbers of ruminal protozoa is associated with a reduction in motility.

Abnormalities of fermentation

Ruminal acidosis

Synonyms for this condition include D-lactic acidosis, grain overload, ruminal overload, grain engorgement and overeating disease. The essential features are depicted in Fig. 7.2.

Acute ruminal acidosis results from the ingestion of highly fermentable carbohydrate, usually in the form of starch-rich grains, fruits, root crops or bakery products. In general the acuteness and severity of the ruminal hyperactivity increases with the quantity and speed of carbohydrate consumption. Animals not properly adapted to a carbohydrate-rich diet are particularly susceptible to the acute form of the disease. A more chronic form is common in dairy and beef cattle fed large quantities of grain for milk production or rapid weight gain.

In the normal rumen, Gram-negative bacteria predominate and many protozoa are present. The fermentation products are approximately 65% acetic acid, 25% propionic acid and 10% butyric acid.

Approximately 2–4 h after the ingestion of excess carbohydrate, Gram-positive cocci (*Streptococcus bovis*) multiply rapidly, utilizing the carbohydrate to produce lactic acid and long-chain volatile fatty acids. The buffering mechanisms in the rumen are rapidly overwhelmed and as ruminal pH falls below 5 (the normal pH ranges from 6.0 to 7.5) protozoa, cellulolytic organisms and lactate-utilizing organisms are destroyed. At pH 4.0–4.5 the streptococci are rapidly overgrown by Gram-positive rods (lactobacilli), whose pH optimum is below 5. By 24 hours these are the most numerous organisms in the rumen and cecum. Additional important microbiologic changes include increased proportions of coliforms and clostridia in the rumen and cecum.

Ruminal motility ceases and the animal becomes anorectic at the time of the initial decrease in ruminal pH. The stasis of the rumen at this time is thought to be due to the

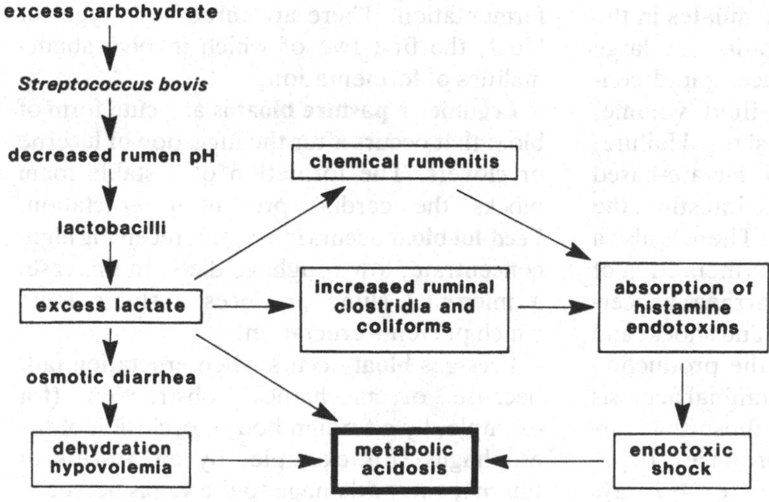

Fig. 7.2. Ruminal acidosis and its consequences as illustrated follows the ingestion of inappropriate amounts of carbohydrates. For details, see the text.

increased ruminal absorption of the long-chain volatile fatty acids rather than to the absorption of lactic acid *per se*. Lactic acid probably provides the acid conditions in the rumen that convert the volatile fatty acids to their more readily absorbable, non-dissociated form. Later in the course of the disease, histamine and endotoxins also play an important role in the inhibition of ruminal motility. Both these substances are produced in the ruminal fluid. Endotoxins are elaborated by the coliforms and clostridia, while histamine is produced by lactobacillary decarboxylation of histidine. Although histamine is poorly absorbed by the normal rumen, significant amounts may cross the acid-damaged epithelium.

Ruminal acidosis is typically associated with a severe metabolic acidosis, dehydration, shock, toxemia and diarrhea. Equal amounts of D- and L-lactate are produced and both forms are slowly absorbed from the rumen. The blood lactate concentration usually reaches a peak 7–24 hours after acute overeating. As the L-lactate is metabolized more rapidly than the D-lactate, the metabolic acidosis is due in large part to the accumulation of the latter.

However, the majority of the lactic acid produced is not absorbed but accumulates in the rumen to cause the sequestration of large quantities of water. This induces rapid contraction of the extracellular fluid volume, dehydration, shock and eventual renal failure. Diarrhea is the result of a lactate-based osmotic overloading of the large intestine, the lactate coming from the rumen. There is also a contribution from the further fermentation of undigested carbohydrate. Diarrhea exacerbates the acidosis, dehydration and shock, and endotoxins also contribute to the production of shock. Animals with acute ruminal acidosis often die in 1–3 days, and, in those that survive, overcompensation may produce a hypochloremic metabolic alkalosis after 4–7 days (see Chapter 4).

A number of additional problems may occur in animals suffering from ruminal acidosis, these being: hypocalcemia resulting from calcium malabsorption, which is in turn caused by the low pH of digesta entering the duodenum in the acute stage of the disease; laminitis resulting from the release of histamine and endotoxins into the circulation; and polioencephalomalacia caused by an induced thiamine deficiency. This has its basis in the destruction of thiamine-producing organisms by the overgrowth of Gram-positive organisms, and in the production of thiaminases in the ruminal fluid. Rumenitis may follow the acute phase of the disease. Various microorganisms, but particularly fungi, readily invade the acid-damaged ruminal epithelium; liver abscesses may develop in animals that have survived the acute phase. Ruminal bacteria, especially *Fusobacterium necrophorum* and *Corynebacterium* spp., reach the portal circulation via the damaged ruminal epithelium and cause the formation of multiple abscesses in the liver. Abscess formation takes weeks or months to occur, and usually causes no clinical evidence of liver disease.

Ruminal tympany or bloat

Bloat occurs when the eructation mechanism fails to expel the gases produced by ruminal fermentation. There are three basic types of bloat, the first two of which involve abnormalities of fermentation.

Legume or pasture bloat is an acute form of bloat that occurs after the ingestion of lucerne or clovers. The formation of a stable foam blocks the cardia, preventing eructation. **Feed-lot bloat** occurs in animals receiving high-concentrate, low-roughage diets. In this case, a microbial slime produces a stable foam which prevents eructation.

Free-gas bloat occurs when eructation fails because of mechanical obstruction (for example, by a foreign body), occlusion of the esophagus (for example, by an abscess or tumor), and/or damage to the vagus nerve.

The basic problem in legume bloat is that feeds such as lucerne and clovers contain cytoplasmic proteins which are adsorbed on to the surface layer of the gas bubbles produced by ruminal fermentation. A stable foam is produced in the rumen, preventing gas bubbles from undergoing the normal process of coalescence, expansion and bursting. The foam blocks the cardia and prevents eructation. The accumulation of gas in the rumen produces a rapid rise in the intraruminal pressure and the bloated rumen causes progressive impairment of cardiopulmonary function by compressing the posterior vena cava, liver and thoracic cavity. Other factors involved in the pathogenesis of legume bloat are the chemical and physical composition of plants and the rate of flow and composition of saliva.

Most legumes are rich in citric, malonic, and succinic acids. These acids react with bicarbonate in the rumen to release large quantities of CO_2 and they also tend to decrease the ruminal pH. Because the isoelectric point of the cytoplasmic proteins is pH 5.4, the foam tends to become more stable as the ruminal pH falls.

Whereas rough, fibrous plant material stimulates recitular motility and the frequency of eructation, legumes, especially the young plants, are soft and relatively non-fibrous. A high intake of legumes will therefore tend to depress the frequency of eructation.

Bloat is less likely to occur in animals with inherently high salivary flow rates as saliva contains bicarbonate, which acts as a ruminal buffer. Bloating is also less likely in animals with inherently high concentrations of salivary mucin, because of its effective antifoaming properties.

Abnormalities of motility

In contrast to other regions of the gastrointestinal tract, the reticulorumen is almost wholly dependent on neural activation for its primary and secondary sequences of contraction. This neural mechanism involves excitatory and inhibitory sensory input from receptors in the reticulorumen, conduction of these signals up the vagus nerves, integration of the signals in the medulla oblongata and conduction of motor impulses back down the vagi to initiate the appropriate contraction sequence. It is therefore convenient to consider the abnormalities of motility on the following anatomical bases: (1) depression of gastric centers; (2) increase in inhibitory reflex inputs; (3) lack of excitatory reflex inputs and (4) blockade of motor pathways (Fig. 7.3).

Depression of gastric centers in the medulla

In addition to their vagal inputs the gastric centers are influenced by other regions of the central nervous system. It is through these influences that depression of gastric centers, and hence of forestomach motility, occurs when ruminants are given general anesthetics or commonly used drugs such as xylazine. This is also the probable basis for the ruminal stasis that occurs when ruminants are in pain, or in a febrile or toxemic state.

Increased inhibitory reflex inputs

Inhibitory reflex inputs to the gastric centers arise from epithelial receptors in the reticulum and cranial ruminal sac. These receptors provoke a transient response to light mechanical stimuli, but a sustained response to extreme distension of the wall and to certain chemicals, the most clinically important of which are acids. The reflex inhibition of forestomach contractions by this sustained response is likely to be an important factor in the ruminal stasis that occurs in severe bloat or in the hyperacidity induced by grain engorgement.

Reflex inhibition of forestomach motility can also be produced by extreme distension of the abomasum or small intestine, as will occur for example when the abomasum undergoes right-sided torsion and becomes distended by fluid and gas.

Lack of excitatory reflex inputs

The most potent reflex excitatory inputs to the gastric centers arise from tension receptors, which are sensory vagal nerve endings lying in the muscle layers of the reticulum, reticulo-ruminal fold, and cranial ruminal sac. They respond to mechanical distortion produced either by distension of the wall or contraction of the adjacent smooth muscle cells. Moderate distension activates the tension receptors both directly and indirectly, through enhancement of the intrinsic contractility of the smooth muscles. When there is anorexia, reduction or absence of these excitatory signals would most likely account for the weak ruminal movements observed clinically.

Vagus indigestion is a relatively common condition in which there is usually irreversible ruminal stasis. The syndrome occurs primarily in cattle and is frequently associated with **traumatic reciculoperitonitis** (also referred to as 'hardware disease'). In this condition, ingested sharp objects such as nails or pieces of wire penetrate the anterior wall of the reticulum and cause peritonitis. The penetrating object may also track through the diaphragm to precipitate pleuritis, pericarditis and abscess formation.

Until recently, it was thought that the ruminal stasis observed in vagus indigestion was always due to vagal nerve damage. However, pathologic investigations of the vagal pathways have failed to demonstrate significant structural changes in the majority of afflicted cattle. It is therefore likely that the signs attributable to vagus indigestion arise because the inflammatory process around the reticulum reduces the distensibility and intrinsic contractility of the smooth muscle in the wall. The excitatory tension receptor drive to the gastric centers would thus be reduced.

Blockade of motor pathways

This is the least common cause of clinical ruminal stasis and is usually secondary to other conditions. Drugs such as atropine block postganglionic cholinergic transmission, thereby preventing vagal motor discharges from causing cyclical contractions of the forestomach musculature. Hypocalcemia has been shown experimentally to block motor pathways, presumably by impairing the release of acetylcholine. This offers a plausible explanation for the ruminal stasis that appears as one of the earlier signs in parturient paresis (milk fever). Blockade of motor pathways would also occur in those cases of vagus indigestion in which a significant number of vagal fibers have been destroyed by an inflammatory process.

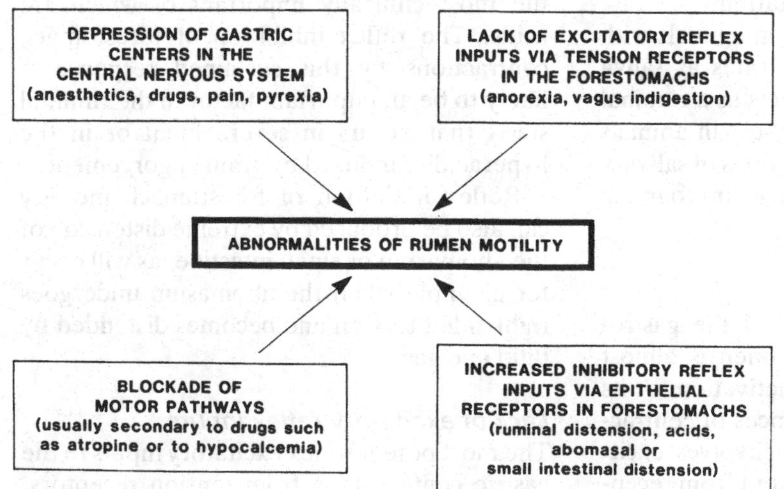

Fig. 7.3. Abnormalities of rumen motility arise from major mechanisms as illustrated. For details, see the text.

The abomasum

Although the abomasum is phylogenetically equivalent to the monogastric glandular stomach, its anatomical position and physiology differ to the extent that its disorders cannot be exactly equated with those of the latter. Clinical abomasitis is virtually confined to one disease entity – **bovine ostertagiosis**. Other abomasal parasites, such as *Haemonchus contortus* and *Trichostrongylus axei* may of course produce severe clinical disease, but these are not manifested clinically as abomasal disorders. Cattle infected with *Ostertagia ostertagi* develop abomasal inflammation and exhibit weight loss, diarrhea and hypoproteinemia. These effects are the consequence of the activities of larval and adult worms in the abomasal mucosa. Third-stage larvae are ingested, exsheath in the rumen, pass to the abomasum and enter the gastric glands, in which they develop to the fourth and fifth stage. During this time, the glands react by metaplastic and hyperplastic transformation to become predominantly mucus-secreting. When immature adults emerge from the glands approximately 20 days after initial penetration, significant functional disturbances are triggered and clinical signs appear. The emergence of the immature adult parasites physically damages the mucosa leading to the induction of inflammation. Large amounts of protein seep into the abomasum in the exudate, producing hypoproteinemia. Glandular metaplasia reduces the secretion of pepsinogen and acid, the luminal pH rises, leading to decreased conversion of pepsinogen to pepsin and to bacterial fermentation. This results in osmotic diarrhea.

The normal intestine

Functional anatomy

The functions of the intestinal tract may be categorized broadly as **digestion**, **absorption** and **secretion**, and an understanding of the physiology of these functions is essential to an understanding of gastrointestinal pathophysiology. This applies particularly when consideration is made of the absorptive and secretory changes that occur in the intestine of animals with diarrhea.

In the small intestine, the functions of digestion and absorption of nutrients are facilitated by a large mucosal surface area, created by the formation of folds, villi and microvilli. The villi are lined by columnar epithelial cells (enterocytes), goblet cells and enteroendocrine cells. All are produced by a population of replicating cells within the crypts of Leiberkuhn (Fig. 7.4).

The enterocytes on the villi are classed as absorptive cells, although they play a major role in digestion. These cells produce an enzyme-rich glycoprotein coating over their luminal surface, known as the glycocalyx, into which they project the microvillous brush border. A layer of mucus, produced by the goblet cells lies over the glycocalyx. Carbo-

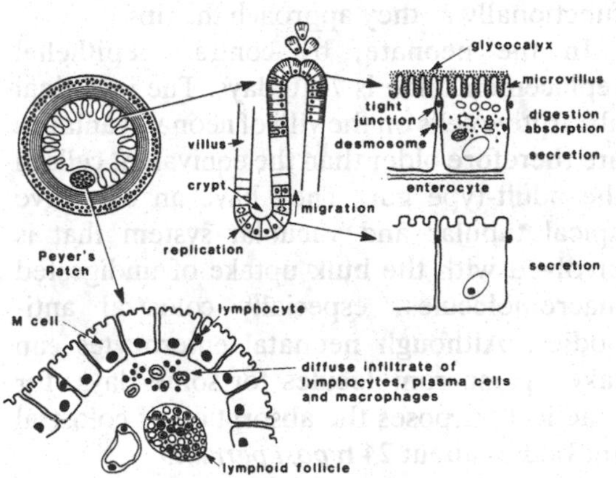

Fig. 7.4. Structure of the mucosa of the small intestine. The surface relief of the intestine is due to the formation of villi and crypts (glands). Villi are pictured as finger-like projections but the shape of villi varies between and within species and may be elongated folds or spatulate rather than finger like. Replication of epithelial cells occurs in the crypts. Epithelial cells migrate distally and mature anatomically and functionally. They are sloughed at an extrusion zone at the tip of the villus.

Villi over Peyer's patches are blunter than other villi and are covered by M cells that have microfolds, not microvilli.

hydrate and protein molecules which have already been subjected to pancreatic and peptic digestion are further digested into smaller components by disaccharidase and dipeptidase enzymes embedded within the cell membrane of the microvilli. The active sites of the enzymes actually project into the glycocalyx, which also contains other enzymes such as ATPase, alkaline phosphatase and intestinal enterokinase. The glycocalyx and cell surface adsorb pancreatic enzymes, which are broken down or, in the case of trypsinogen, converted into the active form trypsin by enterokinase. Nutrients once digested are absorbed into the cell by carrier mechanisms incorporated into the cell membrane.

The adult intestinal epithelium is a highly active cell renewal system in which the cycle of villous cell replacement is normally completed every 2–4 days. Mature senescent cells are extruded and lost into the lumen at the tips of the villi. Replacement cells arise from a pool of proliferating cells in the crypts and migrate out onto the villi, maturing structurally and functionally as they approach the tips.

In the neonate, by contrast, epithelial replacement time is 7–10 days. The intestinal absorptive cells on the villi of neonatal animals are therefore older than the equivalent cells in the adult-type gut. They have an extensive apical tubular and vacuolar system that is involved with the bulk uptake of undigested macromolecules, especially colostral antibodies. Although neonatal enterocytes can take up macromolecules for some days, for practical purposes the absorption of colostral antibodies about 24 h *post partum*.

An 'adult type' gut develops early in life (in the pig, for example, by three to four weeks of age). The normal postnatal development of the mucosa is dependent on the establishment of an intestinal microbial flora. When this flora becomes established soon after birth, the mucosa becomes populated with lymphocytes, plasma cells, macrophages, neutrophils, eosinophils and mast cells. The villi shorten, the crypts lengthen, and the rate of cellular proliferation and peristaltic activity increase. In the adult, intestinal flora is stable, with each type of microorganism occupying its own niche to the exclusion of others. This is an important factor with respect to defense against pathogenic microorganisms.

Intestinal immune systems

As digestive functions mature, immunologically competent cells appear within the lamina propria in response to a wide range of microbial and food antigens. Although detailed discussion of the local immune mechanisms in the gut will not be presented here, it is important to be aware of the essentials, as immune mechanisms are increasingly being recognized as major pathogenetic factors in many enteric diseases. The components of the immune system are the absorptive epithelial cells, specialized epithelial cells known as M cells, and the gut-associated lymphoid tissues (GALT) (Fig. 7.4).

The GALT consists of immune cells in a number of locations including intraepithelial lymphocytes, nodular collections of lymphocytes and plasma cells, including those of the Peyer's patches, and diffuse infiltrates of the same cells throughout the lamina propria. These cells increase in number as the neonatal animal matures normally and also in response to additional stimuli, such as microbial infections.

The M cells cover special dome-shaped villi that lie over Peyer's patches. These cells are more cuboidal than the absorptive cells and have microfolds rather than microvilli. They are actively pinocytotic and appear to sample antigens at the cell surface, absorb them, process them in some way and present them to the intraepithelial lymphocytes. The latter are mainly T cells which have probably already been antigenically primed and they may function in cell-mediated immune reactions and in the handling of antigens that require further processing. It is likely that antigens are transported to the lymphoid nodules of the Peyer's patches by macrophages. Priming of lymphocytes is believed to be initiated within these lymphoid nodules and further differentiation occurs in mesenteric lymph nodes. Primed antibody-producing cells return to the lamina propria via the circulation, where they differentiate into plasma cells. The dominant antibody produced in the lamina propria is the immunoglobulin IgA which is transported across the mucosa only when attached to a 'secretory piece' which is produced by the absorptive epithelial cells. The IgA-secretory piece complex acts at the luminal surface to prevent absorption of soluble antigens and to provide protection against pathogenic microorganisms.

Intestinal absorption and secretion

The epithelium lining the small and large intestine is a hive of activity, with the movement of numerous substances of differing molecular weight and charge **through and between enterocytes**. In many cases it is a two-way flow. Passage through the enterocyte (cellular transport) involves two mechanisms: passive diffusion and carrier-mediated transport. For example, fat is moved by passive diffusion down a concentration gradient, and amino acids, simple carbohydrates and ions are transported by specific transport proteins.

Passage between enterocytes (paracellular transport) via their tight junctions provides substantial access for water and to a lesser extent for electrolytes and small non-electrolytes.

Absorption of carbohydrate, protein and lipids

All these substances are transported through the enterocyte. Lipid and lipid-soluble molecules move through the enterocyte brush border by passive diffusion. This mechanism is unsaturable and absorption is proportional to a concentration gradient across the membrane (Fig. 7.5). In contrast, amino acids, small peptides, mono- and disaccharides are transported by specific transport proteins that are embedded in the microvillus of the cell membrane. The transport proteins are specific

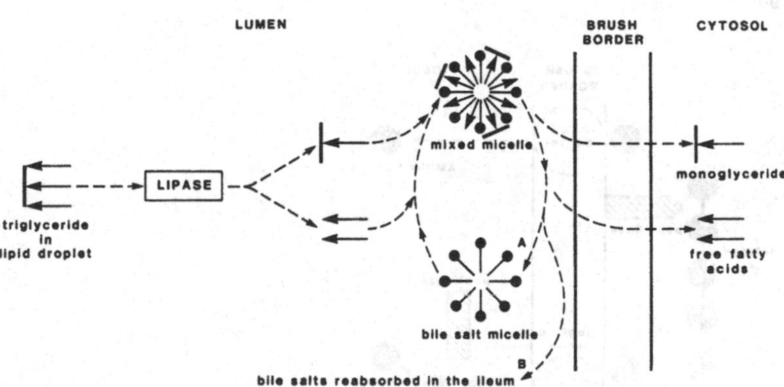

Fig. 7.5. The digestion and absorption of triglycerides. Lipase breaks down triglycerides into monoglycerides (large arrows and bar) and free fatty acids (arrows). For efficient absorption of monoglycerides and free fatty acids, they must be presented to the brush border membrane in micellar form. Bile salts are either reutilized to form further mixed micells or are reabsorbed in the ileum. (Redrawn from Batt, 1980, by kind permission of the author and the British Small Animal Veterinary Association, see Additional reading.)

for a particular substance and may become saturated. Once the carrier mechanism is fully loaded, unabsorbed molecules remain until a site becomes vacant, or they pass to the distal gut. Examples of such transport mechanisms are shown in Figs. 7.6 and 7.7.

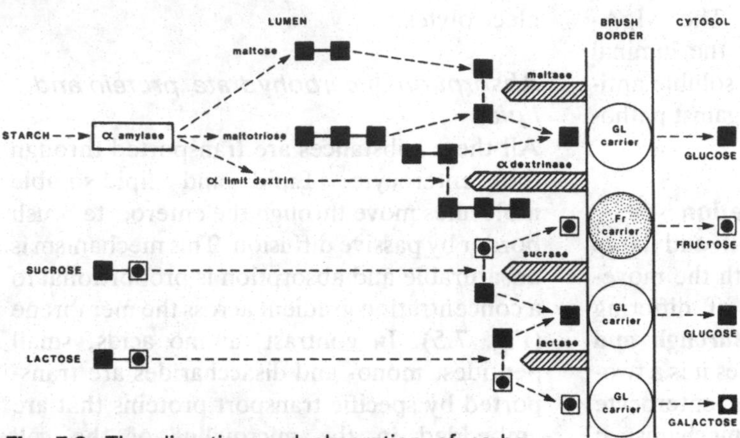

Fig. 7.6. The digestion and absorption of carbohydrate, which must be hydrolyzed into monosaccharides before absorption via specific carrier proteins in the brush border. GL, glucose; Fr, fructose. (Redrawn from Batt, 1980, by kind permission of the author and the British Small Animal Veterinary Association, see Additional reading.)

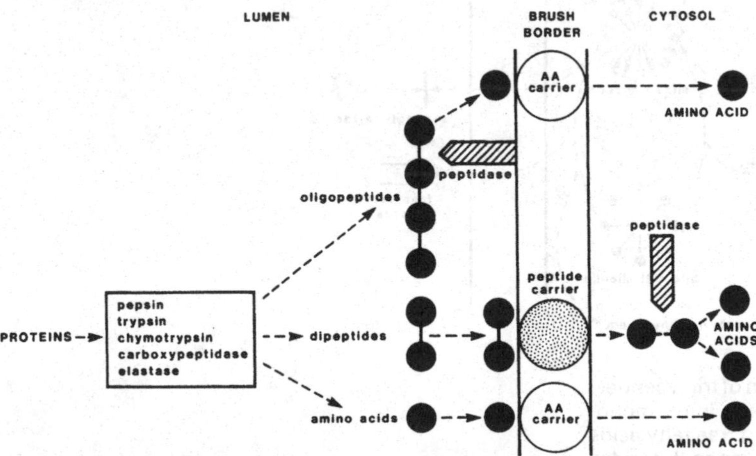

Fig. 7.7. The digestion and absorption of protein follows a pattern similar to that for carbohydrates. There is, however, the ability for dipeptides to cross the brush border. AA, amino acid. (Redrawn from Batt, 1980, by kind permission of the author and the British Small Animal Veterinary Association, see Additional reading.)

Electrolyte and water absorption

Small intestine

In the small intestine, there is net absorption of water, sodium and chloride, along with non-electrolytes such as glucose, amino acids and bile salts. Bicarbonate is the main ion secreted into the small intestine and is important in the buffering of material entering the large intestine.

As stated previously, the two main transport channels are through the enterocyte (cellular) and between the enterocytes (paracellular). The relative importance of both pathways varies with the particular material to be transported. Cellular transport seems to be the important mechanism for ions and paracellular transport for water.

Ion transport across the luminal membrane of the enterocyte may be solitary (uncoupled) or in association with non-electrolytes (coupled). Both are energy dependent. Uncoupled sodium transport involves the active pumping of sodium out of the enterocyte into the interstitial space in exchange for potassium via a sodium–potassium ATPase. Sodium from the intestinal lumen then moves into the enterocyte along a concentration gradient (Fig. 7.8). When coupled, sodium is transported from the intestinal lumen into the enterocyte in association with a variety of non-electrolytes such as glucose, amino acids and bile salts. Once inside the enterocyte, the sodium is actively pumped out and exchanged with potassium via the sodium–potassium ATPase (Fig. 7.9).

Quantitatively, the most important sodium transport mechanism across the enterocyte appears to be that coupled with chloride ion. Two theories for this mechanism prevail. The first proposes transport of sodium and chloride into the enterocyte and the active pumping of sodium from the enterocyte into the interstitial fluid. This mechanism is inhibited by elevations in intraenterocyte cyclic AMP. The second suggests the exchange of hydrogen ion for sodium, and bicarbonate for chloride ion at the luminal face of the enterocyte (Fig. 7.10).

The movement of water is secondary to both osmotic and hydrostatic pressure gradients, established via active solute (ion) transfer. Although water moves through cellular and paracellular pathways, it is the latter which are quantitatively most important.

Movement of water via the paracellular route depends principally upon the active transport of sodium from the enterocyte into the paracellular compartment. Pumping of sodium into the paracellular compartment creates an osmotic gradient across the tight junctions between the intestinal lumen and the

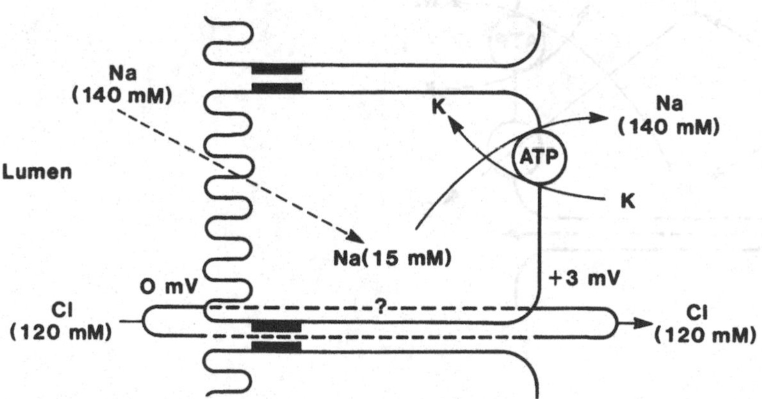

Fig. 7.8. Uncoupled sodium transport across the enterocyte is cellular and involves the energy dependent pumping of sodium from the enterocyte into the interstitium in exchange for potassium. Chloride in this system is transferred paracellularly.

paracellular space across the tight junctions. The paracellular space becomes distended, hydrostatic pressure rises and water flows across into the interstitial spaces and thence into the capillaries (Fig. 7.11).

The net absorption resulting from the activity of all these exquisite transport mechanisms depends to a large degree on the relative permeability of the paracellular pathways. Transported ions tend to diffuse back from the paracellular spaces through the tight junctions and into the intestinal lumen. The 'leakiness' of the paracellular pathway is much greater in the upper small intestine than in the ileum and large intestine. The two counterbalancing mechanisms account for the tremendous bidirectional flow of water and electrolytes, with overall absorption just counteracting back flow.

Mechanisms in the large intestine

In contrast to the small intestine, tight junctions of the large intestine are relatively impermeable to backflow. This permits the avid absorption of sodium, chloride and, most importantly, water.

Sodium is absorbed by an uncoupled, electrogenic mechanism similar to that described for the small intestine but there are two main differences. First, because the epithelium is 'tight', higher potential differences are generated. This provides a greater driving force for chloride absorption. Second, the permeability of the luminal membrane increases (i.e. the ease of entry increases) as the luminal sodium concentration decreases. This enables sodium to be absorbed in spite of very low luminal concentrations. Aldosterone probably plays a role in this mechanism.

Sodium and chloride absorption also occurs by the same mechanism that was described for neutral sodium chloride absorption in the small intestine, with a neutral chloride–bicarbonate exchange mechanism also existing in the large intestine.

The large intestine is similar to the small intestine in that there is a net secretion of bicarbonate, but, in the large intestine, there

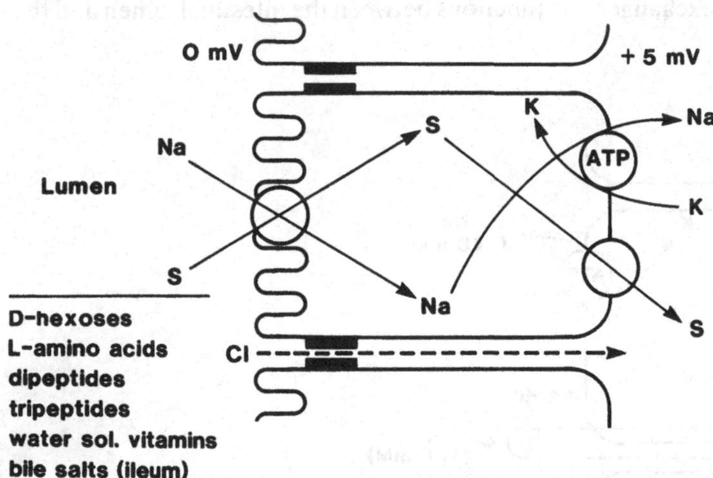

Fig. 7.9. Sodium may be transported coupled with a variety of non-electrolytes such as glucose, amino acids and bile salts. Sodium is actively pumped out of the enterocyte into the interstitium in exchange for potassium. Chloride in this case moves paracellularly. Sol., soluble; S, solute.

is also net secretion of potassium. It is not clear whether the high concentration of potassium in the feces is due to simple diffusion, driven by the transepithelial potential difference, to the more rapid absorption of water compared with that of potassium, or to active potassium secretion.

In the horse, volatile fatty acids (VFAs) are absorbed from the cecum and colon. This is very important because any elevation of VFA concentration in the lumen favors the movement of water into the gut lumen and predisposes to diarrhea. As in the rumen, VFAs are absorbed by diffusion. This process is modified both by the pH of the luminal contents and the metabolism of VFAs by the epithelial cells.

The large intestine is an important site of water absorption in all mammals. Because the epithelium of the large intestine is relatively 'tight', water absorption is very efficient. In this respect, the large intestine has a great reserve capacity, being able to absorb as much as three to four times the volume normally entering from the small intestine. This means that a normal large intestine may mask the net secretion of water and electrolytes by the small intestine.

The importance of water absorption in the large intestine can be illustrated by using the adult horse as an example. In this animal, the daily turnover of water in the large intestine has been shown to be roughly equivalent to the extracellular fluid volume (Fig. 7.12). If one considers that, under normal circumstances, approximately 95% of the water entering the horse large intestine is reabsorbed, the consequences of severe diarrhea become obvious.

Disorders of the intestine

Clinical features of intestinal disease

Diffuse disease of the intestinal tract can make its presence known by giving rise to a wide

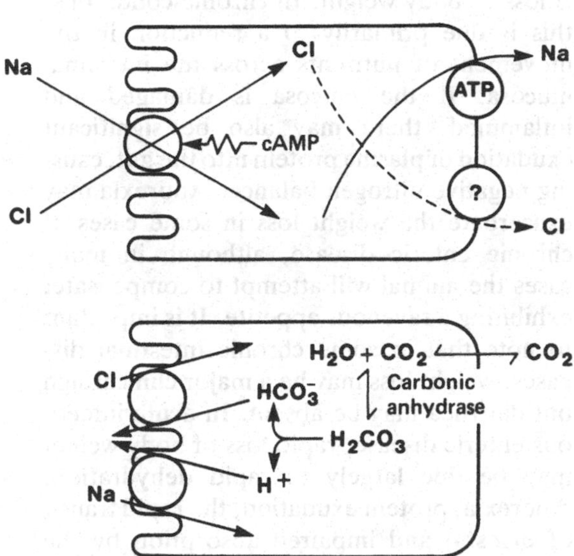

Fig. 7.10. The most important sodium transport mechanism is that coupled with chloride. Two schemes have been suggested, the first via cyclic AMP (c AMP) and the second involving exchange of hydrogen ion for sodium and bicarbonate ion for chloride.

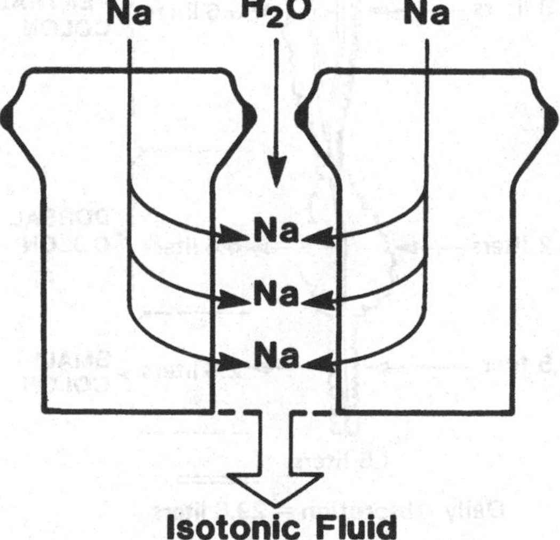

Fig. 7.11. Paracellular transport of water involves an energy-dependent pumping of sodium from enterocytes into the paracellular compartment. This creates an osmotic gradient between the intestinal lumen and the paracellular space across the tight junctions enabling water to move into the paracellular space. The isotonic fluid then moves into the interstitial spaces.

range of clinical signs which may occur in varying combinations. The presence of vomiting and dehydration, electrolyte and acid–base disturbances should alert the clinician to the possibility of enteric disease.

In addition, there is the well-known phenomenon of **diarrhea**, which is defined as an increase in the frequency and liquidity of fecal discharges. It is important to note that diarrhea is not necessarily indicative of intestinal disease, as normal ruminants on lush pastures may exhibit it, as also may very frightened animals. Pathologic diarrhea is associated with an increase in the rate of loss of fecal water and electrolytes and, in the extreme case, the fecal electrolyte content may approach that of plasma. This creates the potential for severe depletion of sodium, potassium, chloride and bicarbonate ions and the development of acidosis, dehydration, shock and renal failure. When diarrhea is the result of acute destructive inflammatory disease, it may contain large amounts of fibrin or even blood, in which case it is referred to as **dysentery**. Depending on its underlying cause, it may be acute and explosive, chronic and intractable or something between those extremes.

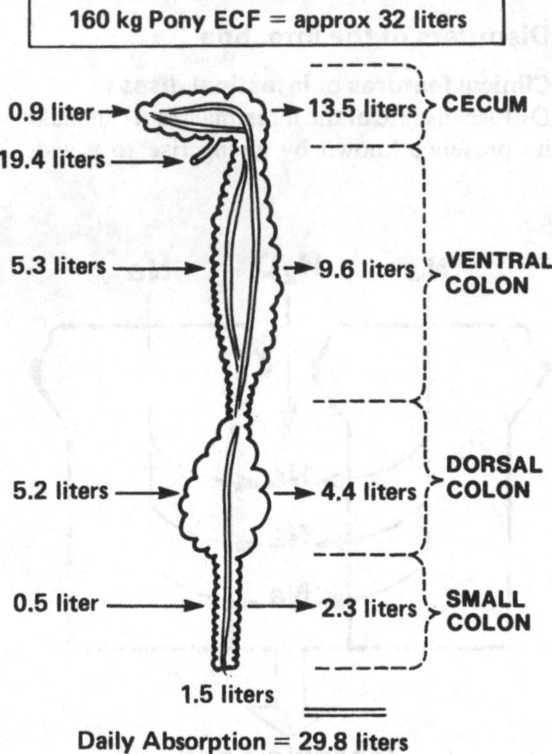

Fig. 7.12. Water absorption from the equine colon. The diagram shows the volume of water delivered each day to segments of the large intestine (volumes shown on the left of the diagram total 31.3 liters). The volume of water absorbed from each segment is shown on the right and totals 29.8 liters. ECF, extracellular fluid. (Adapted from Swenson, 1977, by kind permission, see Additional reading.)

Diarrhea is induced by a number of mechanisms. The most important are an increase in the osmotic pressure in the intestinal lumen, hypersecretion of ions and an increase in intestinal mucosal permeability, all of which cause a net movement of water into the lumen. Finally, and least importantly, in domestic species, diarrhea may be associated with alterations in intestinal motility.

An important clinical sign of enteric disease is loss of body weight. In chronic conditions, this is due primarily to a reduction in the movement of nutrients across the intestinal mucosa. If the mucosa is damaged and inflammed, there may also be significant exudation of plasma protein into the gut, causing negative nitrogen balance. Anorexia may exacerbate the weight loss in some cases of chronic enteric disease, although in many cases the animal will attempt to compensate, exhibiting a ravenous appetite. It is important to note that in some chronic intestinal diseases, weight loss may be a major clinical sign but diarrhea may be absent. In acute infectious enteric disease, rapid loss of body weight may be due largely to rapid dehydration. Anorexia, protein exudation, the rapid transit of digesta, and impaired absorption by the inflammed mucosa may also contribute.

A clinical sign indicative of lower large intestinal disease is tenesmus, which refers to intense straining when passing feces. It can result from stretching of the lower colon or rectum by impacted feces (constipation) or foreign material, or from the irritation induced by an inflammatory lesion.

Finally, abdominal distension can be seen in some animals with intestinal disease. It will usually occur in obstructive disease but may also occur when fluid and gas distend the gut because of an osmotic overload in the lumen.

Pathologic mechanisms in intestinal disease

There are several basic mechanisms which underlie the functional disturbances in enteric disease and ultimately account for the clinical signs expressed. The major ones will be discussed separately, but it should be realized that they are often intercurrent processes.

Villous atrophy

In a number of diseases, the absorptive surface area of the intestinal mucosa is reduced by shortening of the villi. This process is termed villous atrophy and may result either from the excessive loss of mature enterocytes from the villi or from the destruction of progenitor cells in the crypts of Leiberkuhn. If it is due to excessive cell loss, fusion (or bridging) of adjacent villi can occur, further reducing the surface area.

Shortened villi tend to become covered by immature enterocytes in most circumstances. The mere fact of their being short means that replacement cells reach the tips and slough before they have time to develop fully. This is exacerbated if the rate of replication of progenitor cells is increased. The net result of reduced surface area and cellular immaturity is a mucosa with significantly reduced digestive and absorptive functions.

Villous atrophy may also occur in conditions in which there is no obvious destruction of epithelial cells. Although the mechanism is not completely understood, the response appears to be associated with the development of a cell-mediated inflammatory reaction to certain food, microbial, parasitic and other antigens. The reaction provokes hyperplasia of crypt progenitor cells, an increased rate of cell migration and decreased adhesion of villous epithelial cells. The type example of such a disease is gluten enteropathy in humans, which is the result of hypersensitivity to dietary gluten.

Maldigestion/malabsorption

Maldigestion refers to situations in which intestinal absorptive capacity is intact but in which there is incomplete digestive breakdown of large molecules. The undigested molecules are not absorbed and accumulate within the lumen of the small intestine, significantly elevating the intraluminal osmotic pressure. This induces the movement of water into the lumen from the blood. In the large intestine, the large quantities of undigested material entering from the small intestine are rapidly fermented by bacteria. In the case of carbohydrates, poorly absorbed lactate is produced by this fermentive process. This further increases the number of osmotically active particles in the lumen. In addition, the luminal bicarbonate buffer is overwhelmed by the accumulation of lactate and the pH falls. It is probable that this decrease in pH further interferes with absorptive processes in the large intestine. The net result of this osmotic imbalance is the movement of water into the lumen. This causes diarrhea.

One of the best examples of a pure maldigestion syndrome occurs in adult dogs that have an inability to digest lactose (lactose intolerance) and in which osmotic diarrhea is a consequence of ingesting milk. The condition is uncommon but in affected dogs there is a deficit in the activity of the disaccharidase lactase on the brush border. Hand-reared calves fed with poorly digestible milk-replacers commonly develop a similar syndrome, with severe osmotically induced diarrhea as a prominent feature. In the dog and cat, exocrine pancreatic insufficiency is a well-recognized cause of maldigestion (see Chapter 8). Similarly, in hepatic disease the defect will occur if bile salt metabolism or excretion are impaired. It should be remembered that in all the foregoing, the small intestinal mucosa is structurally normal.

Malabsorption refers to situations in which there are structural abnormalities in the intestinal mucosa that result in impairment of the intestinal absorptive capacity. As malabsorption occurs most commonly in diseases in which there is a decrease in the epithelial surface area, it is usually accompanied by maldigestion. These are therefore referred to as **maldigestion/malabsorption syndromes**. The net result is an increase in the luminal osmotic load, enhancement of large intestinal fermentation, and diarrhea.

In general, the clinical signs seen most commonly with maldigestion/malabsorption syndromes are diarrhea and loss of body weight in spite of a normal or ravenous appetite. Affected carnivores may pass voluminous watery or pasty feces, often loaded with undigested fat (steatorrhea). In the horse, because of the enormous water-absorbing capacity of the large bowel, diarrhea may not occur if only the small intestine is involved. Depending upon the degree of large intestinal involvement, the feces may be almost normal, or have an unformed 'cow manure' consistency.

Villous atrophy is the structural abnormality that underlies many diseases that are characterized by maldigestion/malabsorption. As mentioned in the previous section, this contraction and fusion of villi may result from destruction of either the villous enterocytes or the crypt epithelial cells. Both processes lead to a marked reduction in the surface area of the intestinal mucosa.

Destruction of villous epithelial cells is a feature of infections with coronaviruses and rotaviruses, protozoan agents such as coccidia and cryptosporidia, helminths such as *Trichostrongylus colubriformis* and some invasive bacteria.

Intestinal coronaviruses and rotaviruses and coccidia proliferate in, and destroy, villous enterocytes (Fig. 7.13), denuding the villi and causing them to contract and fuse. Crypt epithelial proliferation leads to the covering of the atrophied villi with immature cells. If the lesion is relatively mild, as it is in many rotavirus infections, the mucosa can be completely reconstituted in a fairly short time. Very severe damage, as occurs in some coronavirus infections, can result in permanent villous atrophy. In the case of *T. colubriformis*, destruction of villous cells is caused by the tunneling of the immature parasite into the villous epithelium.

Destruction of crypt epithelial cells is the characteristic feature of infection with the canine and feline parvoviruses (Fig. 7.13) and the togavirus of bovine virus diarrhea. It also results from radiation injury and from the effects of cytotoxic drugs that are used for cancer chemotherapy.

The immediate effect of extensive crypt cell necrosis is rapid collapse, acute maldigestion/malabsorption and osmotic diarrhea. Reconstitution of the mucosa takes longer than after villous cell loss, especially if regenerating cells are being continuously destroyed. Additional severe complications are exudation of protein, hemorrhage into the lumen, and secondary bacterial or fungal invasion of the devastated mucosa.

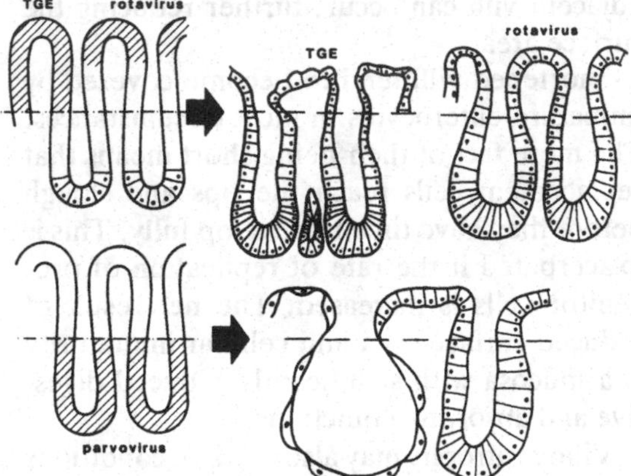

Fig. 7.13. Intestinal coronaviruses (TGE) and rotavirus infections cause villous atrophy by destruction of enterocytes on villi (rotaviruses affecting only the most superficial). This stimulates hyperplasia of crypt epithelial cells. Canine and feline parvoviruses proliferate in the rapidly dividing epithelial cells of the crypts. Destruction of these cells causes villous atrophy as there are less or no cells left to replace those being extruded from the villi. Crypts collapse or become lined by flattened epithelial cells. Surviving cells become hyperplastic.

Hypersecretion

The stimulation of active ion transport by the secretory crypt cells of the intestine can cause severe diarrhea by increasing the effective osmotic pressure of the luminal contents. The stimulation of secretion occurs concurrently with an inhibition of ion absorption by the epithelial cells at the tips of the villi. As the lesion is purely biochemical, no histologic or ultrastructural lesions are apparent in the intestinal epithelial cells.

Hypersecretory diarrheas are characterized by being profuse and watery, and particularly in neonates, rapid dehydration and acidosis occur. The fluid secreted is isotonic, alkaline, low in calcium, magnesium and protein, but high in sodium, bicarbonate and chloride.

Hypersecretory diarrhea is caused by enterotoxins of *Vibrio cholerae*, *Escherichia coli*, *Salmonella typhimurium* and probably other bacteria. It is also caused by prostaglandins and intestinal hormones such as vasoactive intestinal polypeptide (VIP).

The colonization of the human small intestine by *V. cholerae* has provided the basic model of hypersecretory diarrhea. The *V. cholerae* enterotoxin binds to the luminal surface of both the crypt and villous epithelial cells. The production of cyclic AMP is 'switched on' in these cells, stimulating chloride secretion by the crypt cells (secretory effect) and inhibiting neutral sodium chloride absorption by the villous cells (anti-absorptive effect).

The osmotic gradient generated by the luminal accumulation of sodium and chloride causes water to move from the blood to the lumen. There is also a passive net secretion of potassium. The luminal fluid in hypersecretory states also contains large quantities of bicarbonate but the mechanism of secretion is not clear.

As far as is known, the binding of bacterial enterotoxin to the luminal cell membrane of enterocytes is permanent. The natural resolution of the disorder therefore depends on the replacement of affected cells by normal epithelial cell turnover. This takes approximately 3–5 days in the adult. During this period, the patient must be maintained by replacing the water and electrolytes that are lost. An important finding in human cholera was that the mechanism coupling sodium absorption with that of non-electrolytes such as glucose and amino acids remained intact. The oral administration of fluids containing sodium and dextrose therefore plays a very important role in the maintenance of cholera patients. This cheap and simple means of fluid replacement is also used in animals with hypersecretory diarrhea.

The **enterotoxigenic** strains of *Escherichia coli* are the most common cause of hypersecretory diarrhea in both humans and animals. Two enterotoxins are produced by enterotoxigenic *E. coli*. A **heat-labile protein** (LT), which is structurally and functionally similar to cholera toxin, and which stimulates cyclic AMP activity in intestinal epithelial cells. A second, low molecular weight, **heat-stable toxin** (ST) stimulates cyclic GMP activity in intestinal epithelial cells. There are wide (and confusing) species variations in the action of LT and ST. Current evidence suggests that both stimulate secretion of chloride and inhibit absorption of sodium, possibly by increasing the intracellular concentration of calcium, which acts as a 'second messenger' to the cyclic nucleotides in the cell (Fig. 7.14).

Prostaglandins and intestinal polypeptides (e.g. VIP) have also been shown to stimulate secretion in the small intestine in association with increases in the cyclic AMP concentration in the epithelial cells. Prostaglandins, produced during inflammation, mediate the secretogenic effects of invasive bacteria such as some strains of *Salmonella typhimurium*. VIP mediates the secretory diarrhea in some human patients with pancreatic cholera syndrome, which is usually associated with pancreatic endocrine neoplasms.

The degree to which the large intestine is involved in the pathogenesis of hypersecretory diarrhea also needs to be considered. In the previous discussion of normal function, it was noted that the normal large intestine has a considerable reserve capacity. Diarrhea would result if:

1 the quantity of fluid and electrolytes entering from the small intestine exceeded the absorptive capacity of the large intestine, or
2 large intestinal absorption was reduced, possibly in association with a concurrent increase in active ion secretion.

It has been shown that cyclic AMP, prostaglandins, and VIP have the same effect on active ion transport in the large intestine as they do in the small intestine. For example, invasive *S. typhimurium* produces hypersecretion in the large intestine, as well as in the small intestine, mediated by prostaglandin and cyclic AMP.

The pathogenesis of the diarrhea that occurs in infections with enterotoxigenic *E. coli* is less easily explained. The enterotoxin of this organism has no demonstrable effect on ion transport in the large intestine. It is possible that diarrhea occurs because of simple overloading of the large intestine with fluid originating in the small intestine. Another

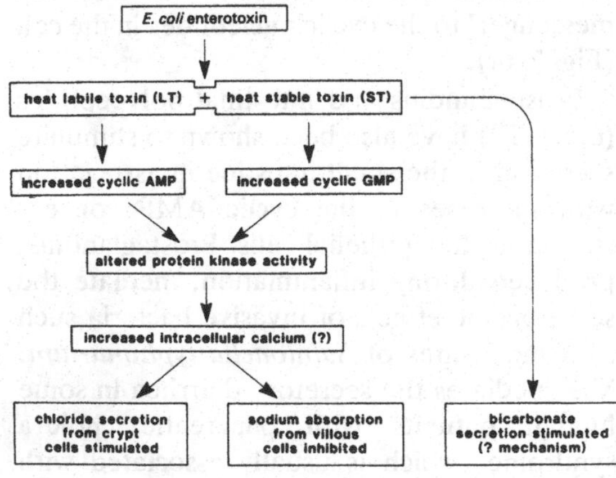

Fig. 7.14. The pathophysiologic effects of enterotoxigenic colibacillosis originate from the action of LT and ST on the enterocyte. Sodium absorption is inhibited and chloride secretion is stimulated. Bicarbonate secretion is also stimulated.

possibility is that there is a qualitative change of the large intestinal contents; for instance, a pH change may be induced by the fluid entering from the small intestine, thereby reducing large intestinal absorption.

Many exogenous and endogenous laxatives cause diarrhea by altering fluid and electrolyte movement in the large intestine. The effects of the common exogenous laxatives such as dioctyl sodium sulfosuccinate (DSS, Coloxyl), phenolphthalein and ricinoleic acid (castor oil) are probably mediated by prostaglandins. The 'endogenous' laxatives, which include bile acids, fatty acids and hydroxy fatty acids are thought to cause diarrhea by a similar mechanism. This accounts for diarrhea in animals with bile salt malabsorption and steatorrhea.

Altered permeability and protein loss

In the preceding review of normal physiology, it was noted that water was transported across the intercellular spaces of the intestinal epithelium down osmotic and hydrostatic pressure gradients. Factors such as increased capillary permeability (as in inflammation), raised venous pressure, lymphatic obstruction, and decreased plasma colloid osmotic pressure are capable of raising the effective tissue pressure in the intestinal lamina propria. This elevation in effective tissue pressure changes the normal osmotic and hydrostatic pressure gradients, so that water tends to move from the blood to the intestinal lumen. This process amounts to an increase in intestinal permeability and it is probable that it involves widening of the tight junctions between enterocytes. The integrity of the tight junctions is very important in preventing backflow of absorbed sodium and water from the intercellular spaces into the intestinal lumen, especially in the ileum and large intestine. Furthermore, large increases in tissue pressure will increase permeability to such an extent that plasma protein molecules pass into the lumen and are lost from the body.

Diseases in which this permeability disturbance is dominant may be characterized by loss of body weight with or without diarrhea. However, diarrhea will almost always be present if the large intestine is affected. If only the small intestine is affected, the large intestine is often able to compensate and the feces will be normal.

When intestinal protein loss is especially severe, concentration of all classes of plasma protein begins to fall and hypoproteinemic edema may develop. Any enteric disease in which this occurs is known as a **protein-losing enteropathy** (or gastroenteropathy should the stomach be involved; gastric protein loss has already been discussed in connection with bovine ostertagiosis, see p. 173). Good examples of such diseases are provided by paratuberculosis in ruminants, equine granulomatous and eosinophilic enteritis and histiocytic ulcerative colitis of Boxer dogs. All these are chronic inflammatory diseases in which the lamina propria of the intestinal mucosa is heavily infiltrated by inflammatory cells (Fig. 7.15). In a rare condition known as intestinal lymphangiectasia, the lymphatics of the lamina propria become greatly distended because of some unexplained disorder of lymph flow. This has been found mostly as a cause of protein-losing enteropathy in the dog.

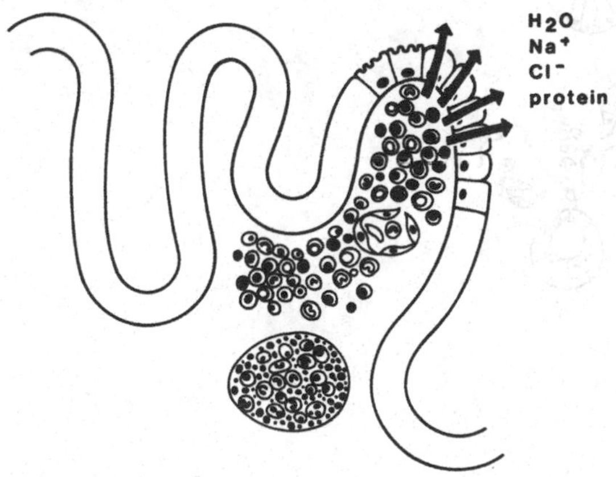

Fig. 7.15. Chronic inflammation. Increased hydrostatic pressure in the lamina propria forces fluid and electrolytes out between cells. Absorption is decreased mainly by some villous atrophy.

Inflammation

The occurrence of inflammation in the gut gives rise to the terms **enteritis**, to specify involvement of the small intestine, **colitis** to specify the colon and **typhilitis** and **proctitis** to specify the cecum and rectum, respectively.

In **acute enteric inflammation**, catarrhal, fibrinous and hemorrhagic reactions may occur, depending on the causal agent. In all types, there is exudation of fluid and cells and, in some cases, necrosis of enterocytes to varying degrees. The amount of fluid loss into the lumen is dependent on the extent of vascular damage and on the presence or absence of hypersecretion. The clinical effects of such disease are often extremely severe and can lead rapidly to death.

Major causes of acute enterocolitis in production animals are the invasive bacteria of the genus *Salmonella* and various strains of *E. coli* (Fig. 7.16). Severe fibrinous reactions are typical of these infections and the gastrointestinal abnormalities are often complicated by the systemic effects of septicemia or endotoxemia. In the case of *S. typhimurium* infection, it has also been shown that diarrhea is caused not only by severe mucosal damage and increased permeability but also by a prostaglandin-mediated hypersecretion.

A very severe necrotizing enteritis is produced in chickens, pigs and foals by *Clostridium perfringens* type C, in which the systemic effects of exotoxins exacerbate the effects of the extremely destructive enteric lesion. **Coccidial** infections can also produce serious intestinal inflammation in several mammalian and avian species (Fig. 7.17).

Swine dysentery is a specific acute colitis involving the combined effects of *Treponema hyodysenteriae* and anerobic bacteria (Fig. 7.18).

In **chronic** inflammatory gut disease, altered permeability and protein loss are major consequences, as has previously been described. There is additionally a high likelihood of maldigestion and malabsorption because villous atrophy is often present, and the lamina propria is greatly expanded by cellular infil-

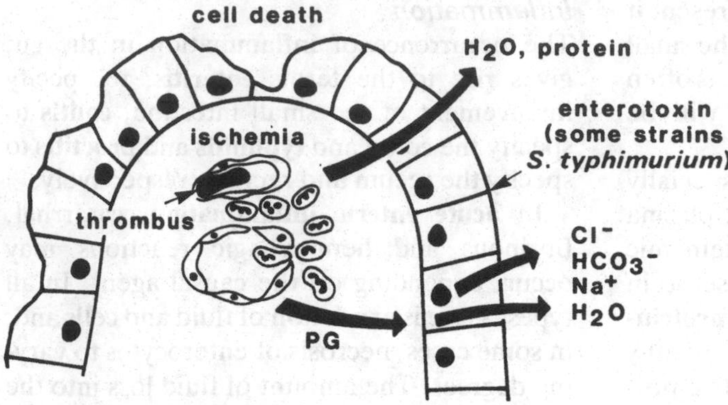

Fig. 7.16. Salmonellae and some strains of *E. coli* induce acute enteritis. Inflammatory fluids are lost through a 'leaky' and/or ischemic epithelium. Prostaglandin (PG)- and enterotoxin-mediated hypersecretion of fluid and electrolytes also occurs.

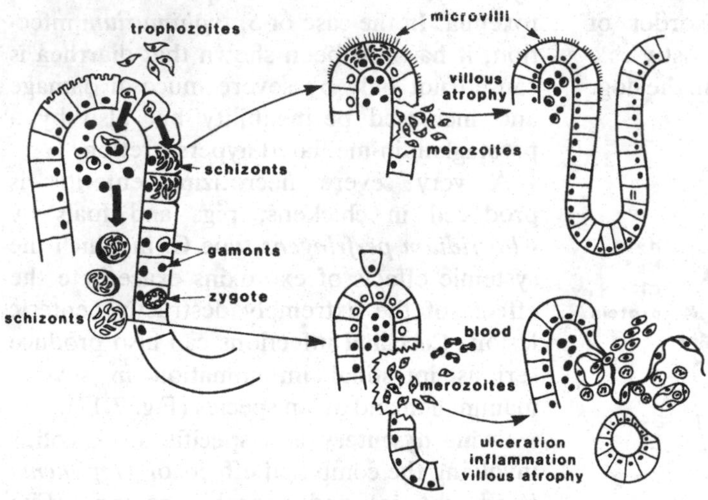

Fig. 7.17. The lesions and clinical disease produced by intestinal coccidia depends on the dose of infective organisms, host immunity and on the site of development of the schizonts and/or gametocytes in the mucosa. Those species that undergo schizogony in superficial epithelial cells produce villous atrophy and its consequences. Those that develop deep within the mucosa destroy the epithelium as they rupture and induce inflammation, hemorrhage and loss of tissue fluids.

tration. The most common of these diseases are granulomatous in nature, featuring the massive accumulation of macrophages and tremendous thickening of the intestinal wall.

Small intestinal bacterial overgrowth syndromes

As far as is known, all animals have a 'normal' small-intestinal bacterial population that is essential for normal function. Syndromes associated with the proliferation of bacteria in the small intestine occur most commonly in humans but are also reported in small animals. These usually occur in patients in which a segment of small intestine is static or hypomotile. In humans, the syndrome is seen commonly in patients who have had a loop of small intestine surgically bypassed or who have primary stasis of the small intestine. This is often referred to as the 'blind loop' syndrome. A loop of small intestine does not always have to be involved as the condition also occurs following partial gastrectomy. Here, the lack of hydrochloric acid in digesta entering the small intestine is thought to promote bacterial overgrowth.

The large intestinal-type bacteria that proliferate in the static or hypomotile segment of small intestine hydrolyze conjugated bile salts, causing bile salt depletion. This results in fat maldigestion and malabsorption in the small intestine. The undigested fat undergoes bacterial hydroxylation in both the small and large intestine. This produces an osmotic diarrhea. In addition, the hydroxylated fatty acids also have a laxative action in the large intestine, possibly by causing prostaglandin-mediated secretion. The end result is that the patient passes loose stools containing large quantities of fat. This is referred to as steatorrhea.

In addition to the fat maldigestion and malabsorption that occurs with bacterial overgrowth in the small intestine there may also be a number of concomitant problems. The most important of these is vitamin B_{12} deficiency, probably related to the binding of the vitamin to the bacteria in the small intestine.

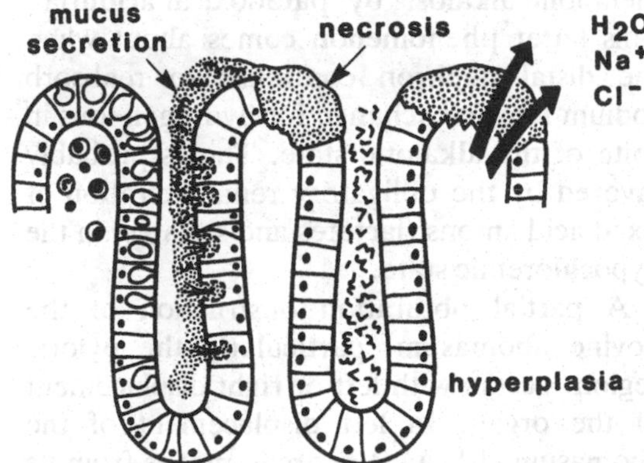

Fig. 7.18. Swine dysentery. *Treponema hyodysenteriae* in association with anaerobic bacteria stimulate mucus secretion and initiate cell necrosis. Necrosis is superficial and 'stimulates' hyperplasia. Fluid is lost through the necrotic and inflammed mucosa and water absorption is reduced.

Changes in intestinal motility

Motility disorders of the small and large intestine are poorly understood. As far as is known, an increase in small intestinal motility *per se* cannot result in diarrhea.

Increased small intestinal motility with decreased digesta retention times have been demonstrated in many hypersecretion disorders. This increase in motility is modulated by the prostaglandin/cyclic AMP stimulation that occurs in these diseases and may represent a defense mechanism whereby the gut attempts to expel the causative agent. Decreased small intestinal motility with prolonged digesta retention times can result in bacterial overgrowth within the lumen. The consequences of this have been discussed in the previous section.

In the large intestine, where motility of the proximal segments may serve to retain digesta for more complete digestion and resorption of ions and water, increased motility may be associated with constipation.

No specific causes of deranged large intestinal motility have been identified in animals, but a condition called 'irritable bowel syndrome' occurs in humans. This disease is characterized by decreased motility, decreased digesta retention in the large intestine, and diarrhea.

Gastrointestinal obstruction

Gastrointestinal obstruction may be defined as a pathophysiologic state in which the aboral passage of ingesta is hindered or completely halted. Although the term as defined usually refers to the intestine, in this discussion it will be applied broadly to include obstruction initiated at any point from the stomach or abomasum and on through the intestine. Disorders of this general type are a common clinical problem in all species and give rise to syndromes which may be acutely fatal if not corrected promptly. They may also give rise to subacute or chronic clinical disease. Once such a disorder is suspected by a clinician, every attempt is made to determine the site at which obstruction has occurred and the nature of the obstructing lesion.

Three basic types of obstruction are defined, and in increasing order of severity these are:

- **Simple obstruction** in which the only problem is interference with the aboral passage of the gut contents.
- **Closed loop obstruction** in which a segment of the gut is isolated by occlusion at two points. This is exemplified by the herniation of a loop of bowel through a restricted opening such as the epiploic foramen.
- **Strangulation obstruction** in which there is interference with the blood flow to or from a closed loop obstruction. An example of this is provided by **volvulus**, which arises from the twisting of a segment of bowel about its mesenteric attachment.

In addition, two major mechanisms of obstruction are recognized, namely mechanical and functional types. Table 7.1 indicates the conditions that may be associated with these two situations.

Mechanisms and clinical features of obstruction

Torsion and displacement of the bovine abomasum

Right-sided torsion of the bovine abomasum is a well recognized clinical entity. The lesion is primarily an obturation obstruction (Table 7.1) with varying degrees of vascular impairment and is brought about by torsion in either direction, about the long axis of the organ.

Under normal conditions, between 30 and 40 liters of hydrochloric acid are secreted into the abomasum each day. Large-scale secretion continues after torsion occurs. This can result in the collection in the abomasum of up to 30 liters of fluid, rich in hydrogen, chloride, sodium and potassium ions, together with a large quantity of gas. The resulting severe distension reduces both motility and the mural blood flow. Torsional tearing of omental vessels may also contribute to the impairment of blood flow. The collected fluids tend to reflux into the rumen, and the sequestration of large quantities of water and ions in the abomasum causes dehydration, hypochloremia, hypokalemia and metabolic alkalosis. The hypokalemia is exacerbated by the movement of potassium into cells in exchange for hydrogen ions, and the metabolic alkalosis by 'paradoxical aciduria'. This latter phenomenon comes about when the distal nephron continues to reabsorb sodium ions in exchange for hydrogen ions in spite of the alkalotic state. This is probably favored by the obligatory renal excretion of fixed acid anions (lactates and sulfates) in the hypochloremic state.

A partial obturation obstruction of the bovine abomasum, particularly the pyloric region, occurs with left or right displacement of the organ. In left displacement of the abomasum (LDA), the organ moves from its normal position on the ventral abdominal floor, slightly to the right of midline, to lodge between the rumen and the left abdominal wall. In right displacement (RDA), the

Table 7.1. *Major mechanisms of gastrointestinal obstruction*

Varieties of mechanical obstruction	Varieties of functional obstruction
1 Obturation – internal blockage of the gut (*a*) Foreign bodies like balls, corn cobs, bones, etc. (*b*) Impaction with ingesta, sand or meconium. (*c*) Internal formation of masses of plant fiber (phytobezoars), hair (trichobezoars) or mineralized masses (enteroliths). (*d*) Intussusception – the invagination of a segment of bowel distally into an ensheathing adjacent segment. 2 Constriction of the lumen (*a*) Intrinsic lesions of the gut wall: Congenital atresia or stenosis. Contraction of scar tissue following injury or neoplasms. (*b*) Lesions extrinsic to the gut wall: Peritoneal and mesenteric adhesions, peritoneal abscess, neoplasms, hernial incarceration, volvulus.	1 Paralytic ileus (failure of peristalsis): (*a*) Causes believed to be potassium depletion and/or excessive sympathetic stimulation. (*b*) May occur after abdominal surgery and in peritonitis, toxemia and electrolyte imbalance. 2 Vascular bowel disease (*a*) Compromised blood supply to a segment of bowel due to thrombosis/embolism in mesenteric vessel. For example, in equine parasitic arteritis. (*b*) Profound shock.

abomasum is dilated with fluid and gas, moving dorsal to its normal position. The abomasal accumulation of water and ions in both conditions tends to cause metabolic changes similar to, but much less severe than, those typical of right abomasal torsion. In a large proportion of cattle with mild LDA and RDA, however, the blood gas and electrolyte picture is often normal.

The etiology of LDA, RDA and right abomasal torsion is not clear, although similar factors affecting motility are probably responsible for all three conditions. The reason for the decreased motility is also not clear, although diets containing large quantities of grain appear to play an important role. It may be that large amounts of ruminal volatile fatty acids in cows receiving grain enter the abomasum and, either directly or indirectly by feedback from the duodenum, produce hypomotility. Many other factors have also been implicated as causing hypomotility. These include subclinical hypocalcemia and concurrent diseases such as ketosis, mastitis and endometritis. An inherited predisposition to LDA and RDA has also been postulated. The acute mechanism whereby the hypomotile abomasum undergoes displacement or torsion is not known, although positional changes of the abdominal organs late in gestation have been suggested as playing a role.

Acute dilation–volvulus of the canine stomach

This syndrome occurs in large, deep-chested dogs such as Great Danes, Bloodhounds, Irish Setters and Saint Bernards. Despite considerable investigation, the pathologic mechanism of the condition remains obscure. Gastric dilation is reported to be due to the combination of rapid eating with the ingestion of air (aerophagia). The accumulated food and gas are thought to contribute to movement of the dilated stomach, which may then undergo torsion (volvulus). When this occurs, the stomach rotates 270°–360° in a clockwise direction when viewed from the caudal aspect. The spleen may also undergo torsion, infarction and rupture as the stomach rotates.

Dry cereal-based dog foods have been implicated in the etiology of acute gastric dilation–volvulus, but their role is controversial. It was thought that the rapid fermentation of these carbohydrate-rich diets pre-disposed to intragastric gas accumulation, decreased gastric motility and delayed gastric emptying. A recent study, clearly showing that these diets do not affect gastric motility or emptying does not support this theory.

It would be expected that the intragastric accumulation of hydrochloric acid in this condition would lead to a metabolic alkalosis. When the disease was produced experimentally, it was found that the dilated stomach caused compression of the vena cava. This compression, in association with a decrease in venous return caused by severe venous congestion of the stomach, rapidly leads to shock and a metabolic acidosis. Clinical surveys of dogs with acute gastric dilation–volvulus have revealed that, whilst all severely affected dogs were in shock, many had almost normal blood gas values. This was presumed to be due to the offsetting effects of the concurrent metabolic acidosis and metabolic alkalosis. Importantly, these clinical surveys also showed that cardiac arrhythmias, possibly initiated by metabolic, neural or humoral factors, were the main cause of death in dogs which underwent surgical correction.

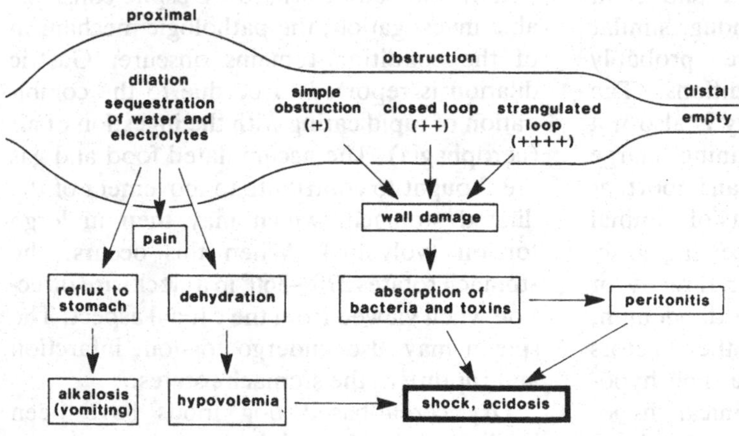

Fig. 7.19. Intestinal obstruction: simple intestinal obstruction produces proximal dilation and the consequences of dilation with little (+) wall damage, whereas closed loop obstruction and strangulation have the added and severe consequences of moderate (++) to severe (+++) intestinal wall damage.

Simple mechanical obstruction of the small intestine

The pathophysiologic changes that occur in simple mechanical obstruction are due to distension, the loss of fluid and electrolytes and bacterial growth within the obstructed segment (Fig. 7.19).

Distension of the intestine proximal to the obstruction is caused by ingested fluid, digestive secretions and intestinal gas. Distal to the obstruction the intestine is usually empty. Intestinal distension is not life threatening, but tends to be self-perpetuating because it increases pressure in the intestinal wall, causing net secretion of water and electrolytes into the distended segment. Later, as increased pressure in the intestinal wall reduces intestinal blood flow, there is also sequestration of edema fluid in the intestinal wall and mesentery. This may lead to transudation of fluid through the wall of the obstructed segment into the peritoneal cavity. In high obstructions of the small intestine, large quantities of fluid may also be lost through reflux into the stomach. In some species, the dog for example, this fluid is lost completely from the body by vomition.

The metabolic consequences of fluid and electrolyte losses in simple mechanical obstruction of the small intestine depend upon how long the obstruction has been present and also on the level of the intestine at which it has occurred. An obstruction may occur in the upper (proximal) small intestine, or lower (distal) small intestine. The more proximal the obstruction the more rapid the onset of clinical signs, and the patient may become extremely ill within hours of the onset.

In **proximal obstruction**, the small intestine tends not to be greatly distended, as there is usually reflux of fluid back into the stomach. Water, sodium, chloride, hydrogen and potassium are lost. This causes dehydration, hypochloremia, hypokalemia and metabolic alkalosis, especially if the refluxed fluid is lost through vomiting. If the obstruction is not relieved, the animal rapidly goes into hypovolemic shock and renal hypoxia and azotemia follow. When this happens, the plasma lactate concentration increases and metabolic acidosis may quickly 'override' the initial alkalosis, especially if refluxed fluid is *not* lost through vomiting.

In **distal obstructions** large quantities of fluid are sequestered within the distended intestine. Electrolyte loss tends to be less severe than in proximal obstructions, but dehydration, hemoconcentration, shock, renal hypoxia and metabolic acidosis create serious problems. These problems are compounded by the proliferation of bacteria within the lumen and the absorption of their toxins into the systemic circulation.

Closed loop and strangulation obstruction of the small intestine

In closed loop and strangulation obstruction, the pathophysiologic changes are due not only to distension and fluid and electrolyte loss but also to vascular changes in the bowel wall and bacterial proliferation in the closed loop.

In a closed loop obstruction, there is distension both proximal to and within the loop. This rapidly reduces blood flow in the wall of the affected loop, which becomes ischemic. Rupture and peritonitis may follow.

Similar changes also occur with strangulation and incarceration but vascular changes are more severe. Strangulation and incarceration obstruct the thin-walled mesenteric veins but blood still passes through the thicker-walled arteries. The affected region of bowel rapidly becomes congested and blood leaks from the wall and into the gut lumen. The wall becomes necrotic and sufficient blood may be lost into the gut to induce hemorrhagic shock.

Intestinal bacteria multiply rapidly in a closed loop of intestine. Toxins elaborated by these bacteria, the bacteria themselves and the by-products of tissue necrosis leak through the degenerate intestinal wall. The presence of this 'strangulation fluid' in the peritoneal cavity causes peritonitis and toxemia. Subsequent absorption of endotoxins from the peritoneum causes rapid deterioration of the patient's condition by inducing profound shock.

Large intestinal obstruction

The pathophysiologic changes associated with simple mechanical obstruction of the large intestine are similar to those described for the small intestine, but the changes are usually less severe and take longer to occur. This means that the onset of clinical signs may occur in days rather than hours following the obstruction.

The obstructed large intestine may become so grossly distended with gas produced by bacterial fermentation, that abdominal tympany occurs. In non-ruminants, the metabolic sequelae of simple mechanical obstruction of the large intestine are dehydration and mild metabolic acidosis. Electrolyte disturbances are minimal. Ruminants, however, tend to develop hypochloremia, hypokalemia and metabolic alkalosis.

In the horse, severe colic, hypovolemic shock and metabolic acidosis are produced when there is obstruction of the large intestine with primary vascular occlusion (thrombo-embolism of mesenteric vessels) or concurrent vascular occlusion (for example, displacement of the colon). In cows, similar signs and metabolic disturbances are associated with mesenteric torsion.

Pathophysiologic basis of clinical signs

Abdominal pain is a prominent feature of intestinal obstruction and is registered by receptors in the mesentery, parietal peritoneum and intestinal wall. It is elicited by stretching, inflammation and ischemia. The clinical manifestations of abdominal pain vary between species. Horses react violently to abdominal pain by sweating, kicking at the abdomen and rolling. Cows exhibit depression and a reluctance to move and dogs show guarding of the abdomen.

Reflux of small-intestinal contents into the stomach occurs because fluid 'backs up' proximal to the site of obstruction, especially when this is high in the small intestine. In horses, this fluid may be regurgitated through the nose, whereas in dogs it is often lost through vomiting. The accumulation of large quantities of fluid in the stomach of horses with intestinal obstruction can result in gastric rupture.

Abdominal distension may occur due to the accumulation of fluid and gas in obstructed segments of intestine. This occurs commonly with large intestinal obstructions. Gas arises from swallowed air, bacterial fermentation and by diffusion from the blood.

Failure to pass feces may occur because of emptying of the intestine distal to the obstruction. Also, obstructions at one site in the intestinal tract cause reflex inhibition of motility in other areas of the gut.

Peripheral circulatory collapse, tachycardia and depression are sequelae of dehydration, acid–base and electrolyte disturbances, shock, toxemia and prerenal azotemia.

Elevations in the white cell count and protein of the peritoneal fluid indicate the presence of devitalized intestine. The cells and protein originate both from the degenerate segment and the inflamed parietal peritoneum. Peritoneal fluid analysis is used routinely in the evaluation of horses with colic.

Partial obstruction of the intestine also occurs and the signs are those of weight loss and diarrhea. The clinical signs are related to maldigestion and malabsorption of nutrients caused by bacterial overgrowth in the lumen proximal to the partial obstruction.

Additional reading

Argenzio, R. A. and Whipp, S. C. (1980). Pathophysiology of diarrhea. In *Veterinary Gastroenterology*, N. V. Anderson (ed.), pp. 220–32. Philadelphia, Lea and Febiger.

Argenzio, R. A., Whitlock, R. H. and Burrows, C. F. (1980). The colon, rectum and anus. In *Veterinary Gastroenterology*, N. V. Anderson (ed.), pp. 523–92. Philadelphia, Lea and Febiger.

Barker, I. K. and Van Dreumel, A. A. (1985). The alimentary system. In *Pathology of Domestic Animals*, vol. II, 3rd edn, K. V. F. Jubb, P. C. Kennedy and N. Palmer (eds.), pp. 1–237. Orlando, FL, Academic Press.

Batt, R. M. (1980). The molecular basis of malabsorption. *J. Small Anim. Pract.* **21**: 555–69.

Binder, H. J. (1981). Colonic secretion. In *Physiology of the Gastrointestinal Tract*, vol. II. L. R. Johnson (ed.), pp. 1003–19. New York, Raven Press.

Blood, D. C., Radostits, O. M. and Henderson, J. A. (1983). Diseases of the alimentary tract, parts I and II. In *Veterinary Medicine*, 6th edn, pp. 132–259. London, Balliere Tindall.

Field, M. (1981). Secretion of electrolytes and water by mammalian small intestine. In *Physiology of the Gastrointestinal Tract*, vol. II, L. R. Johnson (ed.), pp. 963–82. New York, Raven Press.

Gray, G. M. (1978). Mechanisms of digestion and absorption of food. In *Gastrointestinal Disease*, vol. I, 2nd edn, M. H. Sleisenger and J. S. Fordtran (eds.), pp. 241–50. Philadelphia, W. B. Saunders Co.

Harvey, C. E., O'Brien, J. A., Rossman, L. E. and Stoller, N. H. (1983). Oral, dental, pharyngeal and salivary gland disorders. In *Textbook of Veterinary Internal Medicine*, vol. 2, 2nd edn, S. J. Ettinger (ed.), pp. 1126–91. Philadelphia, W. B. Saunders Co.

Hirsh, D. C. (1980). Microflora, mucosa and immunity. In *Veterinary Gastroenterology*,

N. V. Anderson (ed.), pp. 119–219. Philadelphia, Lea and Febiger.

Johnson, J. H., Hull, B. L. and Dorn, A. S. (1980). The mouth. In *Veterinary Gastroenterology*, N. W. Anderson (ed.), pp. 337–71. Philadelphia, Lea and Febiger.

Jones, R. S. (1978). Intestinal obstruction, pseudo-obstruction and ileus. In *Gastrointestinal Disease*, 2nd edn, M. H. Sleisenger and J. S. Fordtran (eds.), pp. 425–36. Philadelphia, W. B. Saunders Co.

Krejs, G. J. and Fordtran, J. S. (1978). Physiology and pathophysiology of ion and water movement in the human intestine. In *Gastrointestinal Disease*, 2nd edn, M. H. Sleisenger and J. S. Fordtran (eds.), pp. 297–335. Philadelphia, W. B. Saunders Co.

Lorenz, M. D. (1983). Diseases of the large bowel. In *Textbook of Veterinary Internal Medicine*, vol. 2, 2nd edn, S. J. Ettinger (ed.), pp. 1346–73. Philadelphia, W. B. Saunders Co.

Merritt, A. M. (1980). Small intestinal diseases. In *Veterinary Gastroenterology*, N. V. Anderson (ed.), pp. 463–522. Philadelphia, Lea and Febiger.

Moon, H. W. (1978). Mechanisms in the pathogenesis of diarrhea: a review. *J. Am. Vet. Med. Assoc.* **172**: 443–8.

Moon, H. W. (1983). Intestine, In *Cell Pathology*, 2nd edn, N. F. Cheville (ed.), pp. 372–91. Ames, IA, Iowa State Univ. Press.

O'Brien, J. A., Harvey, C. E. and Brodey, R. S. (1980). The esophagus. In *Veterinary Gastroenterology*, N. V. Anderson (ed.), pp. 372–91. Philadelphia, Lea and Febiger.

Schultz, S. G. (1981). Salt and water absorption by mammalian small intestine. In *Physiology of the Gastrointestinal Tract*, L. R. Johnson (ed.), pp. 983–9. New York, Raven Press.

Sherding, R. G. (1983). Diseases of the small bowel. In *Textbook of Veterinary Internal Medicine*, vol. 2, 2nd edn, S. J. Ettinger (ed.), pp. 1278–1346. Philadelphia, W. B. Saunders Co.

Swenson, W. J. (1977). *Duke's Physiology of Domestic Animals*, 9th edn. New York, Cornell Univ. Press.

Twedt, D. C. and Wingfield, W. E. (1983). Diseases of the stomach. In *Textbook of Veterinary Internal Medicine*, vol. 2, 2nd edn, S. J. Ettinger (ed.), pp. 1233–77. Philadelphia, W. B. Saunders Co.

Watrous, B. J. (1983). Esophageal disease. In *Textbook of Veterinary Internal Medicine*, 2nd edn, S. J. Ettinger (ed.), pp. 1191–1233. Philadelphia, W. B. Saunders Co.

Way, L. W. (1978). Abdominal pain and the acute abdomen. In *Gastrointestinal Disease*, 2nd edn, M. H. Sleisenger and J. S. Fordtran (eds.), pp. 394–400. Philadelphia, W. B. Saunders Co.

Whitlock, R. H. and Wingfield, W. E. (1980). The stomach and forestomachs. In *Veterinary Gastroenterology*, N. V. Anderson (ed.), pp. 392–462. Philadelphia, Lea and Febiger.

Clive R. R. Huxtable

8 The liver and exocrine pancreas

The liver

The liver is the biochemical core of mammalian metabolism: factory, central depot, clearing house, defense unit, and waste disposal facility for the body, all rolled into one. Well over 1000 biochemical reactions take place in the liver, and its activities influence the farthest reaches of the body. In the face of this biochemical complexity, the challenge has been to identify the clinical signs and biochemical abnormalities that are relevant to the clinician. Clinicians have several tasks to hand when it has been established that liver damage is present. They must decide how significant is the damage, if it is the major cause of the clinical problem, or if it is a relatively insignificant consequence of disease elsewhere.

There is a world of difference between an animal with liver failure and an animal with some liver damage, in spite of the fact that several features might be the same in each case. The diagnosis of liver disease is now sufficiently advanced for the thorough clinician to be able to resolve these questions in most cases, in spite of the variability and nonspecificity of clinical signs. However, it is worth repeating the most important clinical concept: there is a great difference between **liver damage** and **liver failure**. This emphasis arises from the ready availability of biochemical tests which can indicate liver damage. However, positive results from such tests must be kept in perspective.

In the following paragraphs, diagnostic signposts are highlighted against the background of the normal structure–function relationships of the liver, and ultimately against underlying disease processes within liver tissue.

Basic functional anatomy

The liver like most organs is very 'functional', in the sense that almost all its bulk is working parenchyma. Interspecies variation in gross anatomy does occur, with the dog having the most complex version, but the essential structural and functional features can be considered to be the same in all species. The organ is confined by a thin and slightly elastic capsule (Glisson's capsule), which allows expansion under duress. Connective-tissue content within the liver is minimal, consisting of a little collagen around portal triads and hepatic veins. The sinusoids and hepatocyte cords are supported by a very delicate reticular framework, which is scarcely apparent in routine histologic sections. The pig liver is an exception in that distinct collagen septa link portal triads.

The normal liver has a tremendous functional reserve. An animal can maintain adequate function with only 10–12% of its liver intact, provided that the surviving liver is in reasonable anatomical order. A larger mass of liver tissue may be functionally inadequate if the relationship between blood and hepatocytes is perturbed.

Processing of blood

The anatomical design of the liver is directed towards its major mechanical role – to receive and process all the venous blood draining from the digestive tract. The portal blood contains all the small molecular weight products of digestion, plus the metabolic products of the gut microflora. As such, it is a very mixed brew and, while full of good things, it frequently contains some decidedly unwelcome customers. It is the job of the liver to receive the portal blood, to accept the good things and reject or destroy the harmful things as best it can. It also adds its own products to the blood in a major contribution to the metabolic economy of the body.

This role is largely carried out by the parenchymal cells of the liver, the hepatocytes, with assistance from the phagocytic Kupffer cells. To enable an adequate interaction between the hepatocytes and the portal blood, terminal portal venules empty into slit-like sinusoids between plates of hepatocytes – rather like the cavity between two walls (Fig. 8.1). The blood flows fairly slowly and at low pressure through these sinusoids, over a fenestrated endothelial lining. This arrangement provides the hepatocytes with easy access to the blood, and even large protein molecules can diffuse across the sinusoidal wall between the hepatocytes and the blood. The hepatocytes themselves are separated from the endothelium by an additional space, the space of Disse, which also contains reticular collagen fibrils and 'fat storing' or 'Ito cells'. Phagocytic Kupffer cells project from the surface of the endothelium at frequent intervals and sweep the blood clean of any microorganisms or particulate matter.

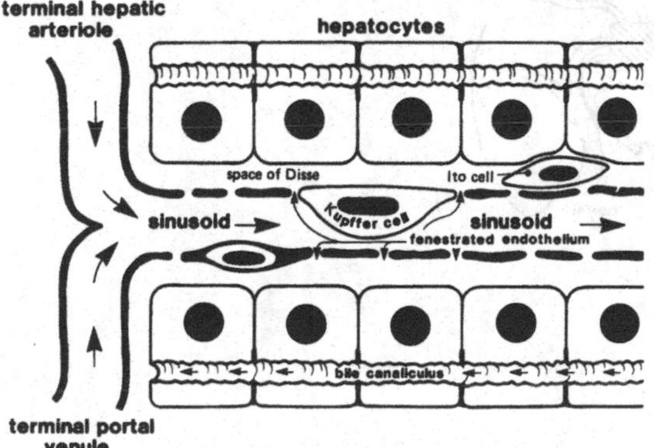

Fig. 8.1. Arterial and venous blood entering a hepatic sinusoid. Hepatocytes, although segregated from the blood, interact vigorously with it across the space of Disse and the fenestrated endothelium.

After traversing a sinusoid, the portal blood enters a terminal radical of the hepatic vein, flows into larger branches, and ultimately leaves the liver via the major hepatic vein on the diaphragmatic surface.

The poorly oxygenated portal blood is supplemented with arterial blood from the hepatic artery as it enters each sinusoid, so that the vitality of the hepatocytes is less threatened by hypoxia. The terminals of the hepatic artery discharge into each sinusoid along with terminal portal venules.

The detailed micro-anatomy of the liver cell plates and sinusoids is well documented and need not be considered here, but some attention needs to be given to the concept of the basic functional unit of the liver – the hepatic 'acinus'. This structure is not clearly defined and is difficult to visualize three-dimensionally. In essence it is a small mass of liver cell cords and sinusoids, at the axis of which are a terminal portal venule, hepatic arteriole and bile ductule (Fig. 8.2). This axis is known as a 'portal triad'. The cluster of sinusoids radiating out from the axis of the acinus drains to at least two terminal hepatic venules ('central' veins) at its periphery.

In histologic sections of liver, hepatocytes adjacent to portal triads can be considered to be at the center of acini, while hepatocytes adjacent to hepatic venules can be considered to be at the periphery of acini. This defines three major reference zones for hepatocytes – **periportal** (around portal triads) and **periacinar** (around hepatic venules; also called the centrilobular zone in older terminology). The intermediate area is termed the **midzone**. The hepatocytes in the different zones, although morphologically similar, are biochemically different and react differently to incoming chemicals and metabolites. This **zonal heterogeneity of hepatocytes** has been convincingly displayed by studying the reaction of liver to various toxic chemicals and will be alluded to again in the pathology section of this chapter (see p. 207).

Conduction of bile

The other major mechanical feature of the liver is the conduit system for transferring bile to the intestine (Fig. 8.3). The bile itself is important in the digestion of fats via the action of the bile salts and acids, and as a route for the excretion and elimination of metabolic end products.

Bile production is initiated in hepatocytes and excretion is first into the bile canaliculi between adjacent liver cells (Fig. 8.1). The canaliculi eventually discharge, via the canals of Hering into the smallest bile ductules in portal triads. Bile flow from hepatocytes is thus in the opposite direction to blood flow from the portal triads.

Bile ductules and ducts conjoin, and bile leaves the liver via the large common duct in most species. In all domestic species, some of the bile is diverted into the gall bladder for storage, except for the horse which has no gall bladder.

Pathophysiologic states most relevant to clinical disease

From the outset it should be kept clearly in mind that clinical signs arising from the liver may reflect merely an inadequacy of some facet of liver function – a problem in one corner of the factory rather than closure of the whole enterprise. This notion of restricted impairment versus total hepatic failure will be maintained throughout this chapter.

Out of the mass of possible malfunctions arising in a damaged liver, those most readily correlated with clinical disease and its diagnosis are as follows.

Disturbance of bilirubin excretion

An excess of circulating bilirubin gives rise to clinical **icterus** (jaundice), but icterus can have several causes. Chemical determination of the proportions of free and conjugated serum

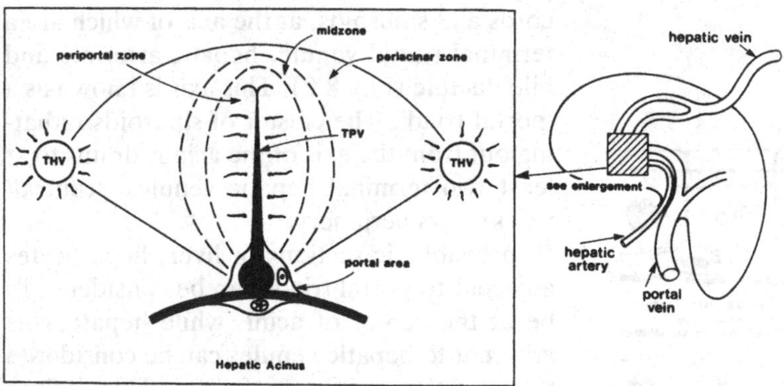

Fig. 8.2. The basic functional unit of the liver – the acinus. Terminal portal venules (TPV) empty directly into sinusoids and the sinusoids drain into two or more terminal hepatic venules (THV). The TPV is at the core of the acinus, the THVs at the periphery. This arrangement delineates three zones of parenchyma in the acinus – the periportal, the midzone and the periacinar.

bilirubin can give a valuable guide to the pathogenesis of icterus.

All bilirubin is derived from the respiratory pigments – hemoglobin, myoglobin and cytochromes. The breakdown of the pigments occurs chiefly outside the liver, producing 'free' (unconjugated) bilirubin which is in fact bound to albumin in the blood. At this point the liver takes over and the relevant steps are:

1 The uptake of free-bilirubin by hepatocytes. This mechanism has a load limit which can be exceeded when large amounts of free bilirubin are circulating. An excessive production of free bilirubin, as in hemolysis, may exceed the capacity of even a normal liver to clear the circulating load. Alternatively, hepatocytes in a damaged liver may lack the capacity to take up free bilirubin produced at a normal rate.
2 The conjugation of bilirubin to glucuronic acid within hepatocytes. This produces a water-soluble molecule of conjugated bilirubin which is suitable for excretion.
3 The excretion of conjugated bilirubin from hepatocytes into the bile canaliculi as part of the general process of bile secretion (see below). In liver cell disease, hepatocytes may not have the ability adequately to conjugate and/or to excrete bilirubin.
4 The conduction of conjugated bilirubin, as a component of bile, to the intestine via the biliary tract. Should the biliary tract be obstructed, conjugated bilirubin will be returned to the blood by the embarrassed hepatocytes, a process called regurgitation.

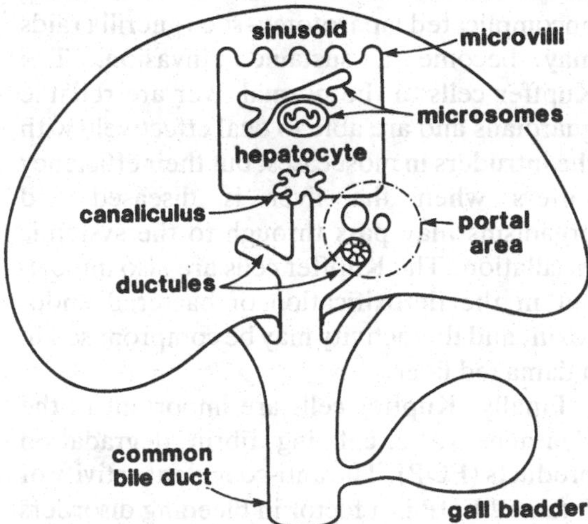

Fig. 8.3. The anatomy of the bile conduit system, showing intrahepatic and extrahepatic components. (Reproduced from Twedt, 1985, by kind permission, see Additional reading.)

Disturbances of bile secretion – cholestasis

Bile is a complex aqueous solution generated by the liver and delivered to the duodenum. Major components are the bile salts and acids which are powerful detergents and are important in the digestion of fats. They are avidly conserved by the body, being resorbed by high affinity receptors in the lower small intestine and returned to the liver in the portal blood. This loop is referred to as the enterohepatic circulation of bile salts. Bile is also the vehicle for the elimination of many metabolites, the most notable being bilirubin and cholesterol.

The process of bile secretion by the liver depends upon a chain of circumstances beginning with biochemical pumps within hepatocytes. Bile formation by hepatocytes is an active process, driven by active transport into the canaliculi, of bile salts and acids and inorganic ions. The movement of these solutes induces the movement of water by osmotic force, so initiating a flow of bile along hepatic cell cords towards a portal triad. Pumping movements of the canaliculi also contribute to the induction of flow. During passage through the ducts, the bile is modified by secretion and absorption via the biliary epithelium. The hormone **secretin** promotes vigorous secretion by the bile ducts.

Any bile that enters the gall bladder is concentrated many-fold by the absorption of water through the mucosa, the water following actively transported sodium and chloride ions. Stored and concentrated bile is expelled by contraction of the gall bladder wall. This contraction is stimulated both by vagal activity and by the hormone **cholecystokinin**, released from the duodenal mucosa by the stimulus of lipids in the lumen.

A failure to excrete bile gives rise to the pathophysiologic state termed **cholestasis**. Not unexpectedly, amongst other things, cholestasis causes a retention of bilirubin, cholesterol and bile acids and may interfere with the digestion of fats. Cholestasis may result from interference with the secretory process at any point, from its origin at the hepatocyte to the termination of the common bile duct.

Within the liver, drugs and toxins may interfere with bile formation at its very source by purely functional means. Alternatively, canaliculi may be obliterated by structural damage to hepatocytes, ranging from cellular swelling, as in lipidosis, to necrosis. In portal triads, widespread inflammatory lesions may disturb bile ductules. All the foregoing scenarios will produce **intrahepatic cholestasis**. **Extrahepatic cholestasis** results from the obstruction of flow in the major bile duct, by whatever means.

Severe cholestasis will usually be signaled by icterus, recognized chiefly by the yellow discoloration of mucous membranes and sclera. Lesser degrees of cholestasis may not be attended by icterus and other means of diagnosis are required. These are discussed in a following section (p. 202).

Inadequate conversion of ammonia to urea

Ammonia produced in the body is a constant toxic hazard. The main source of ammonia is the intestine, where it arises from the action of bacteria on nitrogenous compounds. Ammonia is also produced from the metabolic deamination of amino acids throughout the body. The liver shoulders the task of detoxifying ammonia and does so by converting it to urea for excretion via the kidney. In some chronic liver diseases, the failure of ammonia detoxification is reflected by an elevation of blood ammonia concentration, and the onset of a neurologic disorder known as 'hepatic encephalopathy', which is, in part, ammonia poisoning.

Protein synthetic deficits

The liver is the factory that manufactures and exports a range of essential proteins, the major ones being: (1) albumin; (2) fibrinogen, prothrombin and other clotting factors I, II, V, VII, VIII, IX and XII; and (3) molecules involved in the transport of metals, hormones and lipids, e.g. ceruloplasmin and macroglobulins.

A failure to produce such proteins at a sufficient rate may lead to disorders related to their function, such as edema and bleeding tendencies. Anemia may also be seen because of impaired iron mobilization.

Abnormal carbohydrate metabolism

The liver is the major glycogen store for the body and is the site of gluconeogenesis. It is, therefore, an important contributor to blood glucose concentration. In liver failure, blood glucose values may fluctuate wildly in either direction, but in most cases it is hypoglycemia that occurs.

Failure of Kupffer cell activity

The intestine teems with microorganisms and even in the normal animal there are periodic incursions of microbes into the portal bloodstream. In intestinal disease, or even during uncomplicated laparotomy, such guerilla raids may become a sustained invasion. The Kupffer cells of the normal liver are reliable guardians and are able to deal effectively with the intruders in most cases, but their efficiency suffers when the liver is diseased and organisms may pass through to the systemic circulation. The Kupffer cells are also important in the detoxification of bacterial endotoxin, and this activity may be compromised in a damaged liver.

Finally, Kupffer cells are important in the clearance of circulating fibrin degradation products (FDP). The anti-coagulant activity of retained FDP is a factor in bleeding disorders associated with liver disease.

Abnormal metabolism of chlorophyll

This disorder is of course only relevant to herbivores, but in these species is very much to the fore when the liver is diseased. When large amounts of green feed are ingested, metabolites of chlorophyll are generated, especially phylloerythrin (a photoactive agent). This molecule is normally excreted in the bile and causes no problems but, if its clearance is suppressed, it accumulates in the blood and induces severe photosensitivity in sun-exposed areas of the skin. Retention of phylloerythrin and consequent photosensitization is a classical feature of liver disease in herbivores and is referred to as **hepatogenous photosensitization**.

Portosystemic shunting

This occurs when portal blood bypasses the sinusoids and finds its way directly to the caudal vena cava. Shunts may be acquired or congenital (Fig. 8.4). Acquired shunting may develop in chronic liver disease when there is diffuse fibrosis, distortion and mechanical impedence of blood flow through the liver. The portal blood draining from the gut is then diverted via multiple tortuous collateral channels around the spleen, stomach and kidneys into the vena cava. Shunting may also result from developmental defects in the portal venous system. Various anatomical anomalies have been described. In this case the animal effectively has chronic liver cell insufficiency, as the hepatocytes present are unable adequately to process the portal blood.

Portosystemic shunting directs a host of potentially toxic substances into the systemic circulation, and its major clinical product is **hepatic encephalopathy**.

Portal hypertension

The blood pressure within the portal venous system is normally low and a sustained abnormal increase in this pressure is referred to as 'portal hypertension'. This can be brought about by any condition causing an increased resistance to portal blood flow (Fig. 8.5).

The increased resistance may occur before the blood enters the liver, in which case it is said to be **prehepatic**. This is relatively uncommon in animals but, when present, is usually caused by thrombosis of the portal vein. It will promote portosystemic shunting and congestion of the gut, but does not cause ascites.

Increased resistance to flow may occur within the liver and gives rise to **intrahepatic** portal hypertension, in which there is again a tendency to portosystemic shunting and the development of ascites. Extensive fibrosis and atrophy of the liver is the most common cause in animals. The increased pressure within the sinusoids promotes the generation of edema fluid within the space of Disse. The transuded fluid overloads the lymphatic drainage of the liver and the surplus seeps through Glisson's capsule and accumulates in the peritoneal cavity as ascites. This occurs especially if serum albumin concentration is depressed.

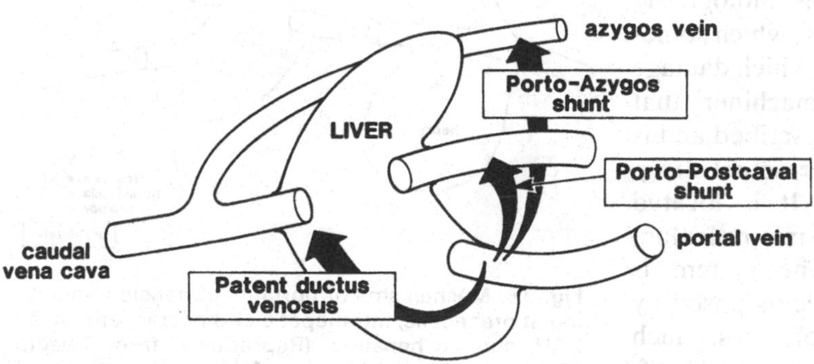

Fig. 8.4. Pathways of portosystemic shunting.

In **post-hepatic** portal hypertension, blood flow is impeded through the hepatic vein or caudal vena cava. The major effects are venous engorgement, enlargement of the liver, and ascites. The most common cause in animals is congestive cardiac failure.

It can be seen that ascites occurs if venous pressure rises *within* the liver. This causes a large-scale movement of fluid into the interstitial compartment. Elevated venous pressure within the portal vein *before* it enters the liver does not generate sufficient fluid movement for ascites to develop.

The question of portal hypertension is important in the clinical evaluation of an animal with ascites. Ascitic fluid emanating from the liver, as described above, will typically be a modified transudate, i.e. will contain a significant quantity of protein, as evidenced by its ability to clot. This is because it is generated under conditions where plasma protein is able to leave the vascular compartment fairly easily through the highly permeable sinusoidal lining. A degree of anoxic damage is also likely to increase permeability.

The metabolism of drugs and chemicals

In life, there is a steady exposure of the body to exogenous chemical agents and metabolically generated endogenous molecules destined for disposal. The liver is primarily responsible for the processing of such materials. An important aspect of this processing is the detoxification of potentially harmful agents, and in many cases this is achieved successfully. However, in some cases the system actually transforms biologically harmless compounds into toxins, which sometimes destroy the liver itself, or which damage other organs. The biochemical machinery that does this work has been well described and is referred to as the **hepatic microsomal drug-metabolizing enzyme system**. It is located largely in the smooth endoplasmic reticulum of periacinar hepatocytes. The system is necessary because of the problems posed by hydrophobic, lipid-soluble molecules. Such molecules by their nature are extremely difficult to eliminate from the body and must be converted to a hydrophilic, water-soluble state, suitable for excretion in bile or urine.

The hepatic drug-metabolizing system processes such molecules in two stages. In phase I, enzymes, by oxidation or reduction, introduce polar groups into the molecules or otherwise modify them. In phase II, the modified molecules are enzymatically conjugated with other chemical groups to become sufficiently water soluble. The best examples are the glucuronide conjugates.

The drug-metabolizing system is remarkably labile and versatile, being competent to process a great variety of molecular types. The activity of phase I reactions may be enhanced or depressed by various chemicals, thereby enhancing or depressing non-specifically the metabolism of a host of other substances. Two classical activators of the system are DDT and phenobarbital. Accordingly, an individual 'normal' animal may vary from its peers, or in its day-to-day response to a drug or toxin. Depending on circumstances, it may be saved or damned. Similarly, an animal with liver

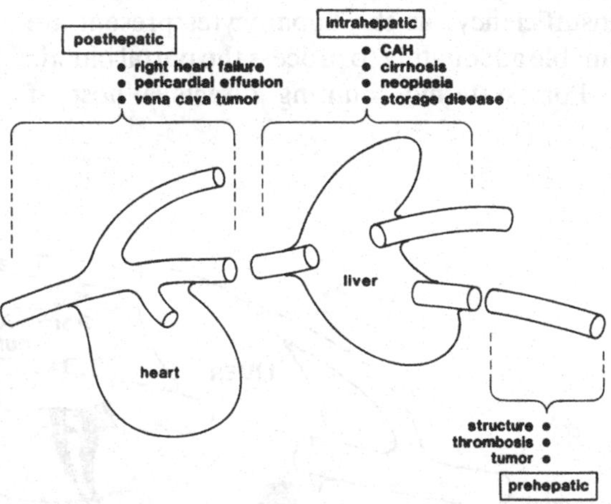

Fig. 8.5. Mechanisms of portal hypertension operating at prehepatic, intrahepatic and posthepatic sites. CAH, chronic hepatitis. (Reproduced from Twedt, 1985, by kind permission, see Additional reading.)

Table 8.1. *Differentiating the three basic causes of icterus*

Pathologic event	Serum	Urine
Hemolysis	Large elevation of unconjugated bilirubin mod. elevation conjugated bilirubin	Large amounts of Hb. Small amount bilirubin
Liver cell disease	Large elevation in both types of bilirubin in roughly equal proportions	Large amount conjugated bilirubin
Biliary obstruction	Large elevation in conjugated bilirubin. Slight elevation in free bilirubin initially – if prolonged obstruction it may rise but not to equal proportions	Massive amount conjugated bilirubin

mod., moderate; Hb, hemoglobin.

disease must be carefully assessed before becoming a recipient of drug treatment, or a candidate for anesthesia.

Interspecies variation is also noteworthy. Cats are notoriously poor at glucuronide, sulfate and glutathione conjugation, and are easily poisoned by several seemingly low risk drugs like salicylates and acetaminophen.

Aberrations in drug-metabolizing ability are probably the cause of 'idiosyncratic' reactions that occur in some individuals. Drugs that are perfectly safe for most recipients will be toxic in such individuals. This is exemplified by the response of some dogs to the anti-convulsant drug Primidone.

Laboratory aids in diagnosis of liver disease and assessment of liver damage

To the clinician, the animal with suspected liver disease can often be frustratingly difficult to assess because of the non-specificity of clinical signs. A number of aids to diagnosis are readily available, but are of use only if the principles of their application are clearly understood and their limitations appreciated.

Biochemical tests can be placed into four categories, often used in conjunction with each other to clarify a particular clinical case. There are those:

1 for the assessment of icterus,
2 for the detection of cholestasis in the absence of clinical icterus,
3 for the detection of damage to hepatocytes,
4 for the assessment of functional liver mass.

The assessment of icterus

The three basic causes of icterus are excessive formation of free bilirubin, diffuse liver cell disease, and obstruction to bile flow. An indication of the pathogenesis of icterus can often be obtained by examining both serum and urine.

In the serum, the total bilirubin concentration will of course be elevated (hyperbilirubinemia), but the proportion of conjugated and unconjugated bilirubin can vary according to the basic cause. In the urine, conjugated bilirubin concentration will increase (bilirubinuria) parallel to the serum concentration. This is because a fraction of the conjugated bilirubin is released from its albumin carrier and filtered by the kidneys. Unconjugated bilirubin is not excreted in urine.

Combining the above, the situation may be stated as follows and is also illustrated in Table 8.1. In **hemolytic icterus**, the great majority of serum bilirubin will be unconjugated, as the massive release of pigment into the plasma simply overwhelms the liver. The liver also suffers some damage due to anemia and shock, and some conjugated bilirubin finds its way back into the blood, mildly elevating its concentration in serum. Glomerular filtration will then result in mild bilirubinuria. Similarly, glomerular filtration of free hemoglobin results in hemoglobinuria.

In **diffuse liver cell disease** the damaged hepatocytes have both an impaired ability to take up free bilirubin and a tendency to leak conjugated bilirubin back into the blood. Therefore, conjugated and free bilirubin concentrations tend to be roughly equal in the serum. Bilirubinuria will be pronounced.

In **obstruction of bile flow**, the major effect is the return of conjugated bilirubin to the blood to produce a massive elevation of serum concentration, with initially only a slight elevation of free bilirubin concentration. Urinary bilirubin concentration directly reflects serum concentration, and bilirubinuria will be intense. Over a period of weeks, the ability of hepatocytes to take up and conjugate free bilirubin is reduced and the serum concentration of free bilirubin will also rise to quite high values.

The detection of cholestasis in the absence of icterus

Early in the course of cholestasis, or in mild cholestasis, serum bilirubin concentration may be only slightly elevated and clinical icterus is not apparent in spite of a significant bilirubinuria. This is particularly so in the dog, which has an extremely low renal threshold for bilirubin and is able to conjugate bilirubin in the kidney. Even ligation of the bile duct will not cause icterus in some individuals.

In addition to the tests mentioned in the previous paragraphs, there are two useful serum enzyme tests available and widely employed. Hepatic **alkaline phosphatase** (AP) is an enzyme produced by both hepatocytes and bile duct epithelium and in cholestasis a fractional amount of it enters the serum to elevate the activity by up to 30 times the normal rate. Serum AP activity may also be elevated by the release of isoenzymes from other sources (e.g. bone, placenta, neoplasms), so on its own it should not be taken to indicate cholestasis. In the cat, AP has an extremely short half-life in the serum. This means that not only is an elevation highly significant, but also that normal values do not rule out moderate cholestasis. In the dog, glucocorticoids and some anti-convulsant drugs can also rapidly induce the overproduction of a hepatic AP isoenzyme, and this should be kept in mind in animals receiving such drugs.

The enzyme **γ-glutamyltranspeptidase** (GGT) originates in the biliary tract and canalicular zone of hepatocytes, and in cholestasis it enters the serum to excess. Serum GGT is a useful and relatively specific indicator of cholestasis on its own. In neonates, however, ingested colostral GGT, originating in the mammary gland, may significantly elevate serum activity.

Detection of damage to hepatocytes

When hepatocytes are damaged, intracellular enzymes escape into the blood and several of these are of great value as diagnostic aids. Even in the case of minor cell injury, without necrosis, increased permeability of the plasma membranes will allow enzymes from the cytosol ('cell sap') to escape to the serum. In the case of more severe injury and hepatocyte necrosis, enzymes from cell organelles, particularly mitochondria, will also escape. A significant elevation in the serum activity of such enzymes is a useful index in the diagnosis and management of liver disease. Once released into the blood, most enzymes have a half-life of 2–4 days, their molecules being too large to be cleared rapidly by glomerular filtration. Serum enzyme assays therefore can be used (1) to detect the presence of acute hepatocyte damage up to about two weeks after its occurrence, and (2) to assess the progress, if any, of ongoing damage. For instance, if serum enzyme activity falls by 50% over 2–4 days, it suggests that damage has ceased. Alternatively, if the activity stays constant or rises, ongoing damage is suggested. To be a perfect indicator, the enzyme, should be easily assayed, and should occur only in the liver. This ideal is not always possible, but is approached in several instances. In addition, the reversibility or otherwise of a disease process cannot be deduced from a single serum assay. It is the repeated series of assays that will help in this regard.

As ever, the clinician must be aware of species difference in the application of these tests. The enzyme assays most commonly employed are the following. **Alanine aminotransferase** (ALT) is a cytosolic enzyme and a specific indicator of liver damage in the dog and cat, but unfortunately of no value in the horse and cow. **Aspartate aminotransferase** (AST) is a mitochondrial enzyme and therefore an indicator of hepatocyte necrosis. As AST serum activity may also derive from muscle, it is wise to co-assay the sample for ALT.

Glutamate dehydrogenase (GLDH) appears to be a useful tool across species, as a specific indicator of hepatocyte damage. **Arginase**, **ornithine–citrulline transaminase** (OCT) and **sorbitol dehydrogenase** (SDH) are good indicators of parenchymal damage, but are seldom used routinely because of technically difficult assays.

It should be stressed that these tests merely indicate hepatocyte damage, but not its cause. Hepatocytes are damaged by any condition causing hypoxia, by direct intoxication, and by the presence of inflammation or neoplasia in the liver. These tests cannot discriminate between such processes.

The assessment of functional liver mass or effectiveness of hepatic blood flow

This style of test is usually employed in advanced chronic liver disease in order to help with prognosis. An animal which fails the test badly is likely to be at the end-stage and is probably untreatable. The traditional test has been the bromosulfonphthalein (BSP) excretion test but, unfortunately, the agent is becoming unavailable commercially. BSP is a dye which, when administered intravenously, is rapidly taken up by the normal liver, conjugated and excreted in the bile. When liver mass is significantly reduced, the BSP is retained in the plasma for a prolonged period. The test is based on determining its rate of disappearance from the plasma with time, thereby deriving a crude index of functional liver mass. Indocyanin green is an excellent alternative dye but is, unfortunately, expensive and therefore restricted largely to small animal use.

The assay of serum **bile acids** shows some promise of being a useful index of functional hepatic mass. The acids tend to accumulate in the blood when the liver is unable to take them up from the portal blood for re-excretion.

The **blood ammonia** concentration can also be used as a rough index of reduced hepatic mass or inadequacy of blood flow. Ammonia absorbed from the intestine is largely cleared from the portal blood by a normal liver, but escapes into the systemic circulation when the liver is compromised. Blood ammonia is also elevated in most animals with congenital portosystemic shunting.

As the liver synthesizes all the **serum albumin**, assay of this compound may be useful on occasions. Reduced serum albumin levels do not occur until 80% or more of the liver mass has been compromised. Thus, if serum albumin values are normal, one can deduce that the liver still has 20% at least of functional mass in terms of protein synthetic capacity. Similarly, a return to normal from a previously low albumin concentration suggests a favorable prognosis. The concurrent application of serum electrophoresis is helpful, because in chronic liver disease there is frequently an elevation of gammaglobulins, resulting from the reduced clearance of antigens from portal blood.

Clinical expression of liver disease

The clinical signs exhibited by an animal with liver disease may not always clearly signal that the liver is the seat of the problem. It may be helpful to analyze the question in basic terms so that clinical logic can be applied to particular cases. In the first place, disease within the liver may be (*a*) predominantly hepatocellular or (*b*) predominantly cholestatic. This statement is valid despite the fact that severe diffuse processes must inevitably involve all components. These two categories may now be further analyzed.

Predominantly hepatocellular disease

This will occur when some agent damages or destroys liver cells in large numbers. This is seen classically in the effects of the multitude of chemical hepatotoxins that reach animals both by accident and design, but infectious agents are equally culpable. Prominent amongst these are canine adenovirus I, *Leptospira* sp. and *Toxoplasma gondii*. Hepatocellular disease may be acute and of short duration or chronic and ongoing, and can be conveniently dealt with from this point of view.

Acute hepatocellular disease

Acute hepatocellular damage will become clinically significant when damage to hepatocytes is extensive enough to make the body 'aware' of the problem, even if most liver functions are intact. In other words there is active liver disease, but not total liver failure. Animals afflicted in this way will produce non-specific, unhelpful signs for the clinician – typically, depression, anorexia, mild abdominal pain and perhaps vomiting in the species able to do so. These signs are explicable on the following basis. The acutely damaged liver swells and places pressure on Glisson's capsule, causing pain, nausea and depression. There may also be some failure of clearance of metabolites from the blood which might affect the vomiting center in the brain directly and contribute to depression. Icterus may appear within several days, with roughly equivalent elevations in serum levels of free and conjugated bilirubin. In the absence of icterus, some retention of conjugated bilirubin is indicated by significant quantities of that compound in the urine. Liver cell damage can be detected by appropriate serum biochemical assays, and these tests provide the most direct evidence of a liver problem, although the enlarged and painful liver may on occasion be detected by palpation.

Such liver disease may well be secondary to another condition (for example severe anemia or cardiac failure) or part of a systemic disease process like toxoplasmosis. This should be kept in mind during the diagnostic work-up. Provided the noxious agent or condition is removed, the prognosis should be excellent in most cases.

Very extensive acute liver cell damage that surpasses the functional reserve of the organ and results in **acute liver failure** is an altogether different situation. Acute liver failure is a metabolic catastrophe, capable of causing death within a short time. There is the sudden onset of severe depression, vomiting (if possible), abdominal pain, weakness and collapse. Polydipsia and polyuria are common signs. Terminally, a comatose state may be reached following other neurologic signs of hepatic encephalopathy. Disseminated intravascular coagulation with resultant bleeding tendencies may be apparent, triggered by the consumption of clotting factors within the damaged liver. This may be contributed to by a defect in the production of serum clotting factors and by failure of clearance of fibrin degradation products. In those animals that survive more than 24 hours, intense icterus develops, with large elevations of both free and conjugated serum bilirubin, and the urine will be darkly discolored by large quantities of conjugated bilirubin. Serum enzymatic evidence of liver cell damage will be overwhelming and hypoglycemia is common.

Even if the patient survives the acute episode, it has to contend with the aftermath, as healing of the damaged organ proceeds. The healing process might be effective, but it can lead to such distortion that the animal is left with permanently defective liver function and may succumb to chronic liver failure.

Chronic hepatocellular disease

In chronic hepatocellular disease, damage to, and destruction of, hepatocytes proceeds over a prolonged period of time and is generally accompanied by increasing fibrosis and distortion of the liver. The prognosis is gloomy. Various aspects of liver function become defective with the passage of time and a fairly typical clinical picture emerges. The animal is generally in poor condition and lethargic and in most cases ascites develops. As the termination is approached, neurologic abnormalities emerge, taking the form of bizarre behavior, progressing to stupor and finally coma. Episodic seizures may occur. In herbivores, photosensitization is likely if chlorophyll intake is high and exposure to sunlight is intense. The clinical signs can be accounted for as follows.

Debility is the result of inefficient digestion of fats and absorption of fat-soluble vitamins, because of reduced quantity and quality of bile. There is also reduced synthesis of a range of serum proteins, including albumin and others important in mineral and hormone metabolism. **Lethargy** is contributed to by the above and perhaps by the persistence of toxic metabolites.

Ascites may be promoted in several ways:

1 By a reduction in serum albumin, causing reduced plasma osmotic pressure.
2 By portal hypertension.
3 By the retention of sodium and water by the kidneys. Failing liver function results in this effect on the kidneys for reasons as yet not clearly understood, but thought to be due in part to reduced catabolism of aldosterone. In end-stage liver disease, there may even be secondary **renal failure** – the so-called **hepatorenal syndrome**. The kidneys are structurally normal but unable to function.

Neurologic signs are the expression of hepatic encephalopathy, an autointoxication caused by accumulation of ammonia and other toxic metabolites (see below).

Gastrointestinal signs – particularly in the dog, gastric dysfunction is common in chronic severe liver disease. Gastric ulceration is probably promoted by alteration of the gastric mucosal barrier, reduced gastric blood flow and increased gastric acid production. These disturbances may be mediated by retained bile acids promoting the secretion of gastrin and by the inadequate hepatic deactivation of histamine, which promotes gastric acid secretion.

Photosensitization is the result of the accumulation of phylloerythrin, a photoactive metabolite of chlorophyll.

Icterus is not a common feature of this kind of disease process until an advanced stage, as the liver has a great reserve capacity to excrete bilirubin. It is characterized in this instance by large increases in both free and conjugated forms of serum bilirubin.

Hepatic encephalopathy

The clinical state known as hepatic encephalopathy deserves special mention. This is a neurologic syndrome resulting from the failure of the liver to clear toxins originating in the gut. It emphasizes the ability of the normal liver to detoxify the portal blood in one pass. The impairment of liver function may occur because of portosystemic shunting (congenital or acquired), hepatocellular disease, and rarely, because of a congenital deficiency in urea cycle enzymes.

The clinical signs are various, often intermittent and often related to eating. Behavioral changes are frequently described, including aimless wandering, head pressing and personality changes. Depression, blindness and stupor may be seen, or by contrast, there may be frenzy, hysteria or seizures.

The underlying neurotoxicity is multifaceted and is due to the combined effects of mercaptans, fatty acids, amines, aromatic amino acids, indoles and skatols. The common diagnostic index is the **blood ammonia** concentration, and ammonia is certainly a contributory toxin. However, it is important to note that the blood ammonia concentration does not invariably correlate with clinical encephalopathy. An animal with significantly elevated blood ammonia may not necessarily have clinical encephalopathy and vice versa. Despite this, blood ammonia assay remains a useful diagnostic tool. When blood ammonia concentration is high, ammonium biurates form and crystallize in the urine as characteristic microscopic crystals or greenish-colored uroliths. These are also useful diagnostically.

In animals with severe hepatocellular disease, encephalopathy will often accompany other liver-related signs like icterus. However, in animals with congenital portosystemic shunting, encephalopathy may be the only overt clinical sign.

Predominantly cholestatic disease

This occurs when the major functional abnormality is the suppression of bile production or bile flow. Clinical icterus should always raise the possibility of cholestatic disease, but it should be remembered that icterus may *not* occur, especially in dogs. The urine, however, will be deeply discolored by the passage of large amounts of conjugated bilirubin.

In herbivores, photosensitization can be the dominant clinical feature.

The blockage of bile flow may be initiated anywhere from the hepatocytes through to the common bile duct, and is usually defined as intrahepatic or extrahepatic. It is worth pointing out that, on clinical grounds alone, it can be very difficult to distinguish between intra- and extrahepatic cholestasis.

Intrahepatic cholestasis is the more common and is usually the result of toxic liver injury, causing concurrent hepatocellular disease. Some toxins directly inhibit the bile secretory mechanism of hepatocytes, thus cutting off bile flow at its origin in the canaliculi. The classic example of such compounds is lantadene, the toxic component of the plant red lantana (*Lantana camara*). Lantana-poisoned ruminants develop acute severe photosensitization and icterus. In man, cholestasis has also been associated with the administration of phenothiazine derivatives or estrogens, but this has not been seen in animals.

Other agents promote intrahepatic cholestasis by causing generalized damage to small and medium-sized bile ducts within the liver. A well-known toxin with this property is the mycotoxin sporodesmin, the cause of ovine 'facial eczema'; so named because of the severe hepatogenous photosensitization that occurs.

Extrahepatic cholestasis is the blockage of bile flow at some point between the liver and the duodenum. In man, 'gall stones' (choleliths) are common, but this is not the case in animals, where inflammatory or neoplastic lesions in the pancreas and duodenum are usually the culprits. For a time, intense icterus is the only significant abnormality resulting from extrahepatic biliary obstruction. If obstruction becomes chronic, the liver begins to suffer and hepatocyte degeneration occurs, with its appropriate biochemical and clinical alterations. This is due to a toxic effect of retained bile.

Pathologic patterns of liver disease and their clinical correlations

In many systemic diseases, evidence of some damage to the liver may be detectable by physical examination (e.g. swelling and tenderness) or laboratory tests (serum enzyme elevation). Because of its nature, the liver cannot avoid becoming distressed in such circumstances, an innocent bystander caught up in the sweep of events. A good example is the hepatic lipidosis that occurs in diabetes mellitus. Another is the hepatocyte necrosis that may occur in severe anemia. This fact should be kept clearly in mind – such circumstances do *not* constitute primary liver disease. In the following paragraphs attention will be focussed on the types of structural lesions that most commonly underlie primary clinical liver disease in animals. It is not the intention to give an exhaustive account of the pathology of the liver – this is dealt with appropriately in pathology texts.

It is also important to note that, while many causes of liver disease are defined, there are frequent occasions when a cause cannot be identified.

Acute hepatic necrosis

This is a disease state characterized by the sudden extensive destruction of liver parenchyma, sufficient significantly to impair total hepatic function, and in the extreme, to cause liver failure. The most common causes across species are drugs and chemicals whose interactions with the liver have previously been described. In addition, certain of the adenoviruses have this ability, especially canine adenovirus type I.

Many hepatotoxins are contained in plants grazed by herbivores, some are known poisons ingested accidentally, and some are commonly used drugs and anesthetics. The list is lengthy and need not be reproduced here. However, in regard to agents directly toxic to the liver itself, it is apparent that two classes may be recognized. Firstly, there are 'predictable' or 'intrinsic' toxins, which can be expected invariably to damage the liver in a dose-related manner. Common examples are phosphorus, carbon tetrachloride and the peptide toxins produced by blue-green algae. Secondly, there are 'unpredictable' or 'idiosyncratic' toxins, which can be harmless to most animals, or toxic to a certain individual on a certain day presumably because of shifts in drug metabolizing activity or hypersensitivity. In this category are numerous agents used for medical purposes in one way or another, making the clinician's life a little more difficult.

All hepatotoxic chemicals arrive in the liver predominantly via the portal blood and their effect is generally distributed evenly throughout the organ, producing a diffuse structural lesion. In rare circumstances a single bolus dose of a toxin may be directed to localized parts of the liver (e.g. one lobe only). This is possible because of laminar flow patterns in the portal vein.

The hepatocytes characteristically react to toxins by rapid degenerative change and necrosis, several patterns of which are recognized. In most acute toxic and adenoviral hepatopathies, inflammatory changes are not conspicuous and the lesions are, therefore, not classifiable as true 'hepatitis'.

Zonal necrosis is the term used to describe a regular diffuse pattern of necrosis related to the acinar units of the liver. The most common pattern is **periacinar necrosis** (also called centrilobular necrosis). It is the pattern produced by many toxic chemicals and by several adenoviruses, most notably canine and avian adenoviruses. In this situation, necrosis of hepatocytes occurs at the periphery of every liver acinus, adjacent to the terminal hepatic venules (central veins). This is largely due to the great drug metabolizing activity of these cells, and is also contributed to by their remoteness from a fresh flow of arterial blood with its oxygen and nutrients. Anoxic periacinar necrosis is also induced in severe congestive cardiac failure or anemia.

Periportal necrosis refers to the selective death of hepatocytes at the center of every liver acinus, adjacent to the portal triads. It is a less common pattern and is caused by a few toxins which directly affect the first cells encountered as they enter the sinusoids. **Midzonal necrosis** describes the death of hepatocytes in the middle of each acinus, with the periportal and periacinar cells surviving. It tends to be seen when animals have had their drug-metabolizing enzyme systems altered by exposure to inducing or suppressing agents.

In any form of zonal necrosis the liver becomes swollen and in many cases mottled, as the areas of necrosis highlight the acinar structure. The necrotic zones are red, as blood pools in the spaces vacated by the liquefied hepatocytes, while surviving zones are often yellowish, due to lipidosis of damaged but still-viable hepatocytes. The red and yellow zones are each a millimeter or so in diameter and form a regular pattern. Severe zonal necroses will be accompanied by clinical signs of acute hepatocellular disease or liver failure, as described above.

A single episode of acute zonal necrosis may resolve satisfactorily with removal of the causal agent and effective therapy. In most cases the stromal scaffold of the acini remains intact, mitotic division occurs in the surviving hepatocytes and perfect regeneration of the liver will occur over 7–20 days. In some cases, stromal destruction will inhibit regeneration and a certain amount of fine scarring will occur, which will, however, usually not prevent the return of normal function (Fig. 8.6).

Massive necrosis can be regarded as a step up in severity from zonal necrosis and may follow a large dose of a toxin that would induce zonal necrosis at a lower dose rate. In the pig, massive necrosis occurs in a complex nutritional deficiency involving vitamin E, selenium and sulfated amino acids. Massive necrosis reflects the wholesale destruction of acini, with surviving acini left in randomly scattered groups amongst the chaos. Sometimes, the process leaves whole lobes of the liver unscathed. Because of the wholesale removal of hepatocytes, large sections of the stroma collapse and the basic architecture of the liver is thrown into disarray. It is for this reason that the liver tends *not* to be swollen in massive necrosis and may in fact be reduced in size, with a flaccid and friable consistency. Fibrin oozes from the devastated liver and precipitates on the capsule, and a bloody exudate may accumulate in the peritoneal cavity.

It is not difficult to appreciate that acute liver failure may develop and death may occur rapidly. Should the animal survive the crisis of the acute disease, it may still have problems ahead. Because of the collapse of stromal scaffold, perfect regeneration is not possible after massive necrosis, and extensive scarring is inevitable. Hepatocyte regeneration from surviving acini produces nodules of parenchyma separated by broad bands of scar tissue ('nodular regeneration', 'post-necrotic scarring') (Fig. 8.7). Although a respectable mass of liver tissue is restored, it is not well aligned with blood flow.

There are two major outcomes possible in this situation. The nodular, scarred liver may reach a new equilibrium, and although distorted and small, may settle down to function adequately with a degree of resorption of

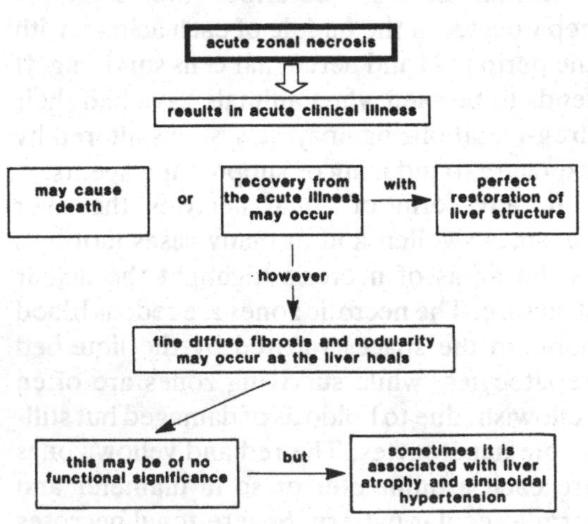

Fig. 8.6. The possible consequences of acute zonal necrosis.

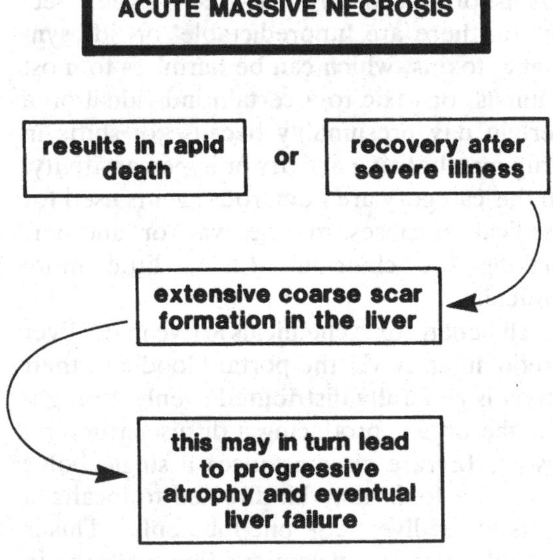

Fig. 8.7. The consequences of massive hepatic necrosis.

collagen. Sometimes, however, there is such distortion that too many hepatocytes have an inadequate relationship with the blood, often being impeded by collagen in the space of Disse. Blood flow itself is obstructed by the haphazard architecture, as is bile flow. Under these conditions, the liver begins to suffer loss of hepatocytes by atrophy, and increasing fibrosis. This is a path of no return, which leads to eventual portal hypertension and chronic liver failure. This may progress over months or years following the initial acute destruction. The result is a shrunken nodular liver.

Acute focal necrosis is the destruction of isolated subacinar masses of liver cells in a random pattern throughout the organ, and is mostly the result of the localization of micro-organisms. It is discussed more fully in relation to hepatitis (see below).

Acute individual cell necrosis (hepatocyte necrobiosis) is the loss by necrosis of single hepatocytes throughout the liver, with no particular zonal location. It is a lesion only definable by microscopic examination, and is non-specifically manifested grossly by swelling, pallor and slight mottling. The clinical expression is of acute hepatocellular disease. This lesion is seen classically in acute lupinosis of sheep, caused by ingestion of the mycotoxin phomopsin.

Chronic hepatic necrosis/fibrosis

This term encompasses a situation in which there is prolonged or repeated exposure to a hepatotoxic agent at dose rates which destroy hepatocytes in small numbers at any one time. There is a progressive loss of hepatocytes and a progressive increase in fibrous tissue, together producing a steady transformation of liver architecture. The liver parenchyma becomes evenly dissected into small (0.5–1 cm) nodules by a network of scar tissue (Fig. 8.8).

Eventually the collagen deposition disturbs the flow of blood and bile and impedes the normal relationship between hepatocytes and the bloodstream. Under these conditions the hepatocytes themselves may produce collagen, so hastening their own demise. As the process continues, the liver will become reduced in size and tough in consistency and clinical signs of chronic hepatocellular disease may emerge. An advanced state of fibrosis, nodular regeneration and ongoing degeneration is sometimes referred to as **cirrhosis** of the liver, and is 'end-stage' liver disease. This type of lesion has been described in dogs as an idiosyncratic response to long-term therapy with the anti-convulsant drug primidone.

An added twist to this forlorn scenario is mitotic blockade induced by several common hepatotoxins. The pyrrolizidine alkaloids are a large group of plant hepatotoxins, and a major problem in many parts of the world. They are usually ingested at low level by grazing animals, partly because of low palatability, and partly because of low alkaloid concentrations in the plants. They are a common cause of chronic liver disease, although quite capable of causing acute clinical effects at appropriate doses. In addition to causing death of some hepatocytes, they prevent regeneration by irreversibly blocking mitotic activity of dividing liver cells in prophase. These cells become greatly enlarged and eventually die. Enlarged hepatocytes are important diagnostic indicators and are known

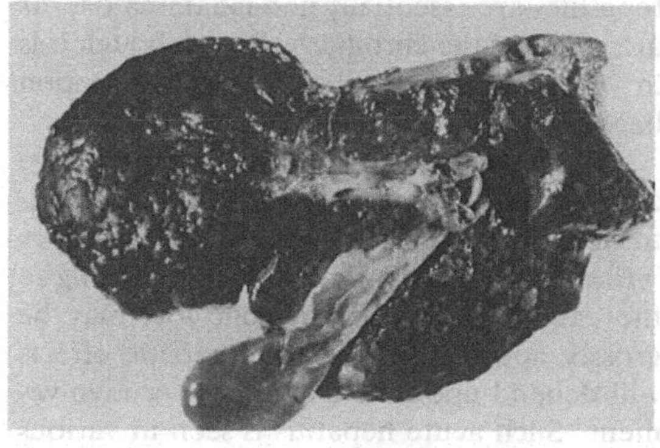

Fig. 8.8. An ovine liver showing the effects of chronic intoxication by pyrrolizidine alkaloids from *Heliotropum europeum*. There are atrophy, nodulation and diffuse fibrosis. Nodules of parenchyma are divided by a fine fibrous-tissue network.

as megalocytes. Some of the aflatoxins also have an anti-mitotic activity and cause megalocytosis.

The mycotoxin phomopsin causes the arrest of hepatocyte mitosis in metaphase and, again, this microscopic feature is useful in the diagnosis of lupinosis.

Clinically, chronic ongoing damage of this type usually reveals itself first as illthrift, although it may be associated with other signs. For instance photosensitization may be the critical manifestation in herbivores. Behavioral abnormalities and other nervous signs may predominate (hepatic encephalopathy). In sheep, chronically damaged livers will avidly store copper if conditions are suitable, and an acute hemolytic crisis may be the first indication of chronic liver disease. Towards its termination, such a chronic process may be associated with a range of clinical abnormalities indicative of liver failure and portal hypertension, as detailed earlier. Clinical signs may occur precipitately, mimicking acute hepatic disease, and the chronic nature of the tissue lesion may emerge only as investigation proceeds.

Hepatitis

The term hepatitis implies the coexistence of hepatocyte degeneration and necrosis with a significant influx of inflammatory cells. As previously discussed, most acute toxic hepatopathies are essentially non-inflammatory, as are many of the chronic variety, although it is in the chronic setting that inflammation becomes significant in toxin-related disease.

Acute hepatitis is usually due to an infective agent and is characterized by random multifocal necrosis attended by an inflammatory infiltrate that will vary in nature according to the causal agent. The necrotic foci may be grossly apparent and clinical hepatic effects will depend upon the extent of liver involvement. Such acute hepatitis is seen in various herpesvirus infections, in salmonellosis, in *Bacillus piliformis* infection (Tyzzer's disease), in toxoplasmosis and other parasitic infestations. As such, it is often part of a systemic infection with multiple organ system involvement. Recovery from such a disease would be expected to leave some focal scarring at the site of large lesions.

In sheep, severe and extensive multifocal acute hepatitis can occur when massive numbers of *Fasciola hepatica* larvae invade the liver. This can produce acute liver failure and the liver itself becomes swollen and congested, with a fibrinous exudate covering the capsule.

Chronic hepatitis has much more interesting associations from the perspective of clinical hepatology. While some infectious diseases like tuberculosis will produce focal chronic inflammatory lesions, major interest is centered on more diffuse chronic inflammatory lesions, sometimes caused by chemical agents. The mechanisms of liver damage encompass either direct conventional toxic damage, or damage inflicted indirectly by hypersensitivity, induced autoimmunity or idiosyncratic reaction. The pathology features ongoing, widespread, focal hepatocyte necrosis, with accumulation of lymphocytes, plasma cells and neutrophils, particularly in portal areas. Liver acini are obliterated and portal areas become connected by bridging fibrosis. In the advanced and terminal stages, cirrhosis of the liver is produced, as described above.

Recently, the diagnosis of **chronic active hepatitis** has been documented in the dog. The term is borrowed from human medicine, where it refers to an autoimmune disease with a list of clinical and pathologic criteria for its diagnosis. Piecemeal necrosis of hepatocytes at the 'limiting plate' (abutting portal tracts) is a major pathologic feature. Some diseases with similar features have been described in animals, and in the future may be shown to be analogous to the human disease, but as yet are probably best classified as chronic hepatitis of unknown pathogenesis. A newly recognized transmissible hepatitis has been described in dogs in the United Kingdom.

Chronic hepatitis, due to **copper storage disease**, has also been well documented in Bedlington Terriers, Dobermans, and West

Highland Terriers. This is a genetically determined metabolic error resulting in accumulation of copper by hepatocytes. A chronic progressive hepatitis leading to cirrhosis is produced, caused by the toxic effect of the stored copper. The copper storage can be assessed chemically and can be visualized by light microscopy with application of appropriate stains.

Cholangiohepatitis

This term denotes an inflammatory reaction within the liver, centered on the biliary duct system and spilling out to involve the periportal liver parenchyma.

Acute cholangiohepatitis may be caused by bacterial infection either ascending from the intestine or descending the biliary tract. It may also be related to the administration of drugs, and exposure to various mycotoxins, notably sporodesmin. The clinical signs will generally reflect acute **cholestasis**, and therefore intense icterus, along with indications of damage to hepatocytes. In cases due to bacterial infection, fever and leukocytosis may be expected. In herbivores, photosensitization is often predominant.

The acutely affected liver will generally be swollen and may be mottled in an arborizing pattern related to bile ducts and ductules. The liver may be diffusely tinged a yellow/green color due to the retention of bile within it. Following resolution of the inflammation, repair will occur with an often explosive proliferation of bile ductule epithelium within portal triads, and a residual scarring process that will link portal triads together ('portal fibrosis'). If the portal fibrosis is sufficiently intense, the liver will be permanently nodular and toughened, but may function adequately. However, progressive fibrosis may ensue, especially if acute insults are repeated.

Chronic cholangiohepatitis, as the name indicates, is a slowly progressive lesion characterized by steady destruction of biliary ducts and adjacent hepatocytes, the accumulation of inflammatory cells in portal triads, portal fibrosis and varying degrees of bile ductule proliferation. The liver is eventually rendered unfit for service, with much of it replaced by scar tissue radiating from thickened bile ducts. In spite of extensive fibrosis, the liver is usually not greatly reduced in size. The classic example of this disease is chronic fascioliasis in sheep. In other circumstances however, the cause remains obscure and it is quite possibly an immune-mediated or autoimmune disease. Idiopathic chronic cholangiohepatitis is reasonably common in the cat, but rare in other species.

Initially, the clinical manifestations in all species encompass a failure to thrive, together with gradually emerging and worsening indications of chronic liver disease. Usually bile flow remains adequate and icterus is not always a feature, although the degradation of bile quality no doubt affects digestion and absorption of nutrients. In ovine liver fluke disease, the clinical effects are largely due to extensive loss of protein and blood from the biliary tract, and then from the body.

Lipidosis and steroid hepatopathy

Most commonly, **lipidosis** ('fatty liver') represents a massive storage of triglyceride by the hepatocytes, and the most common basic cause is the large-scale mobilization of lipids for energy metabolism. This is triggered by the unavailability of other sources of biochemical energy and is the reason for the development of a swollen fatty liver in **diabetes mellitus** (insulin deficiency), when carbohydrate, as glucose, cannot be utilized because of its inability to enter cells. There is a similar state in **ketosis**, and **pregnancy toxemia** of ruminants, or when fat animals are suddenly starved or become anorexic. The liver, confronted by a huge incoming load of mobilized fat, is simply unable to process it, and therefore the fat accumulates. A fat-laden liver still functions reasonably well but, in extreme cases, cellular swelling may impede blood and bile flow, and liver function may suffer a little. Outright liver failure does not occur.

Marked lipidosis may also occur in toxin-damaged liver but tends to be overshadowed by necrosis. In this case, lipid accumulates because damage to hepatocytes impairs their ability to secrete lipoproteins.

Severe hepatic lipidosis will be clinically detectable by enlargement of the liver (heptomegaly). On gross inspection, the liver is yellow/bronze in color and 'greasy' in consistency. The lesion is entirely reversible once metabolic conditions have returned to normal.

Steroid hepatopathy is a pathologic state of the liver induced by a prolonged excess of glucocorticoids, and is recorded only in the dog. The steroids may be endogenous in origin, produced by an inherently abnormal endocrine system, or may be exogenous, administered for therapeutic purposes. Marked swelling and pallor of the liver may develop as a result, being the net effect of hepatocyte swelling. Hepatocytes do not contain much lipid, but are swollen by the accumulation of water and glycogen in the endoplasmic reticulum. Scattered individual hepatocytes become necrotic and attract a few neutrophils. Thus, hepatomegaly and serum-enzymatic evidence of hepatocyte damage may be detected clinically. The pathogenesis of this steroid-induced degenerative change remains unexplained at present, but it appears to be reversible if the hormone excess is corrected.

Developmental vascular anomaly

This is a set of conditions, found to be quite common as a cause of portosystemic shunting and hepatic insufficiency in young dogs and also occasionally in other species. The crux of the problem is the shunting of portal blood directly to the systemic veins, greatly reducing the amount perfusing the liver. Anatomical studies have revealed a variety of anomalous vascular arrangements linking the portal vein to the vena cava or azygos vein. No definite cause has so far been identified, but certain breeds of dog seem to be predisposed. Regardless of anatomical defect, the underperfused liver fails to develop fully and is characteristically undersized, although initially normal in contour and in most other respects. This reflects a deficit in hepatic growth factors normally delivered in the portal blood. With time, the progression of atrophic changes may lead to architectural abnormality.

Affected animals are often stunted, and the underlying anomaly is usually expressed as hepatic encephalopathy – neurologic disturbances frequently related to feeding. Determination of blood ammonia concentration is a useful diagnostic aid, particularly after the application of an ammonia tolerance test.

Neoplastic infiltration

In some neoplastic diseases, especially those of lymphoreticular or bone marrow origin (leukemias), malignant cells may diffusely infiltrate the liver on a massive scale, producing clinical signs of liver disease. The infiltrated liver becomes greatly enlarged, pale and friable. Initially, the neoplastic cells usually crowd into portal triads and then spill out into the parenchyma, displacing and destroying hepatocytes and impeding blood and bile flow.

The progression of clinical signs will vary according to the speed of infiltration and will of course involve other manifestations of the malignancy apart from those relating to the liver.

Primary liver neoplasms that can diffusely invade the liver itself include hepatocellular carcinoma, bile ductular carcinoma and hepatic carcinoid. The last-named tumor arises from the neuroendocrine cells (APUD cells) of the bile ducts.

The exocrine pancreas

Functional anatomy

The exocrine pancreas has a major role to play in the digestion of all classes of food within the upper small intestine. It is a large bilobed and lobulated mass of glandular tissue, adjacent to the duodenum, and consisting overwhelmingly of exocrine acinar cells. These cells secrete into an extensive lobular duct system and ultimately into a large common duct directed into the duodenum.

Pancreatic acinar cells produce a range of enzymes which, between them, suffice to degrade most classes of macromolecules. Enzymes of this potential are obviously hazardous for living tissues and for this reason are generally stored within the acinar cells in an inactive form, in zymogen granules. The acinar cells secrete their stored enzymes when stimulated by the hormone secretin (from the duodenal mucosa) and cholecystokinin–pancreozymin (from the upper intestinal mucosa). These hormones are in turn secreted by intestinal neuroendocrine cells when acid, undigested fats and peptones enter the small intestine. The epithelium of the pancreatic duct secretes a highly alkaline bicarbonate-rich fluid to assist in creating the correct environment for enzyme activity within the intestinal lumen.

The initially inactive pancreatic enzymes are activated in the intestinal lumen where they reduce complex macromolecules to small compounds suitable for absorption across the intestinal wall. Some enzyme activity survives passage through the gut and can be detected in the feces.

Significant enzymes in relation to the genesis and diagnosis of pancreatic disease are lipase, amylase, trypsin, carboxypeptidase, phospholipase, collagenase, elastase, kallikrein and ribonuclease.

From the clinical standpoint, two principles should be kept in mind. The first is that adequate pancreatic function is essential for digestion, and the second is that the pancreas is rather like an explosives store. Should its enzymes be prematurely activated, destruction and chaos ensue.

Clinical, pathologic and pathogenetic features of exocrine pancreatic disease

In veterinary medicine, clinical disease of the exocrine pancreas is almost exclusive to the dog, although pancreatic tissue lesions are often found across species. In sheep, severe pancreatic atrophy has recently been described as a major lesion of zinc toxicity. Clinical exocrine pancreatic disease falls into two distinct syndromes, those of pancreatic insufficiency and acute pancreatitis.

Pancreatic insufficiency

When inadequate quantities of pancreatic enzymes are delivered to the intestine, the digestion of all classes of foodstuffs, but particularly fats and proteins, is severely diminished. The patient virtually begins to starve in spite of eating voraciously. Initially the patient is alert and responsive, but prolonged maldigestion leads to cachexia and debility with a dull lusterless hair coat and dry skin. The clinical signs relate clearly to an inability to utilize food because of maldigestion. Typically, there is much food eaten and large quantities of feces are passed which have an abnormally high content of fat and protein. The passage of voluminous fat-laden feces is called steatorrhea.

Pancreatic insufficiency may arise in adult animals following fibrosis and atrophy of the pancreas (Fig. 8.9). As this process develops, functional reserve is finally exhausted and clinical signs appear. In the dog, this usually follows repeated clinical or subclinical attacks of pancreatic necrosis, while in the cat there

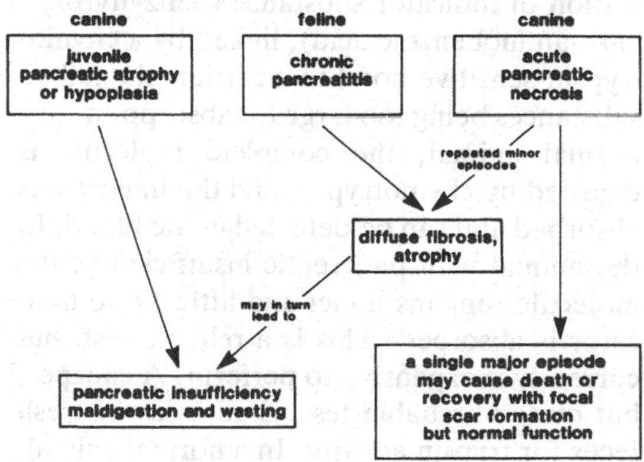

Fig. 8.9. Outline of major disease processes of the exocrine pancreas and their functional consequences.

appears to be a progressive chronic idiopathic pancreatitis, which is possibly immune mediated. By the time clinical disease appears, the pancreas is nodular and tough and greatly reduced in size. However, the surviving pancreas usually contains enough islets of Langerhans to prevent diabetes mellitus and it is rare for this complication to occur.

Pancreatic insufficiency has also been recognized in young dogs, with clinical signs appearing before twelve months of age (Fig. 8.9). There are two suggested mechanisms. One is developmental hypoplasia of exocrine pancreas. Under these circumstances, dogs initially compensate successfully by means of salivary and intestinal enzymes, and decompensate as they grow. The alternative view is that the pancreas is normally formed at birth, but undergoes progressive regression and atrophy, in the manner of an abiotrophy. This term indicates a genetic error causing premature senility of the affected cells. In either case, the pancreas appears as a few scattered lobular remnants in loose connective tissue.

The clinical diagnosis of pancreatic insufficiency is confirmed by using tests to detect both maldigestion and a deficit in enzyme delivery to the gut. Undigested fat and muscle fibers can be detected in feces by simple microscopy and this is a most useful technique. Another test uses the oral administration of indicator substance (benzyltyrosyl-*para*-aminobenzoic acid), linked by a chymotrypsin-sensitive bond to a carrier, the linked substances being too large for absorption. In a normal animal, the complete molecule is digested by chymotrypsin and the indicator is absorbed and can be detected in the blood. In the animal with pancreatic insufficiency, the molecule remains intact and little or no indicator is absorbed. This is a reliable test, but currently is expensive to perform. A cheaper, but quite unreliable test, is an assay of fresh feces for trypsin activity. In a normal animal, trypsin activity should be quite high, whereas in the diseased animal it will be minimal or absent.

Acute pancreatitis (acute pancreatic necrosis)

This is manifested as the sudden onset of dramatic abdominal pain and shock due to acute destruction of pancreatic and adjacent tissues.

In the dog, the only domestic animal afflicted by this disease, the triggering cause is unknown, but the pathogenesis is dependent on the activation and release of pancreatic enzymes into and around the pancreas. The affected tissues undergo autodigestion and a fulminating acute inflammatory reaction is set in train, exacerbated by the activity of pancreatic kallikrein. The typical patient tends to be obese and middle aged and clinical signs usually appear soon after eating.

The lesion usually occurs focally around the head of the pancreas and an early feature is necrosis of fat which appears as intensely white opaque masses surrounded by hemorrhagic inflammation in interlobular septa. The destructive process spreads into the glandular lobules with liquefactive necrosis and inflammation. Sometimes the process is complicated by the invasion of intestinal bacteria via the pancreatic duct. Inflammatory swelling usually extends into the wall of the duodenum and sometimes obstructs the common bile duct. The affected pancreas is swollen, firm, reddened and covered by omental adhesion. It exudes a serosanguineous exudate into the peritoneal cavity.

The explosive nature of the process results in the release of large quantities of pancreatic enzymes into the peritoneal cavity and ultimately the bloodstream, and this is utilized in diagnosis. The activities of **amylase** and **lipase** in serum or peritoneal fluid are commonly assayed for diagnostic and prognostic purposes, but unfortunately are not reliable indicators.

The lesion produces intense pain and severe vomiting and, in some cases, death from shock occurs rapidly. In other cases, clinical signs may subside over several days and the pancreatic lesion resolves by fibrosis to leave a permanent deficit in pancreatic tissue, with omental adhesions and focal mineralization. Provided the area of involvement is not large, the remaining pancreas has the capacity to sustain normal digestion, but repeated attacks may occur, with progressive loss of functional tissue, as the exocrine pancreas has limited capacity for regeneration (Fig. 8.9).

The intense acute destructive inflammation also means that a neutrophilia may be anticipated, especially if secondary bacterial infection has occurred, and biliary obstruction sometimes causes icterus.

As in man, this propensity for pancreatic self-immolation remains an enigma and despite much investigation no etiologic culprit has come to light.

Exocrine pancreatic neoplasia

Pancreatic adenocarcinomas, of ductal or acinar origin, are reasonably common in aged dogs and cats. They usually metastasize extensively to the liver, and clinical signs may be dominated by its destruction. There are, typically, liver enlargement, ascites and icterus, following a period of weight loss and increasing listlessness. In other cases, the neoplasm may obstruct the common bile duct early in the course of disease, and intense obstructive jaundice may be the presenting sign. By the time of diagnosis the prognosis is usually hopeless.

Additional reading

Arias, I., Popper, H., Schaefer, D. and Shafritz, D. A. (1982). *The Liver, Biology and Pathobiology*. New York, Raven Press.

Blood, D. C., Henderson, J. A. and Radostits, O. M. (1979). *Veterinary Medicine*, 5th edn, pp. 200–8. London, Bailliere Tindal.

Coles, E. H. (1974). *Veterinary Clinical Pathology*, 2nd edn. Philadelphia, W. B. Saunders Co.

Duncan, R. J. and Prasse, K. W. (1986). *Veterinary Laboratory Medicine: Clinical Pathology*, 2nd edn. Ames, IA, Iowa State Univ. Press.

Hardy, R. M. (1983). Disease of the liver. In *Textbook of Veterinary Internal Medicine*, vol. 2, S. J. Ettinger (ed.), pp. 1372–434. Philadelphia, W. B. Saunders Co.

Kelly, W. R. (1985). The liver. In *Pathology of Domestic Animals*, 3rd edn, K. V. F. Jubb, P. C. Kennedy and N. Palmer (eds.), pp. 239–312. Orlando, FL, Academic Press.

Rogers, W. M. (1983). Exocrine pancreatic diseases. In *Textbook of Veterinary Internal Medicine*, vol. 2, 2nd edn, S. J. Ettinger (ed.), pp. 1435–55. Philadelphia, W. B. Saunders Co.

Slater, T. F. (1978). *Biochemical Mechanisms of Liver Injury*. New York, Academic Press.

Strombeck, D. R. (1979). *Small Animal Gastroenterology*. Davis, CA, Stonegate Publishing.

Tennant, B. C. and Hornbuckle, W. E. (1980). Diseases of the liver. In *Veterinary Gastroenterology*, N. V. Anderson (ed.), pp. 593–620. Philadelphia, Lea and Febiger.

Thornburg, L. P. (1982). Diseases of the liver in the dog and cat. *Compend. Cont. Educ. Pract. Vet.* **4**: 538–46.

Twedt, D. C. (1985). Liver diseases. *Vet. Clin. N. Am.* **15** (1).

Wisse, E. and Knook, D. L. (eds.) (1977). *Kupffer Cells and Other Liver Sinusoidal Cells*. Amsterdam, Elsevier/North Holland Biomedical Press.

Zakim, D. (1985). Pathophysiology of liver disease. In *Pathology, The Biological Principles of Disease*, 2nd edn, L. H. Smith and S. O. Thier (eds.), pp. 1253–96. Philadelphia, W. B. Saunders Co.

Clive R. R. Huxtable

9 The urinary system

The mammalian urinary system is constructed to generate, store and finally expel the familiar liquid known as urine, a product extracted from the flowing bloodstream by the kidneys. Disorders of the urinary system relate largely to three activities – **production**, **conduction** and **storage** of urine. A failure of any of these activities threatens the health, and ultimately the life of the afflicted animal, which is another way of saying that the urinary system is a vital and indispensible organ system. The total failure of urine formation or expulsion will lead to death in a few days, while lesser degrees of malfunction can give rise to numerous clinical abnormalities.

The kidneys create urine as a vehicle for the elimination of toxic or useless metabolic end products while simultaneously regulating the homeostatic balance of a range of crucial substances. The kidneys are also the site of production of several important hormones and of the degradation of others. The activity of the kidneys is tightly integrated with, and greatly influenced by, the cardiovascular system, the neurohypophysis, the adrenal cortex and the parathyroid gland. Any primary disease in one may be reflected in disturbances of function in the other.

In the lower urinary tract, the storage and voiding of urine are heavily subject to, amongst other things, the state of the autonomic nervous system, spinal cord and pelvic nerves.

In the face of such a diversity of possible clinical ramifications and interactions, the clinician has no choice but to come to grips with the fascinating complexities of the urinary system in health and disease.

The kidney

(Basic relationships between structure and function: principles of dysfunction relating to filtration of the blood and modification of the filtrate; hormone production; functional reserve; pathophysiology of uremia; urinary cast formation)

Although there are some obvious gross anatomical differences between the kidneys of various domestic animals, the basic design is the same. However,most of the available data on function and malfunction derive from studies of the dog and, to a lesser extent, the cat, and may not be entirely applicable across species. This fact of veterinary clinical life is familiar and the deficiencies will no doubt be corrected with time.

The kidneys, as organs, represent the division of the total nephron mass into halves which normally function symmetrically. The kidneys contain several million nephrons and are constructed to ensure the vitally important relationship between them, the blood vessels and the interstitial tissue.

The urine that flows from the renal pelvis into the ureter is the product of two major processes. The first is **the filtration of blood**, the second **the modification of the resulting**

filtrate. These two processes constitute the major part of the kidneys' activity, as they play their role of sensitive servant to the body, organizing the expulsion of the unwanted and the conservation of the needed.

Filtration of the blood

Filtration of the blood is the first step in urine formation and takes place in the renal glomeruli. The highly specialized walls of the glomerular capillaries constitute the filter, providing a large total filtration surface (Fig. 9.1). To this large filtration surface a high filtration pressure is applied, regulated by the tonic balance between the afferent and efferent arterioles of each glomerulus.

Great quantities of water, ions and small molecules easily cross the filter and separate from the plasma, but large molecules are held back. This **filtration barrier** against the larger molecules operates on the basis of molecular size, shape and electrical charge. In general, molecules of molecular weight 50000 and above are large enough that their size alone precludes their passage through the filter. Smaller molecules whose molecular weight is between 10000 and 50000 may be turned back at the filter by virtue of their shape and/or their electrical charge. The filter itself carries a strong net-negative charge and will electrostatically repel negatively charged molecules which might otherwise pass through. The importance of the filtration barrier will become evident in later sections.

The blood filter must do the things described above at a rate sufficient to satisfy the considerable demand created by the normal metabolic activity of the whole body. This is achieved by a large supply of blood to the system, a considerable total filtration surface and an adequate hydrostatic pressure to drive filtration.

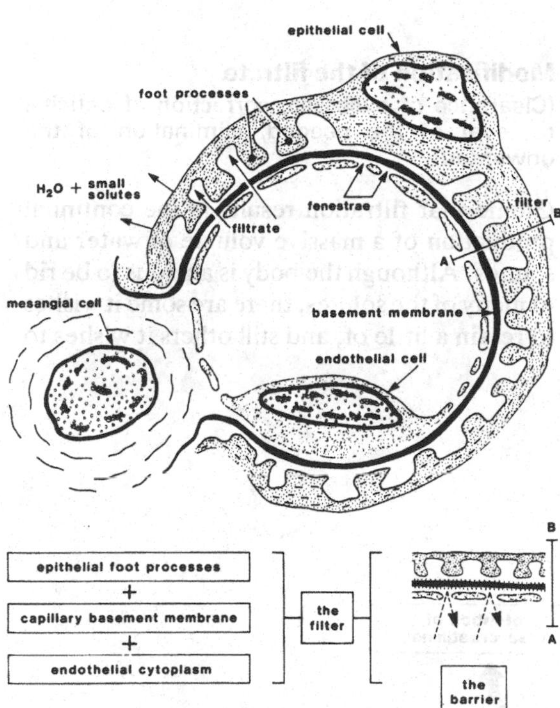

Fig. 9.1. A glomerular capillary in cross-section. The unique structure of the capillary wall provides for the filtration of water and small solutes and the retention of protein molecules.

Disorders of blood filtration
(Azotemia, barrier failure)

The filtration performance of the kidneys can be assessed by determining the **glomerular filtration rate** (GFR), which is a measure of the volume of plasma filtered in a given time. Normally about 20% of the plasma delivered to the kidneys is filtered. Precise measurements of GFR are difficult and not commonly undertaken in clinical practice, but there are some quick and easy ways to obtain a fairly accurate estimate. A reduction in GFR will result in an increase in the plasma concentration of substances cleared by glomerular filtration. Two such substances are creatinine and urea, both small nitrogenous molecules, easily assayed, and derived endogenously from skeletal muscle and liver, respectively. This laboratory finding is referred to as **azotemia**, i.e. a state when blood urea and creatinine concentrations are greater than normal. Thus, a finding of azotemia can be taken to indicate a strong probability of reduced GFR. Be cautioned at this stage, however, that azotemia may sometimes occur with a normal GFR, if for example there is an abnormally high rate of urea formation by the liver. Pitfalls in interpretation are well covered in clinical urology and clinical pathology texts.

In what situations could reduced filtration occur? If the delivery rate of blood to the filter is inadequate, then the GFR will fall, even if the kidneys are basically healthy and the filter normal. This occurs in cardiac or circulatory disorders which reduce renal blood flow and pressure. The resulting azotemia is termed **prerenal** because the basic problem is 'in front of' the kidneys. If the filter itself is damaged because of extensive renal tissue disease, the GFR may fall and produce in this case a **renal** azotemia. The problem this time is centered 'at home', within the kidney tissue.

In addition, the free flow of urine from the lower urinary tract may be interrupted, resulting in an effective decrease of the GFR. In this instance there is **postrenal** azotemia because the problem is centered 'after' the kidneys, in terms of urine flow.

In summary, it can be seen that blood filtration can be reduced by prerenal, renal, or postrenal factors, and that the reduced GFR will produce azotemia, which is measured as the blood concentration of creatinine and/or urea (Fig. 9.2).

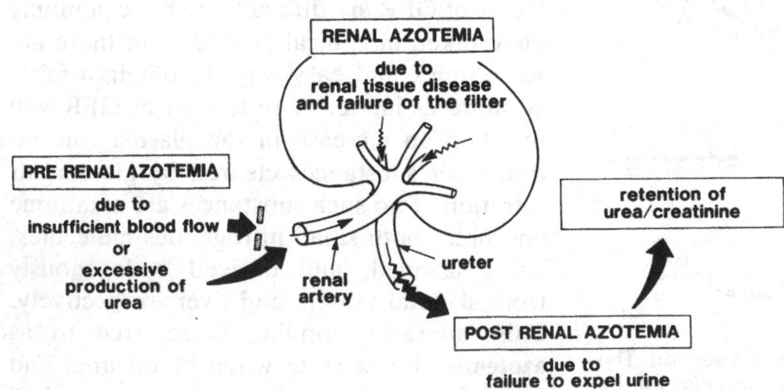

Fig. 9.2. The possible origins of azotemia.

The process of blood filtration can also be disturbed by a failure of the **glomerular filtration barrier** (Fig. 9.1). If the barrier fails, large molecular weight substances, principally plasma protein, will cross the filter and ultimately appear in the urine to produce proteinuria. It is mostly albumin that leaks across, but in severe barrier failure even larger molecules such as globulins will cross the line. This type of filter malfunction is always renal in origin. That is, it is always the result of some type of glomerular lesion. Stated conversely, widespread glomerular disease will always cause a measure of barrier failure and proteinuria.

An important point to note is that severe barrier failure *may not* be accompanied by a significant reduction in GFR. Sometimes, it is possible to see patients with massive proteinuria but no azotemia. Many patients, however, will be both proteinuric and azotemic.

Modification of the filtrate

(Clearance of excesses, correction of deficits, retrieval of the needed, elimination of the unwanted)

Glomerular filtration results in the continual production of a massive volume of water and solutes. Although the body is anxious to be rid of many of the solutes, there are some it wishes to retain a little of, and still others it wishes to conserve avidly. The loss of all the initial glomerular filtrate from the body would be totally unacceptable – the loss of the water alone would necessitate continual drinking. The urinary system therefore requires the means of combating unacceptable losses of water and valued solutes, while retaining the ability to eliminate the unwanted, both endogenous and exogenous. This of course requires scope for further addition of substances from the blood to the filtrate as a 'back up' to glomerular filtration.

Following glomerular filtration, the newly formed filtrate and the newly filtered blood flow in close proximity, allowing for further exchanges of water and solutes between them (Fig. 9.3). These exchanges involve both passive diffusion mechanisms, and active pumping of solutes against concentration gradients. In addition, the ascending limb of the loop of Henle has the special property of being impermeable to water, and provides the facility for active dilution of the filtrate (Fig. 9.4). This is achieved by the active pumping of chloride ions out of the tubule lumen, with concurrent egress of sodium ions by electrostatic attraction. This mechanism serves to dilute the filtrate and helps to concentrate the medullary interstitium. The renal interstitium is vitally involved in nephron function, especially in the medulla where the countercurrent mechanisms of the long loope of Henle (of juxtamedullary nephrons) generate a high interstitial osmolality (Fig. 9.4).

The passage of the collecting ducts through

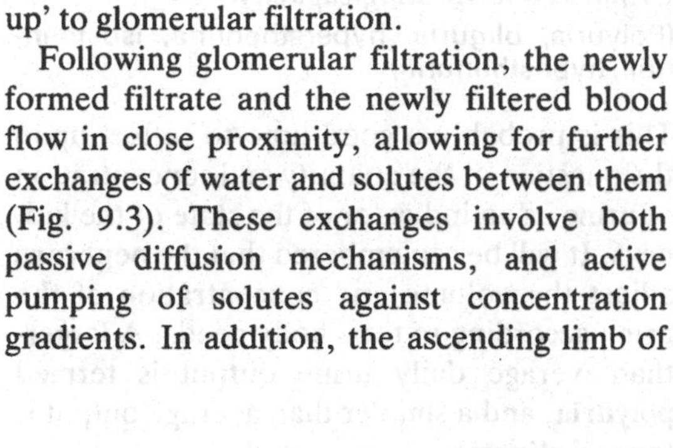

Fig. 9.3. Nephron structure showing the distribution of arterial blood, cortical and juxtamedullary nephrons and the close relationship between renal tubules and postglomerular capillaries. (1) Proximal convoluted tubule; (2) proximal straight tubule; (3) descending limb of Henle; (4) ascending limb of Henle; (5) ascending thick limb of Henle; (6) distal convoluted tubule; (7) connecting tubule; (8) collecting duct.

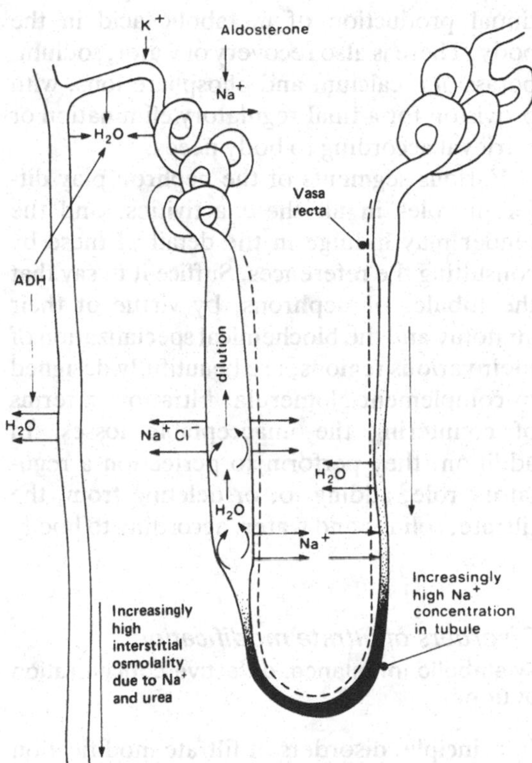

Fig. 9.4. Some urinary concentrating mechanisms in a juxtamedullary nephron. In particular, the active diluting capability of the thick limb of Henle plays a vital role in allowing for either the excretion of free water or the conservation of water via the effects of anti-diuretic hormone (ADH) and a concentrated medullary interstitium.

the concentrated medullary interstitium gives rise to a steep osmotic gradient between the urine in the collecting ducts and the fluid in the interstitium. The wall of the collecting duct, the last part of the nephron, is permeable to water *only* in the presence of anti-diuretic hormone (ADH). In this region, the filtrate can be concentrated if necessary by the osmotic movement of water from the lumen of the tubule into the medullary interstitium.

From a clinical standpoint, the most relevant of these tubular functions are as follows. There is virtually total recovery of glucose and amino acids. There is large-scale elimination of acid in the form of hydrogen ions and organic acids, with a concurrent large-scale recapture of bicarbonate ions (see Chapter 4). This is necessitated by the continual production of metabolic acid in the body. There is also recovery of water, sodium, potassium, calcium and phosphate ions, with provision for a final regulatory elimination or retrieval according to body needs.

Various segments of the nephron play different roles in all these activities, and the reader may indulge in the detail of these by consulting the references. Suffice it to say that the tubules of nephrons, by virtue of their anatomy and the biochemical specialization of their various regions, are beautifully designed to complement glomerular filtration in terms of countering the unacceptable losses. In addition, they perform to perfection a regulatory role, adding to, or deleting from, the filtrate, solutes and water, according to bodily needs.

Disorders of filtrate modification

(Metabolic imbalance, defective concentration of urine)

In principle, disorders of filtrate modification relate to excessive retention, or excessive loss, of particular substances from the body, most notably sodium, potassium, calcium, phosphate, acid and water.

Such a disorder may have a prerenal basis. For instance an excess or deficit in the activities of aldosterone or ADH will compromise the efforts of perfectly normal kidneys. On the other hand, such a disorder may reflect a biochemical or structural renal lesion. An example is the 'Fanconi syndrome', in which the tubules are biochemically incompetent with regard to absorption of glucose, amino acids and phosphate. The massive urinary loss of these produces clinical disease.

Urinary concentrating capacity

(Polyuria, oliguria, hypersthenuria, isosthenuria, hyposthenuria)

This is probably a good time to reflect upon the question of the quantity and concentration of urine as an indicator of the state of the kidneys. It will be remembered that the nephrons adjust the volume and concentration of the urine according to total body needs. A larger-than-average daily urine output is termed **polyuria**, and a smaller than average output is termed **oliguria**.

A normal animal with unlimited access to water will generate daily between 30 and 50 milliliters or urine/kilogram of body weight, and the concentration of the urine will be on average somewhat higher than that of the plasma. If the normal animal drinks, or is given a large load of water, the kidneys will clear this water load by generating a large volume of dilute urine, with a concentration less than that of plasma. Alternatively, if this animal is deprived of water, the kidneys will strive to preserve body water by generating a small quantity of urine with a concentration four to eight times that of plasma.

This knowledge provides the basis for clinical testing of the concentrating ability of the kidneys. The concentration of the urine is routinely estimated by determining its specific gravity and/or its osmolality (Osm). The term **hypersthenuria** describes fully concentrated urine whose concentration is significantly higher than that of plasma. This means a specific gravity of greater than about 1.025, and a urine:plasma osmolality ratio (U_{osm}:P_{osm}) of at least 2 or 3. **Isosthenuria** describes urine whose concentration is about equal to that of plasma, i.e. equal to that of the glomerular filtrate. The specific gravity is in the range 1.008–1.012 and the $U_{osm}:P_{osm} = 1$. **Hyposthenuria** describes urine whose concentration is significantly less than that of plasma. The specific gravity is 1.008 or below, and $U_{osm}:P_{osm} < 1$.

It is apparent from the above that a normal animal can produce urine across the whole concentration range. Impairment of renal concentrating ability can be detected if the water consumption of an animal is controlled. When deprived of water, the animal with severely impaired renal function will continue to make insufficiently concentrated urine. In dogs, if after 24 hours of water deprivation, the urine specific gravity does not reach at least 1.030 (in cats 1.035) and $U_{osm}:P_{osm}$ does not reach 4 or 5, the animal is classified as having defective renal concentrating function. If the urine specific gravity remains in the isosthenuric range and $U_{osm}:P_{osm}$ at 1, then the functional failure is total, as the tonicity of glomerular filtrate has been unchanged. This is called fixation of urine concentration.

The foregoing is an outline of the water deprivation test. Remember that if an animal fails the test it might have any one of the following three abnormalities:

- A defective production of ADH } but essentially normal kidneys
- A problem with tubular receptors for ADH } but essentially normal kidneys
- An insufficient number of nephrons to create and maintain a concentrated medullary interstitium. This means a loss of 65–70% of functional nephrons, due to severe renal disease.

These conditions may be distinguished in a number of ways (Table 9.1). The water-deprived animal with a deficiency of ADH should, of course, respond to injected ADH and concentrate its urine. It could also produce dilute urine if necessary. Similarly, the animal with ADH receptor problems could still actively dilute urine in response to a water load and could, therefore, generate urine with a specific gravity of less than 1.008 or $U_{osm}:P_{osm} < 1$. It would not, however, respond to ADH if water deprived. In neither type of ADH defect would azotemia be expected.

Table 9.1. *Major mechanisms of urinary concentrating defects*

Pathophysiologic state	Result
1 Lack of production or release of ADH	Ability to make dilute urine. Inability to make concentrated urine. Normality restored by ADH. No azotemia
2 Lack or blockade of ADH receptors	Ability to make dilute urine. Inability to make concentrated urine. No response to ADH. No azotemia
3 Insufficient nephrons	Inability either to concentrate or to dilute urine. No response to ADH. Usually concurrent azotemia

ADH, anti-diuretic hormone.

The animal with insufficient nephrons could neither concentrate nor dilute the urine and when challenged by a water load or water deprivation would make urine of inappropriate concentration, somewhere in the range between specific gravity 1.009 and the point of effective concentration for the particular species. In addition it would be expected to have renal azotemia.

It is useful to think of urinary concentration in terms of its being dilute, fixed, partially concentrated or fully concentrated and, above all, appropriate to body hydration. If, for instance, a severely dehydrated, vomiting animal is passing a small quantity of partially concentrated or 'fixed' urine, it is immediately apparent that it has severe renal impairment. If its urine is fully concentrated, it can be assumed immediately that its kidneys are normal. Similarly, any animal passing diluted urine can be assumed to have structurally normal kidneys.

Hormone production

The kidney is the site of production of three hormones – calcitriol, erythropoietin and renin.

Calcitriol (1,25-dihydroxycholecalciferol) is manufactured in the kidney to promote the absorption of calcium from the gut. Any inadequacy in its production can induce a calcium deficit, which often becomes evident in chronic renal disease.

Erythropoietin stimulates the production of red blood cells in the bone marrow. The kidney may not produce erythropoietin directly, but rather a substance called erythrogenin, which then activates a plasma precursor factor to produce erythropoietin. In any event, renal tissue is able to sense a need for more red cells, usually by detecting hypoxia. This is indicated by the fact that most animals with severe chronic renal disease have a non-regenerative anemia because of deficiency in erythropoietin.

Renin is produced by the juxtaglomerular apparatus in response to a fall in systemic blood pressure and causes the activation of the angiotensin system, which induces the release of aldosterone. This promotes the retention of sodium and water and may contribute to edema formation, as in congestive cardiac failure.

Functional reserve

A most important concept is that of the remarkably high functional reserve of the renal tissue mass. In the dog, it has been shown that renal function will continue normally until 65–70% of nephrons are non-functional – meaning effectively the loss of 1.5 kidneys. At this point, the surviving nephrons can continue to perform most functions, but, because they are overloaded with solute, are unable to regulate urine concentration. This dictates that the filtrate from the proximal tubule onwards is changed little in its tonicity and the animal, even if it needs to conserve water, has no option but to pass a large daily volume of urine with a tonicity about the same as the original plasma filtrate. There must be a large daily intake of water in compensation.

When 70–75% of nephrons are not functional, blood filtration becomes inadequate, azotemia develops, and failure of most renal functions is imminent.

In many cases of chronic renal disease, functional reserve may be roughly gauged by assessing the size and shape of the kidneys by palpation and radiography. If it appears that more than 70% of renal tissue mass has been irreparably destroyed, there is no possibility of return to normal function.

Pathophysiology of uremia

When the kidneys are unable to perform the majority of their functions, there are severe systemic metabolic consequences. There are a number of compensating maneuvers that the body can perform, but if all fails a severe **toxemic multisystem disorder** arises which is fatal if uncorrected. This complex multisystem autointoxication is called **uremia** – and a uremic state is the end stage of any disease condition causing renal failure.

Uremia is a pathophysiologic state resulting from factors including metabolic acidosis, electrolyte imbalances and the tissue-toxic effects of numerous retained amines, guanidines, peptides and other nitrogenous compounds. It is exacerbated by the induction of an alteration of bacterial flora in the gut and malabsorption of calcium, glucose and amino acids. There is a negative protein and caloric balance and impaired ability to utilize carbohydrates. The function of platelets is suppressed and there is a tendency for bleeding in the gastrointestinal tract. The life span of circulating erythrocytes is reduced. A number of hormones normally catabolized by the kidneys may accumulate (parathormone, insulin and glucagon), while erythropoietin and calcitriol, produced by the kidneys, are deficient. There may also be a suppression of thyroxin and somatomedins. The combined effects of calcium malabsorption and retention of phosphate by inadequate glomerular filtration will stimulate the secretion of parathormone and result in a state of hyperparathyroidism (renal secondary hyperparathyroidism). This results in removal of calcium from the skeleton and atrophy of the bone (see Chapter 12).

A uremic animal has marked clinical signs and it is useful to reflect upon this in relation to the underlying metabolic disturbances as outlined above. The clinical features include pale mucous membranes, anorexia, lethargy, weakness, muscle wasting, hypothermia and a harsh coat. There is usually vomiting of blood (hematemesis) and passage of blood from the rectum (melena). A non-regenerative anemia and an ulcerative glossitis and stomatitis are frequently present.

Bone disease of varying degrees may be encountered. Initially skeletal demineralization may only be detectable radiographically, but, when advanced, the bones of the maxillae and mandible become fibrotic, swollen and soft, exhibiting classical features of osteodystrophia fibrosa. The combined effects of inadequate calcium absorption and inadequate phosphate excretion are fundamental to this problem.

As uremia approaches its conclusion, central nervous dysfunction becomes prominent, with dullness, drowsiness, tremors and ataxia and finally coma or seizures.

The speed of progression of the uremic state is variable and depends upon the speed of progression of the disease destroying the kidneys. In slowly progressive disease, compensatory mechanisms may maintain life for quite a long period. The animals may have remarkable vitality in spite of marked azotemia and other abnormalities. Conversely, an animal thrown suddenly into a uremic state may become terminal when the measurable azotemia is comparatively low.

Urinary cast formation

The urine voided by a normal animal has several consistent qualities that are of everyday clinical interest. It is largely free of protein, cells and crystals (with the exception of the horse), and contains no glucose or ketones. The pH and electrolyte concentrations are predictable. Urine is generally transparent, clear, and the yellowish color is imparted by urochrome pigments derived from the renal tubules. All these qualities can be changed by prerenal, renal and postrenal factors and the principles of urinalysis are addressed in the final section of this chapter.

There is one urinalysis finding, however, that specifically indicates active renal tissue damage. This is the presence of coarse granular and cellular **urinary casts**. Casts are cylindrical plugs formed in the lumens of distal tubules and collecting ducts. They have a matrix of glycoprotein secreted by these regions of the nephron (called Tamm–Horsfall protein), in which other elements may become enmeshed. Degenerate tubular epithelial cells and inflammatory cells may become enmeshed, giving rise to the **granular** and **cellular** casts. Large numbers of such coarse granular and cellular casts in the urine sediment clearly indicate active severe renal tissue destruction (Fig. 9.5). Note, however, that there are other kinds of casts as well, with other implications.

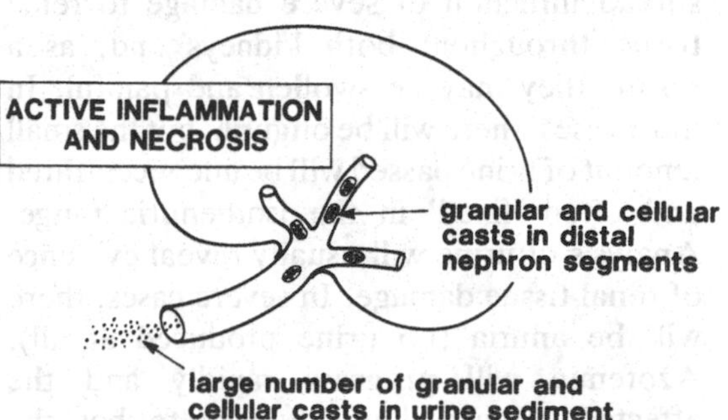

Fig. 9.5. The formation and passage in the urine of granular and cell casts when renal tissue is being actively damaged.

Clinical syndromes of renal malfunction

Renal function may be disturbed in three basic ways.

1 When there is disease within the renal tissue. This is a **primary renal** disorder.
2 When a disorder in some other organ system seriously affects the function of essentially normal, healthy kidneys. Such a situation is, therefore, a **prerenal** disorder. Two examples are:
   (*a*) inadequate production of ADH by the neurohypophysis; and
   (*b*) cardiac failure with underperfusion of the kidneys.
3 When urine flow is obstructed in the lower urinary tract, or urine is released into the peritoneal cavity or subcutis. Such a disorder is described as **postrenal**.

Primary renal disorders

(Acute renal failure, chronic renal failure, protein loss and nephrotic syndrome, nephrogenic diabetes insipidus, Fanconi syndrome, cystinuria, urate uria)

Acute renal failure

The sudden loss of 70–100% of nephrons in a previously healthy animal will produce classical acute renal failure, with serious impairment of all kidney functions.

Acute renal failure implies the sudden widespread infliction of severe damage to renal tissue throughout both kidneys and, as a result, they may be swollen and painful. In most cases, there will be oliguria, but the small amount of urine passed will be unconcentrated and often 'fixed' in the isothenuric range. Analysis of urine will usually reveal evidence of renal tissue damage. In severe cases, there will be anuria (no urine produced at all). Azotemia will progress rapidly and the affected animal will quickly begin to show the effects of acidosis and electrolyte disturbances. The clinical signs will be lethargy, anorexia, weight loss and perhaps vomiting and dehydration. At this point, the **inability to produce a concentrated urine** becomes highly significant as a diagnostic sign (Fig. 9.6).

Oliguria in some cases of acute renal failure has been suggested to have its basis in a defensive response by the nephrons. The potential for sodium loss through badly damaged kidneys is enormous. Thus the surviving nephrons, stimulated by high concentrations of sodium in their distal tubules, drastically reduce their glomerular filtration by adjusting the tonic balance between afferent and efferent arterioles.

It should be noted, however, that some cases of acute renal failure are polyuric. These cases may represent animals with about 75% nephron loss and enough surviving nephrons to continue production of filtrate, but severely overloaded by solute and unable to function effectively.

The important mechanisms that are triggered by renal damage and contribute to the syndrome of acute renal failure in man are thought to be:

1 A reduction in renal blood flow because of intrarenal vascoconstriction.
2 Obstruction of tubule lumens by degenerate and necrotic epithelial cells or crystal precipitates.
3 Back-leak of filtrate from damaged tubules into capillaries.

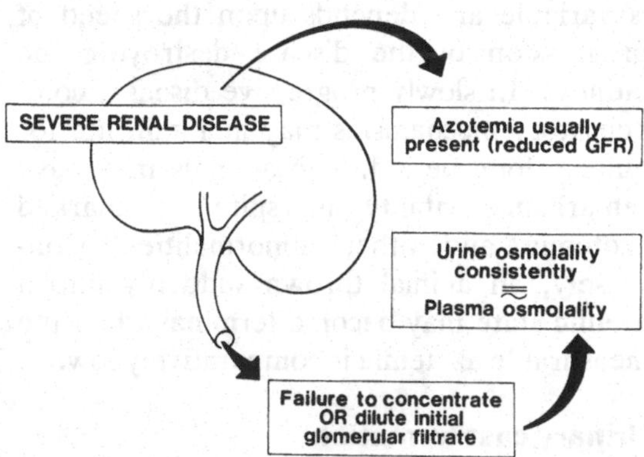

Fig. 9.6. The key clinical disturbances which occur when severe renal disease erodes the functional reserve of the kidneys and causes renal failure. GFR, glomerular filtration rate.

4 A reduction in glomerular permeability.

Similar mechanisms probably operate in domestic animals. The relative importance of each factor will vary with the etiology. The outcome of acute renal failure will depend upon the severity and nature of the underlying disease process. Outright renal failure will result in death from uremia in about a week. In less severe malfunction, clinical signs can persist for some weeks before the patient either begins to recover or succumbs to uremia.

Some diseases, for example acute tubular necrosis, will resolve by regeneration of damaged tissues and restoration of normal function. Typically in these cases, an oliguric state is succeeded by a phase of polyuria, as urine flow increases through regenerating, but still functionally defective, nephrons. After some weeks, they may function perfectly, and improving function can be monitored as the animal corrects its metabolic deficits and excesses.

Some diseases may subside, but leave permanently damaged kidneys, adequate to sustain life, but functionally imperfect. Examples are acute cortical or medullary necrosis and pyelonephritis.

These diagnostic and prognostic variables are the challenges that face the clinician dealing with acute renal failure. Once the diagnosis of acute renal failure is established, it is important to determine as quickly as possible, the nature of the renal disease and its potential reversibility.

Chronic renal failure

The gradual progressive loss of nephrons will lead to the development of chronic renal failure. This process is very common in dogs and cats. The effect of such a process is to throw an increasing burden on a diminishing number of nephrons until the functional reserve of the system is surpassed. As a bigger demand is placed upon surviving nephrons, each must handle a massive load of water and solutes.

The first function to fail noticeably is the regulation of urinary concentration, when about 65–70% of nephrons are lost and the survivors are *unable* to sustain a concentrated medullary interstitium. This, combined with a high solute load per nephron, means that a large volume of unconcentrated urine is passed each day (polyuria), and, to maintain body tonicity, the animal must respond by drinking large amounts of water (polydipsia). At this time a slight reduction in glomerular filtration rate and moderate azotemia can be accommodated by body wide adaptations. For example, parathormone is released to increase urinary shedding of phosphate, thus countering the tendency for phosphate retention. The early stage of chronic renal failure is sometimes referred to as **compensated renal failure**. The animal is secure so long as it has adequate access to water and no additional stress to provoke a uremic crisis.

However, the destruction of nephrons being inevitable, compensation ultimately fails, the patient lapses into overt renal failure and ultimately uremia. These animals generally have a tissue disease causing atrophy, fibrosis and distortion of the kidneys, all of which can be detected by palpation or radiography. Urinalysis may not reveal evidence of intense renal tissue damage because of the slowly progressive nature of the lesion. As will be imagined, the prognosis is usually gloomy, there being little hope of any tissue regeneration.

Protein loss and nephrotic syndrome

The nephrotic syndrome is a clinical state resulting from the loss of plasma protein (mainly albumin) into the urine at a rate which exceeds the capacity of the liver to compensate for the loss. Plasma albumin then falls to levels which seriously reduce plasma osmotic pressure, and hypoproteinemic edema results, exacerbated by a tendency to sodium retention. It is a characteristic set of four clinical findings that defines the nephrotic syndrome in man and most domestic animals – **edema**, **heavy proteinuria**, **hypoproteinemia** and **hypercholesterolemia**. The first three directly reflect massive urinary protein loss, while the last probably reflects the loss of a cofactor of serum lipoprotein lipase. In dogs and cats, the edema is expressed for the most part as ascites, with subcutaneous edema being more prevalent in herbivores.

Animals with the nephrotic syndrome may not have any more than a mild azotemia to indicate that a renal disease is present, as the underlying tissue lesion principally causes a loss of glomerular barrier function. Thus edema and weight loss may frequently constitute the whole clinical picture. In less severe cases, plasma albumin is maintained at adequate levels, edema does not occur and the clinical effect is weight loss and failure to thrive.

Any animal with the nephrotic syndrome will have an underlying glomerular disease of some type (a glomerulopathy) and if this disease is progressive, full blown renal failure can eventuate. The nephrotic syndrome may persist for weeks or months if the degree of edema is not too severe. Due to the excessive loss of plasma anti-thrombin III into the urine, affected animals are prone to thrombosis, and may die as a result of thrombotic obstruction of a major vessel. Sometimes the degree of edema proceeds rapidly to a stage where the animal dies from the effects of hydrothorax in a relatively short time.

Nephrogenic diabetes insipidus

This is an uncommon disorder in which there is no effective action of ADH despite its normal production and release from the neurohypophysis. Affected animals will have essentially normal renal function except for an inability to regulate urinary concentration. Such animals may have a structural lesion within the renal medullary tissues (e.g. amyloidosis or fibrosis) or simply an inability to respond to ADH, because of a genetically determined absence of specific receptors on tubular cells. The clinical features are polydipsia and polyuria, inability to concentrate the urine if water is withheld, and no response to injected ADH.

Fanconi syndrome

This is a rare, usually genetic disorder, which results from the failure of multiple tubular transport mechanisms. It is characterized by the heavy urinary loss of amino acids, glucose and phosphate. These substances are freely filtered at the glomerulus and then actively recovered by specific transport pathways. Classical Fanconi syndrome as a genetically determined defect has been reported in Basenji dogs. It may also result from acquired renal disease and can be seen in the recovery phase of severe tubular necrosis. Affected individuals fail to thrive, are polydipsic and polyuric and have skeletal disorders.

Cystinuria

Cystinuria is an inborn error of metabolism that has been recorded in several breeds of dog. Such animals have a defect in renal tubular resorption of several amino acids and thus their urine contains larger than normal concentrations of these molecules. The defect is of no consequence apart from the fact that one of the amino acids, cystine, is sparingly soluble at the usual urinary pH, and affected individuals are predisposed to the formation of cystine stones.

Urate uria

The Dalmatian breed of dog excretes an unusually high concentration of urates in the urine, owing to a peculiarity of purine metabolism. This is probably largely caused by a limited ability of the liver to oxidize uric acid and partially by defective renal tubular reabsorption of filtered urate. This predisposes Dalmatian dogs to a high incidence of urate-stone formation.

Renal disorders of prerenal origin

(Underperfusion, neurogenic diabetes insipidus, ADH blockade, ADH excess, aldosterone deficiency, aldosterone excess)

Underperfusion

If, for any reason, cardiac output is significantly reduced, GFR will fall and azotemia will develop. Normal kidneys respond by interpreting the situation as a need to conserve water and sodium. Thus, there will be oliguria and the urine will be highly concentrated. It is sometimes a problem to decide if an azotemic, oliguric animal has renal failure or circulatory failure, and the decision is made by reflecting on the above information. There are a variety of conditions in which underperfusion may occur and in which azotemia develops, and a thorough clinical method is required to evaluate any particular case. It should also be remembered that severe prolonged underperfusion can produce ischemic renal damage and primary renal failure as a major complication.

Neurogenic diabetes insipidus

Neurogenic diabetes insipidus occurs when the kidneys are normal, but there is a deficiency in the release of ADH from the neurohypophysis in spite of an increase in the tonicity of body fluids. In the absence of ADH, the animal has no ability to concentrate the urine although it may still dilute it if necessary. Generally, the afflicted animal will have increased water consumption (polydipsia) to compensate for obligatory polyuria. Provided access to water is unrestricted, the urine is usually in the hyposthenuric range. This is probably because of a tendency to overdrink and a consequent necessity actively to dilute the urine. **If water intake is restricted, the urine will become fixed in the isosthenuric range**. If exogenous ADH is administered, the urine will be concentrated normally.

Anti-diuretic hormone blockade

This is a condition that mimicks diabetes insipidus, but is easily differentiated by the accompanying signs. Certain substances have the ability partially to block the interaction between the hormone and its receptors on the distal nephron, thereby preventing adequate urinary concentration from occurring. Quite predictably, there is obligatory polyuria and compensating polydipsia, but a poor response to injected ADH. In veterinary medicine, ADH blockade is recognized typically in canine hyperadrenocorticism, pyometra and hypercalcemia. In the first case, the blockade results from overproduction of adrenocortical glucocorticoids, in the second, by endotoxin absorbed from the pus-filled uterus. Excessive concentration of divalent calcium ions interferes with ADH and reduces the permeability of the collecting duct. In each, the associated clinical signs and laboratory findings overshadow the renal malfunction and point the way to the correct diagnosis. However, the ADH hormone blockade contributes greatly to the clinical profile in each case.

Anti-diuretic hormone excess

This condition, although documented in human medicine, has rarely, if ever, been recorded in animals. In man, it usually results from the secretion of ADH by a functional neoplasm. The patients have an inability to regulate their plasma osmolality. Predictably, they tend, with a normal water intake, to have a high U_{osm} in spite of a low P_{osm}, and this persists in spite of a water load. This has been included here as a useful illustration of an endocrine-induced renal malfunction.

Aldosterone deficiency (hypoaldosteronism)

This defect is most commonly seen in dogs with adrenocortical insufficiency. The most serious disturbance in this disease is the imbalance in potassium metabolism caused by the aldosterone deficit. It will be remembered that this hormone promotes the exchange of sodium for potassium in the distal nephron. In its absence, there is a tendency for urinary loss of sodium and systemic retention of potassium. This predisposes the animal to hyperkalemia (excessively high plasma potassium concentration), which has a profound effect on the myocardium and is frequently the cause of death.

Aldosterone excess (hyperaldosteronism)

An elevation in the activity of aldosterone will result in a high level of urinary potassium loss and a tendency for sodium retention. The most common cause of hyperaldosteronism is congestive cardiac failure. Reduced cardiac output and low blood pressure combine to stimulate the release of renin from the kidneys. Renin, in turn, stimulates the release of aldosterone via the angiotensin pathway. There is, thus, a secondary hyperaldosteronism. In this circumstance, the predominant effect is sodium retention resulting in hypervolemia and edema. As the additional sodium is distributed equally throughout the plasma and interstitial fluid compartments, the plasma sodium concentration is usually normal.

Primary hyperaldosteronism is extremely rare, and would almost certainly be associated with a hormone-secreting neoplasm. In such cases, the clinical picture may be dominated by potassium loss and consequent hypokalemia, producing muscular weakness and cardiac arrhythmia.

Diuretics

Agents which promote the excretion of sodium and water are known as diuretics, and are frequently used for clinical effect. If so desired, the administration of a diuretic may be combined with a regulated water and sodium intake to reduce body water content. Alternatively, a diuretic may be used to promote the flow of urine through nephrons in oliguric renal failure. There are four general types of diuretics: loop diuretics, osmotic diuretics, carbonic anhydrase inhibitors, and aldosterone inhibitors.

Loop diuretics are chemicals which inhibit the chloride pump in the ascending thin limb of Henle's loop and thereby prevent the absorption of sodium in this segment (see Fig. 9.3). They are the diuretics most commonly used in clinical practice, and are exemplified by the drug furoseamide. Because of their effect, large amounts of sodium are delivered to the distal tubular sites of sodium/potassium exchange promoting heavy loss of potassium into the urine. For this reason, animals or persons being treated with loop diuretics must also receive dietary potassium supplementation.

Osmotic diuretics are substances that act by 'holding' water in the renal tubules by their osmotic effect. A commonly used compound of this type is the sugar alcohol mannitol. Similarly, glucose will promote diuresis when present in the urine of patients with diabetes mellitus, provoking obligatory polyuria and compensatory polydipsia.

Carbonic anhydrase inhibitors act by suppressing the exchange of sodium for hydrogen in the proximal tubular segment of the nephron. Patients treated with such drugs may develop a metabolic acidosis as a result, as the body attempts to eliminate acid via exhaled carbon dioxide.

Aldosterone may be inhibited by the drug spirinolactone. In a situation of hyperaldosteronism, the drug may be used to block the effect of the hormone and thereby aid in the elimination of sodium, which would otherwise be retained excessively. The clinical application of this drug is confined to the treatment of congestive heart failure.

Psychogenic drinking

Psychogenic drinking is a behavioral quirk, possibly triggered by boredom, and recognized mostly in dogs. These animals become habitual drinkers, and drink water greatly in excess of the daily requirement. This behavioral polydipsia causes a compensatory polyuria. The urine is in the hyposthenuric range as the body rids itself of the water load. If the animal is deprived of water, the basic health of the kidneys is revealed, as the urine becomes concentrated.

Renal disorders of postrenal origin

All the good work of normal kidneys may be negated if the urine they generate is unable to leave the body because of some malfunction in the lower urinary tract.

Urine recycling

If most of all of the urine leaks from the lower urinary tract into the abdomen, azotemia and other clinical signs of renal failure will appear within a few days, with the added feature of an abdomen swollen by fluid. The cause will most often be traumatic rupture of the urinary bladder. If the lesion can be surgically repaired, there should be a return to normality.

Urethral obstruction

This is a very common and difficult clinical problem in several domestic species. It almost always occurs in males and is usually caused by urinary calculi (uroliths). The obstruction is usually precipitous in onset and results in a great deal of discomfort.

The distress caused by extreme distention of the bladder, coupled with the development of acute renal failure, results in rapid clinical deterioration. There is usually **anuria** and straining. The distended bladder is palpable as a firm smooth mass in the posterior abdomen. Sometimes, the bladder will rupture after 2–3 days, removing the palpable mass and causing abdominal distension. During acute obstruction, renal blood flow and GFR fall significantly and azotemia, electrolyte and acid–base disturbances rapidly develop. Hyperkalemia may become the major life-threatening disturbance. Unrelieved obstruction is usually fatal within 7 days. If obstruction is relieved the return to normality may be slow, with renal blood flow and GFR remaining suppressed for several days. Furthermore, during the post-obstruction phase, there is impaired urine concentrating ability with consequent large-scale loss of sodium and water. The sodium excretion may be due, in part, to the accumulation of natriuretic factors in the circulation. There can also be significant post-obstruction potassium loss and the animal may plunge from hyperkalemia to severe hypokalemia.

Provided the kidneys are basically healthy and the above disturbances correctly managed, renal function will return to normal several days after the relief of obstruction.

Types of renal tissue lesions, their common causes and clinical consequences

(Tubulo-interstitial disease, glomerulopathies, end-stage kidney, neoplastic diseases, developmental diseases)

Renal lesions which do not surpass the functional reserve of the kidneys will not cause clinical signs of renal dysfunction. Indeed, they may produce no clinical signs at all. They may, however, require interpretation at necropsy or surgery. The significance of renal tissue disease will depend on both its nature and extent. In general, destructive lesions need to eliminate about 70% of nephrons (i.e. 1.75 kidneys) before signs of renal failure appear. The more slowly progressive the disease, the more opportunity there is for compensatory mechanisms to be successful and vice versa. Non-progressive disease that has effectively destroyed only one kidney will obviously have no effect on overall renal function. Some examples are the developmental anomalies, such as congenital ureteral obstruction, or unilateral aplasia.

Some lesions may cause no impairment of total renal function, but may cause quite severe clinical disease. For example, a benign neoplasm in one kidney may lead to massive blood loss via the urine, or, alternatively, a bacterial infection in one renal pelvis may cause abdominal pain, loss of appetite and depression.

At this point, the focus is on renal lesions and disease processes that are responsible for clinical signs of renal dysfunction. A clinician evaluating an animal with renal disease will be assisted by knowledge of the exact nature of the underlying tissue disease. There are several irreversible and progressive renal disease processes. Equally, others are potentially reversible and such information is obviously valuable. The clinician may resort to radiography, biopsy and urinalysis in order to determine the nature of renal disease.

Tubulointerstitial disease

(Acute tubular necrosis, renal tissue necrosis, crystal precipitate nephrosis, interstitial nephritis, pyelonephritis, interstitial amyloidosis)

This section focuses on diseases in which tissue injury is initiated amongst the interstitial and tubular elements of nephrons in previously healthy kidneys. The lesions reflect the degenerative, inflammatory and reparative reactions resulting from such injury. In some cases, the causal agents are clearly identifiable, but sometimes the cause is unknown. The causes include a number of toxic and infectious agents and encompass hemodynamic and immunologic mechanisms.

Acute tubular necrosis (ATN, acute 'nephrosis')

In this process, the central event is the sudden death of tubular epithelial cells throughout the kidney and their subsequent shedding into the tubular lumens. The functional consequences of this depend upon the extent of necrosis, but oliguric acute renal failure can be anticipated.

In severe acute tubular necrosis, the kidneys may appear swollen and slightly-to-markedly pale, and in ruminants, perirenal tissues may be edematous. Histologically, disintegrating tubular epithelial cells and amorphous debris can be seen in tubular lumens throughout the kidneys (Fig. 9.7). The interstitium is distended by fluid accumulation, and small to moderate numbers of neutrophils accumulate and migrate into tubular lumens. In most instances, it is the cells of the proximal convoluted tubules that suffer most. Very soon after the onset of necrosis (within 24 hours), surviving tubular cells will begin to divide so that cells in mitosis are frequently seen, together with flattened, basophilic immature cells that re-line previously denuded segments of the tubules. Mineralization of debris, which develops very rapidly, is often a feature.

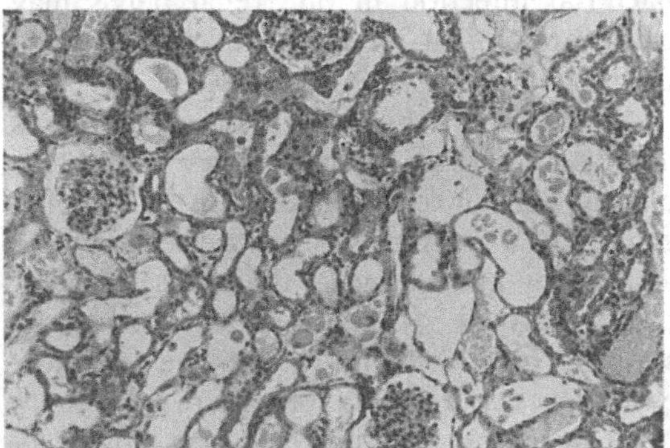

Fig. 9.7. Light micrograph of acute tubular necrosis which shows many dilated renal tubules containing degenerate epithelial cells and necrotic debris.

The potential for regeneration is favorable, provided that the causal agent is removed and the tubular basement membranes remain intact. If not, nephrons will collapse and will be replaced only by fibrous scar tissue. In either case an improvement in renal function must occur within 7–10 days, or the animal will become uremic.

In cases in which full recovery is attainable, oliguric renal failure will begin to subside after a few days and urine flow will increase. This correlates with cessation of necrosis and completion of regeneration. At this stage, if there has been extensive necrosis, the regenerated tubules may lack full functional capacity. Water, electrolyte and pH regulation may remain inadequate for several more days, so that the initial oliguric failure converts to polyuric failure. However, structural and functional normality is eventually re-established.

In less fortunate individuals, the oliguric failure is severe and the animal dies. Sometimes there may be partial recovery, with permanent loss of some nephrons and fibrosis, and the extent of this loss will determine whether or not there are lingering problems with renal function. The flow of events is illustrated in Fig. 9.8.

Acute tubular necrosis can be caused by both exogenous and endogenous agents including some plant toxins. It is a classical lesion in poisoning with heavy metals, especially salts of mercury, uranium, bismuth and calcium. In the case of calcium, the induction of hypercalcemia is well recognized as a basic cause of kidney damage. Hypercalcemia may be induced by dietary excess, excessive calcitriol (vitamin D) intake, hyperparathyroidism and by factors produced by certain neoplasms, notably lymphosarcoma and carcinoma of the apocrine gland of the canine anal sac. Some therapeutic agents can cause acute tubular necrosis, most notably the aminoglycoside antibiotics. Acute tubular necrosis may be associated with heat-stroke and may occur in febrile disease states. Finally, an important cause of acute tubular necrosis is ischemia associated with severe shock from any cause. Ischemic acute tubular necrosis can, therefore, be a serious complication of shock, and is made more serious by the fact that the necrosis is frequently accompanied by extensive tubular collapse and rupture (tubulorrhexis), which impedes regeneration (Fig. 9.8).

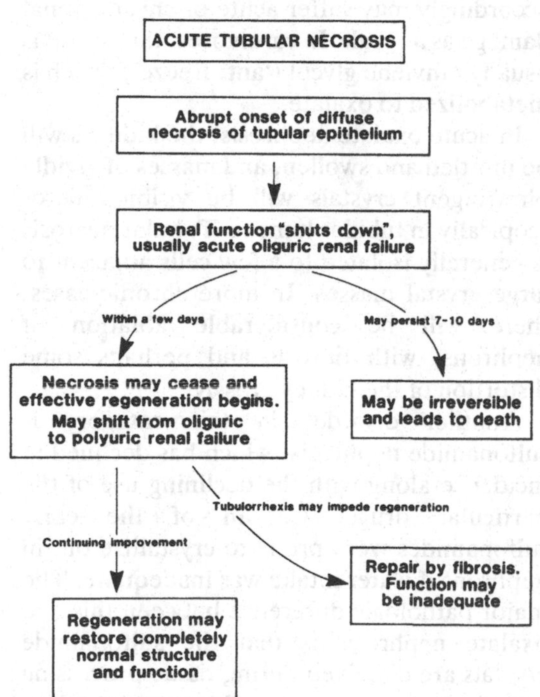

Fig. 9.8. The course and consequences of acute tubular necrosis.

Renal tissue necrosis

A necrotizing process may do more than 'pick off' the renal epithelium from its stromal scaffold. It may be so severe that all the elements of the tissue become necrotic. Such lesions may result from typical infarction following occlusion of an end artery. Renal infarcts, either recent in origin or old and healed are not uncommonly found at necropsy, but are seldom detected clinically, and rarely if ever involve enough tissue mass to cause renal failure.

Occasionally, extensive tissue necrosis may develop which is unrelated to the mechanical obstruction of a single vessel. Necrosis may involve both cortex and medulla and may be patchy or massive (Fig. 9.9). Necrosis of the entire cortex or medulla may occur. Zones of recent necrosis are pale in color and delineated by a narrow zone of hyperemia. Necrotic areas may resolve into cystic spaces in the tissue (Fig. 9.10), with associated scarring and distortion. Extensive cortical necrosis will be associated with severe oliguric renal failure, which will be irreversible because of the nature of the lesion. Extensive medullary necrosis *may* be compatible with continued life, but as the necrotic tissue is permanently

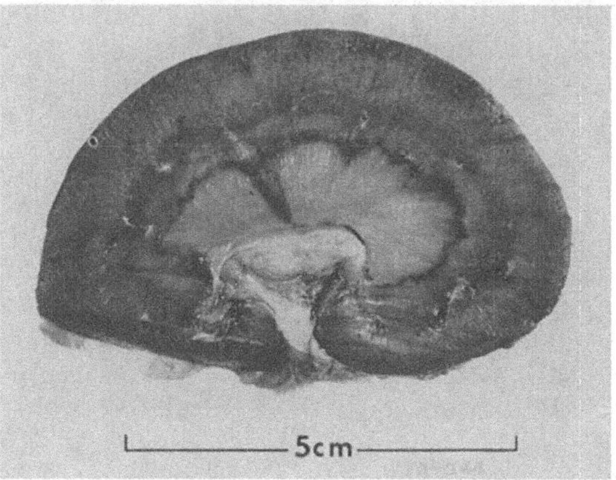

Fig. 9.9. Extensive recent necrosis of much of the renal medulla in a canine kidney. Factors involved in this type of lesion include 'failure of reflow' and depressed synthesis of prostaglandin PGE_2.

lost, residual problems of water and electrolyte regulation may occur.

A prominent cause of renal necrosis of this type is ischemia due to circulatory failure in shock. Under these conditions, blood flow to the kidneys may effectively cease for a time and then restart in part, but not all, of the tissue. This 'failure of reflow' phenomenon is considered to be a major cause of medullary necrosis as the medulla is relatively poorly perfused even in the normal state.

Medullary necrosis may also be provoked by the blockade of synthesis of prostaglandin PGE_2 *in situ*. Recent research indicates that these prostaglandins are important in maintaining vascular patency in the renal medulla. Their synthesis may be blocked by certain non-steroidal anti-inflammatory drugs.

The Schwartzmann reaction is the term used to describe a rare phenomenon thought to be based on a heightened reaction to bacterial endotoxin and it may develop during infections with Gram-negative organisms. During the reaction, renal cortical blood flow stagnates in association with widespread capillary and venular thrombosis, and the entire renal cortex may become necrotic. The process is based on intravascular coagulation. There is

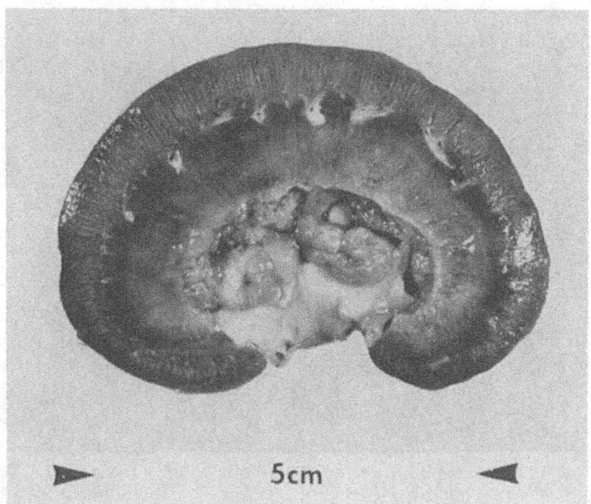

Fig. 9.10. The aftermath of a lesion as depicted in Fig. 9.9. Necrotic tissue has disappeared leaving an extensively 'excavated' medulla.

obviously an irreversible destruction of nephrons, after which no regeneration is possible, and fatal oliguric renal failure ensues.

Crystal precipitate nephrosis

This is a pathologic state resulting from the massive crystallization of solutes within the tubular lumens of nephrons. If this happens suddenly, the mechanical obstruction to urine flow may be sufficient to cause acute oliguric renal failure. If it occurs more chronically, there may be progressive atrophy of nephrons, fibrosis and chronic renal failure. Crystal precipitates do not generally provoke acute tubular necrosis on a large scale, but their persistence does tend to lead to tubular atrophy. The veterinary archetype of this kind of disease is 'oxalate nephrosis'. Herbivorous animals may ingest large quantities of oxalate when grazing plants like *Oxalis* sp., *Chenopodium* sp., sorrel or rhubarb leaves, and accordingly may suffer acute or chronic renal damage as a result. In carnivores, the culprit is usually ethylene glycol ('anti-freeze') which is metabolized to oxalate.

In acute oxalate nephrosis, the kidneys will be mottled and swollen, and masses of vividly birefringent crystals will be visible microscopically in tubular lumens. Tubular necrosis is generally isolated to a few cells adjacent to large crystal masses. In more chronic cases, there will be considerable ablation of nephrons, with fibrosis and perhaps some distortion of the kidney grossly.

Another disorder in this category is sulfonamide nephrosis, which has declined in incidence along with the declining use of the particular drugs. Certain of the early sulfonamides were prone to crystallize out in nephrons if water intake was inadequate. The major pathologic difference between this and oxalate nephrosis is that the sulfonamide crystals are dissolved during tissue processing and thus cannot be seen within the distended, obstructed tubules in routine histologic sections.

Interstitial nephritis

This term refers to a particular pathologic pattern of renal disease in which there is a significant infiltration of reactive lymphoid cells, associated with the destruction of nephrons. The reaction may be acute, subacute or chronic and although numerous causes have been defined, the cause in many cases remains unknown. Causes include leptospiral infection, canine adenovirus infection, mycotoxins (for example, ochratoxin), and in man, a long list of drugs and chemicals. It is also suggested that immune mechanisms are important, and in the case of drug-induced disease, some drugs may act as haptens to convert tissue antigens to auto-antigens. Other drugs may induce hypersensitivity responses.

The classical example of acute interstitial nephritis is *Leptospira canicola* infection in dogs. Following infection, leptospiremia occurs and there is hematogenous infection of the kidneys. The organisms have a selectively destructive effect on the tubular epithelium and an acute cortical tubular necrosis ensues during the first 24 hours or so. This early phase leads on to an intense infiltration of the kidneys by lymphocytes and plasma cells. Many nephrons may be totally destroyed. After 3–4 days, the kidneys are swollen and the cortices irregularly mottled by pale areas of cellular infiltration. As time goes by, increasing fibrosis may alter the shape, size and consistency of the kidneys leading, in severe cases, to considerable contraction and fibrosis. The functional consequences will vary with the severity of the disease, from minimal to severe. In severe acute disease oliguric renal failure can be expected, while later on polyuric failure may occur.

In fact, the great majority of animals with the lesion classified as interstitial nephritis, will have a chronic and slowly progressive disease, with no obvious etiology. Most of these animals will be dogs and cats. Over a period of months to years such individuals will have increasingly impaired renal function. In effect there is a shrinking number of nephrons left to carry the burden of work. The clinical progression will be a phase of compensation, succeeded by polyuric renal failure, and finally uremia (Figs. 9.11 and 9.12). Terminally the fibrotic kidneys reach an end stage, which will be discussed more comprehensively at the end of this section.

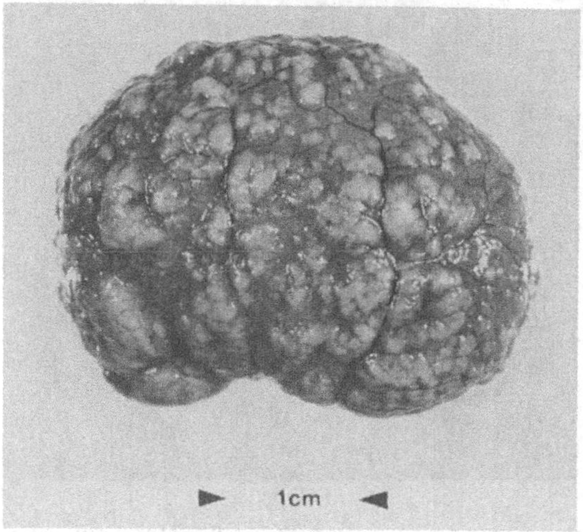

Fig. 9.11. Subacute interstitial nephritis in a feline kidney. Multiple confluent pale nodules are surviving normal tissue amongst depressed atrophic areas heavily infiltrated by inflammatory cells, in this case chiefly lymphoid cells and a few eosinophils.

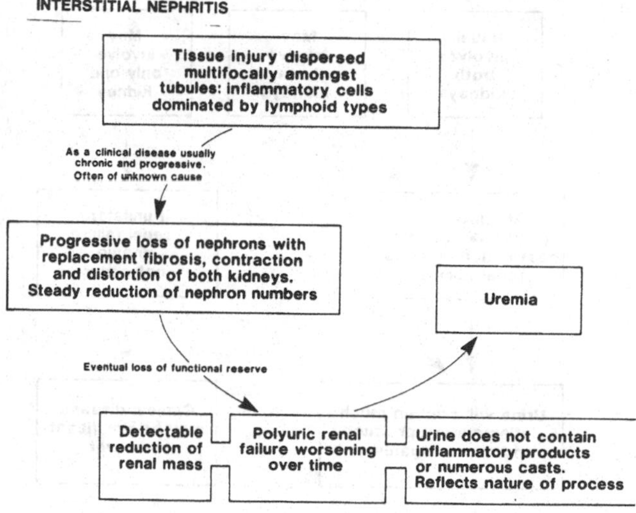

Fig. 9.12. The evolution and consequences of interstitial nephritis.

Pyelonephritis

This is a disease process characterized by relapsing episodes of subacute to chronic inflammation centered in and around the renal pelvis and deep medullary tissues. There is always an infectious agent or agents involved, mostly bacterial but occasionally fungal. The renal tissues in this region provide a favorable environment for microbial growth, as oxygen tension is low and electrolyte and urea concentrations are high. Microorganisms may reach this environment via the bloodstream or by ascending the urinary tract. For this reason any circumstance which impedes the free flow of urine in the lower tract, particularly in females, can predispose to pyelonephritis. Similarly, any disorder which promotes the reflux of urine from the bladder back into the ureters will predispose to pyelonephritis.

The pathology of pyelonephritis is characterized by tissue destruction, inflammation and fibrosis, which may steadily extend from the pelvic area out towards the cortex and capsule (Fig. 9.13). Not infrequently there is considerable pus formation (Fig. 9.14). Histologically, the inflammatory cell infiltration contains neutrophils, plasma cells, lymphocytes and macrophages, and different zones of the lesion will reflect varying phases of active tissue destruction and resolution.

Pyelonephritis, by its nature, may be asymmetrical or even unilateral. Its main clinical effect may be as a focus of toxemia and pain, rather than as a cause of renal failure. However, if the lesion is severe and bilateral, acute or chronic renal failure and uremia will

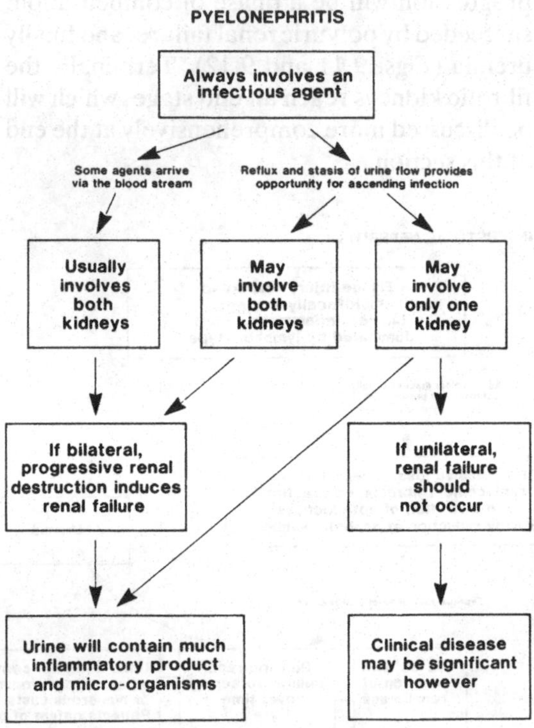

Fig. 9.13. The evolution and consequences of pyelonephritis.

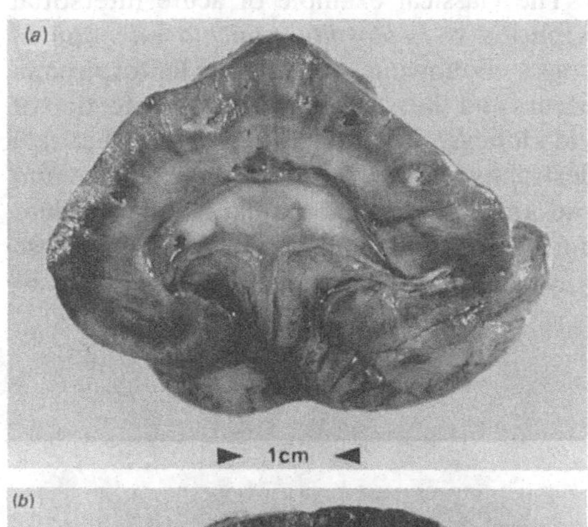

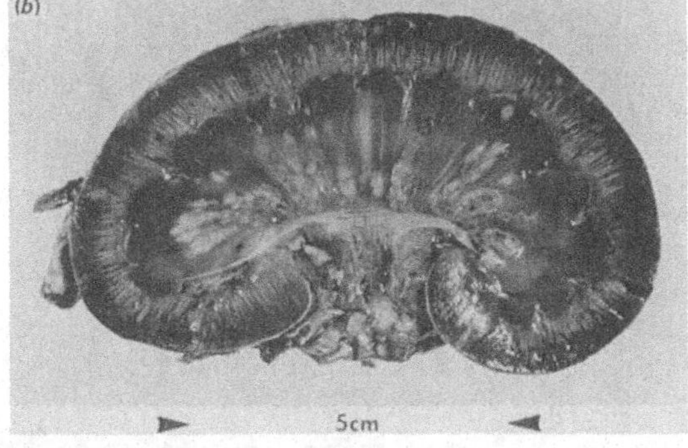

Fig. 9.14. (*a*) Pyelonephritis in a feline kidney, illustrating pus formation and medullary necrosis. (*b*) Multifocal granulomatous pyelonephritis in a German Shepherd dog with *Aspergillus terreus* infection. Lesions appear as pale confluent foci principally in the medulla.

eventuate. In either case, numerous casts, neutrophils, bacteria or fungi are usually found in the urine.

Interstitial amyloidosis

Amyloid deposition within the medullary interstitium may occur in cats on a sufficient scale to cause chronic irreversible renal failure. The deposits result in progressive atrophy of nephrons, with increasing distortion and fibrosis of the kidneys. On gross inspection the kidneys resemble those in advanced pyelonephritis, with retraction and excavation of the renal crest, and radial scarring from medulla to capsule. Histologic examination reveals the amorphous amyloid deposits. The cause is generally unknown, although a high dietary intake of vitamin A seems to predispose cats to amyloidosis and extensive deposits may also occur in other organs like the liver and intestine. In some cats, glomerular amyloidosis may occur concurrently, causing the additional clinical feature of heavy proteinuria.

Glomerulopathies
(Amyloidosis, glomerulonephritis)

Disease processes which originate within glomeruli are classified as glomerulopathies. If all glomeruli are affected, there is said to be a 'diffuse' glomerulopathy, whereas if only some are affected, there is said to be a 'focal' glomerulopathy. If, in any one glomerulus, the lesion involves the whole tuft, it is said to be a 'global' lesion, if it involves only part of the tuft it is said to be a 'segmental' lesion. Glomerular lesions, as has been mentioned previously, may reduce glomerular filtration rate and impair the filtration barrier against proteins.

In severe glomerulopathies there will inevitably be eventual tubular involvement. As the blood supply to the tubules is 'downstream' from the glomeruli, it can be expected that a reduction of blood flow through badly damaged glomeruli will lead to tubular atrophy and degeneration. An animal with severe glomerular disease can be expected to have a persistent heavy proteinuria, and varying degrees of azotemia. Glomerular disease may therefore be associated with acute or chronic renal failure or the nephrotic syndrome (Fig. 9.15). Glomeruli are highly specialized structures, but they can recover from certain levels of injury and some lesions may resolve completely. In other cases, residual scarring may leave glomeruli partially competent. Completely destroyed glomeruli are irreplaceable. An animal may compensate successfully for the loss of up to 70% of its glomeruli and associated nephrons, but the process is one of adaptation and not regeneration. Having established this background, we can now discuss the various glomerulopathies that have been identified.

Amyloidosis

Glomerular amyloidosis, as a cause of clinical disease in the domestic species, is mostly seen in the dog and to a lesser extent in cattle. It is

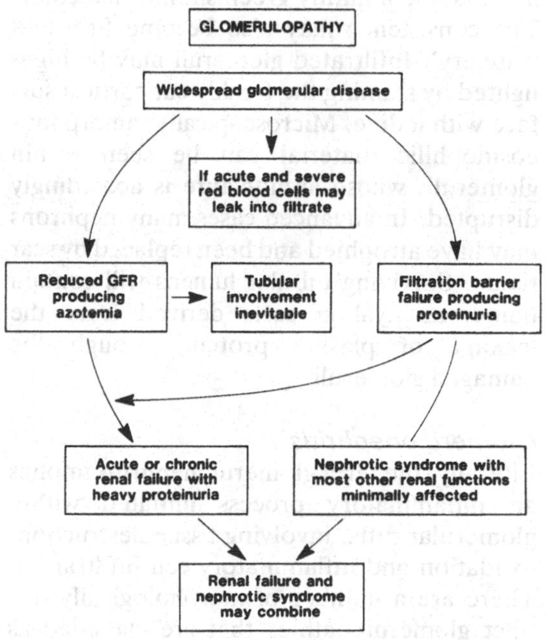

Fig. 9.15. The consequences of glomerular disease. GFR, glomerular filtration rate.

an uncommon disease. The process is relentlessly progressive and there is no known cure to date. In this form of amyloidosis, the disease is confined to the renal glomeruli and there is usually no involvement of any other tissues. Amyloid is progressively deposited in the mesangium, and between endothelial cells and the glomerular basement membrane. As amyloid accumulates, glomerular function becomes increasingly impaired. By most accounts, the usual clinical disease is chronic progressive renal failure although cases of acute renal failure have been described. Other individuals may first exhibit the nephrotic syndrome, but progression to renal failure can be predicted. Glomerular amyloidosis is thus a chronic, progressive, irreversible disease, so that an accurate early diagnosis is desirable. Fortunately, the pathology is sufficiently distinctive to make diagnosis by biopsy fairly easy.

In the early stages of the disease the gross appearance of the kidneys may be unremarkable, but when the lesion is advanced, the kidney surfaces become irregularly depressed, and assume a faintly greenish, mottled color. The consistency may also become firm and 'rubbery'. Infiltrated glomeruli may be highlighted by staining the freshly cut cortical surface with iodine. Microscopically, amorphous eosinophilic material can be seen within glomeruli, whose architecture is accordingly disrupted. In advanced cases many nephrons may have atrophied and been replaced by scar tissue. Surviving tubular lumens will contain numerous hyaline casts derived from the leakage of plasma protein through the damaged glomeruli.

Glomerulonephritis

Literally, the term glomerulonephritis implies an inflammatory process initiated within glomerular tufts, involving tissue destruction, exudation and inflammatory cell infiltration. There are a number of morphologically distinct glomerulopathies that are classified as glomerulonephritis, although some of them have very little 'itis' about them, there being no exudation or inflammatory cell accumulation. However, we shall bow to convention and discuss them as such. As a tissue lesion, glomerulonephritis is a fairly common finding in necropsy material and clinical glomerulonephritis is diagnosed with increasing frequency.

It has been well established that the vast majority of clinically expressed cases of glomerulonephritis have an immunologic basis. The pathogenesis relates to the response of an animal to a specific antigenic stimulus. Sometimes the antibody response to an antigen will result in the formation of soluble antigen/antibody complexes of a type which circulate for a prolonged period in the blood. This is likely to occur in the event of prolonged moderate antigen excess. If these complexes are fairly large, they may be taken into the mesangium of glomeruli, and if they are smaller they may become 'stuck' at some point in the glomerular capillary walls, usually between the endothelium and the basement membrane or within the basement membrane (Fig. 9.16). Alternatively, some circulating antigens may have size and charge characteristics which allow them to become trapped in the filter, subsequently to form complexes with low-avidity antibodies. These complexes, trapped within glomeruli, may be injurious and provoke some reaction in the affected tissue. They are particularly 'toxic' if they fix complement, as this is likely to induce a vigorous inflammatory reaction. In other situations, various reactions occur which produce structural changes within the mesangium and/or capillary loops.

The lesion resulting from this trapping of antigen/antibody complexes in the glomerulus is called **immune-complex glomerulonephritis** and it is really a disorder of the immune system which is expressed as glomerular damage. Sometimes the antigen involved can be characterized, but in most cases this is not so. However, by appropriate laboratory techniques the antibodies within the complexes can be detected in tissue samples, as can deposits of complement. The pattern of such deposition within glomeruli for the various classes of immunoglobulin has been extensively used as a diagnostic criterion in this group of diseases.

True **autoimmune glomerulonephritis**, in which auto-antibodies against glomerular tissue are produced, is an extremely rare condition, and has been studied mainly by experimental induction. The classic model is 'Masugi' nephritis, produced by autoimmunizing rats with kidney tissue. In autoimmune glomerulonephritis, the auto-antibodies can be demonstrated to bind smoothly and evenly along the glomerular capillary walls, rather than in the granular pattern produced in most cases of immune-complex deposition.

Finally, in several species, glomerulonephritis has been found to be a genetically determined disease, due to a hereditary (but generally undefined) metabolic defect. It has been described in Doberman Pinschers, Cocker Spaniels and Landrace sheep.

To restate matters a little from the clinical viewpoint, an animal with acute renal failure, chronic renal failure, or nephrotic syndrome, may have glomerulonephritis (see Fig. 9.15 for a précis). Certain clinical features will suggest a diagnosis of glomerulonephritis in the living animal. It can be taken as a good general rule that heavy persistent proteinuria in the absence of evidence of lower tract inflammation or hemorrhage is strongly suggestive of glomerulonephritis. However, if the glomerular lesions are of a type that is particularly destructive, red blood cells (free or trapped in casts), white blood cells and granular casts may also appear in the urine. Oliguria and azotemia accompany these changes. In the early stages of an acute severe glomerulonephritis, tubular function may still be intact, so that concentrating ability is preserved for a short time. After initial injury, healing and complete recovery may occur. If the injury progresses, there is a decline into severe renal failure and uremia. Throughout the entire course of the disease, proteinuria will be persistent, and at any stage the nephrotic syndrome may appear if protein loss outstrips the synthetic capacity of the liver. A biopsy can both confirm a diagnosis of glomerulonephritis and indicate a certain morphologic type of lesion. At present the following morphologic types of glomerulonephritis have been recognized in animals.

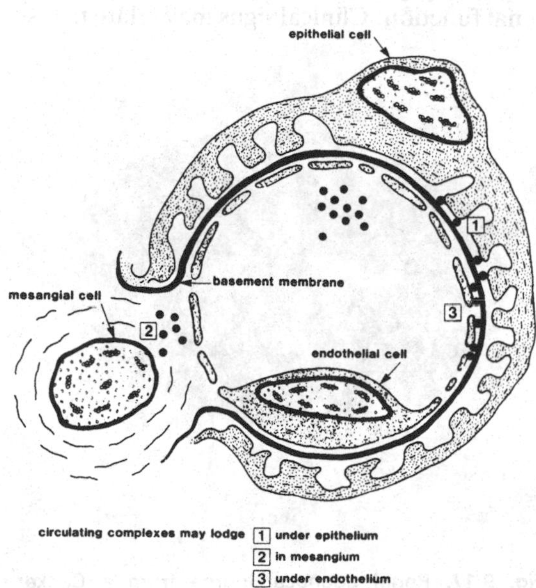

Fig. 9.16. Circulating immune-complex (solid circles) may cause disease when deposited within the glomerulus.

Diffuse membranous glomerulonephritis (also called membranous nephropathy) is probably the most common type of immune-complex glomerulonephritis seen in the dog and cat. It is a subacute to chronic disease and may resolve spontaneously, or progress relentlessly. In the dog it has sometimes been associated with infestation with *Dirofilaria immitis*, and in the cat with leukemia-virus infection. In such cases it is these antigens that are present in the complexes. It may also occur in animals with systemic lupus erythematosis, when the complexes are DNA/anti-DNA.

In this disease, medium-sized, soluble immune complexes lodge within the basement membrane at its outer (subepithelial) aspect (see Fig. 9.16), in all glomeruli. In response, the basement membrane thickens, but there is no cellular proliferation or infiltration. The complexes can be detected by immunodiagnostic techniques and by electromicroscopy.

Initially, affected kidneys may appear grossly normal. Histologically, the lesion is difficult to detect using routine methods and requires special stains and electron microscopy for diagnosis. In time, atrophy of nephrons may lead to a finely nodular fibrotic state, with a microscopic picture of fibrosis, atrophy and lymphoid cell infiltration.

The urine of an animal with diffuse membranous glomerulonephritis will contain excessive quantities of albumin but no debris derived from the large-scale destruction of renal tissue and no inflammatory cells or red blood cells. Problems of urinary concentration will not be evident until late in the disease, when significant tubular atrophy has occurred. These interpretations should all be acceptable if the nature of the lesion is called to mind.

The other forms of glomerulonephritis have not been so regularly associated with clinical renal disease in animals, and are less well understood in clinical terms. Some, however, have been detected in several domestic species fairly frequently in necropsy material.

Membrano-proliferative glomerulonephritis is characterized by membranous thickening of capillary loops and accompanied by mesangial, epithelial or endothelial proliferation. Immunoglobulin deposits may be detected within the mesangium and between the basement membrane and both the endothelium and the epithelium. This may be an acute or subacute disease and in the early stages the kidneys may be grossly enlarged.

Minor types of glomerulonephritis include a **proliferative** type where there are proliferations of epithelial cells in Bowman's space and sometimes endothelial and mesangial cell proliferation. **Exudative glomerulonephritis** has as its main feature neutrophil infiltration of the glomerulus. **Necrotizing glomerulonephritis** is characterized by segmental or global necrosis of glomeruli accompanied by fibrin thrombi in glomerular vessels.

End-stage kidney

Chronic, progressive, destructive disease processes in the kidneys, whatever their initial nature, tend ultimately to result in a common final pathologic state. This is generally referred to as end-stage kidney. The kidneys are reduced in size and distorted in shape due to the combined effects of loss of nephrons and fibrous contraction.

Dystrophic mineralization may be extensive and small cystic spaces may be present, derived either from obstructed surviving nephrons or previous tissue necrosis. An animal with end-stage kidney disease will have chronic renal failure and severe uremia. It must be emphasized that it is frequently difficult or impossible to determine what the initial pattern of disease may have been. The process may have begun as glomerulonephritis, interstitial nephritis or pyelonephritis.

Neoplastic disease

Primary renal neoplasms are relatively uncommon in animals, and when they do occur they generally have no effect on total renal function. Clinical signs may relate to loss

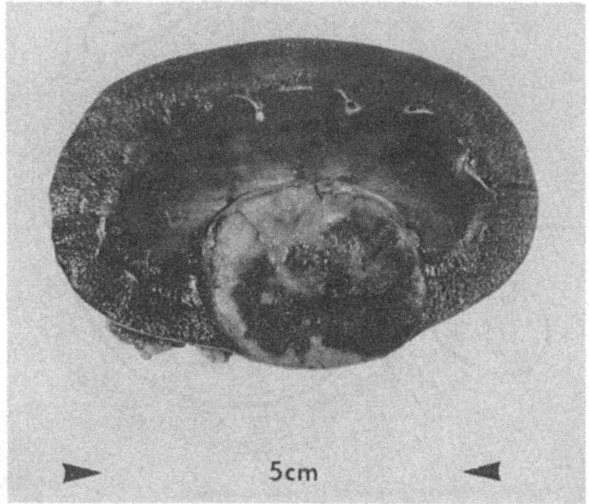

Fig. 9.17. Renal adenocarcinoma from a Cocker Spaniel dog. The initial clinical problem was hematuria but eventually metastases led to euthanasia.

of blood by persistent hematuria, or to sublumbar pain (Fig. 9.17).

A notable exception, however, occurs in cats with lymphosarcoma. Bilateral infiltration of the kidneys may occur early in the disease with the result that renal failure may be the first clinical expression of the underlying lymphoid neoplasm. The kidneys are usually moderately enlarged, with irregularly bulging cortical surfaces (Fig. 9.18). The malignant cells infiltrate in a patchy manner, mostly into the cortex, so that the cut surface has ill-defined solid gray-white foci replacing normal parenchyma. Histologically, the aggressive and destructive crowding of masses of neoplastic cells into the renal tissue is immediately evident. This process is sufficiently rapid for death to occur within a matter of weeks from the onset of clinical signs. This pattern of renal involvement in lymphoid and hematopoietic neoplasms is occasionally seen in other animal species, but feline lymphosarcoma leads the field in this respect. This diagnosis should always be ruled out in any young cat found to have acute renal failure.

Developmental diseases

In the dog, an increasing number of familial renal diseases are being recognized, most of which seem to involve progressive atrophy of nephrons with concurrent fibrosis. The cause is almost certainly a genetic defect, but this has not yet been fully defined. The animals exhibit signs of chronic renal failure at a young age and many die within the first two years of life. At the termination, the kidneys are small, pale and tough, and sometimes nodular and distorted (Fig. 9.19). Formerly they were often considered to be hypoplastic; that is the result of a failure to form an adequate number of nephrons during organogenesis. The evidence now does not support this and it seems that postnatal loss of nephrons is the central feature.

In the Doberman Pinscher, Cocker Spaniel and Samoyed breeds, the disease is reputed to be a chronic progressive glomerulonephritis. In other breeds, progressive interstitial fibrosis and nephron atrophy occur.

The salient point is that, if chronic progressive renal failure is encountered in a young animal, the possibility of a familial disease of this type should be high on the diagnostic list and will have important implications in breeding establishments.

Occasionally, clinical renal disease may result from polycystic dysplasia. This condition results from focal dilations along the

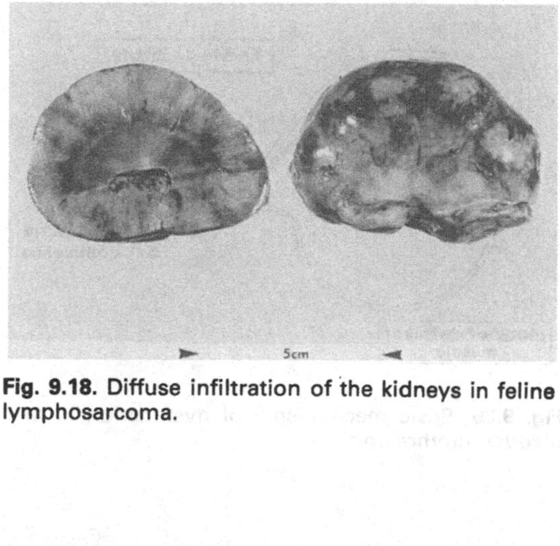

Fig. 9.18. Diffuse infiltration of the kidneys in feline lymphosarcoma.

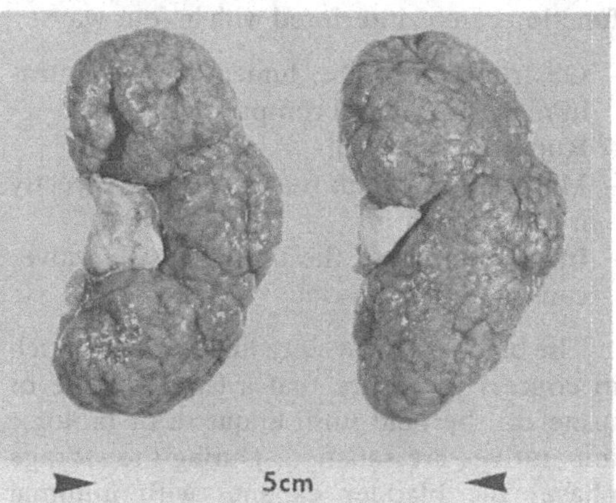

Fig. 9.19. Fibrosis and atrophy of the kidneys in a young dog with familial nephropathy.

nephrons, anywhere from the glomerulus onwards. The renal tissue is transformed into a honeycombed mass of small cystic spaces and there is renal failure at or shortly after birth. The cause of the lesion is probably genetic in most cases, but some chemicals are known to induce it in experimental animals and might be a mechanism in some spontaneous cases.

Function and malfunction in the lower urinary tract

The urine generated by the kidneys is directed to the lower urinary tract and then out of the body. The mechanisms involved in this process are conduction, storage and final voiding. The process involves the ureters, the urinary bladder and the urethra, and requires an integrated functional balance.

Recalling that urine is produced continually, it can be appreciated that the ureters must be in constant action to conduct it away, as the kidneys themselves have no storage capacity. There is thus rapid involuntary peristaltic ureteral contraction. To function normally, each ureter must be a fully patent closed tube, be correctly innervated and implanted obliquely into the neck of the bladder. These requirements allow urine to enter the lumen of the bladder, but not to reflux back up the ureter when the bladder contracts. Ureteral function can be interfered with in four ways:

- Obstruction of the lumen due to intraluminal or external compressive agents.
- Rupture of the wall.
- Malposition, when the duct opens directly into the urethra.
- Incompetence of the ureterovesicle valve, causing reflex of urine.

The bladder and urethra function very much in concert to ensure that a large volume of urine can be held until etiquette or biologic imperatives are satisfied. During the storage phase, the bladder expands with minimal muscle tone to keep hydrostatic pressures low, while urethral muscle tone is high to prevent leakage from the enlarging bladder.

When voiding time arrives the situation is reversed. Parasympathetic stimulation causes contraction of the detrusor muscle of the bladder and simultaneously the muscles of the urethra relax. The most obvious cause of voiding dysfunction is mechanical obstruction of the urethra, classically by calculus. However, problems of urine voiding may have other underlying causes. This has been fairly simply stated as a failure to store or a failure to empty.

A failure to store can be analyzed as the result of reduced bladder capacity, hyperactivity of the detrusor muscle, or urethral incompetence (Fig. 9.20). The consequence of failure to store is a continuous or frequent discharge of small quantities of urine. A continuous dribbling of urine is referred to as **incontinence**, while frequent discharge is termed **frequency**. The bladder is generally small in these situations.

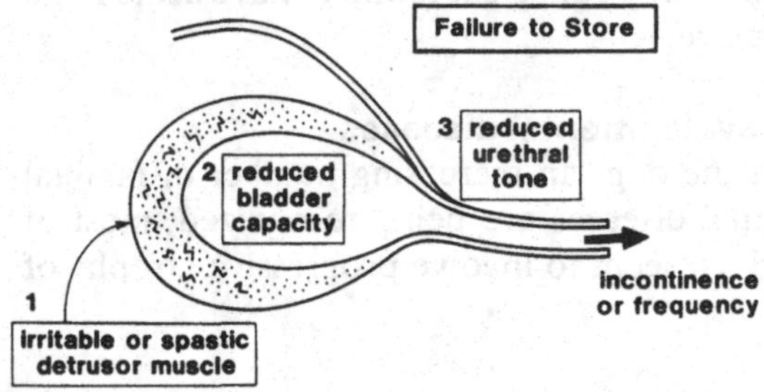

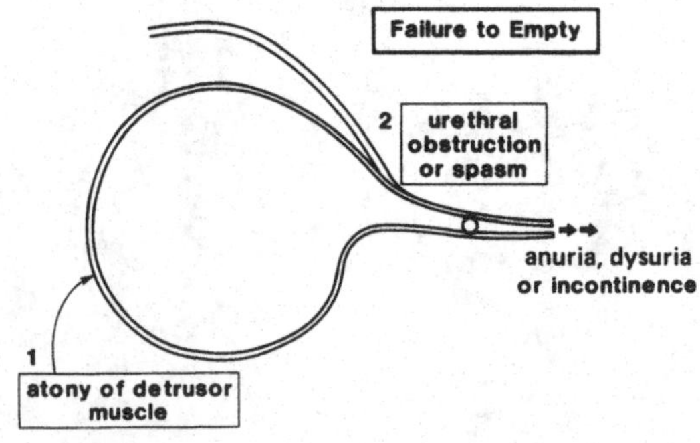

Fig. 9.20. Basic mechanisms of dysfunction in the bladder–urethra unit.

Failure to empty results from bladder atony and/or urethral obstruction or spasm (Fig. 9.20). The result is urinary retention, and this may lead to postrenal azotemia and renal failure. If there is bladder atony, the bladder will become greatly distended and eventually there will be 'overflow incontinence'. Urethral obstruction will, of course, result in a greatly distended bladder and there will be straining and discomfort. If the obstruction is incomplete, small quantities of urine may be passed during straining. Such problems may result from lesions within the organs themselves or from lesions interfering with their innervation and muscle tone. The ultimate nervous control of micturition is at the level of the lower spinal arc, but voluntary control of the act depends on centers in the brain. Lesions within the spinal arcs may produce varying states of spasticity or atony of the bladder and urethra (see Chapter 13). Current understanding of the autonomic innervation of the bladder and urethra has allowed a rational approach to treatment of some of these problems, by using drugs that stimulate or block sympathetic or parasympathetic pathways.

Common disorders of the lower urinary tract

Cystitis

Cystitis is inflammation of the urinary bladder and it is safe to say that bacterial infection via the urethra is always a major component of the problem. Other factors, however, may predispose to such infections, either by damaging the mucosa of the bladder, or by causing interference to free urine flow. Cystitis is much more common in females than males, because the short, wide female urethra offers an easy pathway to bacteria with ascending aspirations. There is little doubt that stasis of urine and ascending infection are the two major factors in the initiation of cystitis. This accounts for the high incidence of cystitis in animals with neurogenic bladder atony.

In a normally innervated bladder, **acute cystitis** provokes irritability of the wall, causing 'failure to store' and frequency of urination. The act of urination, or palpation of the bladder, may be painful. The inflammatory process in the bladder mucosa leads to large numbers of red cells, neutrophils and transitional epithelial cells being shed into the urine. The urine is, therefore, cloudy, frequently blood-tinged and contains numerous bacteria.

In **chronic cystitis** the wall of the bladder may become permanently thickened by proliferative changes in the mucosa and this may be detected radiographically. In a **dysinnervated bladder**, there is frequently 'failure to empty', overfilling and incontinence. The presence of cystitis in this case is indicated by turbid bloody urine with an offensive ammoniacal odor caused by bacterial urease activity.

The bladder has excellent powers of healing and if basic causes can be removed, cystitis is readily reversible. Uncomplicated bacterial cystitis can, therefore, be expected to respond well to chemotherapy.

Neurogenic bladder–urethral malfunction

Any lesion interfering with the innervation or neural control of the bladder and urethra may cause problems of urinary storage and voiding, as described above (see also Chapter 13).

The most common disease process in this context is degeneration and prolapse of canine lumbosacral intervertebral disks. This results in direct traumatic damage to the spinal cord and/or autonomic nerves. Trauma in this area from motor vehicle accident also ranks highly in small animals. Neoplastic or inflammatory lesions are uncommon. This type of situation classically causes detrusor atony and failure to empty. Any vesicourethral malfunction occurring in such cases will usually become complicated by bacterial cystitis. The outlook will depend on the result of neurologic examination and assessment of the extent of lesions and the prospects of regeneration. Attempts at pharmacologic therapy are based on cholinergic stimulation.

Peripheral bladder denervation may on

occasion produce a spastic bladder and failure to store. Regeneration of damaged nerves will depend on the type and severity of the basic disease process. Spasticity of the bladder may be approached pharmacologically by treatment with anti-cholinergics and anti-spasmodics.

Other neurogenic disorders have been found to be functional in nature and not associated with structural lesions. They may be regarded as imbalances of muscle tone. A well-known example is incontinence in old, desexed bitches. In such cases incontinence during rest or sleep is thought to be due to reduced alpha-sympathetic tone and urethral incompetence. The disorder usually responds well to estrogen therapy, but the relationship between hormone levels and autonomic function is not clear. An alternative pharmacologic approach is to provide alpha-adrenergic stimulation.

Ectopic ureter

This is a developmental anomaly seen most frequently in young bitches in which one or both ureters opens directly into the urethra, causing persistent incontinence. However, if the defect is unilateral, there will still be regular filling of the bladder and micturition. A bilateral defect results in no urine entering the bladder, which therefore never fills, and the animal never micturates but constantly dribbles urine. Abnormal ureters are usually dilated (megaloureter) and pyelonephritis is a not uncommon complication.

Patent urachus

This is another developmental anomaly, most prevalent in foals. The fetal connection between the bladder and the umbilicus persists and urine dribbles continuously from the umbilicus.

Mechanical urethral obstruction

Possibly, the most common clinical urologic problem encountered in veterinary medicine is acute urethral obstruction. In contrast to the state of affairs relating to cystitis, urethral obstruction is infinitely more common in the male animal. This predisposition has its basis in the anatomy of the male urethra – generally long and narrow and, according to species, complicated by sharp bends, enveloping bones or narrow terminal processes. By far the most common cause of urethral obstruction is **urinary calculus**. The pathophysiology of calculus formation (urolithiasis) will be discussed in a following section, but for now our interest centers on the mechanical effect of uroliths in the urethra. Such uroliths develop in the bladder from where they may be flushed into the urethra during micturition. One or more of these, or a fine crystalline mass of calculi may obstruct the urethra at various points, or throughout its length. In this regard there is considerable interspecies variation. In the cat, for example, it is common for the entire urethra to be plugged with a sand-like mass of crystals. In the dog, several individual calculi may lodge in the region of the os penis. Bulls and rams frequently become obstructed at the sigmoid flexure and ischiatic arch and, in the ram, the urethral process is also a common site of obstruction. In all cases the effect is total, or near total, cessation of urine flow through the urethra, in the face of continuing delivery of urine to the bladder. The result causes considerable discomfort. Initially the animal is preoccupied with the discomfort and constantly strains to urinate, and small quantities of urine may be passed if the blockage is incomplete. In time acute renal failure will eventuate, and unrelieved obstruction is usually fatal in 3–5 days. Additional complications may include bacterial urethritis and cystitis and/or rupture of the bladder or urethra.

Affected animals will have a tense, painful abdomen, distended bladder and sensitive kidneys, and will exhibit other signs associated with uremia. Some temporary relief of discomfort occurs if the bladder ruptures but of course uremia persists. Pathophysiologic events during obstruction and after the relief of obstruction have been previously discussed.

Urethral urine flow may be partially

obstructed by prostatic enlargement and by neoplasms of the bladder neck. The clinical defect is one of **dysuria**, with urination being prolonged and accompanied by straining. Prostatic enlargement may be the result of hyperplasia (most common in canines), neoplasia or inflammation. Neoplasms of the bladder neck are frequently malignant and virtually impossible to treat.

Urolithiasis

(In association with Prof. William T. Clark)

Urolithiasis is a pathologic state characterized by the precipitation of solids within the urinary tract. Such precipitation usually takes place in the renal pelvis or the urinary bladder, but the precipitate may be flushed into the ureter or urethra. The precipitates originate mainly from solutes normally present in the urine (see below), although they include small amounts of organic material which may come from the lining of the urinary tract.

The precipitated material may occur as a sludge of small particles, usually crystalline, or as discrete firm masses known as uroliths, urinary calculi or simply 'stones'. These can reach quite a startling size (Fig. 9.21) and can be present in large numbers. In veterinary medicine, the clinical significance of urolithiasis relates largely to obstruction of the male urethra, and to mechanical irritation of the bladder mucosa. In the latter case, there will be frequent urination and, sometimes, hematuria. In contrast, urethral obstruction is generally a life-threatening emergency and is a common problem in dogs, cats, sheep and cattle.

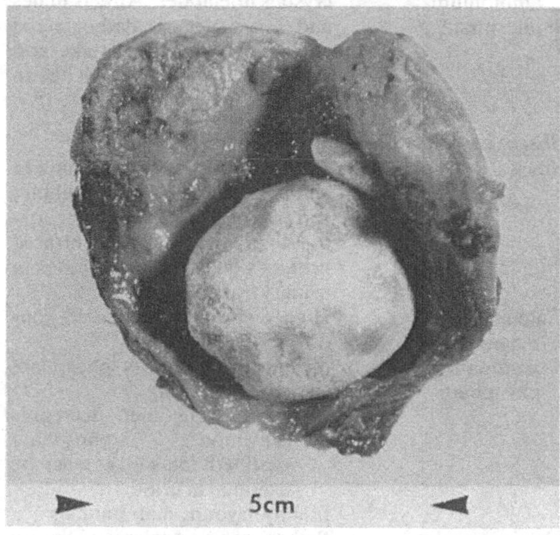

Fig. 9.21. Large calculus in the bladder of a bitch.

The process of urolith formation

The fundamental problem in urolithiasis is the oversaturation of the urine with a particular solute or solutes. The situation is complex in that the urine normally contains numerous solutes in concentrations above the normal solubility limits. The normal urine thus has the capacity to maintain supersaturated solutions of various solutes. Its ability to do this is not fully understood and the literature contains conflicting theories on this remarkable physicochemical phenomenon. It should also be noted that the urine in many species normally contains small crystals, intermittently or consistently (as in the horse), but these are insufficient to cause any problems. Also, while excessive precipitation may lead to deposits, they may sometimes redissolve completely. The process becomes clearly pathologic when precipitation is initiated and the crystals aggregate to form large masses. Indeed, it is crystal aggregation and growth that are the key events in urolithiasis.

When solute oversaturation occurs and precipitation begins, there must be either a marked excess of the solute present in the urine, or a reduced capacity of the urine to maintain supersaturation. The former situation is illustrated by silicate urolithiasis in range cattle. Certain pastures contain high concentrations of soluble silicates, which leads to a high concentration of silicic acid in the glomerular filtrate. The further concentration of the filtrate by the renal tubules, leads to a 100-fold increase in silicic acid concentration in the urine. At such concentrations, it precipitates in association with organic material to form silica uroliths.

Similarly, in dogs, there may be excessive

urinary solute excretion due to a genetically determined biochemical defect. In this case, the problem occurs without a dietary excess of the solute. The Dalmatian breed is noted for a predisposition to urate urolithiasis, and other canine breeds for cystine urolithiasis.

Reduced capacity to maintain supersaturation is well exemplified by magnesium ammonium-phosphate urolithiasis in the dog. This occurs if urinary pH is abnormally elevated. Under these conditions, the solute load normally present comes out of solution and uroliths form. The common predisposing cause is staphylococcal infection of the bladder. The bacterial urease activity splits urinary urea, releasing ammonia, which raises the urinary pH.

It has also been suggested that disturbance of the physicochemical balance of the urine occurs when organic debris is shed from the urinary epithelium. Such debris may provide a nidus for the crystallization of solute. Urinary infections, bacterial and viral, have been proposed as inducers of urolithiasis in this way, but experimental evidence is meager.

In spite of an energetic research effort over the years, the general phenomenon of urolithiasis largely remains an enigma and a challenge for future effort. Unfortunately, a clear-cut relationship between urolithiasis and a high dietary intake of solute is not obvious in most instances. Many solutes simply cannot be absorbed to excess, while others are metabolized before excretion. Nor is altered urinary pH involved in most cases. There is also little evidence that urolithiasis is provoked by generation of highly concentrated urine in prolonged water deficiency. However, the stimulation of water consumption and urinary dilution has been shown to help to prevent urolithiasis in some circumstances. It seems that the key to the puzzle lies in defining the physical chemistry of urine as a supersaturated solution.

Urolithiasis across species

Whenever urolithiasis is encountered as a clinical problem, the precipitated material should always be submitted for chemical analysis. In general, the chemical identity of the uroliths provides a useful clue to pathogenesis, prognosis and prophylaxis.

The most important types of uroliths are as follows:

Chemical nature	Pathogenesis and clinical features
The dog	
Magnesium ammonium phosphate	Forms in the bladder when staphylococcal urease generates ammonia and elevates urine pH. Most common in females
Calcium oxalate	Pathogenesis unclear. Mainly in old males
Cystine	Inadequate renal tubular resorption from glomerular filtrate leads to high urinary cystine concentration. Occurs only in males as a sex-linked genetic defect. Uroliths may form in renal pelvis and bladder in some individuals. Clinical effects first evident at two to four years of age
Ammonium urate	High urinary concentration of urates due to hepatic and renal tubular defect in urate transport. Mainly in Dalmatians, which cannot convert urate to soluble allantoin. Sometimes in other breeds. Uroliths form in renal pelvis and bladder
The cat	
Magnesium ammonium phosphate	Forms as a heavy sludge of small crystals in bladder. Affects males and females. Pathogenesis unclear. Diet, water intake and viral infection have all been proposed
Sheep, cattle and goats	
Silica	High intakes of soluble silicates in animals eating cereal stubble, certain pasture grasses or oat grain. Highly concentrated urinary silicic acid precipitates in urinary tract
Calcium-magnesium-ammonium phosphate	High dietary phosphate in concentrates: (*a*) increases urinary phosphate concentration (*b*) concentrate diet decreases salivation – phosphate destined for saliva must be excreted in urine. Mainly in young male animals
Calcium oxalate	Pathogenesis unclear. Dietary oxalate not directly related

Organic complex	Phytoestrogen intake from certain clovers causes epithelial metaplasia and desquamation. Debris deposits as softish mass known as 'clover stone'. May cause urinary obstruction in male sheep
The horse	
Calcium carbonate	Pathogenesis unclear. Normally large quantity of $CaCO_3$ crystals secreted in urine. Only rarely do uroliths form in the bladder

Principles of urinalysis

As the urine represents the culmination of the multiple efforts of the kidneys in their service to the body, it is to be expected that the urine will reflect disorders and disasters in the urinary tract itself, and in the body at large. Examination of the urine in this context is referred to as urinalysis and is based upon the predictable qualities of normal urine. When combined with a thorough clinical examination, it is a most useful diagnostic tool, complementing other clinical and laboratory tests, radiography and biopsy.

In urinalysis, the focus is upon the quality of the urine. The **quantity** and **concentration** of urine produced must be related to body water status, as has already been discussed (p. 220). In this discussion, emphasis will be upon the routine factors that are 'everyday in nature' but are often not fully appreciated. The following paragraphs will attempt to provide a conceptual framework, as the materials and methods required for adequate urinalysis are well documented in clinical pathology tests.

The qualitative characteristics that are routinely assessed are as follows:

Color	– normally pale, clear, yellow. This is imparted by urochrome pigments of renal origin. The exception is the horse in which the urine is opaque and cloudy due to carbonate crystals and mucin.
Glucose	– normally free.
Ketones	– normally free.
Protein	– normally free or a trace of albumin.
Bilirubin	– normally free (or a trace in the dog).
pH	– normally slightly acid in carnivores and alkaline in herbivores.
Cell content	– normally only a *few* sloughed cells from the urinary tract, mostly transitional and squamous epithelial cells. In males, sperm.
Micro-organisms	– normally free of bacteria and fungi. Crystals normally very few except in the horse where calcium carbonate crystals are found. Dalmation dogs – urates.

Under special circumstances, the urine may be examined for the presence of enzymes, amino acids, electrolytes, snake venom, heavy metals and various macromolecules, for instance glycosaminoglycans and oligosaccharides in lysosomal storage diseases.

Urine abnormaliies of prerenal origin

Urine quality may be altered by prerenal disturbances, even when the urinary tract is perfectly normal. In such instances, in the absence of concurrent urinary tract disease, the urine abnormality is dominated by the presence of molecules coming from the blood through the glomerular filter and causing some measurable change in urine quality. The following list identifies the classical culprits and indicates their effects.

Diabetes mellitus

Insulin deficiency results in hyperglycemia. The filtered load of glucose exceeds the capacity of renal tubular transport mechanisms and the urine, which should be entirely free of glucose, contains small to massive amounts (glycosuria). The urine specific gravity and Osm will be moderately high because of the glucose molecules and the diuretic effect will be indicated by polyuria and a rather pale color of the urine. If the patient has diabetic keto-acidosis, the urine will also react positively to a ketone test. In the absence of complications, the urine should appear otherwise normal.

Intravascular hemolysis

The rupture of red blood cells in the circulation releases free hemoglobin, which is able to cross the glomerular filter and appear in the urine (hemoglobinuria). The pigment will change the color of the urine to a deep red/brown in severe cases and will result in a positive protein test (proteinuria). In severe hemolysis there is frequently renal tubular necrosis which will produce additional abnormalities (see below).

Rhabdomyolysis

Acute degeneration or necrosis of skeletal muscle can release large quantities of free myoglobin into the blood, which can then appear in the urine (myoglobinuria). All comments pertaining to hemoglobin apply.

Cholestasis

Diseases which result in the regurgitation of conjugated bilirubin from the liver back into the blood will cause plasma concentration of the pigment to rise. A proportion of conjugated bilirubin is freely filtered by the glomerulus and will appear in the urine (bilirubinuria). Unconjugated bilirubin is more tightly bound to albumin and cannot be filtered. The urine becomes discolored to a dark olive green and reacts positively to a bilirubin chemical test.

Exogenous pigments

These may derive from ingested food or administered drugs and may change the color of the urine.

Plasma cell neoplasia (myeloma)

Neoplastic plasma cells may produce immunoglobulin light chains which are small enough to cross the glomerular filter and cause proteinuria. These protein molecules can be identified for what they are by appropriate tests and can thus be distinguished from normal plasma proteins in the urine. They are sometimes referred to by their old name of Bence-Jones protein.

Portocaval shunt

In congenital or acquired abnormalities which cause portal blood to bypass the liver, one of the metabolic consequences is the accumulation of ammonium biurates and their excretion in the urine. Such patients will have characteristic crystals of ammonium biurates in their urine.

Urine abnormalities of renal origin

Active renal tissue disease will be reflected in the quality of the urine produced, even if total renal function is normal. In this discussion, the major categories of renal disease presented earlier will be used to highlight the associated urine abnormalities.

Glomerulopathy

In a situation where large numbers of glomeruli are suddenly damaged severely, there will be marked glomerular barrier failure with perhaps sufficient damage to cause hemorrhage. Urine abnormalities will be dominated by heavy proteinuria (mainly albumin), the presence of red blood cells (hematuria), and perhaps some neutrophils. Red blood cells may be found in cylindrical casts. The excessive protein will produce fine granular casts and purely amorphous hyaline protein casts may also be expected, as the proteinuria promotes the precipitation of Tamm–Horsfall mucoprotein.

When glomerular damage is less severe, the major abnormalities will be as above, but red cell casts and neutrophils will not be present.

Acute tubular necrosis

When there is active destruction of renal tubular epithelium, there is an associated inflammatory reaction, with exudation and mild cellular infiltration. Necrotic tubule cells, exudate and inflammatory cells will therefore find their way into the urine. Extensive destruction of proximal tubule cells will also prevent the resorption of glucose and any albumin that is filtered.

The expected urine abnormalities on routine analysis will be proteinuria (less marked in glomerulopathy), glycosuria (usually moderate), renal tubular cells and neutrophils (often in casts), granular casts and perhaps a few red cells. Neutrophils are never in massive numbers, and the urine color is usually little changed. In the phase of regeneration, these abnormalities will resolve, although renal function may be abnormal.

Acute interstitial nephritis

Urine abnormalities will be generally similar to acute tubular necrosis, with perhaps more inflammatory cells, and in the case of leptospirosis, spirochetes may be demonstrable.

Chronic interstitial nephritis

In this disease, there is very little active destruction of renal tissue at any particular time, so that the abnormalities described above are not obvious. Proteinuria is slight, casts are few and inflammatory cells absent.

Pyelonephritis

In this disease involving bacterial infection, inflammation and renal tissue destruction, there will be proteinuria, neutrophils, cellular and granular casts, and often bacteria. Neutrophils may be present in such numbers that the urine is turbid and cloudy (pyuria).

Renal tumor

Renal neoplasms not infrequently bleed into the urine. The urine will therefore contain red blood cells and white blood cells in appropriate proportions, and of course, protein. The blood may or may not discolor the urine, depending on its quantity. A small amount of blood may cause a clouding rather than red discoloration.

Urine abnormalities of postrenal origin

The most common postrenal conditions that alter urine quality are inflammatory lesions and proliferative lesions with a tendency to bleed.

Cystitis

In cystitis there is frequently a rather destructive bacterial inflammation of the mucous membrane of the bladder, with a hemorrhagic exudate and considerable sloughing of the epithelium. The urine is turbid and often blood tinged, and will contain protein. The sediment contains masses of neutrophils, transitional epithelial cells, red blood cells and bacteria. In many instances bacterial urease activity will release ammonia from urea and render normally acid urine alkaline. The picture will resemble that seen in pyelonephritis, except that there will be no casts and no renal epithelial cells.

Lower tract hemorrhage

Neoplastic, hyperplastic and traumatic lesions in the bladder and urethra may produce hematuria, with characteristics similar to those described for renal tumor.

Additional reading

Blood, D. C., Henderson, J. A. and Radostits, O. M. (1979). *Veterinary Medicine*, 5th edn, pp. 276–93. London, Bailliere Tindall.

Bovee, K. E. (ed.) (1984). *Canine Nephrology*. Media, PA, Harwal Publishing Co.

Breitschwerdt, E. B. (1981). Clinical abnormalities of urine concentration and dilution. *Compend. Cont. Educ. Vet. Pract.* **3**: 414–21.

Brenner, B. M. and Rector, F. C. (1976). *The Kidney*. Philadelphia, W. B. Saunders Co.

Cameron, S. C. (1981). *Kidney Disease: The Facts*. New York, Oxford University Press.

DiBartola, S. P. (1980). Acute renal failure: pathophysiology and management. *Compend. Cont. Educt. Vet. Pract.* **2**: 952–8.

Duncan, J. R. and Prasse, K. W. (1986). *Veterinary Laboratory Medicine*, 2nd edn. Ames, IA, Iowa State Univ. Press.

Ettinger, S. J. (ed.) (1983). *Textbook of Veterinary Internal Medicine*, 2nd edn. Philadelphia, W. B. Saunders Co.

Kaneko, J. J. (ed.) (1980). *Clinical Biochemistry of Domestic Animals*, 2nd edn. New York, Academic Press.

Kaysen, G. A., Myers, B. D., Couser, W. G., Rabkin, R. and Felts, J. M. (1986). Mechanisms and consequences of proteinuria. *Lab. Invest.* **54**: 479–98.

Jamison, R. I. and Kriz, W. (1981). *Urinary Concentrating Mechanisms*. New York, Oxford Univ. Press.

Leaf, A. and Cotran, R. S. (1980). *Renal Pathophysiology*, 2nd edn. New York, Oxford Univ. Press.

Maxie, M. G. (1985). The urinary system. In *Pathology of Domestic Animals*, vol. 2, K. V. F. Jubb, P. C. Kennedy and N. Palmer (eds.), pp. 343–411. Orlando, FL, Academic Press.

Michell, A. R. (1984). Ins and outs of bladder function. *J. Small Anim. Pract.* **25**: 237–47.

Osborne, C. A. (1983). Azotemia: a review of what's old and what's new. Part I. *Compend. Cont. Educ. Pract. Vet.* **5**: 497–510.

Osborne, C. A. (1983). Azotemia, a review of what's old and what's new. Part II. *Compend. Cont. Educ. Pract. Vet.* **5**: 561–74.

Osborne, C. A., Low D. G. and Finco, D. R. (1972). *Canine and Feline Urology*. Philadelphia, W. B. Saunders Co.

Papper, S. (1980). *Clinical Nephrology*, 2nd edn. Boston, MA, Little, Brown and Company.

Their, S. O. (1985). The kidney. In *Pathophysiology: The Biological Principles of Disease*, 2nd edn, L. M. Smith and S. O. Their (eds.), pp. 799–920. Philadelphia, W. B. Saunders Co.

Wayne F. Robinson and Susan E. Shaw

10 The endocrine glands

Hormone structure, function and regulation

Before embarking on a discussion of the abnormalities affecting endocrine glands, a number of facets of the normal state will be highlighted.

Hormones are chemical messengers. They originate in one part of the body and act on other, often distant, parts. Hormones are selective in their action, influencing only those organs, tissues or cells that are receptive to them.

Although the variety and metabolic consequences of hormones are remarkable, they may be classified into three groups according to the tissues upon which they act. There are those that act directly on non-endocrine target tissues (effector hormones); those that control the synthesis and release of effector hormones (tropic hormones); and those that control the synthesis and release of tropic hormones (releasing hormones).

Hormones may also be classified by their molecular structure into three chemical groups; polypeptides, steroids and amino acid derivatives. This latter classification carries with it some of the fundamental concepts of biochemical mechanisms of hormone action.

Polypeptide hormones

Of the endocrine glands considered in this chapter, the hypothalamus, pituitary, endocrine pancreas and the parafollicular (C cells) of the thyroid produce polypeptide hormones. The hormones within this group vary tremendously in size, from simple molecules to complex polypeptides of up to 190 amino acid residues in length. However, the mechanism of action is basically the same, regardless of the size or amino acid composition of the hormone.

All act by binding to membrane receptors specific for the particular hormone. Their effects may be mediated either by an alteration of membrane permeability to various substances, or by an alteration in the activity of membrane proteins. Most polypeptide hormones induce a change in the activity of the membrane proteins adenyl cyclase or guanyl cyclase. A notable exception to this is insulin, which appears to act via a different membrane glycoprotein.

The binding of polypeptide hormones to adenyl cyclase or guanyl cyclase catalyzes the conversion of ATP and GTP to cyclic AMP and cyclic GMP, respectively. The discovery of the cyclic derivatives of ATP and GTP revolutionized endocrinology and gave rise to the concept of second messengers. Through this mechanism, the information carried by many polypeptide hormones is transferred to the interior of a cell (Fig. 10.1).

Cyclic AMP and cyclic GMP modify the activity of protein kinases, resulting either in an immediate response, such as the activation of enzymes concerned with glycogen breakdown, or a delayed response, such as the induction of a new family of enzymes via RNA transcription and protein translation. Often one polypeptide hormone provides both an immediate and delayed response; the immediate response becomes apparent within minutes, and the delayed response within hours to days. An example of such dual action is the response of the thyroid gland to thyrotropin. There is an immediate release of preformed thyroid hormone, but at the same time the synthesis of enzymes involved in the slower process of thyroid hormone production is enhanced.

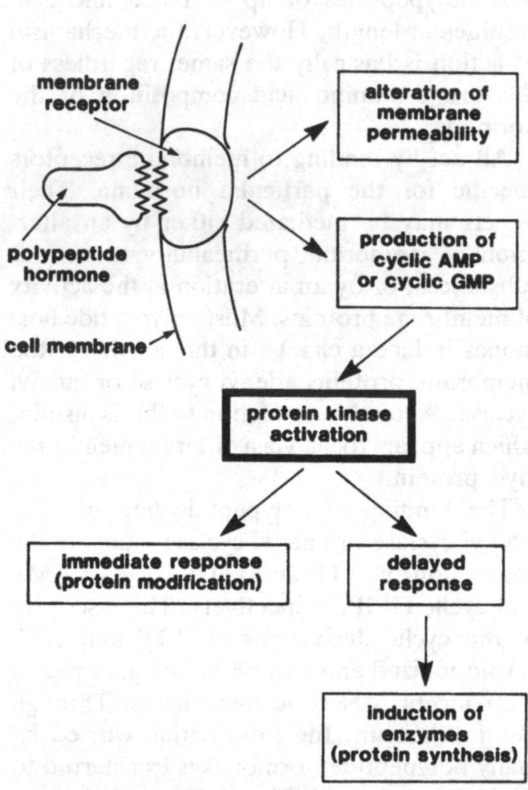

Fig. 10.1. Polypeptide hormone action: the dominant actions of polypeptide hormones on target cells are membrane mediated. The responses may be immediate or delayed, depending on whether (1) the activity of enzymes already present is modified or (2) induction of one or a group of enzymes occurs.

Steroid hormones

These hormones are produced, modified and secreted by a number of organs, including the adrenal cortex, the kidney, the ovary and testis. While the basic structure of steroid hormones is similar, the actions of a particular steroid are related to small modifications in side chains added to the basic structure. The exquisite specificity produced by side-chain modification has been utilized to produce synthetic steroids in which a particular activity is greatly enhanced.

The mechanism of action of steroids is quite unlike that of polypeptides. Whereas polypeptides are shunned by the interior of the cells, steroids, like long-lost friends, are welcomed. Steroids pass easily through the cell membrane and bind to a hormone-specific cytosolic receptor. The steroid–receptor complex is then transported to the nucleus. Once within the nucleus, the steroid binds to a nuclear acceptor, modifying the transcription of particular RNA types. This is followed by the enhancement or suppression of the production of a protein or a group of proteins. The altered complement of proteins within the cell modify the cell function (Fig. 10.2). For example, in the liver, glucocorticoids induce the formation of tyrosine aminotransferase, an enzyme intimately involved in gluconeogenesis. Only those cells with cytosolic receptors specific for a particular steroid will have their function modified, reinforcing the concept of target organs or tissues.

Amino acid derivatives

In this group are the thyroid hormones and the catecholamines produced by the adrenal medulla. Catecholamines are unusual in that their most significant effect on the body is not hormonally mediated, but occurs through their release at sympathetic nerve endings. Catecholamines act both by membrane modification, enhancing the flux of ions such as calcium, and by adenyl cyclase to produce a variety of intracellular responses. This is not so with thyroid hormones; their effect on target organs is massive and somewhat general, inducing a variety of enzymes by modifying transcription.

Hormone regulation

The regulatory mechanisms for hormones fall within two general patterns. In one, the rate of secretion of the hormone and its plasma concentration undergoes wide fluctuations. For instance, the rate of secretion of aldosterone and anti-diuretic hormone (ADH), reflects the state of the body at that particular time. The second pattern is characterized by constancy, where the hormonal needs of the body do not vary on a day-to-day basis.

For the first group, the hormone acts as the **controlling element**, the rate of secretion of the hormone keeps the substance to which it is sensitive (the controlled variable) within narrowly defined limits. For instance, the plasma levels of insulin change, sometimes markedly, in an endeavor to keep blood glucose levels constant (Fig. 10.3).

For the second group of hormones, the plasma level of hormone is kept constant, and the hormone becomes the **controlled variable**. In this case the controlling element (which is usually another endocrine gland such as the adenohypophysis or hypothalamus) is sensitive to the levels of the hormone that acts on non-endocrine target tissues. The regulation of thyroid hormone is an example of such a mechanism. Cells in both the hypothalamus and the adenohypophysis respond to the plasma levels of thyroid hormone and secrete thyrotropic hormone releasing factor (TRF) and thyrotropin (thyroid-stimulating hormone, TSH) respectively (Fig. 10.4). A list of the control mechanism for each endocrine gland will be found in Table 10.1.

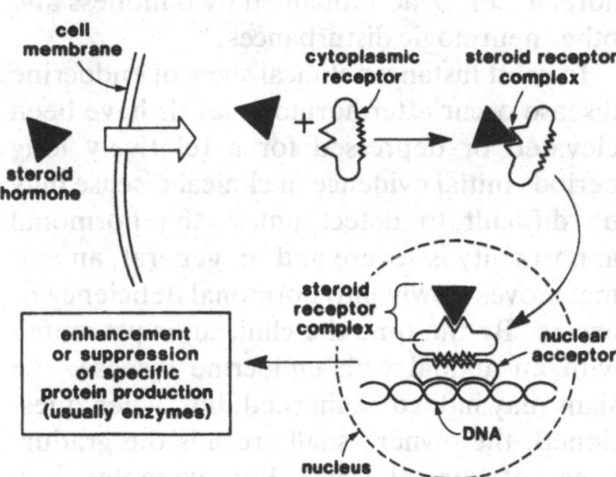

Fig. 10.2. Steroid hormone action: steroid hormones pass easily through the cell membrane and bind to specific receptors in the cytoplasm of target cells. The steroid–receptor complex moves to the nucleus, binding to chromatin. RNA transcription is modified to enhance or suppress production of particular enzymes.

Principles of endocrine dysfunction

Clinical signs due to endocrine disease are the result of either one or both of the following circumstances. The first, and most common, is an **exaggeration of, or deficiency in, endocrine function**. The second and less common is the

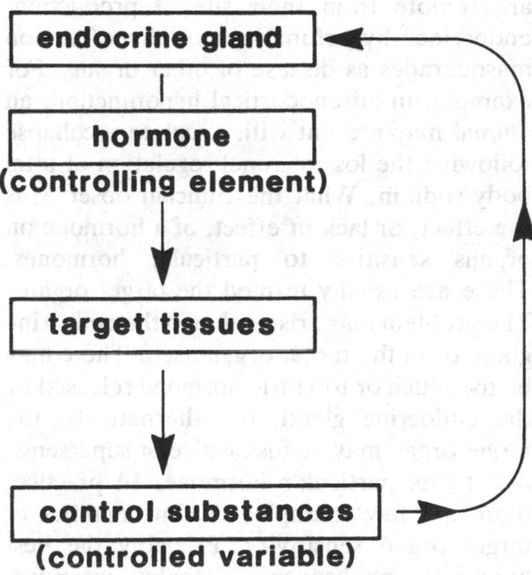

Fig. 10.3. Hormones as controlling elements: in this regulatory pattern it is the plasma level of the control substance (the controlled variable such as blood glucose) that is kept within strict limits. The hormone (the controlling element such as insulin) undergoes wide fluctuations.

Table 10.1. *Endocrine control mechanisms*

Gland	Hormone secreted	Controlling element	Controlled variable
Adenohypophysis	Growth hormone	Growth hormone	Somatomedins
Neurohypophysis	ADH	ADH	Plasma osmolality
Thyroid	Thyroid hormone	TSH and TRH	Thyroid hormone
Adrenal cortex (inner zones)	Glucocorticoids	ACTH and CRF	Glucocorticoids
Adrenal cortex (glomerulosa)	Aldosterone	Aldosterone	Sodium
Endocrine pancreas	Insulin	Insulin	Glucose

ADH, anti-diuretic hormone; TSH, thyroid-stimulating hormone; TRH, thyrotropin-releasing factor; CRF, corticotropin releasing factor.

destruction of tissues adjacent to endocrine glands by invasion or compression.

The clinical recognition of an exaggeration or deficiency of endocrine function does not depend on knowing the anatomical location of an endocrine organ. Rather, it is dependent on knowledge of the metabolic effects of a particular hormone. As the effects of hormones are remote from their site of production, endocrine hyperfunction or hypofunction masquerades as disease of other organs. For example, in adrenocortical hypofunction, an animal may present with circulatory collapse following the loss of renal regulation of total body sodium. What the clinician observes is the effect, or lack of effect, of a hormone on organs sensitive to particular hormones. These are usually termed the target organs. The problem may arise either in the endocrine gland or in the target organ itself. There may be too much or too little hormone released by the endocrine gland, or, alternatively, the target organ may be insensitive or supersensitive to its particular hormone. In practice, there are few examples of abnormalities of target organ sensitivity. Probably the best known is nephrogenic diabetes insipidus, where the kidney is insensitive to the effects of ADH.

Clinical signs referable to destruction of tissues adjacent to an endocrine gland are almost always due to neoplasia of the gland. These local effects are no different to those produced by any other space-occupying mass. Pituitary and thyroid neoplasms are of most clinical significance in this context.

Finally, there may be clinical signs reflecting *both* a disturbance in endocrine function and an occupation of space. For example, a pituitary neoplasm may not only produce excessive amounts of corticotropin (adrenocorticotropic hormone, ACTH) but also invade or compress the adjacent central nervous system. These will cause excessive stimulation of the adrenal cortex accompanied by blindness and other neurologic disturbances.

In most instances clinical signs of endocrine disease occur after hormone levels have been elevated or depressed for a relatively long period. Initial evidence of clinical disease may be difficult to detect unless the hormonal abnormality is severe and, in general, an animal moves slowly into hormonal deficiency or excess. By the time the clinician is presented with an animal with endocrine disease, the signs may indeed be marked, but, when questioned, the owner usually recalls the gradual onset of clinical signs. For example, with thyroid deficiency in dogs, obesity and a decrease in activity are often interpreted by the owner as signs of approaching old age in the dog.

Because of the diversity of hormone actions, it is difficult to generalize about the effects of endocrine disease on target tissues, but, in the short term, hormone deficiency or excess in many instances does not produce a dramatic change in target tissue size. There is no accompanying inflammation or fibrosis, the change being limited to a gradual enlargement or atrophy of the involved target tissue. Indeed, in some cases there is no observable gross or microscopic change. However, the resulting metabolic changes can have a profound effect on the animal. For example, a loss of sodium regulation with aldosterone deficiency quickly leads to electrolyte disturbances, dehydration, cardiac dysrhythmias and death. Similarly, the loss of ADH becomes life threatening when the affected animal is deprived of adequate amounts of water and is unable to concentrate urine.

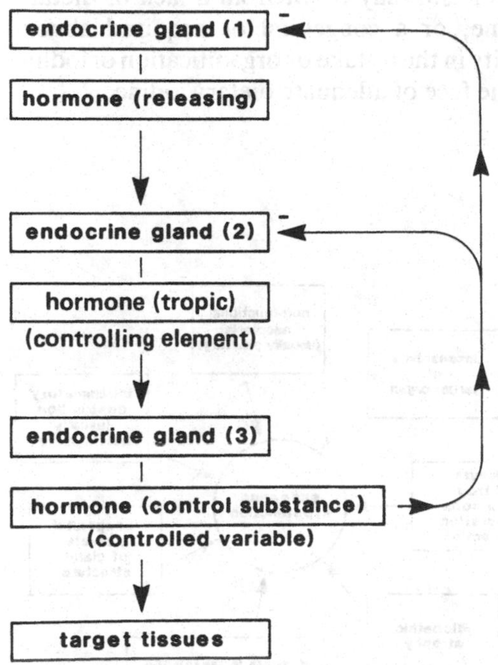

Fig. 10.4. Hormones as the controlled variable: in this case plasma levels of the hormone acting on the target tissue are kept constant (the controlled variable). The controlling element in these circumstances is usually another endocrine gland. Plasma levels of thyroid hormone are regulated in this way.

The level of the functional deficit or excess

One of the difficulties faced by a clinician when confronted with a case of endocrine disease is that only the end product is seen; that is, an abnormality of the target organ. There is usually no indication of the level or site at which the primary disease is occurring. In cases where the endocrine pancreas is producing too much or too little insulin, the problem is a relatively simple one, as there are only two possibilities, the beta cells themselves or the target tissues. However, the difficulties are compounded when there is a complex regulatory mechanism. One that is particularly relevant clinically is dysfunction of the glucocorticoid mechanism. The final clinical signs will be quite similar, but the abnormality may be in one of four sites: the hypothalamus, the pituitary, the adrenal cortex, or the non-endocrine target tissues. At least these are the possibilities, but, practically, only the first three need be considered. Target tissue insensitivity or supersensitivity to glucocorticoids has yet to be recognized in animals.

The need to differentiate between potential sites is at this stage partly academic and partly a clinical necessity. If, for example, the cause is a functional tumor in the adrenal cortex, it may be removed surgically, but a lesion in either the hypothalamus or the pituitary is for all practical purposes inaccessible.

A final possibility is the production, ingestion or injection of hormones or hormone-like substances. There are isolated examples of non-endocrine tumors, such as lymphosarcoma, producing hormonally active compounds, mimicking the effects of certain hormones. Similarly, herbivores may ingest plants containing hormone analogs, such as those with vitamin D activity. Also, a major problem in grazing areas in Australia is the ingestion of subterranean clovers containing estrogen-like compounds.

The basis of endocrine hypofunction

Consistent with the above discussion, endocrine hypofunction will be dealt with according to the functional nature and level of the abnormality. There is a marked species difference in the frequency of particular patterns of dysfunction, and reference will be made to this fact as each type is mentioned.

Endocrine hypofunction most commonly follows primary disease of endocrine glands that are the source of hormones acting on non-endocrine target tissue. In some cases it may be secondary, as the end product of a depression or lack of tropic hormones or releasing factors from the pituitary or hypothalamus. As mentioned previously, it occasionally follows end-organ unresponsiveness.

The abnormality may include inflammatory destruction, non-functional neoplasms, defects in the pathway of hormone production or release, congenital defects in gland structure, idiopathic atrophy and failure of tropic hormone secretion or action (Fig. 10.5).

Destruction by inflammation or non-functional neoplasia

Inflammatory destruction is common, at least for the thyroid and adrenal cortex and in most instances is probably autoimmune in origin.

A similar mechanism may also affect the endocrine pancreas. In all cases, the suspicion of autoimmune involvement is based on the microscopic appearance of the affected gland, accompanied by evidence of humoral or cell-mediated immune attack directed against the particular gland. Hypofunction in the case of the endocrine pancreas may also follow necrosis or inflammation of the exocrine pancreas. Non-functional neoplasia occurs most commonly in the adrenal and thyroid glands, but is of little clinical importance with respect to endocrine function because these glands are paired. The remaining normal gland is sufficient to maintain their endocrine function. Non-functional neoplasia may assume significance because of the possibility of the tumor metastasizing. In contrast to the paired endocrine organs, non-functional neoplasms of the pituitary assume importance because of the progressive destruction of the remaining normal anterior and posterior pituitary. In such cases of destruction, there may be secondary atrophy and hypofunction of the adrenals, thyroids and gonads.

Defects in the pathway of hormone production or release

The ultimate hormonal product of any endocrine gland is the result of a number of metabolic steps. The requirements for each of these steps varies with the gland. With some hormones only the protein-synthetic machinery is essential, whereas in others, organic or inorganic precursors are also needed.

The defect may be congenital, where it usually takes the form of a single enzymatic abnormality, or it may be an acquired inhibition of one or more enzymes. A lack of substrate or an inhibition of substrate uptake is another mechanism. Probably the best example of both occurs in the thyroid. Hypothyroidism may result from a lack of dietary iodine, or a congenital or acquired abnormality in the uptake or organification of iodine in the face of adequate dietary iodine.

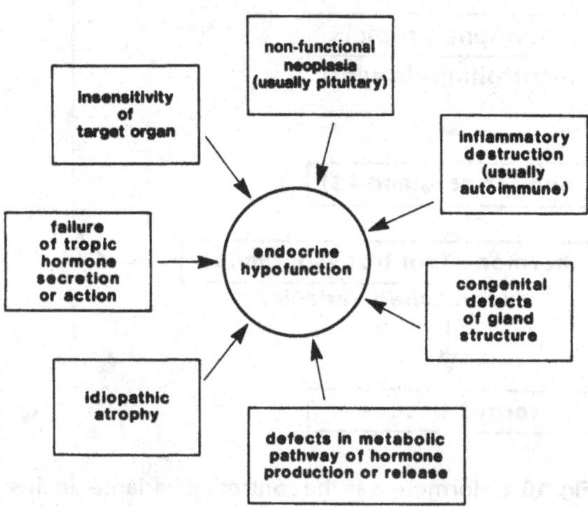

Fig. 10.5. Endocrine hypofunction may follow a number of congenital or acquired disorders of gland structure, hormone production and action.

Congenital defects of gland structure

As with endocrine neoplasia it is usually only structural defects of the single glands that are of clinical importance. Severe hypoplasia or aplasia usually only affects one of the paired organs. There are examples of gross developmental defects in the pituitary, particularly in certain breeds of dog and cow and there are instances of pituitary abnormalities in lambs where the ewes have grazed on pastures containing teratogens. There are also isolated examples of endocrine pancreatic hypoplasia in conjunction with exocrine pancreatic hypoplasia.

Idiopathic atrophy

This most unsatisfactory title describes those cases of hypofunction where the gland is grossly and microscopically atrophied for no apparent reason. There is some evidence for the atrophy being immunologically mediated, particularly in the case of the thyroid gland.

Failure of tropic hormone secretion or action

Structural or functional defects of the pituitary may result in hypofunction of those glands subservient to it. Those involved include the two inner zones of the adrenal cortex, the follicular cells of the thyroid and the gonads.

The basis of endocrine hyperfunction

The broad patterns of endocrine hyperfunction are encompassed in four categories: those tumors that produce excessive amounts of hormone (functional neoplasia), the iatrogenic administration of hormones, feedback regulation problems and hormone-like substances (Fig. 10.6).

Functional neoplasia

All the endocrine glands considered in this chapter may be the origin of functional neoplasms. Most commonly the pituitary gland, but also the thyroid, adrenal gland and endocrine pancreas may be involved. Neoplasia in these cases results in a loss of feedback regulation on the neoplastic cells. They become more or less autonomous.

Iatrogenic hyperfunction

The administration of exogenous hormones will sometimes lead to signs of hyperfunction. This is most commonly seen with the overzealous use of corticosteroids.

Interference with feedback regulation

Whilst not proven, there is some suggestion that clinical signs of hyperfunction may result from the alteration or 're-setting' of feedback regulation. This is particularly so for the hypothalamus and pituitary.

Hormone-like substances

There are increasing examples of compounds in this category and they fall into two broad categories; those produced endogenously, usually from non-endocrine tumors, and those of plant origin. The latter are by far the most common and important, at least as far as the economic animals are concerned.

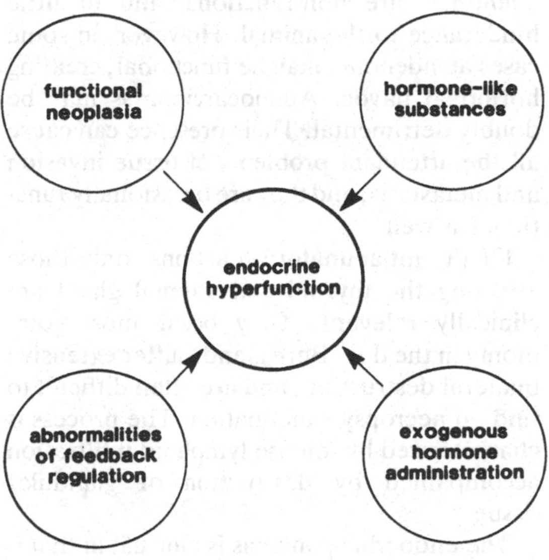

Fig. 10.6. Endocrine hyperfunction may be divided into four major categories. All are acquired abnormalities.

The pathologic features of endocrine disease

Endocrine glands are reactive. As well as releasing hormones, the endocrine cells as a whole are sensitive to their particular stimulators or inhibitors. They are usually reactive in both the directions, of hyperplasia or atrophy, depending on the level of stimulation or inhibition. The change of hyperplasia or atrophy follows the call for more or less hormone, and the cell is the servant of that call. Under these circumstances the hyperplasia or atrophy is restrained and controlled. Once the drive to enlarge or to atrophy has passed, then normality ensues. In short, it is reversible. This type of hyperplasia or atrophy, at least for the paired organs, is diffuse and affects both equally.

Occasionally, hyperplasia or atrophy occurs independently of an obvious driving force. Such cases may be focal or diffuse. While nodular hyperplasia is of little clinical significance, idiopathic atrophy, particularly of the thyroid gland, is important.

Neoplasia, a renegade, a deviant from the main stream, is a relatively common affection of endocrine glands. Fortunately many adenomas are non-functional and of little hinderance to the animal. However, in some cases an adenoma may be functional, creating hormonal havoc. Adenocarcinomas may be doubly detrimental. Their presence can cause all the attendant problems of tissue invasion and metastasis, and they are occasionally functional as well.

Of the inflammatory reactions, only those affecting the thyroid and adrenal gland are clinically relevant. They occur most commonly in the dog. Both glands suffer extensive bilateral destruction, and are often difficult to find on necropsy examination. The process is characterized by intense lymphoid infiltration accompanied by destruction of glandular tissue.

The endocrine pancreas is unusual in that it rarely appears to undergo physiologic atrophy or hypertrophy. However, individual cells within the islets are often vacuolated in diabetes mellitus. The endocrine pancreas may also be destroyed because of its intimate association with the larger and comparatively overbearing exocrine pancreas. In the dog, necrosis of the exocrine pancreas, especially when repetitive, unavoidably destroys islet tissue.

Table 10.2. *Relative prevalence of clinical endocrine disease*

Endocrine gland		Species					
		Dog	Cat	Cow	Sheep	Horse	Pig
Pars distalis	↑	+++	+	–	–	–	–
	↓	+	–	+	+	–	–
Neurohypophysis	↑	–	–	–	–	–	–
	↓	+	–	–	–	–	–
Pars intermedia	↑	+	–	–	–	+++	–
	↓	+	–	–	–	–	–
Adrenal	↑	+++	+	–	–	–	–
	↓	++	+	–	–	–	–
Thyroid	↑	–	+++	–	–	–	–
	↓	+++	–	+	+	+	+
Pancreas	↑	+++	–	–	–	–	–
	↓	+++	++	+	–	–	–

The minus sign indicates rarity or absence of the condition; ↑ indicates hyperfunction, ↓ indicates hypofunction; + gives a guide to the prevalence with which endocrine disease occurs.

Species prevalence of endocrine disease

With the exception of the dog, which seems to suffer from most endocrine disorders, other species only sporadically exhibit disease due to endocrine dysfunction. The notable exception is hypothyroidism following iodine deficiency in ruminants, the horse and pig, which is limited to those areas of the world where soils are iodine deficient. In addition, the cat is the only species in which hyperthyroidism is commonly diagnosed. Table 10.2 displays the relative prevalence of clinically significant endocrine disease.

The pituitary gland

The hypothalamus and the pituitary gland together form a functional endocrine unit which controls the status of a number of other endocrine glands including the thyroid, the two inner zones of adrenal cortex, the ovary, and the testis. It also regulates extracellular tonicity and in young animals skeletal and somatic growth. The pituitary is divided anatomically and physiologically into two distinct units, the adenohypophysis and the neurohypophysis.

The **adenohypophysis** is subdivided into three areas: the pars distalis, pars tuberalis and the pars intermedia. Of these, the pars distalis predominates and the contained cells secrete most of the tropic hormones. Cells within the pars distalis are classified according to the staining qualities of their cytoplasmic granules.

Acidophils manufacture and secrete growth hormone (GH) and prolactin (LTH); basophils synthesize and secrete luteinizing hormone (LH), follicle-stimulating hormone (FSH) and TSH; chromophobes, so named because they contain few cytoplasmic secretory granules, synthesize and release ACTH and melanocyte-stimulating hormone (MSH). The pars tuberalis surrounds the infundibular process and functions as a framework for the hypophyseal portal capillaries. The pars intermedia lies between the pars distalis and the pars nervosa (posterior lobe) of the pituitary and lines the residual lumen of Rathke's pouch. Cells from the pars intermedia secrete MSH and, in the dog, ACTH.

The release of tropic hormones from the adenohypophysis is governed by appropriate 'factors' synthesized in neurons in the hypothalamus and released into the hypophyseal portal circulation. Each releasing factor is named for the tropic hormone it regulates.

The neurohypophysis is composed of neuronal cell bodies in the hypothalamus, the respective axons of which travel via the infundibular stalk of the pituitary and terminate in the pars nervosa. The hormones ADH and oxytocin are synthesized in the hypothalamic neurons, and travel down the axons attached to a carrier molecule (neurophysin). Under the appropriate stimulus, they are released into the circulation from the terminating axons.

Because of its complexity, the potential for multiple hormonal disasters emanating from the pituitary may seem enormous. In fact, they are comparatively rare. Fortunately, **hyperfunction is limited to the excessive secretion of ACTH, or a substance with a similar effect**. It may be the result of an ACTH-secreting tumor or an altered regulatory mechanism for the hypophyseal–pituitary–adrenal axis. In all cases the result is hyperadrenocorticism. In the case of functional pituitary tumors, there may also be additional clinical signs associated with compression of adjacent tissues.

Conversely, hypofunction reflects an inadequacy of either adenohypophyseal tropic hormones alone, or of ADH emanating from the pars nervosa dysfunction. In some cases, there is both adenohypophyseal and neurohypophyseal hypofunction. The time of onset is critical, as congenital or juvenile hypofunction often has the added feature of abnormalities of gestation and the growth problems associated with a depression or lack of growth hormone and TSH.

Because of these differences, adenohypophyseal dysfunction is divided into congenital hypofunction and adult onset neoplasia. Altered regulation is discussed under hyperadrenocorticism.

Adenohypophyseal hypofunction

A number of breeds of dog, including the German Shepherd and the Carolean Bear Dog, exhibit failure of the oropharyngeal ectoderm of Rathke's pouch to differentiate into cells of the pars distalis. The resulting pituitary is composed of multiloculated cysts and a variable number of tropic-hormone-secreting cells.

Affected puppies appear normal at birth but

subsequently develop growth abnormalities because of insufficient growth hormone. A variable number of puppies also show clinical signs of hypothyroidism.

Congenital defects of the adenohypophysis also occur in a number of breeds of cattle, including Guernseys and Jerseys. The defect is inherited and the fetus is often of the cyclopian type, but, in some, only the pituitary is absent. All result in prolonged gestation.

Abnormalities of pituitary development also occur in sheep grazing pasture containing the teratogenic plant *Veratrum californicum* in the United States. Affected fetuses are often cyclopian, with a displaced pituitary, and are delivered after a prolonged gestation (Fig. 10.7).

There is an interesting contrast between the consequences of anomalies of the pituitary in dogs compared to sheep and cattle. Onset of parturition is normal in bitches producing affected offspring, whereas in cattle and sheep

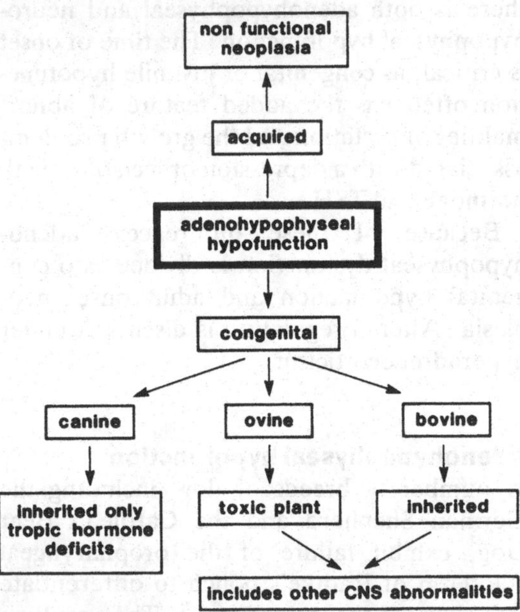

Fig. 10.7. Adenohypophyseal hypofunction may be congenital (inherited or caused by ingestion of toxic plants) or acquired (usually non-functional neoplasia). CNS, central nervous system.

the rule is prolonged gestation. To induce parturition at least in sheep, there needs to be an intact fetal hypothalamus–pituitary–adrenal axis, which implies that in the dog there are sufficient ACTH and circulating corticosteroids for normal parturition to occur. There may be a difference in single births versus multiple births or of the number of fetuses affected. As the canine pituitary defect is an autosomal recessive trait there are usually only one or two affected puppies per litter. The affected puppies invariably survive, probably because the defect is limited to the pituitary, whereas with affected cattle and sheep they are either born dead or die shortly after birth as they usually have multiple neurologic abnormalities.

Acquired adenohypophyseal hypofunction is really only of clinical significance in the dog and rarely the cat. Non-functional tumors of the pars distalis or pars intermedia compress and destroy the remaining normal areas leading to panhypopituitism. Such animals exhibit clinical signs following the loss of tropic hormones such as ACTH, TSH, growth hormone and the gonadal hormones FSH and LH. Additional neurologic clinical signs follow invasion of the optic chiasm and adjacent hypothalamus. There is also in some cases a loss of ADH because of destruction of the pars nervosa.

Adenohypophyseal hyperfunction

Functional adenohypophyseal neoplasia is only of clinical importance in the dog and horse. Adenomas tend to be functional and adenocarcinomas not. Compression or invasion of the adjacent tissues may also occur, particularly with adenocarcinomas.

In dogs, functional tumors arise from either the pars distalis or the pars intermedia and are usually composed of chromophobes. Functional adenomas of the pars intermedia are more commonly seen in dolicephalic breeds, whereas tumors of the pars distalis are more commonly seen in the brachycephalic breeds: Boxers and Boston Terriers. **The clinical syndrome is that of adrenocortical hyperfunction**

because of excessive production of ACTH by the tumor (Fig. 10.8). For details of adrenocortical hyperfunction refer to the description under the adrenal gland (p. 269).

Functional tumors also occur in old horses, are derived from the pars intermedia in most instances, and are more common in females. The clinical signs were originally thought to be the result of excessive production of ACTH, but there is now some doubt about this. Plasma cortisol levels may or may not be elevated. However, many affected horses do not exhibit a normal diurnal variation in plasma cortisol levels. The most consistent finding is persistently high plasma MSH levels.

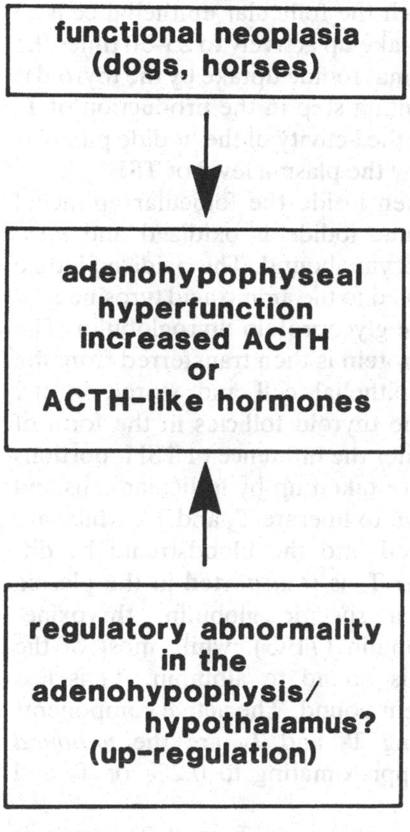

functional neoplasia (dogs, horses)
↓
adenohypophyseal hyperfunction increased ACTH or ACTH-like hormones
↑
regulatory abnormality in the adenohypophysis/ hypothalamus? (up-regulation)

Fig. 10.8. Adenohypophyseal hyperfunction may be the result of either functional neoplasia or abnormality in hypothalamic/adenohypophyseal regulation.

MSH has some ACTH-like effects. A number of other ACTH-related peptides have been shown to be raised both in the tumor and in plasma.

Clinical signs in both dogs and horses are referable to excessive plasma corticosteroid levels and its catabolic effects, which include muscle atrophy, weight loss and ravenous appetite, and in some cases compression of the hypothalamus and posterior pituitary. Polydipsia and polyuria are also frequently seen. These are not ADH responsive and are probably due to corticosteroid blockage of hypothalamic osmoreceptors, although the neurohypophysis is often invaded by the tumor. Horses may exhibit the classical hematologic pattern of corticosteroid excess, which includes relative neutrophilia, lymphopenia and eosinopenia.

There is a marked species difference with respect to the effects of excessive corticosteroids on hair growth. In dogs alopecia is the rule, whereas in man and horses hirsuitism develops when corticosteroid levels are excessive. In human females, the development of hirsuitism in adrenocortical hyperfunction is considered to be due to excessive sex steroid production by the adrenal cortex.

Neurohypophyseal hypofunction

As ADH is required for water resorption from dilute urine in the distal convoluted tubules of the kidney, a lack of ADH leads to the production of excessive amounts of dilute urine. ADH acts via an adenyl cyclase–cyclic AMP mechanism, enhancing the phosphorylation of a lumenal, membrane-bound protein kinase. The natural stimulus for ADH control is the excitation of osmoreceptors located in the supra-optic nucleus.

The failure of action of ADH produces the clinical syndrome **diabetes insipidus** (Fig. 10.9). This is due most commonly to inadequate plasma levels of ADH, or more rarely to an insensitivity of the kidney to the effects of ADH. The latter is termed **nephrogenic diabetes insipidus**. Inadequate circulating levels of ADH may follow the destruction of the neurohypophysis by neoplasia, but in many instances no lesion can be found to explain the lack of ADH.

The clinical signs of diabetes insipidus are polyuria and compensatory polydipsia. The urine is consistently dilute with a specific gravity of less than 1.008 (hyposthenuria). The production of excessive amounts of dilute urine indicates that the kidney is functioning normally, at least up to the phase of final concentration by the action of ADH.

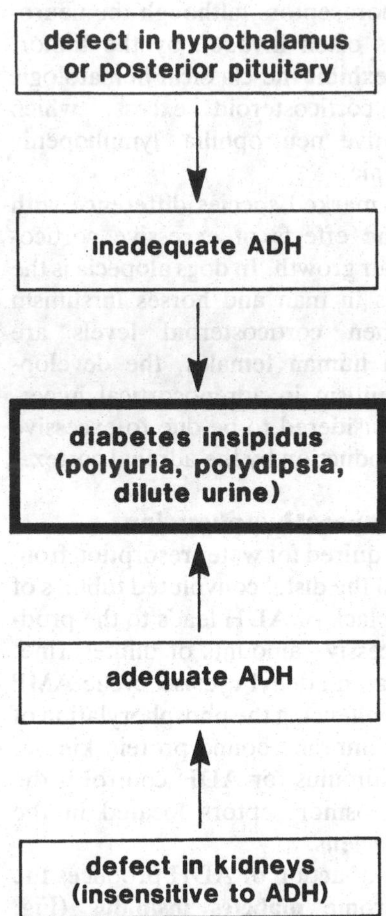

Fig. 10.9. Neurohypophyseal hypofunction is almost always reflected by a lack of adequate plasma levels of anti-diuretic hormone (ADH). Target organ insensitivity to ADH (nephrogenic diabetes insipidus) simulates neurohypophyseal hypofunction (see also Chapter 9)

The diagnosis of diabetes insipidus of pituitary origin depends on the demonstration of the inability of the animal to concentrate urine in the face of water deprivation, and the ability to concentrate urine following the administration of exogenous ADH. Nephrogenic diabetes insipidus is diagnosed when an animal continues to produce dilute urine following ADH administration.

The thyroid

Normal thyroid function

Iodine is an absolute requirement for the production of the two active hormones produced by the thyroid gland, tetra-iodothyronine (thyroxine or T_4) and tri-iodiothyronine (T_3).

Iodine is normally ingested in the form of iodides which the follicular epithelial cells of the thyroid take up actively to 25–30 times the level in plasma. Iodide uptake by the thyroid is the rate-limiting step in the production of T_3 and T_4, and the activity of the 'iodide pump' is controlled by the plasma level of TSH.

Once taken inside the follicular epithelial cell, inorganic iodide is oxidized and most probably enzyme bound. The oxidized iodine is then coupled to the amino acid tyrosine contained in the glycoprotein thyroglobulin. The iodinated protein is then transferred from the follicular epithelial cell and stored in the lumen of the thyroid follicles in the form of colloid. Under the influence of TSH, portions of colloid are taken up by follicular cells and broken down to liberate T_3 and T_4, which are then released into the bloodstream by diffusion. Most T_4 is transported in the plasma bound to a specific globulin, thryoxine-binding globulin (TBG), while most of the remainder is bound to albumin. T_3 is less avidly protein bound. The active components of circulating T_3 and T_4 are the *unbound* moieties, approximating to 0.2% of T_4 and 50% of T_3.

The maintenance of T_3 and T_4 levels is governed by the level of TSH, a glycoprotein produced by the basophils in the adenohypophysis. TSH output is in turn controlled by a hormone produced in the hypothalamus TSH releasing factor (TRH). This rather complex mechanism is controlled by the negative feedback of T_3 and T_4 on the adenohypophysis and the hypothalamus (Fig. 10.10).

The thyroid hormones have a marked general effect on the cellular activity of most tissues of the body. They stimulate the metabolic rate, and in the young have a marked effect on growth and development. This applies particularly to epiphysial growth areas of the skeleton.

The mechanism of action of thyroid hormones is still one of the great mysteries of endocrinology. They freely cross plasma membranes and bind specifically to nuclear protein acceptors. There is no cytoplasmic receptor similar to that seen for steroid hormones. There is good evidence that T_3 is the only metabolically active thyroid hormone, T_4 probably being de-iodinated to T_3 upon entry into the cell. The consequence of T_3 binding to nuclear acceptors is an increase in RNA and protein synthesis, but the specific nature of the effect is obscure.

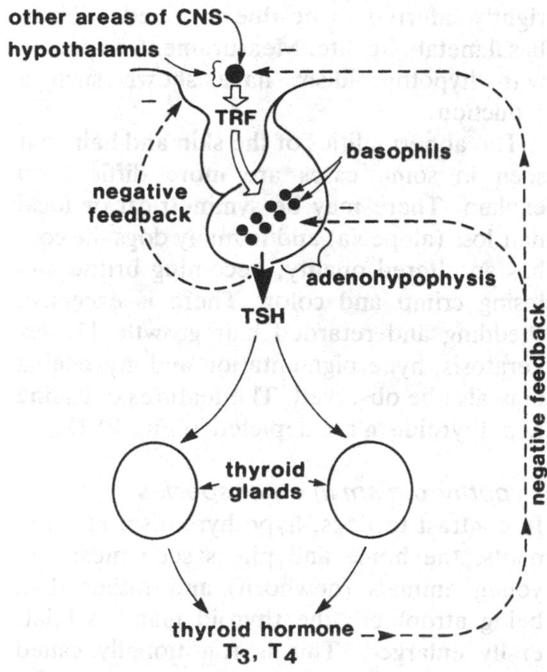

Fig. 10.10. Regulation of thyroid hormone is achieved by a negative feedback on the hypothalamus and adenohypophysis controlling the release of thyrotropin-releasing factor (TRF) and thyroid-stimulating hormone (TSH). CNS, central nervous system

Thyroid hypofunction

With the exception of the cat, and very rarely the dog, we are concerned with hypofunction, and mostly with hypofunction at the level of the thyroid. In young ruminants, horses and pigs, the predominant etiology is iodine deficiency, which may be primary or secondary. In dogs, hypothyroidism is mostly a disease of middle to old age and follows inflammatory destruction or idiopathic atrophy of the thyroid gland. Congenital and acquired disorders of the adenohypophysis occasionally produce secondary hypothyroidism in the dog through a total or partial deficiency of TSH.

Because of the disparity between the age of onset and the etiology of the majority of cases of hypothyroidism in the dog compared to the grazing species, each will be considered separately.

Canine hypothyroidism

Primary hypothyroidism is, with rare exceptions, a disease of adult dogs and there are two major pathologic patterns. The first is progressive immunologic destruction of follicles. The gland is atrophied and contains many lymphocytes. A high proportion of these dogs have circulating antibodies to thyroglobulin (Fig. 10.11). Almost equally common is atrophy of the thyroid unaccompanied by any evidence of inflammation. This so-called idiopathic thyroid atrophy may be the end stage of lymphocytic thyroiditis. A rare form of congenital primary hypothyroidism is seen in Scottish Deerhounds.

Secondary hypothyroidism follows congenital or acquired abnormalities in the pituitary gland. Whilst basically similar to the clinical signs seen in primary hypothyroidism, the animal may exhibit abnormalities referable to deficiencies of other pituitary hormones such as ACTH, growth hormone and gonadal hormones. Congenital secondary hypothyroidism is seen, for example, in German Shepherd dwarfs. Acquired secondary hypothyroidism is almost always due to pituitary neoplasia.

Special clinical features
In hypothyroidism the clinical signs observed follow a depression in circulating T_3 and/or T_4 levels. The most accurate means of measuring T_3 and T_4 is radioimmunoassay. Enzyme-linked immunosorbent assay (ELISA) methods are currently being evaluated and may become the standard method in the future. A normal resting T_3 or T_4 level effectively rules out a diagnosis of hypothyroidism. The finding of a low T_4 or T_3 may be indicative of hypothyroidism, but a TSH stimulation test is necessary to confirm the diagnosis. The TSH stimulation test provides evidence of the ability of the thyroid gland to secrete T_3 and T_4. Serum T_3 and T_4 levels are most commonly measured before, and up to 12 hours following, the administration of TSH. It also provides a method for distinguishing between primary and secondary (adenohypophyseal) hypothyroidism. A thyroid biopsy may be used to classify morphologically the process affecting the thyroid.

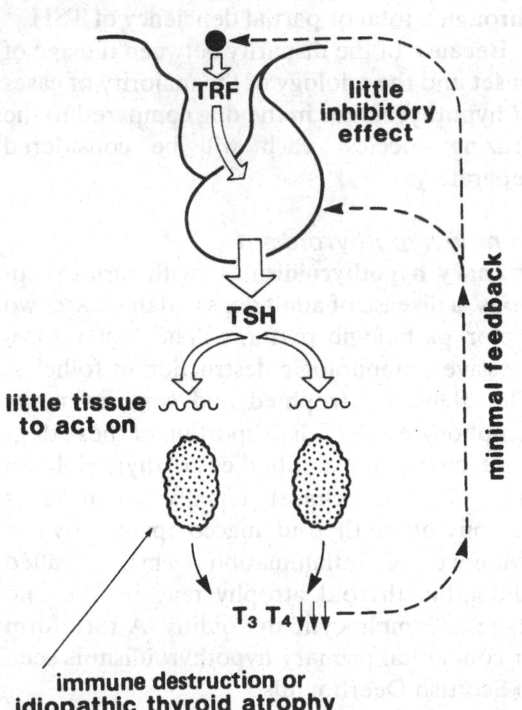

Fig. 10.11. Primary hypothyroidism in the dog most commonly follows the destruction (immune mediated) or idiopathic atrophy of the thyroid glands. For abbreviations, see Fig. 10.10.

Thyroid hormones maintain the basal metabolic rate with a very narrow range. Dogs with hypothyroidism are usually easily fatigued, sleep more, seek warmth, have reduced mental activity and have difficulty in maintaining normal body temperature. There is often an increase in body weight in spite of a reduced food intake. All these clinical signs may be rightly inferred to be due to a reduction in basal metabolic rate. Measurements in people with hypothyroidism have shown such a reduction.

The abnormalities of the skin and hair coat seen in some cases are more difficult to explain. There may be symmetrical or focal hair loss (alopecia) and in many dogs the coat has an altered quality, becoming brittle and losing crimp and color. There is excessive shedding and retarded hair growth. Hyperkeratosis, hyperpigmentation and myxedema may also be observed. The features of canine hypothyroidism are depicted in Fig. 10.12.

Hypothyroidism in other species
In contrast to dogs, hypothyroidism in ruminants, the horse and pig is seen mostly in young animals (newborn) and rather than being atrophied, the thyroid gland is bilaterally enlarged. This is traditionally called goiter. Hypothyroidism in these species is the result of iodine deficiency, either from inadequate dietary intake or from a block of either iodine uptake or organification by the thyroid gland (Fig. 10.13). Iodine deficiency

occurs world wide, usually in areas where there is a high rainfall, offshore winds and soils lacking organic matter. Historically, clinical iodine deficiency was an important disease, but it is now quite uncommon following the institution of simple control measures. Diets rich in *Brassica* spp. sometimes induce hypothyroidism in ruminants. The plants contain glucosinolates, which are converted to thiocyanate in the rumen and inhibit the linking of iodine to tyrosine. A glucoside in linseed meal may also cause hypothyroidism, by a similar mechanism. Congenital organification defects have been reported to occur in Merino sheep and goats. In all cases, an enlarged thyroid is a prominent feature.

It is important to note that, while some cases of overt hypothyroidism may occur, there is a much higher prevalence of animals with thyroid enlargement and normal thyroid function (euthyroid).

Special clinical features

A high incidence of stillbirths and weak newborn animals is most common. In calves and lambs, partial or complete alopecia is the rule, often accompanied by palpably enlarged thyroid glands. Neither alopecia nor thyroid enlargement is seen in foals (Fig. 10.14).

Goiter in each instance mentioned thus far is the end product of low T_3 and T_4 levels, which fail to regulate the level of TRF and TSH secretion. The resulting high levels of TSH stimulate the thyroid to undergo hypertrophy and hyperplasia.

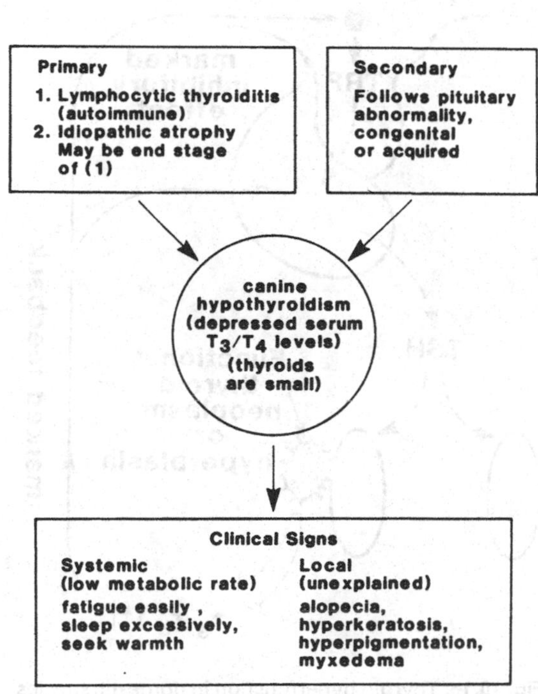

Fig. 10.12. Canine hypothyroidism may be primary or secondary. It is important to note that not all clinical signs need be present in any individual case.

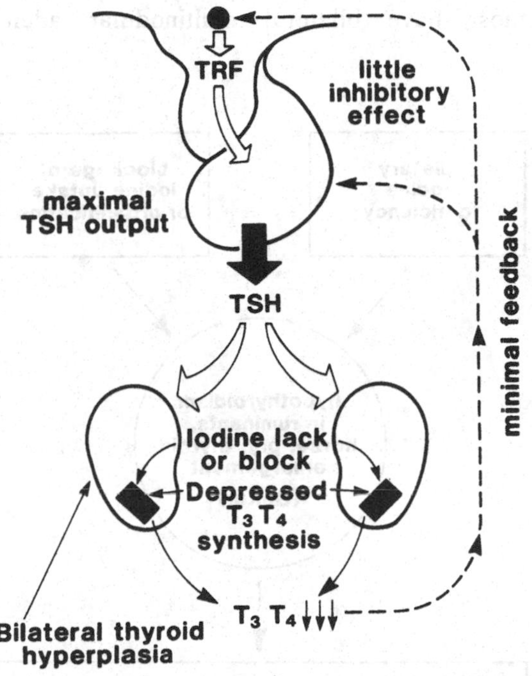

Fig. 10.13. Hypothryoidism in economic animals may follow (1) dietary iodine lack, (2) a block in iodine uptake, or (3) a block of iodine organification. Because of lowered thyroid hormone levels there is little feedback leading to increased plasma TSH levels. Consequently a major feature is bilateral enlargement of the thyroid glands (goiter). For abbreviations, see Fig. 10.10.

Thyroid hyperfunction

In veterinary medicine, hyperthyroidism has so far been reliably recognized only in the cat. It is an uncommon disease, occurring in late middle-aged to old animals. Clinical features may include weight loss in spite of a sometimes ravenous appetite, frequent defecation with stools of soft or fluid consistency, polydipsia and polyuria. Most cats show an increase in physical activity, being restless, excitable and constantly pacing.

The most important physical findings are a palpable enlargement of one or both thyroids, pyrexia, tachycardia or other cardiac arrhythmias, and the presence of systolic murmurs. The clinical diagnosis is confirmed by demonstrating elevated serum T_3 and T_4 levels.

Whilst a minority of cats have adenocarcinomas as the basis of the hyperthyroidism, most have bilateral multinodular adenomatous hyperplasia, the pathogenesis of which is unclear (Fig. 10.15). One can only presume that the clinical features of this disease are the result of an increase in metabolic rate.

C (parafollicular)-cell hyperfunction

Thyroid C (parafollicular)-cell tumors are a common tumor of adult to aged bulls and can be associated with neoplasia of other endocrine cell populations of neural crest origin such as the adrenal medulla and pituitary gland. C-cell tumors are known to contain calcitonin and many affected bulls have higher than normal levels of calcitonin in plasma. Many bulls, possibly because of elevated calcitonin levels, develop extensive vertebral osteophytes and osteosclerosis.

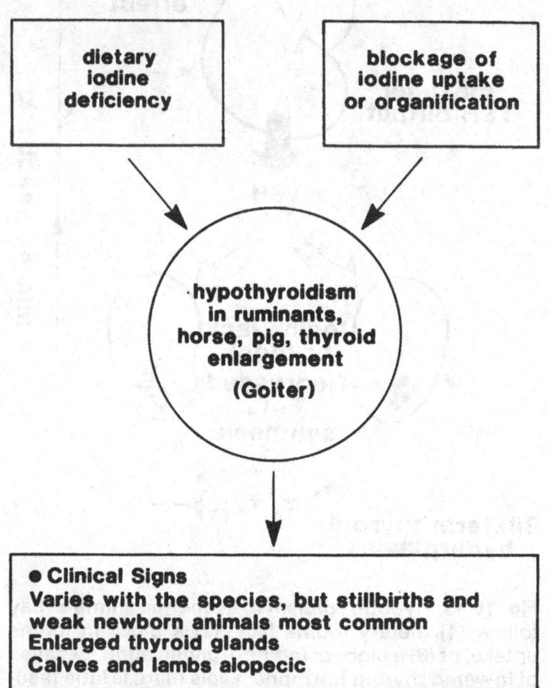

Fig. 10.14. The clinical signs of hypothyroidism in economic animals vary with the species affected.

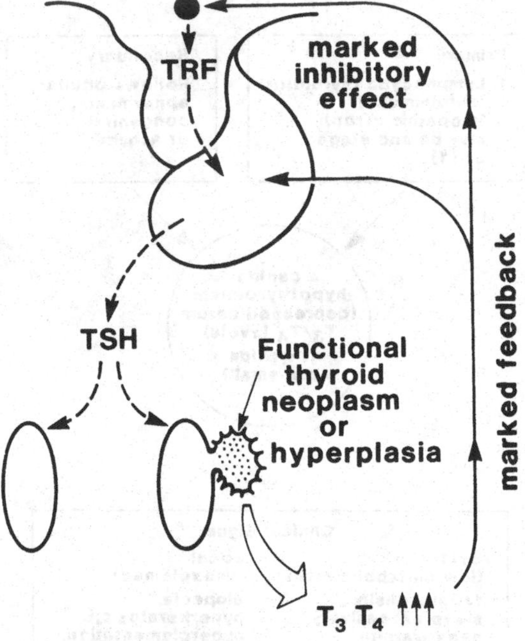

Fig. 10.15. Thyroid hyperfunction in domestic species appears limited to the cat and is the result of thyroid neoplasia or adenomatous hyperplasia. The result is marked feedback inhibition and little TRF or TSH release. For abbreviations, see Fig. 10.10.

Non-functional thyroid neoplasia

Although the functional aspects of thyroid disease have been discussed at some length, just as clinically important at least in the companion animals are follicular cell thyroid tumors that are locally invasive and prone to metastasize. The clinical signs are (1) those of a palpable enlargement in the ventral laryngeal region, which may, in the case of malignancy, be accompanied by signs of invasion of the jugular vein, and (2) respiratory signs following pulmonary metastases.

The adrenal gland

With adrenocortical dysfunction there are two metabolic syndromes, the first is the result of a deficiency primarily of adrenal mineralocorticoids, the second an excess of adrenal glucocorticoids. Adrenocortical hypofunction is almost exclusively adrenal in origin and independent of pituitary influence, whereas adrenocortical hyperfunction follows adrenal, pituitary or possibly hypothalamic disease.

The metabolic effects of glucocorticoids and mineralocorticoids are not dissimilar. As with all corticosteroids, there is some overlap, but particular modes of action are either accentuated or depressed.

Glucocorticoids

From the name glucocorticoid, one can surmise that there is some involvement with glucose, and indeed there is. The glucocorticoid effect is to spare the utilization of glucose by peripheral tissues, and to mobilize precursors such as amino acids from skeletal muscle, for conversion to glucose in the liver. Such gluconeogenesis may lead to a mild hyperglycemia. There is also peripheral mobilization of fat. Glucocorticoids also have a marked negative effect on the immune system, the inflammatory response, and wound healing.

Glucocorticoid secretion is controlled by ACTH, which is in turn controlled by corticotropin-releasing factor (CRF) from the hypothalamus. Glucocorticoids released from the inner zones of adrenal cortex feedback on the pituitary and hypothalamus inhibiting the release of ACTH and CRF, respectively (Fig. 10.16). At physiologic levels it is a slow phenomenon, requiring at least 24 hours for an inhibiting effect to be demonstrated.

As for many of the pituitary and hypothalamic hormones, the release of ACTH and CRF is intermittent. Glucocorticoid secretion is, in most species, greater in the early morning and at its lowest at night. This intermittent release is also observed for growth hormone and the gonadotropins.

Glucocorticoids are metabolized in the liver and excreted bound to glucuronic acid, largely in the urine. The presence of glucocorticoids in the urine allows the estimation of total glucocorticoid output, but this test is usually impractical in domestic animals.

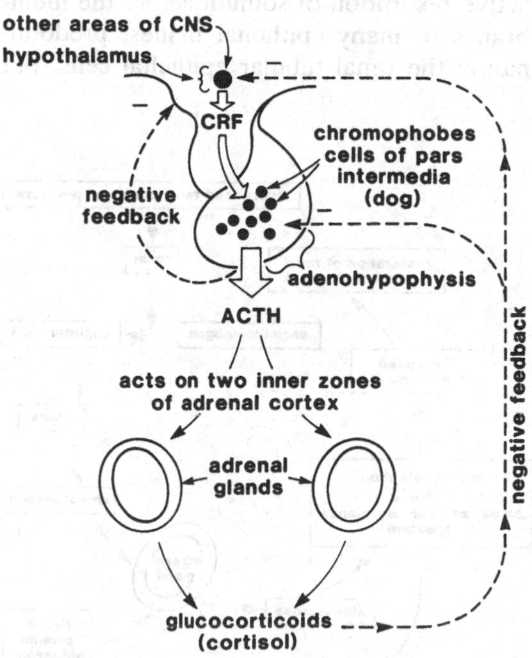

Fig. 10.16. Glucocorticoid regulation is similar in many respects to thyroid hormone regulation with involvement of the hypothalamus and the pituitary. CRF, corticotropin-releasing fctory; ACTH, corticotropin; CNS, central nervous system.

Mineralocorticoids

Although similar in structure to glucocorticoids and produced within the same gland but from the zona glomerulosa, the regulation of the mineralocorticoid aldosterone is for practical purposes independent of ACTH. The synthesis and release of aldosterone is controlled via the renin–angiotensin system, and by the plasma levels of potassium.

In the renin–angiotensin system, release of renin from the juxtaglomerular apparatus in the kidney is stimulated by a drop in renal perfusion pressure. Within the plasma, renin converts angiotensinogen to angiotensin I, which is then activated to angiotensin II in the lung. Among other effects, angiotensin II stimulates the release of aldosterone from the zona glomerulosa of the adrenal cortex. The second and equally potent stimulus for aldosterone release is a rise in plasma potassium levels (Fig. 10.17).

The action of aldosterone is to promote the active resorption of sodium across the membranes of many epithelial tissues, predominantly the renal tubular epithelial cells, but also the salivary glands, sweat glands and intestine. It also promotes simultaneously the urinary excretion of potassium. For a complete description of the regulation of sodium balance, see Chapter 9.

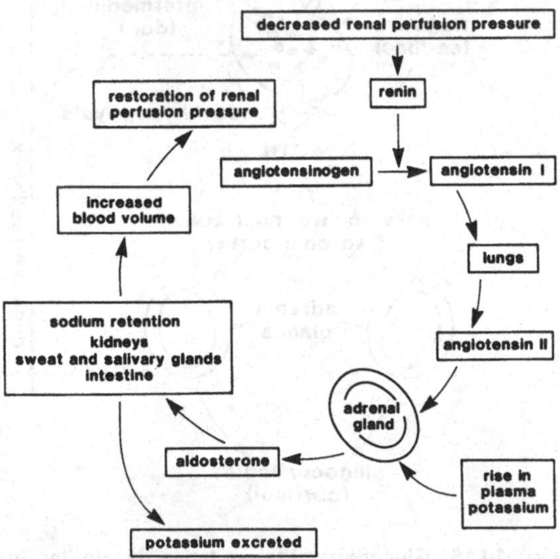

Fig. 10.17. Mineralocorticoid regulation is governed by two mechanisms: (1) the renin–angiotensin system, and (2) the level of potassium in the plasma.

Adrenocortical hypofunction

(Addison's-like disease)

The often severe clinical signs seen with adrenocortical hypofunction are a consequence of mineralocorticoid deficiency. Glucocorticoid levels are also depressed, but contribute little to the clinicopathologic picture. In dogs, where the disease is by far most commonly diagnosed, the adrenal cortex *in toto* undergoes idiopathic atrophy, or is destroyed by extensive lymphocytic invasion. This is probably autoimmune in origin (Fig. 10.18).

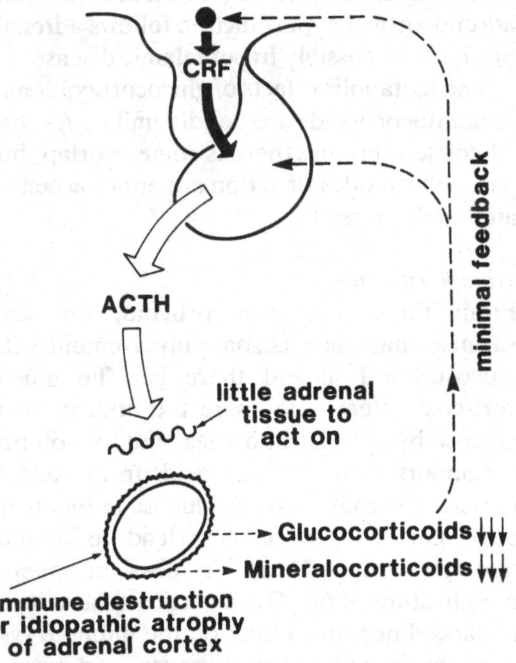

Fig. 10.18. Adrenocortical hypofunction follows destruction of all zones of the adrenal cortex. Although plasma levels of both mineralocorticoids and glucocorticoids are depressed, clinical signs are due to mineralocorticoid lack. For abbreviations, see Fig. 10.16.

Table 10.3. *Physical examination findings in 137 dogs with hypoadrenocorticism*

Finding	Percentage tested
Depression	97
Weakness	77
Dehydration	52
Bradycardia	34
Weak pulse	23
Melena	10

From Feldman and Peterson, 1984, by kind permission, see Additional reading.

Table 10.4. *Historical signs in 136 dogs with hypoadrenocorticism*

Sign	Percentage tested
Lethargy/depression	90
Decreased appetite	85
Weakness	75
Vomiting	69
Diarrhea	37
Weight loss	34
Waxing-waning course	33
Shaking	27
Polyuria/polydipsia	22

From Feldman and Peterson, 1984, by kind permission, see Additional reading

Once sodium regulation by aldosterone is impaired or lost, large amounts of sodium and water are lost in the urine, with an associated retention of potassium and hydrogen ions. Affected animals become hyponatremic and hyperkalemic, but often it is better to compare the sodium:potassium ratio. If this falls below 23:1, then hypoadrenocorticism should be considered. Chloride ions accompany the sodium loss in the urine. Sodium loss also occurs across the intestinal mucosa. Such a drop in extracellular sodium concentration leads to hypovolemia and circulatory shock. The hyperkalemia often leads to cardiac dysrhythmias, predominantly a sinus bradycardia, and occasionally ventricular escape beats. The T waves are usually enlarged.

Special clinical features

The predominant clinical signs are depicted in Table 10.3 in descending order of prevalence: all are probably due to hypovolemia. The bradycardia seen in a minority of cases is due to hyperkalemia. Intermittent diarrhea, polydipsia and polyuria are the result of excessive sodium and chloride loss into the intestine and urine, respectively (Table 10.4). Death from adrenocortical insufficiency may occur within 24–28 hours. The cause of vomiting is obscure but it may also be due to chloride loss via the gastric mucosa.

In chronic relapsing cases, weight loss, intermittent diarrhea, vomiting and anorexia are features of the disease. It is worth emphasizing that there is little observable clinical effect due to glucocorticoid insufficiency. A direct effect of low plasma levels of glucocorticoids is the removal of inhibitory effect on ACTH synthesis and release. Because of its MSH-like activity, excessive levels of ACTH may lead to hyperpigmentation. The sequence of events and clinical features are depicted in Fig. 10.19.

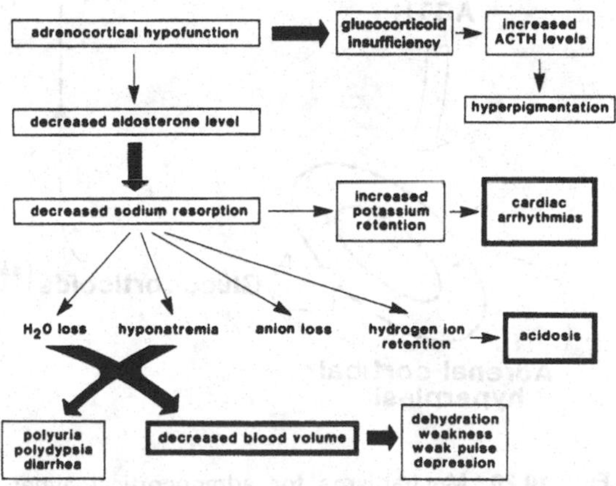

Fig. 10.19. The effects of decreased aldosterone levels dominate the clinical feature of hypoadrenocorticism. All relate back to a loss of sodium and potassium regulation

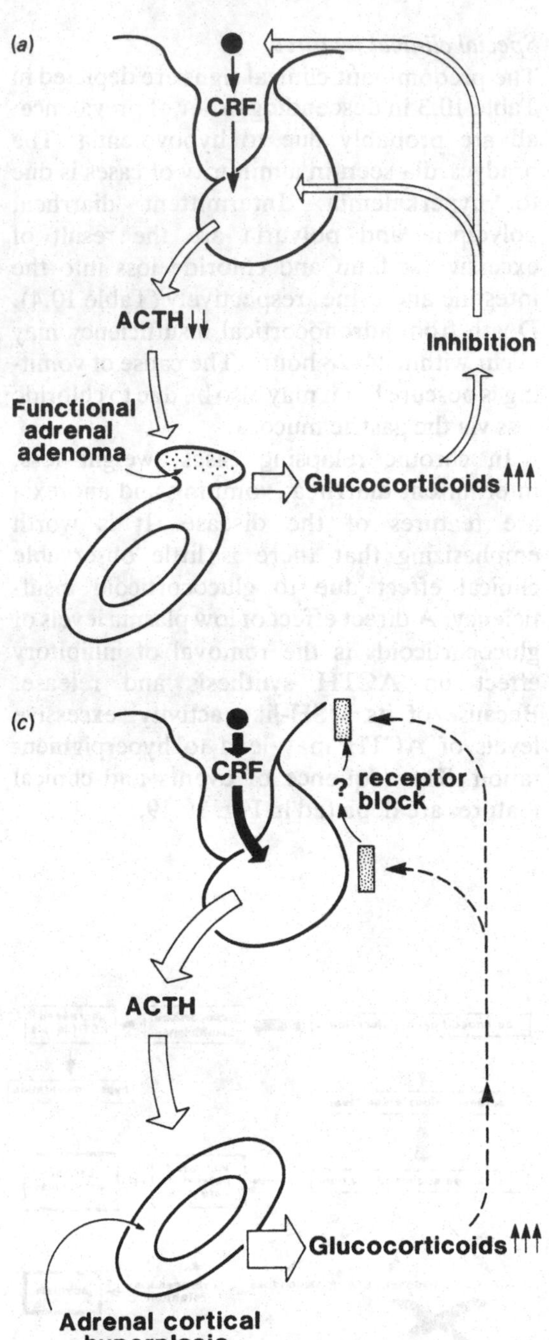

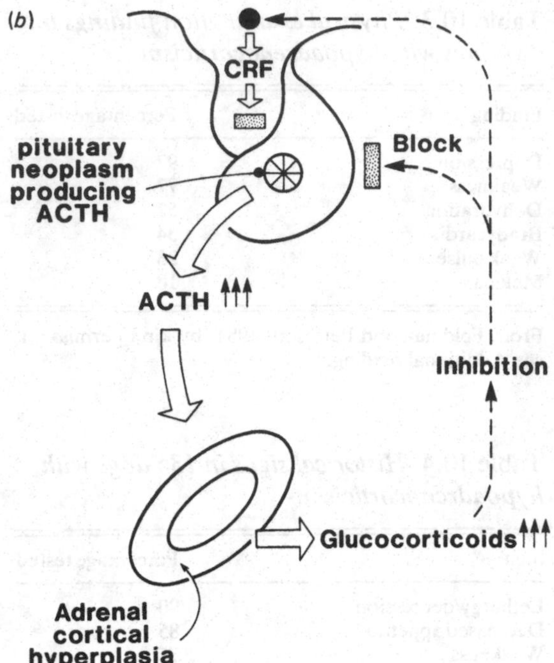

Fig. 10.20. Mechanisms for adrenocortical hyperfunction: (*a*) following a functional adrenal tumor, (*b*) a functional ACTH-producing pituitary tumor, or (*c*) insensitivity of hypothalamic and possibly pituitary receptors to glucocorticoid feedback (upregulation). For abbreviations, see Fig. 10.16.

Adrenocortical hyperfunction
(Cushing's-like disease)

In contrast to hypoadrenocorticism which is primarily adrenal in origin, an excess of corticosteroids may be of adrenal, pituitary or possibly hypothalamic origin (Fig. 10.20). The clinical signs observed are the result of excessive levels of glucocorticoids. There are rare examples of mineralocorticoid excess, in which clinical signs are related to the presence of hypokalemia.

Canine hyperadrenocorticism

Hyperadrenocorticism is often due to hormone-secreting neoplasms. About 10% of cases in the dog are the result of steroid-secreting adrenal tumors, while the majority, and there is some argument about this, follow stimulation of the adrenal cortex by ACTH-secreting adenomas of the pituitary gland. Published accounts vary in describing from 20% to 85% of cases as the result of either adenomas of the pars distalis or pars intermedia. There is speculation that the remaining 5% to 70%, depending on the figures, follow excessive CRF release from the hypothalamus (the so-called idiopathic cases). The remaining possibility for clinical signs of hyperadrenocorticism is iatrogenic disease following exogenous steroid administration. This latter variant of hyperfunction is becoming increasingly recognized.

In dogs, the disease is most common in the middle-aged, of small breeds such as Dachshunds, Toy and Miniature Poodles. Boxers appear to be overrepresented in the pituitary neoplasm group.

Special clinical features

Many of the clinical signs of hyperadrenocorticism are the result of the action of glucocorticoids on the intermediary metabolism of protein, carbohydrate and fat. Also in most cases, there is the contributory effect of excessive levels of ACTH such as hyperpigmentation. Only when the disease is due to a functional adrenal adenoma is there no effect of ACTH. Excessive levels of glucocorticoids also affect a variety of other processes. The most commonly seen clinical signs are listed in Table 10.5, and include polyuria and polydipsia. The urine is dilute with a specific gravity of less than 1.008. The mechanism for this is probably interference with the action or secretion of ADH. It has been suggested that the action of ADH is blocked by glucocorticoids at the renal collecting tubule level. In the dog, the ADH block is not absolute as affected animals can concentrate urine following water deprivation. Alternatively, there may be an inhibition of ADH release from the pituitary. It is known that in man, glucocorticoids raise the threshold of osmoreceptors in the hypothalamus.

Table 10.5. *Clinical signs in 300 dogs with hyperadrenocorticism*

Sign	Percentage of dogs
Polyuria/polydipsia	82
Pendulous abdomen	67
Hepatomegaly	67
Hair loss	63
Lethargy	62
Polyphagia	57
Muscle weakness	57
Anestrus (69 females)	54
Obesity	47
Myscle atrophy	35
Comedones	34
Increased panting	31
Testicular atrophy (128 males)	29
Hyperpigmentation	23
Calcinosis cutis	8
Facial nerve palsy	7

From Peterson, 1984*b*, by kind permission, see Additional reading.

The clinical signs of abdominal enlargement, muscle atrophy and weakness are related to the catabolic effects of glucocorticoids, inhibiting protein synthesis and enhancing protein breakdown. Abdominal enlargement is probably also contributed to by an increase in abdominal fat stores and hepatomegaly. The catabolic effect of glucocorticoids is also probably the prime reason for the osteoporosis and the skin fragility. Both bone and skin contain abundant collagen. The osteoporosis is also contributed to by the effect of glucocorticoids on the action of vitamin D on the intestines. The mechanism for polyphagia is unknown, but it may result from a stimulation of appetite by glucocorticoids. Obesity, a common presenting complaint, is probably due to the redistribution of body fat stores.

The often spectacular cutaneous findings of alopecia and atrophy, are difficult to explain mechanistically. There is bilaterally symmetrical pilosebaceous atrophy accompanied by follicular plugging with keratin and very poor hair growth. Dermal atrophy appears to be due to a loss of collagen and elastic fibers. In a small percentage of cases, mineralized dermal plaques are noticed, the pathogenesis of which is obscure. The cutaneous hyperpigmentation seen in about 20% of cases is the result of excessive amounts of ACTH and β-lipotropin, both of which have some MSH-like activity.

Dogs with hyperadrenocorticism are prone to infection and may have delayed wound healing. This follows the marked anti-inflammatory effect of glucocorticoids, accompanied by a suppression of immune function, particularly cell-mediated immunity. For example, glucocorticoids block the effect of one of the lymphokines, macrophage inhibition factor, on macrophages. Glucocorticoids also inhibit the release of arachidonic acid from phospholipids, decreasing the formation of prostaglandins and related compounds. Glucocorticoids delay wound healing primarily by the inhibition of mitosis in fibroblasts.

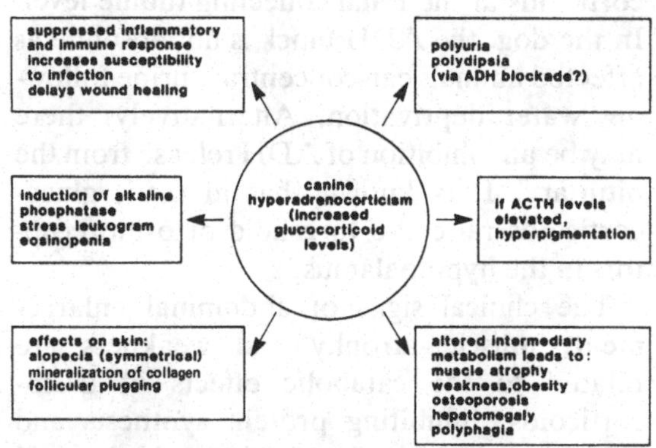

Fig. 10.21. Increased plasma glucocorticoid levels in canine hyperadrenocorticism produces a considerable variety of clinical effects. All the clinical signs depicted in the diagram may not be present in an individual case. The most commonly observed clinical signs are shown in Table 10.5.

A complete blood count in affected dogs reveals a typical 'stress' response: neutrophilia and monocytosis follow a decrease in the marginated pool; eosinopenia because eosinophils are sequestered in the bone marrow; and lymphopenia following necrosis of lymphocytes.

Increased gluconeogenesis and a decreased peripheral utilization of glucose results in a mild hyperglycemia which, if severe enough, may exceed the renal threshold for glucose, leading to glycosuria.

Serum alanine aminotransferase is usually mildly elevated following hepatocyte vacuolar change. Alkaline phosphatase (AP) is elevated mainly through the induction of a specific hepatic AP isoenzyme. The major clinical features of canine hyperadrenocorticism are depicted in Fig. 10.21.

The adrenal medulla

The adrenal medulla is of clinical significance because of tumor formation. Such tumors reside under the rather exotic title of pheochromocytoma. They occur most frequently in dogs and cattle and may be benign or malignant. The overwhelming majority of pheochromocytomas are non-functional, but the minority which are functional are spectacularly so. The great surges of catecholamines released from such tumors induce tachycardia, edema, cardiac hypertrophy and hypertension.

The endocrine pancreas

Unlike most other endocrine organs which are discrete, the endocrine pancreas is scattered within, and hidden from the naked eye by the exocrine pancreas. However, its hidden nature does not mask the powerful influence that hormones from the endocrine pancreas have on carbohydrate, fat and protein metabolism. Of the hormones produced, insulin, secreted by the beta cells, directly or indirectly affects the structure and function of every organ in the body, particularly adipose tissue, the liver and muscle. Insulin is a conserver, facilitating anabolic reactions, and inhibiting catabolic reactions.

The endocrine pancreas also produces glucagon and somatostatin, which are secreted from the alpha and delta cells, respectively. Glucagon promotes insulin secretion, but at the same time stimulates glycogenolysis and lipolysis, so-called 'anti-insulin' actions. Excessive secretion of glucagon may play a role in the pathogenesis of diabetes mellitus. Functional tumors of the glucagon-producing islet cells have not been documented in animals.

Somatostatin inhibits the release of growth hormone and ACTH from the pituitary. It depresses the action of TRF and the release of insulin and glucagon from the pancreas. Abnormalities of somatostatin secretion producing syndromes of clinical significance in animals have not been documented.

We are concerned with the understanding of two disorders, firstly **diabetes mellitus**, a complex disease, defined as an absolute or relative deficiency of insulin, and, secondly, with functional neoplasia of beta cells, producing inappropriate amounts of insulin. Once again the dog is the species most commonly affected and the cat less so. Diabetes mellitus occurs rarely in other domestic species.

Very rarely, certain islet cells of the endocrine pancreas (G islet cells) become neoplastic and secrete excessive amounts of the hormone gastrin. Gastrin-producing tumors may also arise in the gastroduodenal mucosa. Gastrin stimulates production of gastric acid by the oxyntic cells of the stomach and overproduction results in gastric hyperacidity. Clinical signs associated with this include vomiting, diarrhea, anorexia and gastrointestinal bleeding secondary to gastric and duodenal ulceration. In man, this is known as the Zollinger–Ellison syndrome and a similar disorder has been reported to occur rarely in dogs with endocrine pancreatic neoplasia.

Metabolic effects of insulin, glucagon and growth hormone

Insulin is a two-chain polypeptide whose synthesis and release from beta cells is controlled primarily by the levels of glucose in the bloodstream. Its release is governed to a lesser extent by plasma amino acid and free fatty acid levels. Insulin has a profound effect on glucose, lipid and protein metabolism. Insulin also enhances lipogenesis and inhibits lipolysis. The final major action of insulin is the enhancement of protein synthesis by facilitating the movement of both glucose and amino acids across the cell membrane.

Insulin influences the metabolism of three major organs – the liver, muscle and adipose tissue. In the liver, insulin influences glucose metabolism within the hepatocyte, in which it promotes the formation of glycogen, and suppresses glycogenolysis and gluconeogenesis.

Insulin effectively antagonizes the hepatic effects of glucocorticoids, catecholamines and glucagon. The actions of insulin within the hepatocyte are numerous. There is an increase in the rate of synthesis of glucokinase, phosphofructokinase and pyruvate kinase, and a corresponding decrease in the synthesis of enzymes involved in gluconeogenesis, such as glucose 6-phosphatase, phosphoenolpyruvate carboxylase and pyruvate carboxylase. The activity of some enzymes is also increased, for instance glycogen synthetase.

In muscle, insulin stimulates the active transport of glucose and amino acids across the cell membrane. Similarly in adipose tissue, insulin stimulates the membrane transport of glucose and, additionally, stimulates the phosphorylation of glucose to glucose 6-phosphate. Lipoprotein lipase activity in adipose tissue is also stimulated. All these activities lead to the enhancement of lipogenesis.

The mechanism of action of insulin,

although intensively investigated remains an enigma. It certainly does not appear to act via the adenylate cyclase–cyclic AMP system, nor indeed by the guanylate cyclase–cyclic GMP system. There is some suggestion that changes in calcium flux mediate the membrane effect of insulin, but this is not universally accepted. There is as yet no unifying hypothesis to explain the many and varied actions of insulin.

Glucagon is a polypeptide which is manufactured and secreted by alpha cells primarily in response to a reduction in blood glucose levels. Its action is via the stimulation of the adenyl cyclase–cyclic AMP system, resulting in an increase in glycogenolysis, gluconeogenesis and lipolysis. These effects are basically opposite to those of insulin. Glucagon acts primarily on the liver and adipose tissue.

As **growth hormone** has assumed greater relevance in regard to diabetes mellitus in recent times, its non-skeletal actions will be reviewed. Growth hormone is manufactured by, and released from, the adenohypophysis. It stimulates both RNA and protein synthesis in the liver, and the transport of amino acids across cell membranes. These are the anabolic effects of growth hormone. Growth hormone also has an influence on carbohydrate metabolism, which involves both decreased peripheral utilization of glucose and augmented production of glucose via hepatic gluconeogenesis. Growth hormone also stimulates the release of fatty acids from adipose tissue.

Endocrine pancreatic hypofunction

(Diabetes mellitus)

Diabetes mellitus is a common disease of the dog, and to a lesser extent the cat. Most cases are seen in middle-aged to old dogs, where females are affected twice as commonly as males. Diabetes mellitus is distinctly uncommon in other domestic species.

Although initially thought to result from absolute deficiency of insulin, it has now been shown that not all cases of diabetes mellitus, at least in the dog, have low circulating levels of insulin. The hormonal characterization of diabetes mellitus is in its infancy as it is only recently that assays to measure plasma hormone levels have become available to assess the hormonal status of a particular patient. The investigation, particularly in dogs where diabetes mellitus is most common, have followed findings in man, where there are two major subgroups. The first (type I) is usually juvenile in onset and is **insulin dependent**; the second (type II) is usually adult in onset, is usually seen in obese patients, and is usually **insulin independent**. Type I diabetes is often controled by dietary adjustment alone, or in combination with oral hypoglycemic agents. Dogs with diabetes mellitus have been placed into two major groups.

1 A minority of cases that are hyperinsulinemic, have elevated growth hormone levels, and mild clinical signs (mainly ketotic).
2 A majority of cases that are hypoinsulinemic. This group has been subdivided into two subgroups:
   (*a*) Mildly ketotic.
   (*b*) Severely ketotic.

Group 1 is similar to mature-age onset diabetes in man, whilst group 2 is similar to juvenile onset diabetes in man.

The metabolic state in diabetes mellitus

Once an animal has an absolute or relative lack of insulin, a number of key metabolic processes are decreased in activity, and another group of processes are enhanced.

1 The ease of entry of glucose into many tissues, such as adipose tissue and muscle, or the conversion of glucose into glycogen in the liver is impaired. The net result contributes to an increase in blood glucose levels (hyperglycemia).
2 The entry of amino acids into cells, especially muscle cells, is impaired. There is also an increase in protein degradation. Surplus amino acids are converted to glucose by the gluconeogenic pathway in the

liver and kidney. Glucose production by this method contributes further to the hyperglycemia.

3 Lipogenesis is impaired because lipoprotein lipase activity is decreased, and lipolysis is enhanced. This leads to an increase in plasma free fatty acids and triglycerides. The excess free fatty acids are converted in the liver by beta oxidation to acetyl CoA. As much more acetyl CoA is produced than is required, the excess is converted to the ketone bodies acetoacetate, β-hydroxybutyrate and acetone. In diabetes mellitus, the ketone bodies produced, whilst capable of being utilized peripherally, overwhelm the system. This leads to ketonemia and to ketonuria.

Special clinical features

The consequences of hyperglycemia are based on the renal threshold for glucose being exceeded. In the dog, the threshold is in the range 10–12 millimoles/liter. Once this concentration is exceeded, glucose appears in the urine (glycosuria) and by osmotic attraction causes the obligatory retention of water in the urine. An osmotic diuresis ensues, resulting in polyuria. Compensatory polydipsia follows.

Although hyperglycemic, the animal reacts to the relative inability of glucose to move into cells, or to be phosphorylated once in the cell, by consuming more food. The mechanism for this is apparently central and the animal thinks it is starving. It thus becomes polyphagic.

Three 'polys' – polyuria, polydipsia and polyphagia – are often observed in diabetes mellitus. Weight loss is common due to muscle atrophy and increased lipolysis.

In many cases ketosis and ketoacidosis develop. Ketones are readily excreted in the urine, usually as salts of sodium or potassium. Water is also required for their excretion, which tends to exacerbate the already severe polyuria. Within the bloodstream, ketone bodies exist in an ionized form, where the free hydrogen ion donated by the keto-acids is neutralized by bicarbonate and other buffer systems. If ketosis is sufficiently severe, the animal becomes acidotic and attempts to compensate by blowing off carbon dioxide via the lungs. A dog with keto-acidosis therefore develops a deep and rapid type of breathing. It is worth emphasizing that not all diabetic animals are severely keto-acidotic. Similarly, not all animals invariably exhibit glycosuria.

A clinical feature is the development of bilateral lenticular opacity, the pathogenesis of which is thought to be due to the conversion of glucose to sorbitol in the lens after the glycolytic pathway is saturated with glucose. As sorbitol is unable to leave the lens, an osmotic gradient it set up requiring water to move into the lens. The lens swells, and some lens fibers rupture and opacity results.

On physical examination, affected animals may exhibit hepatomegaly. The hepatic enlargement is the result of extensive lipidosis within hepatocytes following peripheral mobilization of fat. The chief features of diabetes mellitus are depicted in Fig. 10.22.

Three major pathologic changes are responsible for the development of diabetes; it may follow extensive exocrine pancreatic necrosis, but it is uncommonly associated with inflam-

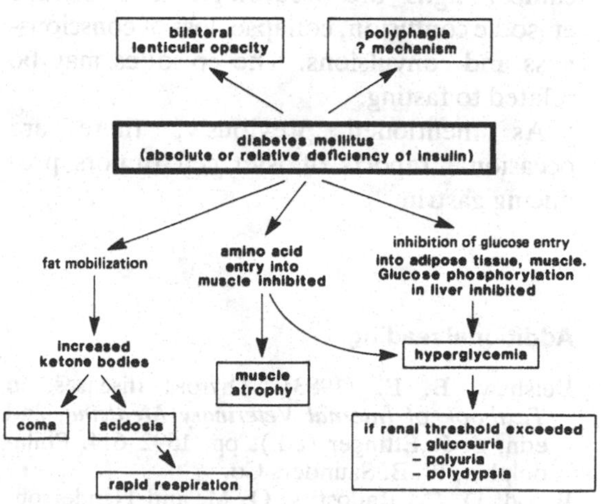

Fig. 10.22. Diabetes mellitus: an absolute or relative deficiency of insulin results in clinical signs that are dominated by polyuria, polydipsia and polyphagia and the metabolic effects of fat and muscle breakdown.

mation of the exocrine pancreas. Amyloidosis of the islets in cats is a prominent finding. Lastly, islets may contain a scattering of lymphocytes, suggesting an autoimmune basis.

As would be expected from the metabolic derangements, a consistent finding is a spectacularly yellow fatty liver; there is also glycogen nephrosis of the loops of Henle and the distal convoluted tubules. The glomeruli may also exhibit a characteristic thickening of glomerular basement membrane. In many cases in the dog, there is vacuolar change in the islets and exocrine pancreatic duct epithelium.

Endocrine pancreatic hyperfunction (Insulinoma)

Tumors of the beta cells of the islets may be either benign or malignant and are often functional. They are uncommon, but not rare in the dog, are usually solitary but may be multiple.

The clinical signs are directly referable to the tumor releasing inordinate amounts of insulin and insulin-like peptides into the circulation, causing profound **hypoglycemia**. The clinical signs are neurologic and include episodic confusion, collapse, loss of consciousness and convulsions. The episodes may be related to fasting.

As mentioned previously, there are occasional reports of islet cell tumors producing gastrin.

Additional reading

Belshaw, B. E. (1983). Thyroid diseases. in *Textbook of Internal Veterinary Medicine*, 2nd edn, S. J. Ettinger (ed.), pp. 1592–614. Philadelphia, W. B. Saunders Co.

Blood, D. C., Radostits, O. M. and Henderson, J. A. (1983). *Veterinary Medicine*, 6th edn. London, Bailliere Tindall.

Burns, T. W. (1979). Endocrinology. In *Pathologic Physiology – Mechanisms of Disease*, 6th edn, W. A. Sodeman Jr and W. A. Sodeman (eds.), pp. 1002–58. Philadelphia, W. B. Saunders Co.

Capen, C. C. (1985). The endocrine glands. In *Pathology of Domestic Animals*, vol. 3, 3rd edn, K. V. F. Jubb, P. C. Kennedy and N. Palmer (eds.), pp. 237–303. Orlando, FL, Academic Press.

Capen, C. C. and Martin, S. L. (1983). Diseases of the pituitary gland. In *Textbook of Internal Veterinary Medicine*, vol. 2, 2nd edn, S. J. Ettinger (ed.), pp. 1523–49. Philadelphia, W. B. Saunders Co.

Crispin, S. M. and Langslow, D. R. (1980). Diabetes mellitus in the dog. In *Physiological Basis of Small Animal Medicine*, A. T. Yoxall and J. F. R. Hird (eds.), pp. 3–25. Oxford, Blackwell Scientific Publications.

Eigenmann, J. E. and Peterson, M. E. (1984). Diabetes mellitus associated with other endocrine disorders. *Vet. Clin. N. Am.* **14**: 837–58.

Feldman, E. C. (1983). Diseases of the endocrine pancreas and the adrenal cortex. In *Textbook of Internal Veterinary Medicine*, vol. II, 2nd edn, S. J. Ettinger (ed.), pp. 1615–96. Philadelphia, W. B. Saunders Co.

Feldman, E. C. and Peterson, M. E. (1984). Hypoadrenocorticism. *Vet. Clin. N. Am.* **14**: 751–66.

Hoenig, M. (1983). Hyperthyroidism in cats. In *Veterinary Annual*, 23rd issue, C. S. G. Grinsell and F. W. G. Hill (eds.), pp. 281–6. Bristol, John Wright and Sons.

Kashgarian, M. and Burrow, G. N. (1974). *The Endocrine Glands*, Structure and Function in Disease Monograph Series. Baltimore, The Williams and Wilkins Co.

Ladds, P. W. (1985). The female genital system. In *Pathology of Domestic Animals*, vol. 3, 3rd edn, M. V. F. Jubb, P. C. Kennedy and N. Palmer (eds.), pp. 305–407. Orlando, FL, Academic Press.

Lavelle, R. B. and Yoxall, A. T. (1980). The adrenal glands. In *Physiological Basis of Small Animal Medicine*, A. T. Yoxall and J. F. R. Hird (eds.), pp. 39–68. Oxford, Blackwell Scientific Publications.

Mehlhaff, C. J., Peterson, M. E., Patnaik, A. K. and Carrillo, J. M. (1985). Insulin producing islet cell neoplasms: surgical considerations and general management in 35 dogs. *J. Am. Anim. Hosp. Assoc.* **21**: 607–12.

Peterson, M. E. (1984*a*). Feline hyperthyroidism, hyperadrenocorticism. *Vet. Clin. N. Am.* **14**: 809–26.

Peterson, M. E. (1984*b*). Hyperadrenocorticism. *Vet. Clin. N. Am.* **14**: 731–49.

Rogers, K. S. and Luttgen, P. J. (1985). Hyperinsulinism. *Comp. Contin. Educ. Pract. Vet.* **7**: 829–41.

Siegel, E. T. (1977). *Endocrine Diseases of the Dog*. Philadelphia, Lea and Febiger.

Clive R. R. Huxtable and Susan E. Shaw

11 The skin

The skin, being the outward manifestation of the body, is often admired, but frequently misunderstood by the everyday observer. The 'real' organs, it would seem, are wrapped in the dark mysteries of the interior while the skin is somehow set apart – a sort of bloodless, hairy and inert covering, useful for turning into shoes, handbags or rugs.

The well-trained biologist knows better; the skin is a marvellous adaptation of living tissue directly confronting the stresses and hazards of the environment. It is one of the largest organs in the body and no other tissue has the same problems to overcome. Consider the requirements for its design. It must be strong but elastic, pliable and resilient, and must be able to repair itself quickly. It must be able to resist desiccation – a particularly difficult accomplishment for an exposed system of living cells. It must protect against onslaughts chemical, physical, thermal and microbiologic. It must provide for the reception of countless sensory inputs needed for adequate proprioception and it must provide visual, olfactory and tactile stimuli for sexual and special species-related behavior. In wild animals this includes camouflage.

Far from being inert, the skin is very much alive in spite of its large non-living keratin component. It is interlocked with general body metabolism and has especially strong links with the endocrine and nervous systems. Its construction ensures that it is able to carry out these special functions. The basic structure is the same throughout, but there are areas of specialized modification where differences are either grossly or microscopically obvious.

The skin is a combination of connective tissue and epithelial elements, the connective tissue making up the greater part of its bulk. The deepest layer is the **hypodermis**, a fibro-fatty pad abutting the superficial body muscles and fascia. The hypodermis acts as an insulator, mechanical protectant, body energy store, and deep anchor for the skin. It is also referred to as the **panniculus** layer.

The **dermis** is a tough, pliable, compliant and elastic sheet of interwoven fiber bundles containing about 90% collagen fibers and 10% elastin fibers. The fibers are woven into a specific pattern to produce these physical properties.

The blood supply to the skin is generous, with cutaneous arteries coursing through the hypodermis to terminate in extensive capillary beds in the dermis. In most species there are two dermal capillary plexuses, one superficial and the other deep. The total capacity of cutaneous vessels is enormous, but many of the capillary beds are normally shut down at any given time. This rich blood supply optimizes conditions for metabolic activity, immune mechanisms and healing, and is important for thermoregulation in some species. Lymphatic vessels are also numerous.

Cutaneous nerve supply is abundant and the senses of touch, pain and pressure are vital for survival. There are various classes of specialized sensory receptors in the dermis and epidermis for these modalities. The degree of sensory innervation is also reflected in the clinical phenomenon of **pruritus** (itching).

Autonomic innervation relates to vasomotor activity, pilimotor activity (the movement of hairs) and the secretory activity of the sweat glands. Hair movement is powered by small straps of muscle attached to large hair follicles – the arrector pili muscles.

The dermis also contains a diffuse population of mast cells and histiocytes which figure prominently in several disease states, especially allergy in the case of mast cells. Thus the dermis, far from being mere shoe leather, can be seen as a richly endowed foundation upon which the epithelial elements of the skin depend for their survival.

The **epithelial** components of skin are organized into three major systems of cells:

Keratinocytes,
Melanocytes,
Glandular cells.

The function of the **keratinocytes** is to produce keratin, a complex fibrous protein that constitutes the non-living outer surface of the body. It comes in two major forms, a very thin surface sheet and a long cylindrical hair. Keratin is produced by a continual process of cell division, maturation and final degeneration to give the end product. The process is still not completely understood, but is influenced by nutritional and endocrine factors which will be mentioned in a clinical context. Suffice to say that the continual formation of the protein keratin is a major synthetic operation for the body.

The **melanocytes** are closely linked with keratinocytes. They carry out the synthesis of melanin from tyrosine via the copper-dependent enzyme tyrosinase. Melanocytes are neuroepithelial cells that migrate into the skin during organogenesis. Melanin in granular form (melanosomes) is transported via the long dendritic processes of the melanocytes, and taken up by keratinocytes. It is finally incorporated into the keratin of the surface layer and the hairs. The major biologic role of epidermal melanin in man, and perhaps in other relatively hairless animals, is to protect the skin from ultraviolet (UV) radiation which can both damage the skin and induce neoplastic transformation in epidermal keratinocytes and melanocytes. The melanin granules, by effectively absorbing UV light, act as a biologic sun-screen. In most domestic animals, the hair coat effectively protects the skin from UV radiation. The amount and distribution of epidermal melanin in an individual animal is genetically determined, but can be modulated by hormones, nutritional factors and local skin inflammation, which will be discussed later.

The melanin in hair shafts determines the color of the coat, and in wild animals this has immense implications for survival. In domestic species it has mainly an esthetic influence on owners.

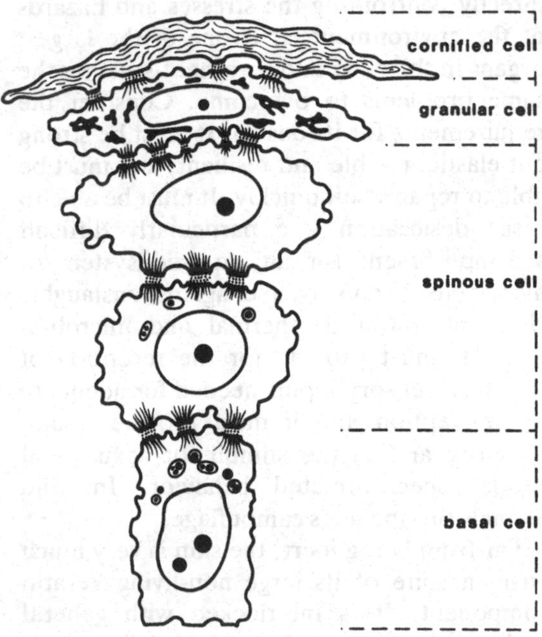

Fig. 11.1. Representation of the stages of epidermal keratinocyte maturation (Adapted from Ackerman, 1978, by kind permission, see Additional reading.)

There are also variable numbers of dermal melanocytes in animal skin. Their biologic role is obscure and they may represent heterotopia, or cells that became 'lost' during developmental migrations.

The **glandular cells** occur in two types of gland, the sebaceous and sweat glands. In most parts of the skin of animals, their secretions enter the infundibulum of the hair follicles. Sebum is important both as a lubricant for maintaining the quality of hair and for some bacteriostatic properties. The secretion of sweat plays a role in heat regulation in some species like the horse, but in general its purpose in animals is poorly understood.

The special epithelial structures of the skin

The epithelial cells of the skin are organized into the following structures:

The epidermis.
The hair follicles.
The sebaceous glands.
The sweat glands.

The last three are often collectively referred to as the adnexae.

The epidermis

The epidermis is the exterior cellular layer of the body, exposed directly to the external environment and, seemingly, a fragile barrier against the hostile world. In animals, it is generally no more than three to five cells in thickness. It consists predominantly of **keratinocytes**, but also contains **melanocytes** and **Langerhans** cells. The last of these are immigrant cells of bone marrow origin and have a macrophage-like function in the processing of antigen. The epidermis is a flat, avascular sheet of cells, depending entirely upon diffusion from the dermis for its nutrition and oxygen supply. It is separated from the dermis by a basement membrane and, in animals, is anchored to the dermis largely by the hair follicles and by a system of fibers and fibrils adjacent to the basal lamina.

The keratinocytes of the epidermis form an active cell-renewal system with a turnover time of 20–30 days (Fig. 11.1). The germinal layer is the basal layer of cells situated along the basement membrane. In sparsely haired skin, about 80% of cells in this location are undifferentiated keratinocytes with up to 10–20% melanocytes, although melanocytes may be scarce in densely haired skin. Basal keratinocytes undergo mitosis and a proportion of daughter cells move outwards into the next zone, where they lie with their cytoplasm interdigitating and connected by well-developed desmosomes. These cells, with their desmosomes appearing 'spinous' when viewed with the light microscope, are also called 'prickle cells' or 'acanthocytes' (Fig. 11.1). The dendritic processes of melanocytes, stretching from parent cells in the basal layer, ramify amongst the prickle cells and it is here that the keratinocytes acquire melanin (Fig. 11.2). It is here also that the Langerhans cells are located amongst the keratinocytes. The prickle cells move outward again and change from rounded to flattened cells when they begin to produce granules in their cytoplasm (keratohyalin). Microscopically, this zone appears as the 'granular layer'.

The final phase occurs when the maturing cells degenerate to yield a layer of flattened keratin scales which form an adherent sheet, the stratum corneum or 'horny layer'. The keratin flakes are connected by vestigial desmosomes, but disintegrate and are shed at a steady rate in the normal animal. Under normal circumstances the stratum corneum is barely visible to the naked eye.

The hair follicles

Hair follicles are cylindrical sleeves of cells which project down into the dermis from an opening in the epidermis. The basal layer of the epidermis reflects down to form the basal layer of the follicle. Hair is formed by keratinocytes at the base of the follicle, which occur in a cluster called the hair matrix. Again, melanocytes are present with the keratinocytes. The intense production of keratin sheets and melanin in the hair matrix is nourished and modulated by a tuft of vascular connective tissue which invaginates the base of the follicle and is called the dermal papilla (Fig. 11.3). The hair shaft, when first formed, is tightly sheathed by multiple layers of specialized follicular cells, the details of which need not concern us here. In the upper third of the follicle, the infundibulum, the hair shaft is free of the follicular wall, which is lined by conventional keratinocytes only. In most areas of the skin, the ducts of sebaceous and sweat glands open into the infundibulum (Fig. 11.3).

At the skin surface, the hairs emerge from the follicles either singly, or in multiple fashion, with up to 15 hairs emerging from one surface pore. Such an arrangement is called a compound follicle and is seen typically in the dog.

The **growth of hair** is an immensely complex process. In some species the hair coat is repeatedly renewed throughout life. In others, like sheep or Poodle dogs, there is continual growth of the coat, with no cyclical shedding. In wild animals, coat renewal often occurs in a synchronized 'wave' pattern, starting at one end of the body and spreading symmetrically over it. In many domestic species, hair growth for coat renewal seems to have a 'mosaic' pattern, occurring continually all over the body.

In those animals with cyclical hair growth, activity in each follicle is discontinuous (Fig. 11.4). In simple terms, the active production of hair by the matrix is termed the stage of **anagen**; keratin sheets are produced and compacted to build the hair shaft, which steadily elongates. Melanin is incorporated into the shaft in the correct manner according to the genetic program. It is remarkable to think that the diversity of animal coat colors and patterns is achieved by the varying concentrations of just two kinds of melanin. After the hair has grown for the genetically specified time, matrix activity ceases and the base of the follicle retracts towards the surface. This regressive stage is known as **catagen**. The follicle then lapses into a resting inactive state called **telogen**, with the hair shaft detached from its sheath and easily removed by friction. After the genetically determined time, an

Fig. 11.2. The relationship between epidermal keratinocytes and melanocytes. The melanocyte cell body in the basal layer projects its processes amongst maturing keratinocytes. (Adapted from Ackerman, 1978, by kind permission, see Additional reading.)

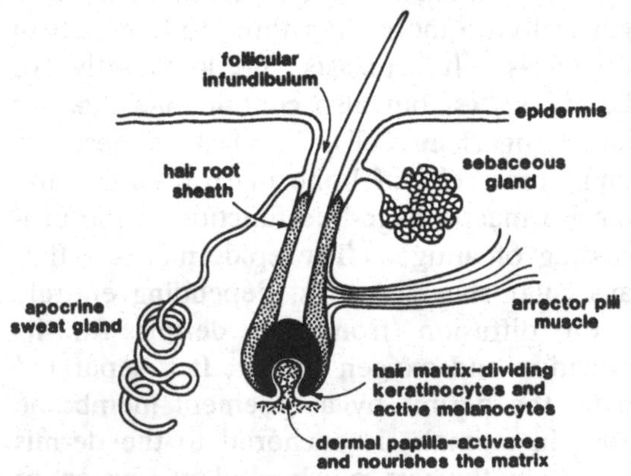

Fig. 11.3. Representation of a pilosebaceous unit with a hair in the active, anagen growth phase.

active matrix develops and a new hair shaft is formed that pushes up to dislodge the old shaft, which is shed. The full round of activity is called a **hair growth cycle** (Fig. 11.4). Each follicle has its own genetic program which it obeys, even if transplanted to another part of the body. However, the hair growth cycle is influenced by some hormones, and disturbances of hair growth are a feature of some endocrine disorders.

The sebaceous and sweat glands

Sebaceous glands are composed of solid clusters of lipid-laden cells which eventually disintegrate to become the waxy secretion of the gland (holocrine secretion). Over most of the skin they discharge via a duct into the follicular infundibulum (Fig. 11.3). Their activity is stimulated by androgens and inhibited by high doses of estrogens. Apart from these examples, there is total ignorance regarding the regulation of their normal activity. Sebum diffuses through the stratum corneum, sealing the margins of the squames and forming a physical and bacteriostatic barrier.

In animals, most of the **sweat glands** are of the apocrine type and discharge into the infundibulum of primary hair follicles after coiling their way up from the deeper dermis (Fig. 11.3). They are innervated by adrenergic and cholinergic sympathetic fibers and probably function mainly as scent organs. However, the regulation of their activity is poorly understood.

Eccrine sweat glands, discharging directly to the skin surface and independent of hair follicles, are found only in the footpads of dogs and cats, but are widespread in the horse, where they are important in heat regulation. They are most sensitive to parasympathetic stimulation. Upon stimulation, they flood the surface of the skin with a watery secretion.

Reactions of the skin in disease and their clinical interpretation

In dermatology the clinician is in the unique situation of having the whole organ system readily available for inspection. Finding the lesions is a small problem. The major challenge lies in interpreting them. The aim should be to decide how much of the skin is affected, what is the distribution pattern of affected skin, what type of pathologic process is indicated, and what elements of the skin are primarily affected.

It should always be kept in mind that the skin is not set apart from the rest of the body, but is an organ interacting with other organ systems. It can become secondarily involved in disorders of other organ systems or disturbed

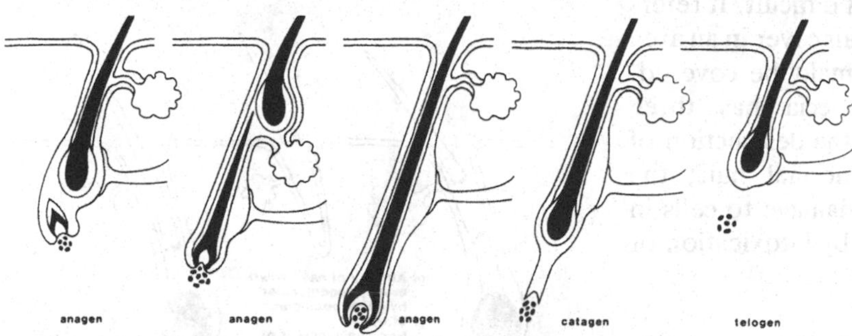

Fig. 11.4. The phases of the hair growth cycle. Over the course of anagen (left) a new matrix develops and a new hair shaft progressively displaces the old. (Adapted from Ackerman, 1978, by kind permission, see Additional reading.)

by systemic upsets. There are numerous dermatologic disorders that are secondary to internal diseases, with the skin often acting as the distress signal. A thorough dermatologic investigation, therefore, must always include a total physical examination and history. The skin in disease is able to express a range of appearances which is based upon the responses of its component parts. Clinical skill will be enhanced if there is an understanding of these components and their responses. In the following paragraphs, the responses of the components will be examined individually for ease of understanding. In life, of course, things will often happen in complex combinations, but the basic disturbances should still be apparent to the skilled clinician. The complexity can be frustrating, but it adds to the fascination and the challenge.

Reactions of the hair coat

Disorders of the hair coat encompass deficiency in hair cover, surplus hair cover, and/or changes in hair quality. These abnormalities may be localized or extensive.

Deficiency in hair cover

In any severely destructive lesion of the skin, hair may of course be lost along with all the other skin elements. There are situations, however, in which hair loss is the primary feature and this is what concerns us here. The lack of hair cover is referred to as **alopecia** and its clinical recognition is not difficult. It refers to a partial or total lack of hair cover in an area of the skin that would normally be covered. The pathogenesis of alopecia has three possible bases (Fig. 11.5): the destruction of the shafts of essentially normal hair; the regression of hair growth; damage to cells in the matrix of growing hair by intoxication or metabolic injury.

Destruction of normal hair shafts

The destruction of normal hair shafts by self-mutilation is inevitable in any area of skin that is intensely irritating or itchy. Animals will rub, lick, chew or scratch any such lesion and can rapidly remove hair cover. Most lesions of this type are inflammatory in nature and so the alopecia will be accompanied by indications of inflammation. Conversely, any very itchy area of skin will become inflamed as it is traumatized by the frantic patient. Successful treatment of the underlying problem will soon result in restoration of the hair coat, unless there has been deep and severe inflammation with destruction of the follicles. In this case the skin will heal by forming a hairless scar.

Microorganisms may also cause the destruction of hair shafts and consequent alopecia. Prominent in this category are the dermatophyte fungi, and the demodectic mange mites. The **dermatophytes** use keratin as their substrate for growth and actively invade and destroy hair. The **demodectic mites** live within the follicles and when they are present in large numbers, the contained hairs are damaged and dislodged. In both dermatophyte and demodex-induced lesions, alopecia may be accompanied by other changes related to epidermal reactions and inflammation, but in neither case is itching usually a feature.

Regression of hair growth

The maintenance of the hair coat depends upon normal cyclical activity in the hair follicles. If the hair cycle fails for any reason, all hairs will go into the resting telogen phase.

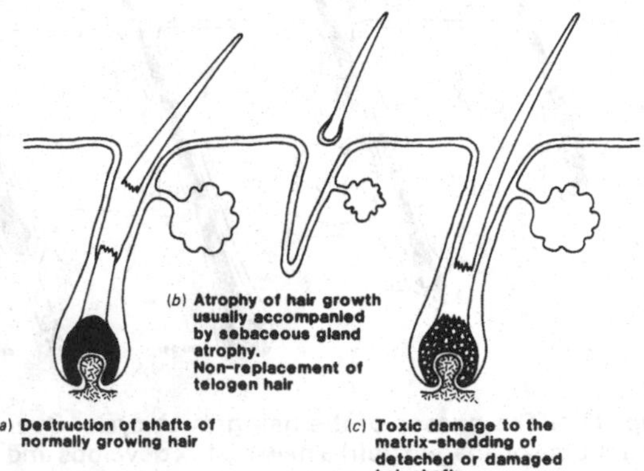

Fig. 11.5. The three basic mechanisms of alopecia.

Any hair removed by mechanical friction or clipping will not be replaced. Conditions so affecting the hair cycle are systemic metabolic disturbances particularly endocrine disorders. The best understood are hypothyroidism, hyperadrenocorticism and hyperestrogenism, but a longer list will be found in clinical dermatology texts.

Beyond stating the obvious, that active hair growth is greatly influenced by the endocrine system, little can be said to explain the mechanisms involved. The outstanding general dermatologic feature of these disorders is symmetrical hair loss, chiefly from areas of sustained frictional contact, for example, the collar region and thighs. In extreme cases the hair may be lost from most of the body, but the head and feet are characteristically spared (Fig. 11.6). In most, but not all endocrine dermatoses, the skin lesions are non-pruritic and non-inflammatory, and the alopecia may be accompanied by hyperpigmentation. More information is to be found in Chapter 10. In most instances, correction of the endocrine imbalance will result in the resumption of normal hair growth and restoration of the coat.

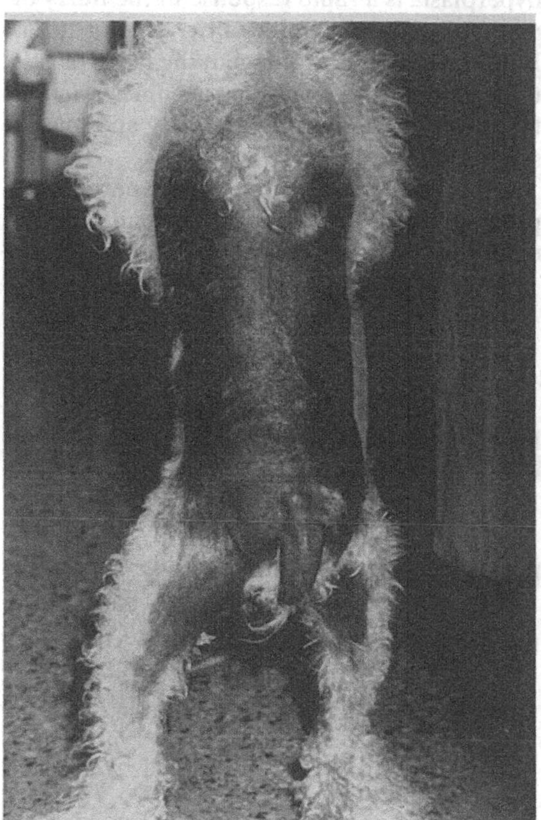

Fig. 11.6. Extensive symmetrical alopecia in a dog with hyperadrenocorticism. Atrophy of hair growth results in non-replacement of telogen hair removed by mechanical friction.

In some individuals, hair growth in some regions of the body may irreversibly cease for no apparent reason and the baldness in these cases is assumed to be genetically determined. Genetically determined baldness has been actively selected for in some breeds of dogs, for instance the Chinese Crested dog. Increasing hairlessness may also be a feature of senility or debility.

Toxic insult or metabolic disturbance

The growth of hair depends on intense cellular proliferation in the matrix and, as is the case with any such system, these cells are especially vulnerable to acutely cytotoxic agents and to certain other toxins.

Acutely cytotoxic agents (which may be used in cancer therapy), such as radiation, cause an abrupt dislocation of hair growth in the matrix and separation of the matrix from the shaft. Within a very short time the hair can be easily removed. This cytotoxicity has been used in experiments for the chemical shearing of sheep, but is hampered by the side effects.

More chronic intoxication, rather than abruptly interrupting hair growth, may cause the production of defective, fragile hairs, which are easily broken off or shed. This is exemplified by chronic selenium poisoning, when there is a qualitative defect in keratin. Toxic chemicals like thallium and arsenic can also affect hair growth and result in the shedding of damaged hair. As the growth of hair requires up to 30% of the daily protein requirement of the body, it is understandable that alopecia is seen in severe protein malnutrition and cachetic states in general. In these toxic and metabolic disorders, the skin tends to be diffusely and symmetrically affected.

Surplus hair cover

Surplus hair cover reflects the production of an abnormally long hair coat (hirsutism or hypertrichosis) due to a prolonged anagen phase of the growth cycle. The cause is usually a systemic metabolic disorder, and a well-known entity is equine hyperadrenocorticism, resulting from pituitary neoplasia. In several species, notably cattle, cachetic debility can lead to hirsutism because of cessation of seasonal shedding.

Changes in hair quality

Pre-senile generalized changes in coat color and/or texture are usually seen in states of malnutrition. The classic example is the effect of copper deficiency in black-wooled sheep. Melanization of the wool fails because the activity of tyrosinase, the melanin-forming enzyme, is copper dependent. In addition, the nature of the keratin is altered and the wool loses its crimp. In copper deficient cattle, depigmentation of hair is often particularly noticeable around the eyes and the coat generally may become harsh and lusterless.

The pigmentation of the hair coat may diminish and the texture may become coarse in any animal with severe protein malnutrition, and these are also common senile changes in many species. The depigmentation and graying of hair is called **leukotrichia**. In localized areas of skin where there has been severe inflammation, the hair will sometimes regrow with a loss of pigmentation because melanocytes have failed to regenerate. This has been exploited in the freeze-branding technique.

Conversely, in Siamese and Burmese cats, hair regrowing first on an area that has been clipped, as for surgery, will often be hyperpigmented. The reason for this melanocyte activation is obscure, but may be related to the effect of skin temperature on heat-sensitive, melanin-synthesizing enzyme systems.

Adaptive responses of the epidermis

In those situations where the skin is diseased, but still essentially intact, the appearance of the lesions will be governed by the responses of the epidermis and the hair coat. Excluded from consideration here are severe inflammatory reactions sufficient to destroy epidermal structure. The adaptive responses of the epidermis can be analyzed according to the following scheme.

Hyperplasia

The conventional hyperplastic response of the epidermis is an increase in the number of cells in the stratum spinosum or acanthocyte layer of keratinocytes, leading to the hyperplastic state being termed **acanthosis**. When inflammatory events occur in the dermis, epidermal hyperplasia is a rapid response to the flurry of metabolic changes induced by inflammation. If inflammation is prolonged, the hyperplastic thickening of the epidermis becomes a major feature of the lesion, both grossly and microscopically.

In areas where the stimulus to hyperplasia is prolonged and intense, the thickening of the epidermis becomes spectacular and the skin is

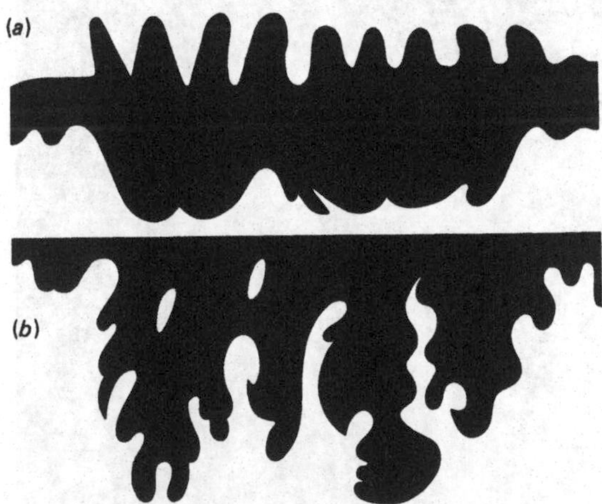

Fig. 11.7. Representation of (*a*) pseudopapillomatous and (*b*) pseudocarcinomatous epidermal hyperplasia. (Adapted from Ackerman, 1978, by kind permission, see Additional reading.)

thrown into ridges and folds and may become deeply fissured. This is referred to as **lichenification**.

In advanced hyperplasia, the epidermis may push downward into the dermis, forming 'rete pegs'. In extreme cases, the intradermal downgrowth mimicks neoplasia and is termed pseudocarcinomatous hyperplasia. Conversely, it may push projections outwards from the skin surface, as frond-like papillae, and be termed pseudopapillomatous hyperplasia (Fig. 11.7).

Epidermal hyperplasia should be recognized for what it is – a common non-specific response to metabolic alteration in the skin found in a wide variety of dermatoses. It will often be accompanied by alterations of the surface keratin layer, as will now be described.

Alterations in keratinization

An increase in the keratinized layer of the epidermis is referred to as **hyperkeratosis**, which may be further classified as orthokeratotic or parakeratotic (Fig. 11.8), although both may occur together.

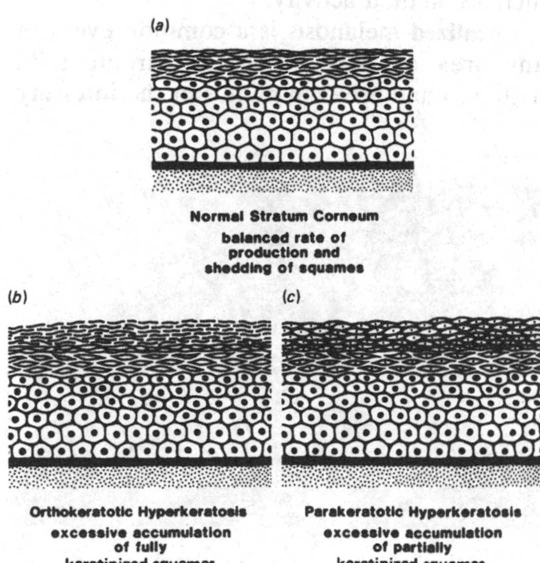

Fig. 11.8. Representation of (*a*) normally keratinized epidermis, (*b*) orthokeratotic hyperkeratosis, and (*c*) parakeratotic hyperkeratosis.

Orthokeratotic hyperkeratosis describes an increased thickness in the surface layer of normal mature keratin squames, and reflects either an increased rate of production of normal stratum corneum or an abnormal retention of the squames. The reaction may produce tough and horny thickening of the skin classically exemplified by the callus that develops at points of prolonged frictional stimulation, or it may be more subtle. In the dog for instance, focal orthokeratotic hyperatosis may occur in superficial staphylococcal infection to produce circular 'collarettes' of scaling at the margins of the lesions.

Also in the dog, marked hyperkeratosis of the footpads ('hard pad') and/or the nose can be an interesting accompaniment to distemper virus infection. This is a useful diagnostic sign, but how it relates to the virus infection is not understood. It may also occur spontaneously with no obvious cause.

Generalized or widespread hyperkeratosis suggests a systemic metabolic disturbance, the best-known examples of which are endocrine imbalances. In such cases, there may also be severe accompanying hyperkeratosis within hair follicle infundibulums, and the mouths of the follicles become plugged with keratin (Fig. 11.9). These keratin plugs are called **comedones** and are visible grossly.

Parakeratotic hyperkeratosis is a more common phenomenon. It results from increased turnover of epidermal keratinocytes, with an increased rate of entry of keratinizing cells into the stratum corneum. The result is that the stratum corneum contains only partially keratinized cells, which still contain their nuclei (hence parakeratosis). The sheets of parakeratotic cells accumulate on the skin surface as a layer of grayish brown, bran-like flakes and scales. This reaction is likely to develop in any localized area of mild skin irritation and inflammation. The inflammatory events in the underlying dermis incite the hyperplastic response in the overlying epidermis. It is seen in conditions with a variety of causes, infectious and non-infectious.

A tendency to generalized or extensive parakeratosis suggests a systemic metabolic disturbance. The classic example is zinc responsive dermatosis in pigs (Fig. 11.10) and dogs. Normal keratinization has a high requirement for zinc, and a dietary deficiency or imbalance of zinc can lead to extensive parakeratotic scaling of the skin. Parakeratosis is also a feature of vitamin A inadequacy and diffuse ectoparasitism, sarcoptic mange for example.

Seborrheic dermatosis is a term embracing those skin conditions whose dominating feature is the disorder called seborrhoea, implying an increase in sebum excretion. This classification is inappropriate because the real basis of the disorder is altered keratinization, with or without excessive sebum secretion. In seborrheic dermatosis, there is sometimes, dry, or greasy, scaling and flaking of most of the skin surface due to marked ortho- and/or parakeratotic hyperkeratosis, as described above. The surface lipid has an increased content of free fatty acids, and an abnormally high population of bacteria develops on the skin surface, which commonly gives rise to skin infections and inflammation (seborrheic dermatitis). The seborrheic state is commonly secondary to systemic states of malnutrition, cachexia and hormonal imbalance, or to environmental agents like mange mites and dermatophyte fungi. Primary idiopathic seborrhoea is recognized in some breeds of dog, for example, the American Golden Cocker Spaniel. This appears to be a genetically determined fault in epidermal keratinization.

Abnormalities of epidermal pigmentation

It should be realized that the epidermal melanocytes are intimately associated with the keratinocytes but are independent of the melanocytes in the hair bulbs. In any animal, the normal activity of epidermal melanocytes is governed by its own genetic program, which defines the pattern of skin pigmentation. This activity can be modified by factors operating both locally and systemically.

An increase in pigmentation is called **melanosis** (or melanoderma) and a decrease is called **leukoderma**. The recognition of either depends upon knowledge of the normal pigmentation pattern of the subject animal. In acquired melanosis there is probably actual hyperplasia of melanocytes as well as an increase in their activity.

Localized melanosis is a common event in any area of skin subject to chronic mild irritation and inflammation, but the intensity

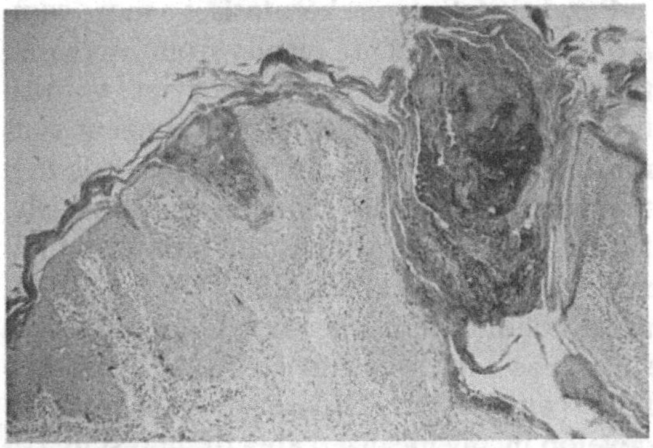

Fig. 11.9. Light micrograph illustrating a large plug of keratin within the ostium of a hair follicle, from a dog with severe parakeratosis. Note also pseudocarcinomatous epidermal hyperplasia and epidermal parakeratosis. The animal had a vitamin A-responsive dermatosis.

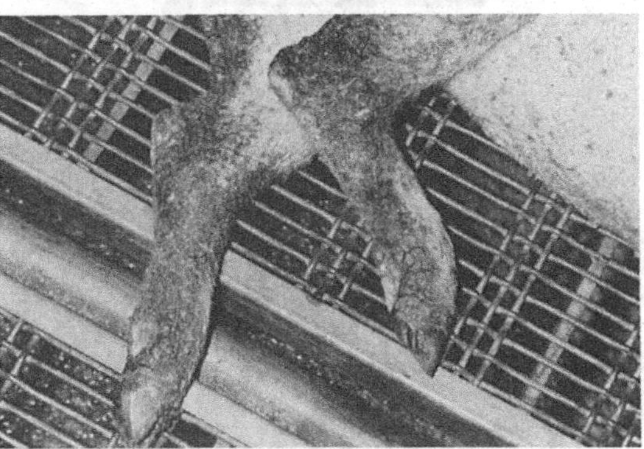

Fig. 11.10. Photograph of the hind legs of a pig with zinc-responsive dermatosis. The skin surface is covered by adherent scales of parakeratotic material. With appropriate treatment the lesion rapidly resolved.

of the response varies between individuals. Because of their intimate association with keratinocytes (Fig. 11.2), the local melanocytes are stimulated by the metabolic changes associated with inflammation. Examples are the lesions of chronic contact and flea-bite hypersensitivities in the dog, which in time may become thickened and darkened areas of skin. Localized patches and plaques of hyperpigmentation should also alert the observer to the possibility of melanocytic neoplasia.

Generalized melanosis implies a systemically acting stimulus and reflects hormonal action on melanocytes. Non-inflammatory melanosis may be seen in certain sex hormone endocrinopathies and in hypothyroidism. When melanosis occurs in these diseases it is usually in a multifocal 'blotchy' pattern, associated with alopecia. Pituitary adrenocorticotropic hormone (ACTH) is melanocyte-stimulating in some individuals, if secreted to excess for a long time, and is the reason some dogs with Cushing's-like disease have generalized melanosis (Fig. 11.6). The pituitary also produces a specific melanocyte-stimulating hormone, which may be overproduced in pituitary disease.

Congenital epidermal **hypopigmentation** (leukoderma) is seen of course in true albinism, in which melanocytes are devoid of tyrosinase activity and are therefore biochemically incompetent.

Localized acquired leukoderma usually reflects the permanent loss of melanocytes from an area of skin following severe damage. While keratinocytes regenerate well in the epidermis, melanocytes may not. In the dog, the lesions of discoid lupus erythematosis may feature marked depigmentation of the nasal skin.

Disorders of the sebaceous and sweat glands

If sebaceous secretion is deficient, the hair tends to become coarse and lusterless, and theoretically at least, the skin becomes more vulnerable to bacterial growth. In dogs, a dietary imbalance of certain fatty acids can lead to sebaceous hypofunction and this will of course occur in malabsorption/maldigestion syndromes. Sebaceous gland atrophy is also a feature of several endocrine dermatoses.

Excessive sebaceous secretion leads to a state known as **seborrhoea oleosa**, in which the surface of the skin is coated with a thick greasy layer of sebum. This could arise when there is excessive androgen production, for instance if there was a functional gonadal or adrenal tumor. In many cases, however, a cause cannot usually be identified in afflicted individuals.

The sweat glands usually have only secondary but sometimes important involvement in skin lesions. Hyperkeratotic and hyperplastic reactions often result in obstruction of sweat ducts, which leads to dilation and sweat retention. This can delay healing, especially if the cystic glands rupture and leak sweat into the dermis.

Inflammation in the skin

Inflammation in the skin has all the classical macroscopic features of the process, as well as some features related to the special character of the organ. Its histologic classification is intricate and will not be considered here. Clinicians often request histologic examination of skin biopsies but the interpretation of these is the province of a specialist dermatohistopathologist. The interested reader should refer to Additional reading, below.

For the clinician, inflammation of a mild character does not produce much exudate, but rather stimulates hyperplastic epidermal responses of the types described in the previous section. The dominant feature will be scaling or scurf production, the result of parakeratosis.

Dermatitis

More overt dermatitis is an exudative reaction, involving dermis and epidermis, with exudate drying and crusting over the surface of the affected area, causing the hair to matt. There are several possible degrees of severity.

In the *least* severe type of exudative

reaction, the epidermis remains largely intact and the exudate oozes through it, perhaps accumulating here and there to produce intra-epidermal blebs. Small blebs, a few millimeters in diameter are called **vesicles** and larger ones **bullae**. If neutrophils are numerous, the vesicles and bullae become **pustules**. The passage of exudate through the epidermis separates the epidermal cells somewhat from one another, and this, in histopathologic terms, is called spongiosis. The separated cells of the prickle cell layer adhere to one another only at their desmosomes.

In an acute lesion, the affected skin will appear reddened, with a covering of crusted exudate, matted hair and droplets of recent exudate. The potential for complete regeneration from this grade of lesion is excellent, provided the cause can be identified and removed. If this type of reaction persists for weeks, the epidermis is stimulated to become acanthotic and parakeratotic and a thickened, fissured melanotic area of skin may result.

In the next degree of severity, exudation may lift the epidermis off the dermis, removing it from sustenance and, in the process, forming large bullae. Epidermal necrosis then leads to ulceration. The ulcerated areas are usually covered by an adherent crust of dried exudate. The potential for healing is again good if the process is reversed early. The epidermis regenerates from the edges of ulcers and the mouths of hair follicles.

The most severe type of dermatitis occurs when ulceration is accompanied by deep necrosis of the dermis and adnexae. The deep ulcers must heal by granulation (Fig. 11.14). Considerable scar formation and loss of hair follicles are the sequelae.

Folliculitis

Folliculitis is a pattern of inflammation originating within the follicles, and is often due to staphylococcal infection. In simple acute folliculitis, exudate and inflammatory cells accumulate within the follicle and distend the follicular infundibulum. This, together with perifollicular swelling, may produce an appearnce of small red lumps, or papules (Fig. 11.11).

Folliculitis increases dramatically in severity if the swollen follicle ruptures and releases its contents into the deep dermis. This event produces **furunculosis,** an angry, destructive, deep dermatitis. The violence of the reaction is due not only to bacteria and inflammatory products, but also to the release into the dermis of keratin, sebum and sweat. A foreign body reaction is provoked by these materials and the whole area becomes swollen, red and painful, and may eventually discharge at the skin surface. Widespread furunculosis is well exemplified when canine demodectic mange is complicated by secondary staphylococcal infection.

Cellulitis

Cellulitis is the term used to describe a deep acute inflammation that spreads underneath the skin in the panniculus layer. The underrun skin undergoes extensive necrosis and numerous sinus tracts may discharge at the surface. A good example of this process is the lesion that occurs in cats after bite wounds. Cat bites are very effective for deep inoculation of pathogens into the skin and cellulitis is a common sequel. After severe cellulitis, large deep ulcers can follow sloughing of necrotic skin,

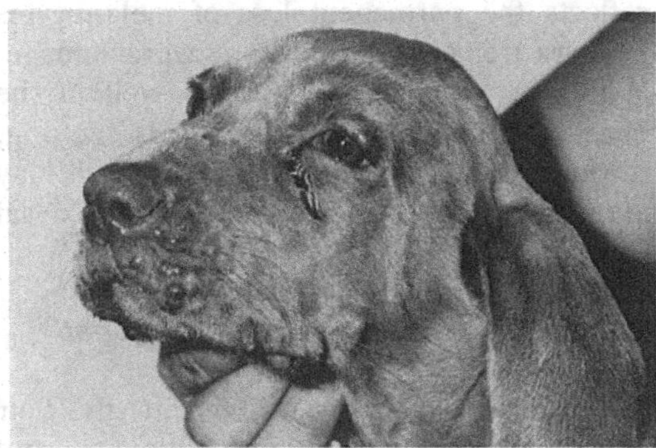

Fig. 11.11. The cluster of papules and pustules around the muzzle of this dog is the result of bacterial folliculitis.

and these can only heal by granulation and scar formation.

Panniculitis

This denotes a localized inflammatory reaction arising within the hypodermal fat layer (panniculus adiposis). Whatever the initiating cause, the reaction tends to assume a character determined by the breakdown of adipocytes and the release of their contained lipids. Constituents of these lipids are potent inducers of an inflammation characterized by focal suppuration with a concurrent intense macrophage response – a so-called pyogranulomatous reaction. This response frequently results in focal liquefactive necrosis of adipose tissue. As a result, panniculitis presents grossly as deep-seated confluent dermal nodules, often quite painful, which may rupture and discharge an oily exudate on the skin surface. In many cases, the cause of panniculitis cannot be determined, but it has been associated with deep bacterial infections, traumatic injury and foreign body penetration. In some instances, the disease appears to be immune-mediated (including autoimmunity) and, in dogs, certain breeds seem predisposed to develop spontaneous panniculitis of this general type. In the cat and dog, panniculitis has been associated with infection by various non-tuberculous mycobacterial species. These are ubiquitous soil organisms and are introduced by trauma or foreign body penetration.

Granulomatous inflammation

This reaction in the skin results in one or more **nodules** within the dermis, or hypodermis, which may ulcerate through the overlying epidermis to form a discharging sinus or a granulating surface. Such lesions can easily be confused with neoplasms, and they often originate as panniculitis or furunculosis, so there is considerable overlap between these categories. Depending on cause, nodules may be quite small (a few millimeters) to enormous, as in the case of some deep fungal infections. Various fungi, mycobacteria and actinomycetes are among the well-known infectious causes. In the horse, the nematode larvae of *Habronema* spp. are commonly involved.

Vasculitis

The term vasculitis denotes a situation in which an inflammatory process is centered on the walls of blood vessels, which consequently undergo varying degrees of damage and destruction. In the skin, the blood vessels involved are usually those of the superficial or middle dermal vascular plexus. As would be expected, the clinical manifestations of dermal vasculitis are multifocal hemorrhage and/or ulceration and, in the dog, the lesions are frequently concentrated over the extremities of the body (the lips, ear pinnae, feet and tail). The true nature of the disease process is only revealed by histologic examination and the recognition of lymphocytic or neutrophilic infiltration around and into the walls of damaged vessels. In many instances, a specific cause cannot be elucidated but it is generally considered that immune mechanisms, particularly type III hypersensitivity reactions, are responsible. In some cases, this has been shown to be the result of hypersensitivity to specific drugs or to staphylococcal products.

In porcine erysipelas severe widespread cutaneous vasculitis is the result of bacteremia with localization in cutaneous vessels.

The implications of cutaneous vasculitis depend upon the extent and severity of the lesions and upon the degree of involvement of internal organs, particularly the kidneys, heart and central nervous system.

Mineralization in the skin

Mineralization of the skin does not occur commonly, but two forms are seen in the dog. Focal areas of mineralization of dermal collagen are sometimes seen in canine hyperadrenocorticism (Cushing's-like syndrome) and appear as flat hard plaques about 20 millimeters in diameter. When present, they are a good diagnostic indicator but their pathogenesis is obscure. Likewise in dogs, deep dermal or hypodermal nodules of mineralization can occur in a lesion termed calcinosis circumscripta. Putty-like masses of calcium salts are surrounded (circumscribed) by a fibrous capsule containing macrophages. These lesions may be quite large and are often located close to limb joints. The pathogenesis is generally unknown, although some lesions appear to arise via the calcification of cystic dilated apocrine glands.

Dermal edema

Acute dermal edema or urticaria is usually seen after insect stings and sometimes in hypersensitivity. It may be precipitated by drugs, food additives, physical or thermal injury, or emotional stress. It is mediated by histamines causing a massive increase in vascular permeability. Edematous areas frequently take the form of lines or papules, known clinically as 'wheals' or 'hives', respectively. Severe urticaria, in which very large localized areas become affected, is sometimes referred to as angioneurotic edema. Urticarial lesions appear suddenly as areas of soft fluid swelling in the skin which may regress rapidly, or persist for weeks, developing and regressing over the body at different sites.

Neoplasia in the skin

Cutaneous neoplasms are common in all species and can arise from most of the epithelial and connective tissue elements; some cutaneous neoplasms can metastasize and kill the animal. Metastasis to the skin by malignant tumors of other organ systems is not common in animals, although malignant lymphoid neoplasms may, on occasion, preferentially invade the skin, having an apparent trophism for it. The neoplastic lymphoid cells may either concentrate within the dermis or seek out and infiltrate the epidermis and adnexal structures. The latter pattern is analogous to a disease in man called 'mycosis fungoides', a further example of an old and inappropriate term. The lesions appear as plaques and ulcers, which progress and fail to heal.

Some kinds of skin neoplasms are locally invasive only, but provide a problem by recurring locally after surgical removal. Cutaneous neoplasms can become troublesome by becoming ulcerated and infected, by growing to a large size, or by interfering with normal structures.

Most cutaneous neoplasms are appreciated much earlier in their development than those of internal organs and the majority develop as a firm nodule or plaque. However, some may be cystic and some may develop as spreading areas of ulceration, with little development of mass. This is a typical feature of squamous cell and basal cell carcinomas, and as mentioned above, the cutaneous lesions of lymphosarcoma also take the form of ulcers and ulcerated plaques rather than large nodules.

Pruritus

The sensation of itching is familiar to everyone, but probably few reflect that it is confined to the skin, anus, nose, vulva and conjunctiva. The possibility of itching in internal organs is awful to contemplate. Pathologic pruritus is a major pathogenetic factor in many skin diseases, because it promotes ongoing tissue damage inflicted by the animal itself (self-trauma). It is particularly important in cutaneous **hypersensitivity reactions** and many **inflammatory conditions**. The itch sensation is initiated in naked nerve endings in the dermis, transmitted via small unmyelinated nerve fibers to the spinal cord, traveling via the thalamus to the cerebral cortex. The sensations of itching and pain are closely related, but distinct. In the dog and cat the major pruritic stimulus appears to be triggered by proteases and leukotrienes released during inflammatory reactions from stimulated mast cells and inflammatory cells. In man and other species, mast cell histamines are involved. This is why pruritus is such a feature of hypersensitivity reactions.

The mechanical effect of scratching temporarily abolishes itching, but it quickly recurs, often intensified by the damaging effects of the self-trauma and is referred to as the itch/scratch cycle. Equally important is psychogenic itching; the animal perceives an itch where one does not apparently exist. There are several well-recognized conditions considered to have this basis in animals, of which anxiety is seemingly a predisposing factor. Therapy is directed to the central problem. Appreciation of the nature, cause and control of pruritus is of prime importance in clinical dermatology.

Etiologic and pathogenetic factors in skin disease

While the diseased skin is readily accessible for inspection, and cutaneous lesions can be appreciated from the earliest moments of their development, clinicians are frequently confronted by things they cannot explain. There has been enormous progress in dermatology over the last couple of decades, but the etiology and pathogenesis of numerous disorders remain an enigma. We can be encouraged, however, by a healthy and rapidly growing balance on the credit side of the ledger.

The skin is often damaged by agents operating directly from the external environment, but their ability to cause disease can depend upon a complex of predisposing factors which may be difficult or impossible to unravel.

What follows is an outline of the major etiopathogenetic factors in skin disease, organized into their various categories. More detailed information is accessible via the reading list (see Additional reading).

Allergy

Hypersensitivity is an extremely important mechanism in canine dermatology and is significant in the cat and horse as well. It involves an immune reaction in which the outcome is detrimental rather than beneficial to the host. The clinical expression of hypersensitivity that most concerns us in dermatology is allergic dermatitis, the trade mark of which is pruritus. The antigens involved may reach the skin by direct external exposure, or systemically via the bloodstream following inhalation, injection or ingestion. Biting insects and parasites of the skin, including fleas, lice and mites are of course a source of allergens. The antigens themselves may either be complete, or may act as haptens which bind to host tissue to create complete allergens.

In general, hypersensitivity reactions have been classified into a system which, although not totally satisfactory, is widely accepted (Fig. 11.12; and Chapter 2). Most applicable to veterinary dermatology are those reactions resulting in the clinical syndromes of allergy. The immediate, or type I reaction, is based on the reaction of potential allergens with IgE antibodies bound to mast cells, while the Arthus, or type III reaction, is initiated when allergens combine with circulating complement-fixing antibodies, the complexes forming in the walls of blood vessels. The delayed, or type IV reaction, is based on the sensitization of T lymphocytes by allergens and the subsequent amplification of the response by the release of lymphokines.

Type I reactions rapidly follow inhalation, ingestion or injection of the culpable allergen. **Allergic inhalant dermatitis** and **food allergy dermatitis** are important clinical entities in the dog and cat, respectively. In this context,

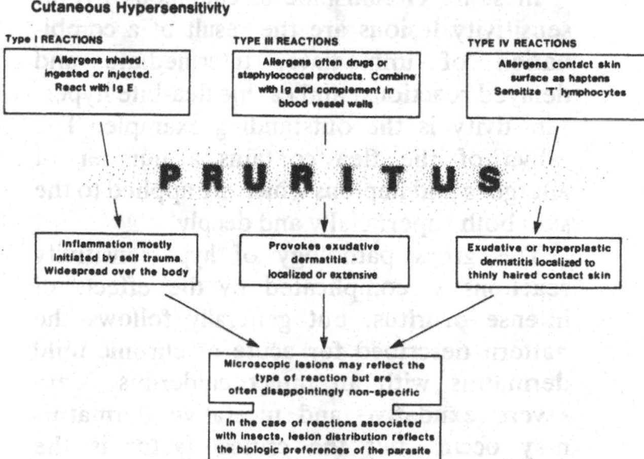

Fig. 11.12. Representation of pathologic mechanisms in cutaneous hypersensitivity reactions.

pollens are important inhaled allergens and certain fish proteins are important ingested allergens. As would be expected in a systemic disorder, the pruritic skin reaction is widespread but exacerbated locally by self-trauma. Dogs with allergic inhalant dermatitis, for example, are notorious 'foot lickers' and 'face-rubbers'. The cutaneous reaction may be accompanied by respiratory or gastrointestinal signs such as sneezing or diarrhea.

Type III reactions have been described following the administration of drugs and as a response to staphylococcal products. The lesions appear as exudative dermatitis, which may be localized or extensive but always pruritic.

Type IV reactions are produced in **contact allergic dermatitis**, when a chemical hapten directly contacts the skin surface, is absorbed through the epidermis, and combines with a tissue molecule to form the complete allergen. Well-known contact allergens are wool carpet dyes, topical drugs such as neomycin, plastics in flea collars and food dishes, and plant constituents. Lesions will occur on sparsely haired skin, in areas where contact with the offending allergen produces a pruritic and inflammatory response. Type IV reactions may also have a systemic basis, when administered drugs act as haptens and sensitize the skin in the manner described above.

In some circumstances, cutaneous hypersensitivity lesions are the result of a combination of immediate, intermediate and delayed reactions, and canine flea-bite hypersensitivity is the outstanding example. The saliva of the flea contains a number of allergens and haptens which are applied to the skin both superficially and deeply.

The gross pathology of hypersensitivity reactions is complicated by the effects of intense pruritus, but generally follows the pattern described for acute or chronic mild dermatitis with an intact epidermis. Very severe exudative and ulcerative dermatitis may occur, but the critical factor is the *distribution* of the affected areas of skin. Contact hypersensitivity, for instance, will occur only on fairly hairless areas accessible to the suspected allergen. Insect-bite hypersensitivity will occur at the favored feeding site of the particular insect.

Histologically, the lesions may contain a pattern of cell types consistent with the type of reaction, for example a predominance of dermal mast cells and eosinophils in type I reactions, and lymphocytes in type IV reactions. However, the histologic picture is often non-specific, with a mixed population of inflammatory and other cells.

Autoimmunity

This is an area which has received increasing attention in the last decade and diseases in this group are now diagnosed more frequently. For major diseases in this group, the common pathogenetic feature is damage to the epidermis caused by auto-antibodies directed against its constituents (Fig. 11.13). The damage is accompanied by an inflammatory reaction and the lesions usually have a distribution characteristic for the particular disease. Autoimmune skin diseases tend to be chronic and relapsing, and the clinical manifestation can be severe.

In the **pemphigus** group, the auto-antibodies are directed towards the intercellular substance binding the keratinocytes. The bound antibody may be demonstrated by immunodiagnostic techniques. The antibody–antigen reaction causes separation of these cells and the creation of intraepidermal vesicles or **bullae**, which rapidly rupture. (The term pemphigus is derived from the Greek for blister.) The breakdown of adhesion between epidermal keratinocytes is called **acantholysis** and is accompanied by an inflammatory reaction. In animals, because initial bullae are rapidly broken down, the lesions are characterized by crusting and shallow ulceration. The most common form is **pemphigus foliaceous**, while less common are **pemphigus vulgaris**, **pemphigus erythematosus** and **pemphigus vegetans**. They are differentiated on the grounds of clinical severity, distribution of lesions and the site of acantholysis within the epidermis. For instance, in pemphigus foliaceous, lesions occur typically mostly on the face, ears and feet and are characterized by acantholysis in the subcorneal layer of keratinocytes. In pemphigus vulgaris, the lesions tend to involve all mucocutaneous junctions and the oral cavity, and are characterized by acantholysis of keratinocytes immediately above the basal layer. Pemphigus vulgaris is generally a much more severe disease than pemphigus foliaceous. The various forms of pemphigus have been recognized most commonly in the dog and cat, and also in the horse and goat.

In bullous **pemphigoid**, auto-antibody is directed against the basement membrane zone of skin and mucosae, resulting in separation of the whole epidermis from the dermis in focal areas of the skin with similar lesions in the oral cavity. Initial bullae rapidly become crusted ulcers, mostly distributed in axillae and groin, at mucocutaneous junctions and in the mouth. Clinically, bullous pemphigoid closely resembles pemphigus vulgaris.

In **systemic lupus erythematosis** (SLE), the pathogenesis of cutaneous disease has a different basis, in that auto-antibodies are not directed primarily against epidermal components, but against nucleotides and nucleoproteins in general. Complexes of antibody and such nucleoantigens produce tissue damage in the manner of a type III hypersensitivity reaction, and the skin is but one organ system that can be involved. In the skin, this reaction occurs in the region of the epidermal basement membrane (Fig. 11.13). It is still unclear whether preformed circulating complexes diffuse into the tissue from blood vessels, or if circulating antibody leaves vessels to complex with antigen released

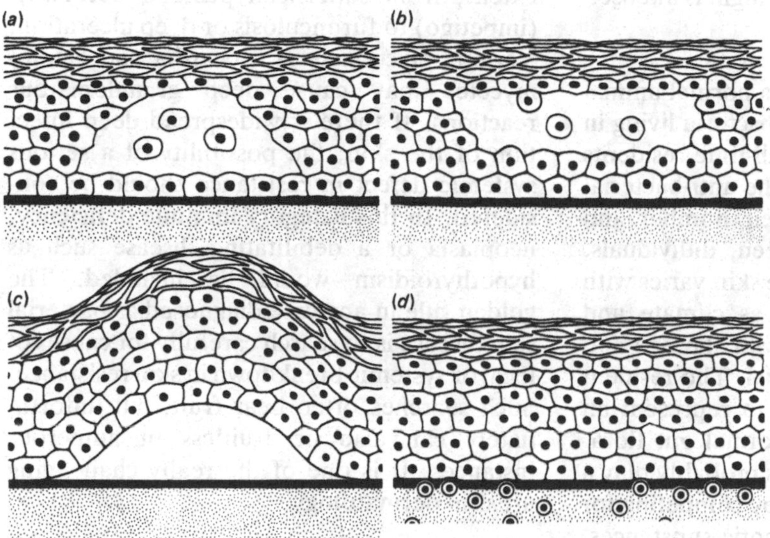

Fig. 11.13. Representation of the major sites of autoimmune attack on the epidermis in several different diseases. (*a*) In pemphigus vulgaris, acantholysis is initiated immediately above the basal layer of keratinocytes. (*b*) In pemphigus foliaceous acantholysis is initiated in the keratinocytes immediately below the stratum corneum. (*c*) In bullous pemphigoid there is separation of the epidermis from the basement membrane. (*d*) In lupus erythematosis, immune complexes (•) become localized in the basement membrane zone.

locally. It is known that injury induced by UV light exacerbates cutaneous SLE lesions, supporting the second alternative. The result of the immunopathologic process is inflammatory skin disease which may be multifocal or generalized and commonly involves face, ears and distal limbs. The inflammatory response may range from mild seborrheic dermatitis, through ulceration, to panniculitis, making the diagnosis of SLE difficult on clinical grounds. The diagnosis depends upon immunologic techniques, particularly the demonstration of circulating anti-nuclear antibodies. SLE is a rare disease and has been documented mainly in the dog and cat. In the dog, a benign variant occurs which is confined to the skin and is termed **discoid lupus erythematosis** (DLE). The typical lesion develops as a depigmenting, scaling, crusting dermatitis spreading over the nose. DLE is a reasonably common disease, especially in climates where exposure to sunlight is intense.

Bacterial infection

The skin and/or hair coat of a normal animal has a resident population of bacteria living in symbiotic harmony. These full-time residents are able to exist in spite of the anti-bacterial devices of the skin, but their number and diversity vary in, and between, individuals. The microenvironment of the skin varies with environmental factors such as climate and husbandry conditions, and with the age and nutritional status of the animal. The resident bacteria are not considered to represent an infection, but are akin to normal gut flora. Many of the normal residents probably play a role in inhibiting pathogens by competing for nutrients or producing antibiotic substances. In addition to the normal resident flora, many bacteria *transiently* populate the skin and potential pathogens amongst these, most notably strains of staphylococci, will seize any available opportunity to invade and destroy.

The normal skin is wonderfully resistant to bacterial infection, mainly due to attributes of the stratum corneum, notably its physical nature and its content of sebum derivatives, fatty acids and immunoglobulins.

In most cases bacteria are able to invade the skin because of a local breakdown in resistance, especially if the bacterial population is large at the time. Areas of skin that are continually moist, mechanically damaged, covered with matted hair, exudate or seborrheic scale are likely to become infected. Ovine footrot is a classical example of such a situation, when local resistance is degraded in the presence of a transient pathogen. Constant wetness and coldness allow bacterial infection of the interdigital skin, but true footrot will develop only if *Fusiformus nodosus* is present in the environment. The pathogenetic properties of this organism allow it to destroy the tissues of the hoof, causing separation, a most debilitating condition.

In general, cutaneous bacterial infections may induce lesions of varying severity and extent, from superficial pustular dermatitis (impetigo) to furunculosis or deep ulceration. Various mycobacteria, nocardia and actinomycetes may cause deep granulomatous reactions. If there is widespread deep infection of the skin, the possibility of a serious systemic defect in resistance should be considered. In this category immune deficiency, neoplasia or a debilitating disease such as hypothyroidism would be included. The golden rule in any case diagnosed as bacterial skin disease is to search carefully for causes of local or systemic breakdown in skin resistance. Such searches often bear fruit, but unfortunately can also be fruitless in numerous instances. It is one of the really challenging areas in dermatology.

Fungal infection

Like bacteria, fungi are ubiquitous in the environment and may reside on the skin and hair of animals, and have a similar relationship with it.

The fungal infections of the skin fall into two main groups, the dermatophytoses and the deep mycoses.

The **dermatophytoses** are caused by several fungal genera that invade and destroy keratin rather than living tissue. These infections frequently give rise to lesions known as 'ringworm', an old and inappropriate term that is unfortunately still in popular use. The dermatophytes tend to cause alopecic, scaly lesions, because of fracture of hair shafts and epidermal parakeratotic hyperkeratosis. Inflammatory changes are generally minimal but there is the occasional exception to this, when exudative dermatitis and even furunculosis can result. These more violent reactions are apparently triggered by soluble fungal products diffusing into the dermis or sometimes by secondary bacterial infection. The common dermatophytes are species of the genera *Microsporum* and *Trichophyton*. The reader is referred to Muller, Kirk and Scott (1983) for an excellent review (see Additional reading).

Deep mycoses are fungal infections of the living tissue and are frequently characterized by nodular or deeply ulcerative and granulating dermal lesions (Fig. 11.14). In most cases they result from a localized by-passing of skin defenses, for instance by the penetration of a foreign body. Numerous fungal species have been documented as causing these lesions, for example *Pithium* spp. and *Basidiobolus* spp., which commonly cause deep mycoses in horses and occasionally in dogs.

Viral infection

A number of viruses attack the skin, having a direct tropism for the epidermal keratinocytes. The poxvirus group has the skin as its major target organ, as do papillomaviruses and some papovaviruses.

Poxviruses infect the skin hematogenously and cause multiple papules and pustules which begin as focal intraepidermal necrosis. After the rupture of pustules, an elevated scab results from the vigorous proliferation of underlying basal cells, together with an inflammatory infiltrate. Common poxvirus diseases include contagious ecthyma of sheep ('scabby mouth') (Fig. 11.15), cowpox and pseudocowpox, fowlpox and swinepox.

The papilloma- and papovaviruses on the other hand stimulate tumor-like proliferation of keratinocytes, the lesions being referred to as infectious verrucae or warts. These lesions are common in cattle, horses and dogs and usually spontaneously regress, succumbing to an immune response by the host.

The multisystem viral infections that produce skin lesions include the **vesicular**

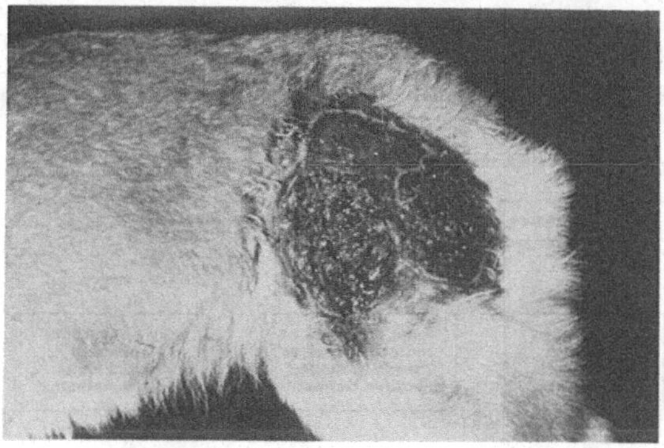

Fig. 11.14. An advanced deep mycosis (*Aspergillus* spp.) on the lateral thigh of a dog. Note the extensive area of deep ulceration, which first appeared as an area of a few square centimeters in extent and relentlessly progressed.

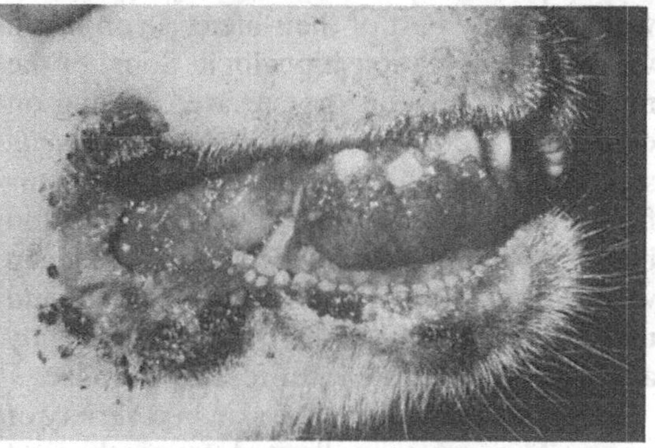

Fig. 11.15. Fully developed lesions of contagious ecthyma about the lips and gums of a lamb. The nodules result from the hyperplastic and inflammatory reactions provoked by the poxvirus infection.

diseases, such as foot and mouth disease, vesicular exanthema and vesicular stomatitis. The lesions are characterized by vesicle and bulla formation, leading to ulcerations. In bovine mucosal disease, and malignant catarrh, inflammatory skin lesions may also develop. Canine distemper virus infection occasionally results in marked hyperkeratosis of the footpads ('hard pad') and nose, but the mechanism of this change is not clear.

One interesting and unusual viral lesion of the skin occurs in ovine Border disease, now known to be caused by the virus of bovine mucosal disease. Lambs infected *in utero* may suffer the damaging effects of the virus on wool follicles and myelin-producing cells (oligodendroglia). Such lambs have a hairy birth coat and exhibit a severe tremor syndrome due to hypomyelinogenesis. As a result, they are often referred to as 'hairy shakers'.

Parasitic infestation

This small title encompasses a vast collection of agents of major importance in skin disease and belonging to the helminth, arthropod and protozoan orders. It would be reasonable to claim that the arthropods constitute the most important group and include the fleas, ticks, flies, lice and mites that can make life miserable across species.

The parasites of the skin may spend the whole or only part of their life cycle on it and vary in their relationship with it. Some of the mites live solely on the surface, feeding on epidermal debris, for example *Cheyletiella* spp. Others, like fleas, move about on the surface, but puncture it intermittently to feed on blood or lymph. Some helminths live deep within the skin. Biting flies, midges and mosquitos may live away from the skin, only alighting to puncture it for feeding purposes.

Parasites may injure the skin in a variety of ways. This may be limited to mechanical injury, irritation, and the provocation of an inflammatory reaction. However, often there is induction of a hypersensitivity reaction with all its consequences. The epitome of this situation is canine flea-bite hypersensitivity, the saliva of the flea being the major source of antigens.

Various parasitic dermatoses have a characteristic distribution pattern relating to the biology of the parasite, and these patterns are diagnostically very important. This fact re-emphasizes the importance of recognizing lesion distribution patterns in dermatologic diagnosis. Some of the parasites of the skin are very small and difficult to find, and, in a hypersensitive host, a relatively small number may provoke a large response. A good example of this is sarcoptic mange infestation in the dog. Multiple skin scrapings are often necessary to demonstrate its presence as the causative agent. However, knowledge that the parasite has a predilection for the elbows and pinnae of the ears should direct the clinician to concentrate on these areas.

Ultraviolet light and photosensitization

In some parts of the world, UV radiation is a significant factor in skin disease, acting directly, or indirectly, via photosensitizing agents (Fig. 11.16). The melanin of the skin has the key role in protection from UV light, especially where hair cover is sparse or absent. If epidermal melanin granules are not

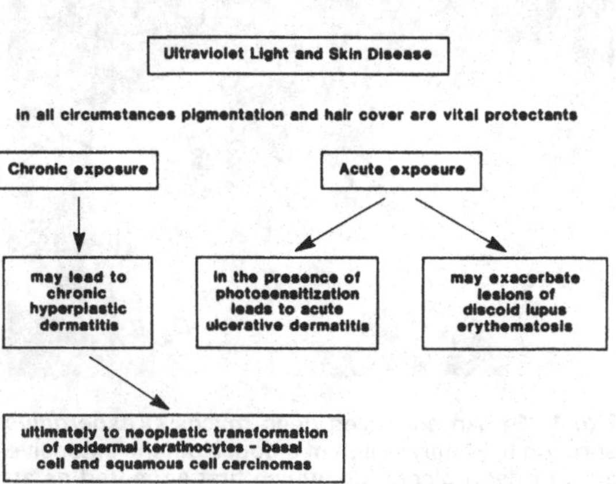

Fig. 11.16. The role of ultraviolet radiation in skin disease.

sufficiently numerous in sun-exposed skin, prolonged exposure leads first to tissue damage and then to a chronic superficial dermatitis. Eventually the chromatin of keratinocytes or melanocytes may suffer to the extent that neoplastic transformation occurs and malignant tumors may result. The most common are basal and squamous cell carcinomas, with melanoma being less common in animals than in Caucasian persons.

Photosensitization is a reaction brought about by the presence of photodynamic chemicals in the dermal tissues. These are agents that absorb UV light and emit infra-red and heat radiation. Such a reaction is potentially devastating for the skin. Photodynamic agents may be ingested directly as plant toxins or drugs, or may be produced as metabolites by the liver. The second mechanism is important in herbivores because chlorophyll is a precursor of a potent photosensitizer, phyloerythrin. An animal with a damaged liver, which is also ingesting large quantities of chlorophyll, is at great risk for such hepatogenous photosensitization when exposed to intense sunlight. The lesions are concentrated in poorly pigmented, thinly haired areas.

Photosensitization produces an acute, severe, deep dermatitis with extensive necrosis and sloughing, owing to the directly tissue toxic effect of the emitted energy.

Systemic intoxication and malnutrition states

Deficiencies or imbalances of vitamin A and zinc can cause dermatoses, principally by disturbing the process of keratinization and producing a hyperkeratotic response in the epidermis. In the case of fatty acid deficiency, quality of sebum is diminished and the quality of the hair degraded. Copper and protein deficiencies may affect pigmentation and hair quality, as has been previously described.

Chronic intoxication with thallium, arsenic and selenium can produce severe dermatoses with shedding of damaged hair and inflammatory changes. Although rare, these conditions illustrate how skin lesions can reflect a systemic, rather than a topical cause.

Trauma and contact irritants

These agents inflict direct damage to the tissues of the skin, inducing an inflammatory response. The severity of lesions will depend entirely on the character and intensity of the insult, as will the distribution of lesions. In such cases therefore, the damage inflicted on the skin can vary from mild to horrendous.

Developmental defects

Developmental defects do not occur with any great frequency in the skin but often produce a great deal of attention from veterinary dermatologists when they do. There may be varying degrees of failure of skin formation (epitheliogenesis imperfecta), fragile skin that tears easily (cutaneous asthenia), anomalies of hair growth (the color mutant alopecias), or grossly thickened abnormal skin (icthyosis). Details of the various entities may be found in pathology and dermatology texts. The cause of some of these conditions has been found to be genetic, while the cause of many remains unknown.

Behavior disturbances

Behavior disturbances are now well recognized as a cause of self-inflicted skin damage and as a mechanism through which skin damage caused by other etiologic agents is exacerbated. Injury is caused by excessive licking, sucking, biting or scratching at the skin. Both anxiety and boredom have been implicated in initiating self-trauma. Anxiety may occur when social behavior patterns are disturbed. Severing of attachments (human or animal) through death or separation, disruption of a stable family unit by a newcomer (another pet or a new baby) or invasion of the home range or territory by a competing individual such as the next door neighbour's cat, are all situations which will induce 'stress' in susceptible animals. Certain breeds of cat, such as the Burmese and Siamese appear to be prone to these disturbances. Boredom is a problem induced by the lack of sufficient interaction with other individuals and often is more of a problem in dogs, which are 'pack animals'. Boredom may result from physical isolation (being left at home all day), lack of work to do in those breeds of dog which have a genetic predisposition towards cooperative interaction with humans (such as Collies, Dobermans), or where the ability to interact is hampered by a lack of mobility, such as obesity or musculoskeletal diseases. The overweight, middle-aged Labrador dog with hip dysplasia is a prime candidate for developing a psychogenically mediated skin disease.

Initially, licking or biting may be an appropriate response to the discomfort of a preexisting skin lesion, for example, a superficial abrasion. Removal of the stimulus through treatment allows healing of the lesion. However, excessive grooming may be stimulated by scar formation and the behavior is reinforced through repetition. The animal orientates readily to this site of previous injury especially when stressed or insufficiently occupied and will resort to self-trauma.

The clinical expression of self-inflicted trauma varies with the individual and also the species. Anxiety and boredom in the cat most commonly causes an increase in grooming behavior. The excessive licking and chewing of hair results in partial patchy alopecia which may be associated with mild superficial nonspecific inflammation secondary to abrasion. This syndrome is often referred to as **neurogenic** or **psychogenic** alopecia or dermatitis. Because grooming follows a definite pattern in the cat, the distribution of the alopecia may appear bilaterally symmetrical. However, on close examination, the hair shafts are obviously broken off and 'tattered' and the hair is not easily epilated, as in endocrine dermatoses.

In contrast, the dog tends to self-traumatize focal areas of skin. These areas are usually easily accessible and continually visible, the most common site being the distal portion of the fore- and hindlimbs. The dog's biting and chewing initiates focal skin lesions, which may range from a superficial exudative dermatitis to the formation of a thick oval ulcerated plaque, often referred to as **acral lick nodule** or **'granuloma'**.

Before behavior disturbances are incriminated as the sole cause for the self-inflicted skin damage, underlying organic causes of pruritus and pain must be ruled out. These include allergy (especially flea allergy dermatitis in the cat), infection (bacterial, parasitic and fungal), and chronic skeletal diseases or neurologic disturbances causing paresthesia.

Endocrine disturbances

Several hormones have a potent metabolic influence on the skin which is revealed clinically when the endocrine system malfunctions. This has been previously mentioned and is discussed in Chapter 9.

The endocrine dermatoses are well documented in companion animals, but are rarely encountered and poorly documented in production animals. Endocrine dermatoses are usually non-inflammatory and nonpruritic, but in some cases secondary inflammatory changes may become superimposed. An excellent and detailed account is available in Additional reading (see Muller, Kirk and Scott, 1983).

The hormones most concerned are thyroxine, the glucocorticoids, estrogens and androgens. Recently, growth hormone abnormalities have been incriminated in cases of canine endocrine dermatosis. In the case of thyroxine and growth hormone, the skin reactions relate to a deficit in hormone activity, in the case of glucocorticoids to an excess in activity, while in the case of the sex hormones either may occur.

In general, endocrine imbalances affect hair growth, follicular and epidermal keratinization, sebaceous gland activity, epidermal melanocyte activity and the metabolism of the dermal connective tissue. Common clinical and histologic signs include symmetrical alopecia due to suppression of anagen, orthokeratotic epidermal hyperkeratosis, follicular hyperkeratosis, dilation and plugging of follicles and sebaceous gland atrophy. Epidermal melanosis is a reasonably common response.

The character of the dermis may be significantly altered as in hypothyroidism where it may become thickened by excess glycosaminoglycan ground substance in the connective tissue. In hyperadrenocorticism, the dermis becomes extremely thin due to atrophy of collagen, and develops focal mineralization (calcinosis cutis).

The symmetrical distribution of skin lesions in endocrine dermatosis is occasionally lacking in canine hypothyroidism, when lesions may be solitary or asymmetrically multifocal.

Additional reading

Ackerman, A. B. (1978). *Histologic Diagnosis of Inflammatory Skin Diseases*. Philadelphia, Lea and Febiger.

Ackerman, L. J. (1984). Canine nodular panniculitis. *Compend, Cont. Educ. Pract. vet.* **6**: 818–24.

Ackerman, L. J. (1985). Canine and feline pemphigus and pemphigoid. Part I, Pemphigus. *Compend. Cont. Educ. Pract. Vet.* **7**: 89–97.

Baker, B. B., Maibach, H. I. *et al.* (1973). Epidermal cell renewal in the dog. *Am. J. Vet. Res.* **34**: 93–4.

Bennett, D., Lauder, I. M., Kirkham, D., McQueen, A. (1980). Bullous autoimmune skin diseases in the dog. 2. Immunopathologic assessment. *Vet. Rec.* **106**: 523–5.

Fukushima, K. (1982). Pathogenesis of pemphigus vulgaris in dog and man – a review. *Can. Vet. J.* **23**: 135–7.

Halliwell, R. E. W. (1974). Pathogenesis and treatment of pruritus. *J. Am. Vet. Med. Assoc.* **164**: 793–6.

Halliwell, R. E. W. (1977). Dermatologic disease. In *Current Veterinary Therapy*, vol. VI, R. W. Kirk (ed.), pp. 493–579. Philadelphia, W. B. Saunders Co.

Halliwell, R. E. W. (1979). Skin diseases associated with autoimmunity. *Vet. Clin. North Am.* **9**: 57–71.

Halliwell, R. E. W. (1980). Dermatologic disease. In *Current Veterinary Therapy*, vol. VII, R. W. Kirk (ed.), pp. 432–503. Philadelphia, W. B. Saunders Co.

Ihrke, P. J. (1981). Canine pyoderma: diagnosis and management. *Proc. Am. Anim. Hosp. Assoc., 48th Ann. Meet.*: pp. 61–70.

Ihrke, P. J. (1983). Vitamin A-responsive dermatosis in the dog. *J. Am. Vet. Med. Assoc.* **182**: 687–90.

Kirk, R. W. (1983). Dermatologic diseases. In *Current Veterinary Therapy*, vol. VIII, pp. 454–529. Philadelphia, W. B. Saunders Co.

Manning, T. O., Scott, D. W., Smith, C. A. and Lewis, R. M. (1982). Pemphigus diseases in the feline: seven case reports and discussion. *J. Am. Anim. Hosp. Assoc.* **18**: 433–43.

Montagna, W. and Parakkal, P. F. (1974). *The Structure and Function of the Skin*, 3rd edn. New York, Academic Press.

Muller, G., Kirk, R. W. and Scott, D. W. (1983). *Small Animal Dermatology*, 3rd edn. Philadelphia, W. B. Saunders Co.

Nesbitt, G. H. (1978). Canine allergic inhalant dermatitis: a review of 230 cases. *J. Am. Vet. Med. Assoc.* **172**: 55–60.

Nesbitt, G. (1983). Canine and feline dermatology: a systemic approach. Philadelphia, Lea and Febiger.

Nesbitt, G. H. and Kedan, G. S. (1985). Differential diagnosis of feline pruritus. *Compend. Cont. Educ. Pract. Vet.* **7**: 163–72.

Nesbitt, G. H., Kedan, G. S. and Caciolo, P. (1985). Canine atopy. Part I. Etiology and diagnosis. *Compend. Cont. Educ. Pract. Vet.* **6**: 73–84.

Nesbitt, G. H. and Schmitz, J. A. (1977). Chronic bacterial dermatitis and otitis: a review of 195 cases. *J. Am. Anim. Hosp. Assoc.* **13**: 442–50.

Parker, W. M. (1981). Autoimmune skin diseases in the dog. *Can. Vet. J.* **22**: 302–4.

Reedy, L. (1984). Pruritus in small animals. *Compend. Cont. Educ. Pract. Vet.* **6**: 95–102.

Scott, D. W. (1980). Immunologic skin disorders in the dog and cat. *Vet. Clin. North Am.* **8**: 641–64.

Scott, D. W. (1981). Observations on canine atopy. *J. Am. Anim. Hosp. Assoc.* **17**: 91–100.

Scott, D. W. and Lewis, R. M. (1981). Pemphigus and pemphigoid in dog and man: comparative aspects. *J. Am. Acad. Dermatol.* **5**: 148–67.

Scott, D. W., MacDonald, J. M. and Schultz, R. D. (1978). Staphylococcal hypersensitivity in the dog. *J. Am. Anim. Hosp. Assoc.* **14**: 766–79.

Scott, D. W., Wolfe, M. J., Smith, C. A. and Lewis, R. M. (1980). The comparative pathology of non-viral bullous skin diseases in domestic animals. *Vet. Pathol.* **17**: 257–81.

Werner, L. L., Brown, K. A. and Halliwell, R. E. W. (1983). Diagnosis of autoimmune skin diseases in the dog: correlation between histopathologic, direct immunofluorescent and clinical findings. *Vet. Immunol. Immunopathol.* **5**: 47–64.

Yager, J. A. and Scott, D. W. (1985). The skin and appendages. In *Pathology of Domestic Animals*, 3rd edn, K. V. F. Jubb, P. C. Kennedy and N. Palmer (eds.), pp. 408–59. Orlando, FL, Academic Press.

Wayne F. Robinson, Robert S. Wyburn
and John Grandage

12 The skeletal system

The skeletal system, a remarkably diverse arrangement of bones and joints, combines strength with mobility, providing structural support for the body and, in conjunction with the nervous system and muscle, movement and locomotion.

Thick, encasing, and virtually indestructible, bones are fitted to protect vital organs, but appearances can be deceptive. Although the bone's shape and form remain after death, the features necessary for life are lost. It is the cells of bone that are required for its continued existence and its mutability during life.

The cells of bone also provide the mechanism for the other major function of bone, that of a mineral reservoir. The body's requirement for mineral is constant, but the dietary availability is not. During periods of deprivation, mineral can be extracted from bone. Because of the absolute requirement of the rest of the body for minerals, particularly calcium, bone is subservient to these needs. Bone therefore is structurally strong, but also dynamic. In normal adult bone these two needs are balanced by continual remodeling. There is, in the adult animal, little net loss or gain, but a continual turnover of the components of bone.

In growing animals there is an additional need for the net accumulation of bone. In this regard, any dietary deficiency of minerals or vitamins necessary for adequate bone growth will lead to a more rapid development of structural deformity. This chapter discusses the diseases that interfere with the strength and mobility of the skeletal system, and bone as a protector and mineral reservoir. But first, there is a brief review of the structure and function of normal bone.

Anatomic considerations

Individual bones come in a remarkable variety of shapes and sizes, but, taking this variation into account, each bone has a similar basic structure. The architecture of a typical bone can be appreciated by examining a longitudinal section of a bone such as the humerus. There are two arrangements of bone, cortical or compact bone of which the cortex and epiphyses are composed, and cancellous bone, which consists of fine trabeculae crossing the medullary cavity from cortex to cortex. Cancellous bone is particularly apparent in the extremities of long bones, and is arranged along lines of stress.

Each bone is invested with a tough external covering, **the periosteum**, of which the innermost cellular layer has the capacity to resorb or form bone when appropriately stimulated. A less readily observable layer, **the endosteum**, covers all internal bone surfaces. It also has the capacity to resorb or form bone. Each bone has a number of identifiable landmarks, and Fig. 12.1 reviews these.

Blood vessels to each bone supply two general areas. The periosteal arteries subdivide and supply the periosteum and outer cortex. The majority of the cortex and medullary cavity, in the case of larger bones, is supplied by the nutrient artery, which enters via the nutrient foramen.

The blood supply to growing bone deserves special mention. As blood vessels do not traverse the cartilage of the physis, the epiphysis receives its blood supply from the epiphyseal–metaphyseal arcade (Fig. 12.2). The importance of the vascular supply to both adult and growing bone will become apparent when fracture repair, bone necrosis and osteomyelitis are discussed.

Nerves usually follow the blood vessels supplying the periosteum and medullary cavity. There is no nerve supply to the cortex.

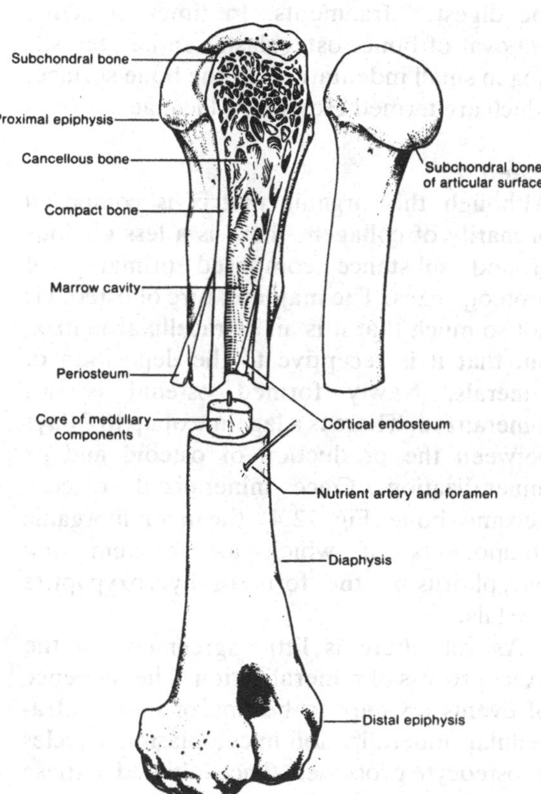

Fig. 12.1. The anatomical landmarks of a typical long bone. (Reproduced from Banks, 1981, by kind permission, see Additional reading.)

The tissue bone

There are three basic components of mature bone.

1 The cells that are involved in the formation of bone and the removal of bone.
2 The primarily collagenous framework (osteoid).
3 The mineral deposited on the osteoid.

The cells of bone

The cells involved in the production and removal of bone are sufficiently distinct in

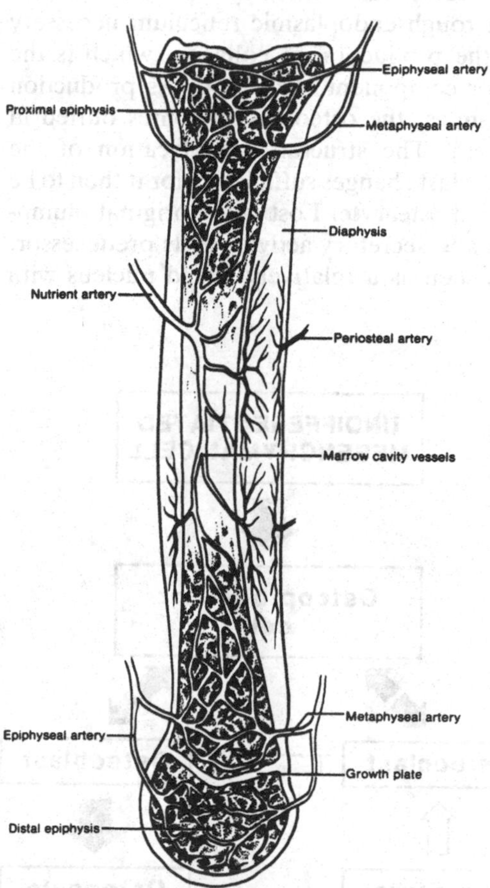

Fig. 12.2. Diagram of the typical blood supply to a long bone. Before growth plate closure, the epiphyseal and metaphyseal vasculature is separate. After closure the vessels anastomose. (Reproduced from Banks, 1981, by kind permission, see Additional reading.)

their mature state to be recognizably different, but there is controversy with regard to their origin and their capacity to revert to another cell type. The consensus is that osteoblasts arise from osteoprogenitor cells. The same consensus does not apply with respect to the suggestion that osteoblasts may arise from osteoclasts, or whether some osteoclasts arise from the macrophage–monocyte system (Fig. 12.3).

Osteoblasts are elongated, rather plump cells which cover most surfaces of bone. They produce the organic matrix of bone (osteoid), and govern the mineralization of the matrix (Fig. 12.4). As such, they are well endowed with rough endoplasmic reticulum necessary for the production of collagen, which is the major component of osteoid. As production continues, the osteoblast becomes buried in osteoid. The structure and location of the osteoblast changes sufficiently for it then to be termed **osteocyte**. Lost is the original plumpness and secretory activity of its predecessor. It is seen as a relatively naked nucleus with numerous cytoplasmic projections that extend through the bone, occupying the canaliculi.

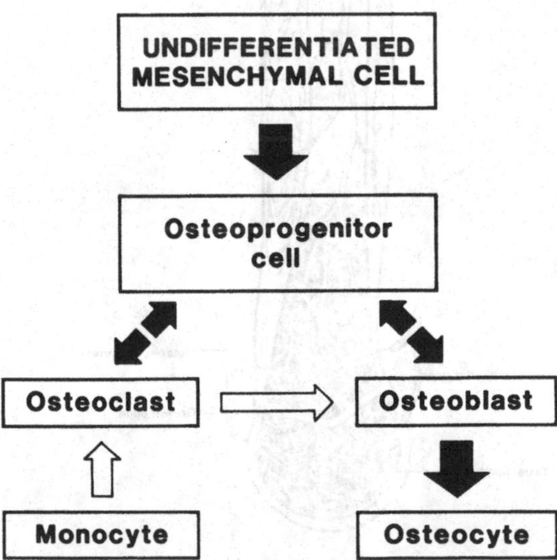

Fig. 12.3. The origin of the cells of bone. The solid arrows indicate confirmed relationships between cells. The open arrows indicate a less well-defined pathway.

Osteoclasts are geared for the destruction of bone. Each osteoclast is a large cell with abundant cytoplasm and many nuclei (Fig. 12.5). An osteoclast is most probably derived from a preosteoclast or from blood monocytes. The cytoplasm of osteoclasts is rich in lysosomes, which contain a variety of hydrolases. When activated, they lie in close proximity to an exposed bone surface. There are numerous short projections (the brush border) in intimate contact with bone. Osteoclasts will only remove mineralized osteoid (bone); they appear not to have the capacity to remove unmineralized osteoid. The process by which bone is removed involves both the initial secretion of enzymes to dissolve mineral and osteoid and the subsequent phagocytosis of the digested fragments. In times of active removal of bone, osteoclasts can be seen sitting in small indentations in the bone surface, which are termed Howship's lacunae.

Osteoid

Although this organic matrix is composed primarily of collagen, there is a less obvious ground substance composed primarily of proteoglycans. The major feature of osteoid is not so much that it is an extracellular matrix, but that it is receptive to the deposition of minerals. Newly formed osteoid is not mineralized. There is a lag time of up to 7 days between the production of osteoid and its mineralization. Once mineralized, osteoid becomes bone (Fig. 12.4), the major inorganic components of which are calcium and phosphorus in the form of hydroxypaptite crystals.

As yet, there is little agreement on the exact process of mineralization. The sequence of events appears to be one of initial intracellular mineralization in cytoplasmic vesicles of osteocyte processes. Once initiated in these vesicles, mineralization extends beyond the vesicles and into osteoid. Crystal growth then continues until all osteoid is mineralized. The central feature of this hypothesis is that initiation of mineralization is under osteoblastic/osteocytic control.

Types of bone

Two types of bone are recognized, woven or immature bone, and lamellar or mature bone.

In **woven bone**, the collagen fibers are arranged in many different planes. Osteocytes are arranged in the same pattern. In the adult animal, woven bone is found at the sites of initial fracture repair and it is the first bone that forms in the fetus.

Lamellar bone is distinguished from woven bone by its lamellar, or orderly arrangement of the collagen fibers. Osteocytes are fewer and appear to be more mature than in woven bone. There are two types of arrangement of lamellar bone. In the internal (endosteal) and the external (periosteal) limits of a bone, bone is deposited in a circumferential pattern, giving rise to the endosteal and periosteal lamellae. Between these limits, bone is arranged in concentric units, termed osteons or Haversian systems. These may be primary or secondary. Primary osteons are formed on the periosteal surface. Secondary osteons form in response to various stimuli and form in any tunnel within bone. The spaces between osteons are filled by interstitial bone (Fig. 12.6).

Bone growth

The growth of bones is the result of an interaction between a number of factors. Among the most important are the genetic potential of the animal, the nutritional status, and hormonal influences.

Little value can be ascribed to the genetic potential of the animal except that it is the limiting factor. Both the nutritional state of the animal and the hormones involved in bone growth will allow the animal to reach only the limit of its own genetic potential.

The morphology of bone growth

The differing forms individual bones assume from birth to maturity are quite variable, but all are the result of only two basic patterns. The first depends on a cartilage template (endochondral ossification), the other on a rather loose fibrous framework (intramembranous ossification). The majority of

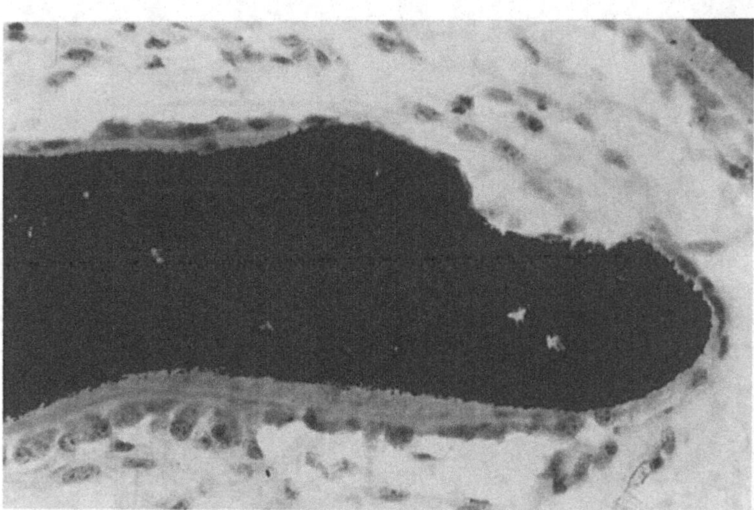

Fig. 12.4. Light micrograph of normal undemineralized trabecular bone. The black area is bone, the grey area surrounding the black area is osteoid and the cells lining the osteoid are osteoblasts. Stains: Von Kossa, hematoxylin.

bones grow in length and width by endochondral means, the minority (some parts of the skull and the mandible) by intramembranous means. What influences the growth of the latter is unknown.

With endochondral ossification, the shape and position of the cartilaginous growth plate determines the direction of growth. The rapidity of growth of each bone also varies. For example, the ulna has only one growth plate at the distal end, whereas the radius has two plates, one proximal and one distal. The ulnar growth plate elongates at twice the rate of each radial growth plate.

There are numerous descriptions of the process of endochondral ossification and the reader is encouraged to refer to texts listed in Additional reading and to Figs. 12.7 and 12.8. However, some features of endochondral ossification are pertinent to the study of skeletal disease. In normal animals, an orderly division and maturation of cartilage occurs and the ground substance of the growing cartilage must be mineralized before osteoblasts will lay down osteoid on the framework.

The factors influencing the remodeling of bone to the correct size and shape during growth are unknown. Suffice it to say that there is constant and active remodeling during growth by intense osteoclastic resorption and by osteogenesis. The metaphyseal area of long bones is wide compared to the diaphysis. In the metaphysis, lies the so-called 'cut back' zone, where there is periosteal resorption and endosteal deposition. In the diaphysis the opposite occurs, that is periosteal deposition and endosteal resorption (Fig. 12.8). There is also remolding or maturation of bones from an infantile shape to that seen at maturity.

Hormonal influences on bone growth

Growth hormone, thyroxin, insulin, sex hormones and corticosteroids influence bone growth. Growth hormone as well as having an influence generally on protein, carbohydrate and fat metabolism, also has a profound indirect effect on endochondral ossification. Growth hormone induces the formation and release of a group of peptides called sometomedins, which are synthesized in the liver, and stimulate endochondral cartilage growth. In appears, at least in man, that growth hormone is effective only from about four years of age onward. In growth hormone-

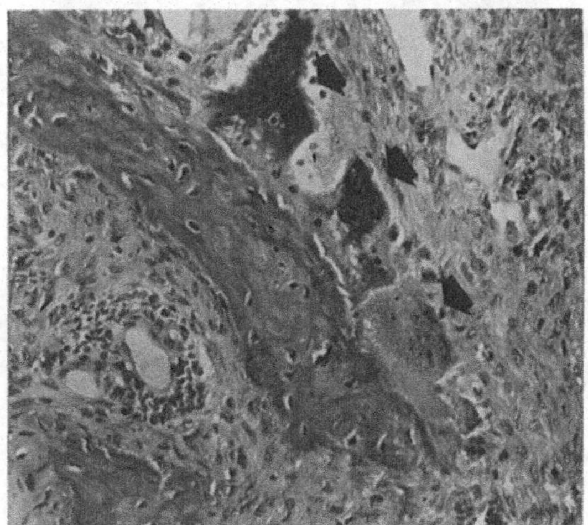

Fig. 12.5. Light micrograph of osteoclasts (arrows). The cells are large, multinucleate and have abundant cytoplasm.

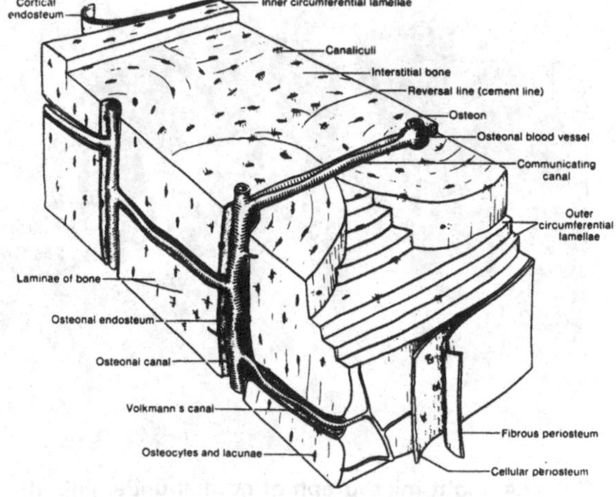

Fig. 12.6. The arrangement of diaphyseal lamellar bone. (Reproduced from Banks, 1981, by kind permission, see Additional reading.)

deficient humans, growth is normal up to that point. A similar pattern is seen in German Shepherd dwarfs. The pups are indistinguishable from their littermates up to two months of age.

Thyroxin also has a marked influence on bone growth, particularly on maturation. Insulin and androgens have a stimulatory effect on bone growth, whereas corticosteroids are inhibitory.

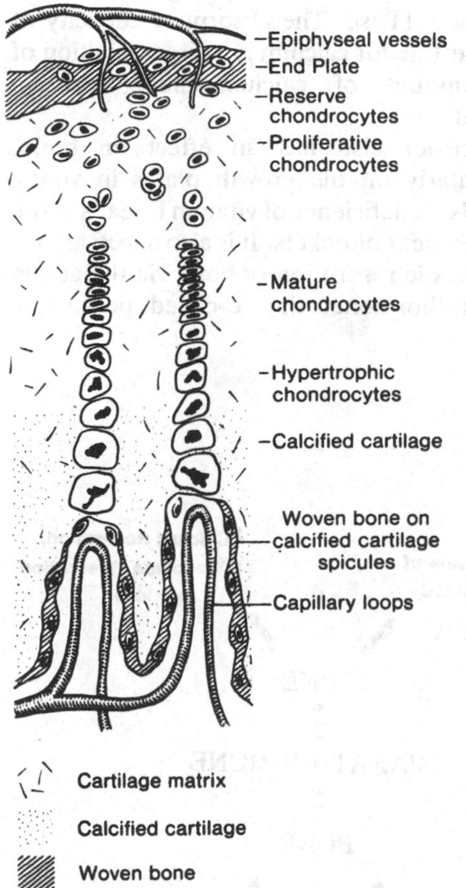

Fig. 12.7. The appearance of the cartilagenous growth plate. Endochondral ossification begins with the proliferation of chondrocytes on the epiphyseal side and concludes with the invasion along the calcified cartilage spicules by capillary loops and osteoblasts. (Reproduced from Banks, 1981, by kind permission, see Additional reading.)

Nutrition and bone growth

A diet that lacks adequate protein or calories will result in poor general growth, including bone growth. The mechanism of attenuation of bone growth has been partially elucidated in rats. It appears that fasted, immature, hypophysectomized rats do not respond to the administration of exogenous growth hormone. Somatomedin levels remain low. However, on refeeding, somatomedin levels are quickly elevated to normal. So it seems that the poor skeletal growth seen in protein/calorie deficiency is partly the result of a shutdown of somatomedin synthesis or release. It may also be due to release of somatomedin inhibitors.

Vitamins and minerals also influence bone growth and will be referred to in the discussion of bone atrophy, below.

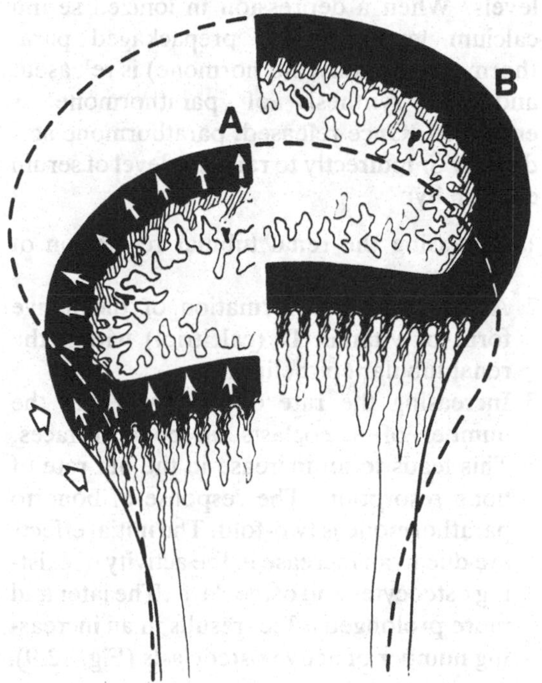

Fig. 12.8. Remodeling of the epiphyses of long bones during growth. The broad open arrows indicate the 'cut back' zone, where there is periosteal resorption of bone. Interstitial growth of the growth plate and articular cartilage (white arrows) moves them from A to B. (Reproduced from Banks, 1981, by kind permission, see Additional reading.)

Bone and calcium metabolism

To the detriment of bone, the rest of the body has first call on the available calcium reserves in the body. Calcium is required for a remarkable variety of metabolic activities. It is essential for cardiac, skeletal and smooth muscle contraction, neuromuscular transmission, hemostatis, and a number of other enzymatic processes.

Consequently, the serum calcium level is maintained within strict limits through an elaborate control mechanism. Bone is an essential part of that mechanism. The major influencing factors on bone are parathormone, vitamin D and calcitonin.

Parathormone

The cells of the parathyroid gland are the major sensors of variation in serum calcium levels. When a depression in ionized serum calcium levels occurs, prepackaged parathormone (parathyroid hormone) is released, and the synthesis of parathormone is enhanced. Once released, parathormone acts directly or indirectly to raise the level of serum calcium by:

1 Increasing the renal tubular resorption of calcium.
2 Accelerating the formation of the active form of vitamin D (calcitriol) within the renal tubular epithelium.
3 Increasing the rate of osteolysis and the number of osteoclasts on bone surfaces. This leads to an increase in the net rate of bone resorption. The response of bone to parathormone is two-fold. The initial effects are due to an increase in the activity of existing osteocytes and osteoclasts. The later and more prolonged effect results in an increasing number of active osteoclasts (Fig. 12.9).

Vitamin D

Vitamin D is now considered to be a hormone and the generic name for the most active form 1,25-dihydroxycholecalciferol is now calcitriol. The precursor of calcitriol is either dietary in origin in the form of cholecalciferol, or is synthesized in the epidermis from substances like 7-dehydrocholesterol. Two processes are then required. Cholecalciferol is first converted to 25-hydroxycholecalciferol in the liver and then to calcitriol in the kidney (Fig. 12.10). This latter process is facilitated by parathormone.

The biologic effects of calcitriol are primarily to enhance the absorption of calcium and phosphorus from the intestine. As a steroid hormone, calcitriol induces the formation of two major proteins by the intestinal epithelial cells: calcium-binding protein, and calcium ATPase. The absorptive capacity of the intestine for calcium is a direct function of the amount of calcium-binding protein present.

Calcitriol also has an effect on bone, particularly on the growth plates in young animals. A deficiency of vitamin D leads to the development of rickets. It is also necessary for osteoclastic resorption of bone via the action of parathormone, the so-called permissive effect.

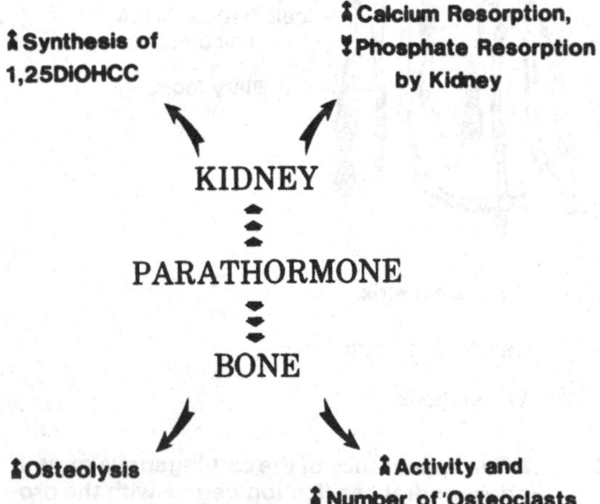

Fig. 12.9. Parathormone exerts multiple effects on kidney and bone. 1,25DiOHCC, 1,25-dihydroxycholecalciferol.

Calcitonin

A peptide hormone secreted by the parafollicular cells of the thyroid gland, calcitonin has effects that run counter to those exhibited by parathormone. Calcitonin depresses serum calcium and phosphorus levels and is released in response to an elevation in serum calcium level. The hypocalcemic effects of this hormone are due to an inhibition of the effects of parathormone on bone resorption. Calcitonin also has an inhibitory effect on osteoclastic activity.

Experimentally, calcitonin induces decreased renal tubular resorption of phosphate and calcium as well as sodium and chloride ions. However, these effects in normal animals are probably minimal, the major renal regulator being parathormone.

From the preceding discussion, it can be seen that there are a number of osteoid mineralization processes that may be interrupted. Deficiencies in dietary calcium, uncontrolled retention of phosphorus through chronic renal failure, or a deficiency of calcitriol precursors all result in defects in mineralization and increased bone resorption.

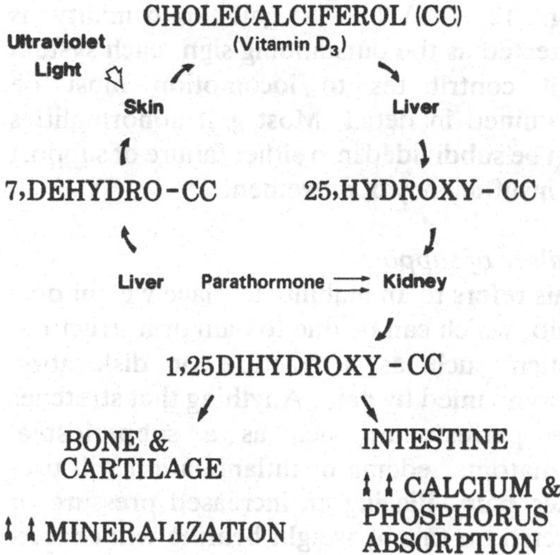

Fig. 12.10. Diagram of the steps involved in the production of 1,25-dihydroxycholecalciferol (calcitriol) and its metabolic effects. Production commences with the liver synthesizing 7-dehydrocholecalciferol.

Stress and bone

Much is known about the importance of hormonal effects on bone, but equally as important with respect to bone remodeling is the everyday stress applied to it. The state of bone in the normal abimal relies on gravity and the muscle tension applied to it. The effect of stress on bone is probably a more common reason for lameness than most other diseases combined.

It is pertinent to introduce a number of terms. **Stress** refers to the force exerted, and **strain** to the effects of the stress. There may be both **tensile** (stretching) and **compressive** strain. Collagen fibers impart tensile strength to bone, whereas the mineral imparts resistance to compression.

The application of stress to bone is encompassed in Wolff's Law which states, 'Every change in the form and function of bones, or their function alone is followed by certain definitive changes in their configuration in accordance with mathematical laws'. The use of Wolff's Law is encompassed in the Flexure–Drift hypothesis. When a flexural stress is applied to a bone without fracturing it, there are two outcomes. A compressive strain occurs along the concave surface and a tensile strain on the convex surface. If the stress is maintained the bone 'drifts' toward the concave surface by an increase in osteoblastic activity on the concave surface and an increased osteoclastic activity on the convex surface. Bone is therefore deposited on the surface where the stress is applied, and removed from the surface where the stress is weakened.

Clinical features of skeletal disease

The clinical appearance of skeletal disease will vary according to the anatomic location of the affected area. For example, an animal may have difficulty with prehension, mastication, defecation or breathing. These may be the result of abnormality of the premaxillae, mandibles, pelvis or ribs, respectively. There may also be neurologic abnormalities if the vertebrae or skull are involved. But the majority of cases of clinical disease involve abnormalities of locomotion or movement, and in most cases the appearance of the deficit is due to pain. The exhibition of pain varies in intensity from slight lameness, observable only during or after prolonged exercise, to a severe lameness, where the animal will not bear weight on the affected limb under any circumstances. In bone disease the origin of the pain can only arise from the periosteum, as it is only the periosteum that contains pain receptors. There seems little doubt that there are no pain receptors in cortical bone and the medullary cavity.

To cause pain, the periosteum must be either stretched or compressed. The intensity of pain depends to some extent on the rapidity of onset of periosteal compression or stretching. A slowly developing periosteal stretch through new bone formation does not appear to be painful, whereas a rapidly developing subperiosteal hematoma is painful. Periosteal lifting in neoplasia or osteomyelitis may or

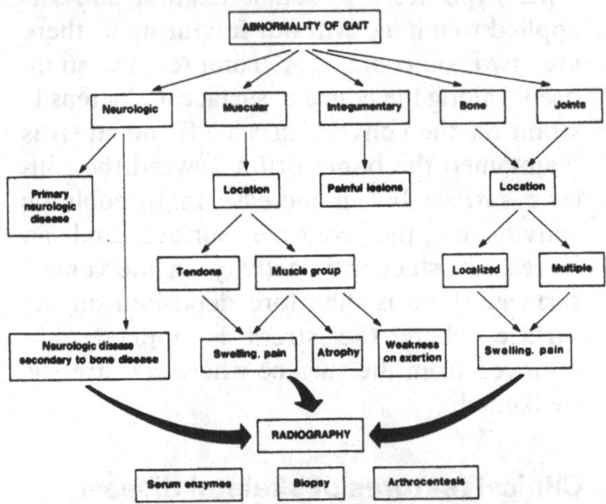

Fig. 12.11. Flow chart emphasizing the organ systems that may be involved in an abnormality of gait. Included also is the general approach used to delineate a problem and the relative importance of radiography to include or discount skeletal disease.

may not be painful. Pain associated with bone disease is not just limited to the appendicular skeleton as mentioned before. Localized pain will arise from any bone when there is periosteal involvement.

There are also abundant pain receptors in the joint capsule which extends to just below the synovium. Damage to articular surfaces can be severe without being accompanied by pain, but when the joint capsule is stretched, pain can be marked.

Disturbances of locomotion or movement may not be manifested as pain only, but may be seen as an abnormality of form or a restriction of movement.

Recognition of locomotor involvement

The clinical signs which suggest disease associated with locomotion revolve around two major alterations from the normal state: abnormality of gait and abnormality of form. They may coexist or occur separately.

Abnormality of gait

Normal locomotion is a complex integration of a number of systems and includes the nervous system, muscle, bones, joints and integument. A method for localizing the lesion is shown in Fig. 12.11. When a gait abnormality is detected as the outstanding sign, each system that contributes to locomotion must be examined in detail. Most gait abnormalities can be subdivided into either failure of support or insufficiency of movement.

Failure of support

This refers to an inability to place weight on a limb, which can be due to pain or a structural deficit such as a fracture or dislocation accompanied by pain. Anything that stretches the periosteum, such as a subperiosteal hematoma, edema or inflammation will produce pain. Placing an increased pressure or stretch on this by weight bearing will exacerbate pain. Similarly, stretching of the joint capsule by an increase in intra-articular pressure or by fluid accumulation within the capsule will exacerbate pain. Both events result in failure of support.

Insufficiency of movement
Textbooks have been written on the differential diagnosis of lameness in all its various forms, and the reader is encouraged to consult these texts at the appropriate time. However, the generality remains that locomotor disease is often manifest by a restriction in the movement of a limb. Once again, the restriction may be due to pain, particularly with joints. It also may be a physical restriction of movement, unaccompanied by pain, such as with a fibrosed or ankylosed joint.

Abnormality of form
Everyday contact with animals results in the building of a mental picture of the 'normal' appearance of an animal. The variation in normal external contour from animal to animal may be narrow or wide, depending on the species, and the limit of 'normal' may be somewhat arbitrary. What is evident is a change from a previously recognized pattern. This may be subtle and only detectable on palpation, or it may be obvious on first glance.

The abnormality of form may be perceived as a swelling, the deviation of a limb or an unusual stance. The abnormality may or may not be painful. Animals with congenital bone disease may exhibit local or generalized alterations in the proportions of the body, or the absence for example, of limbs or digits. The presence of these does not necessarily mean that bone disease is present, but it requires an investigation of the skeletal system as part of the examination.

Acquired swelling may or may not be accompanied by pain and may feel hot on palpation. Acute infectious arthritis is typically hot, swollen and painful, whereas chronic periarticular fibrosis is not.

Combinations
Abnormalities of gait and form are not always mutually exclusive but are frequently a combination of the two. For example, an osteosarcoma of the distal radius may present as a swelling (abnormality of form) and the animal may not bear weight on the limb (abnormality of gait, failure of support). Suffice it to say that the two are recognized.

Localization of the lesion
Most skeletal lesions present clinically as localized changes. However, in some cases it may be part of a generalized process. For example, a kitten with nutritional secondary hyperparathyroidism will often present with a unilateral hindlimb lameness, probably due to a folding fracture of the femur. The pathologic femoral fracture is a consequence of generalized fibrous osteodystrophy.

Localization of the lesion allows the formulation of a number of diagnostic possibilities. Many types of disease have predilection sites, such as a mandible in bovine actinomycosis, all metaphyses in canine metaphyseal osteopathy, and the thoracolumbar vertebrae in spondylosis of bulls.

Characterization of the lesion
Once the abnormality has been anatomically localized, ideally the next step is to define the underlying structure of the lesion followed by the elucidation of the etiology, or a prediction of the etiology. The delineation of the morphology has a number of advantages, particularly with respect to treatment and prognosis.

The reader will notice that a central feature of Fig. 12.11 is the use of radiography to assist in the differentiation of skeletal lesions. Exquisite detail in the form of normal and abnormal bone can be obtained by radiography. No adequate investigation of bone disease can proceed without this diagnostic aid. However, it must be remembered that, in many bone diseases, there is a lag time between the onset of the disease and the radiologic appearance of the disease. The use of radiographic patterns to predict the morphology of the lesion follows.

Other diagnostic aids include biopsy of bone lesions and arthrocentesis if a joint is involved. Culture of joint fluid or material from a lesion in bone may also be indicated to confirm or deny the presence of bacteria or other types of microorganisms.

Radiographic patterns in bone disease

The image seen on a radiograph is formed by shadows cast in an X-ray by the structures through which that beam has passed. The degree of contrast any particular structure has from any other depends on differences in thickness, density and atomic number. Because bone is dense and contains high levels of calcium and phosphorus, which have high atomic numbers, it is clearly demonstrated on a properly exposed and processed radiograph. This makes it an ideal tissue for radiographic examination and most pathologic changes affecting bone are reflected in radiologic changes.

The detection of disease implies the recognition of variations from normal. This requires a sound knowledge of the normal and the use of radiography is no exception to this. **The single most important factor influencing the ability to interpret radiographs is a familiarity with normal patterns.** Therefore, before attempting to look for radiologic signs of disease, it is necessary to know the anatomy of the region being examined and to understand how images are formed on a radiograph.

Normal bone

It is beyond the scope of this text to give a description of the normal radiographic appearance of each bone in each species. However, there are many features common to most bones which can be seen on a radiograph. The femur of a medium-sized, adult dog is shown in Fig. 12.12. Because a radiograph compresses three dimensions into two, the hollow cylinder formed by the cortex appears as two bands of dense bone where the X-ray beam has passed through the walls of the cylinder at a tangent. That part of the cortex which is either close, or at right angles, to the X-ray beam is not clearly depicted. This is one of a number of reasons for examining at least two radiographs in planes 90° apart. On the radiograph, the trabeculation visible between the cortices is most pronounced toward the epiphysis and gradually decreases towards the midpoint of the diaphysis. In the diaphyseal region, the outline of the cortex is smooth because there are no muscle, ligament, tendon or joint capsule attachments. Where there are attachments, such as over the greater trochanter, the cortical outline is roughened. A nutrient foramen seen on the radiographs of most bones depends, for its appearance, on its plane relative to the plane of the X-ray beam. Seen 'end on' it appears as a circular defect in the cortex and side on as a channel running through the cortex.

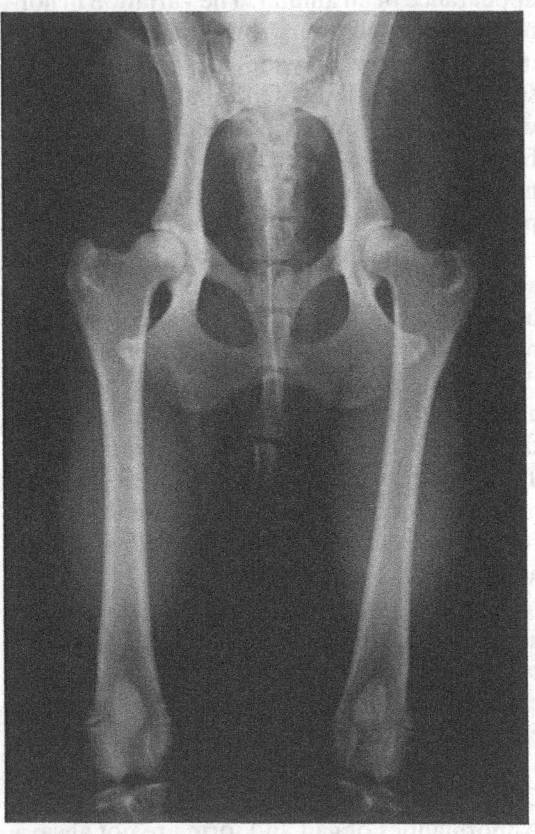

Fig. 12.12. Radiograph of a femur from a normal mature dog.

A radiograph of a femur of young dog with active growth (physeal) plates is illustrated in Fig. 12.13. As cartilage does not have the high mineral content of bone, it has approximately the same opacity to X-rays as most soft tissue. The physeal plate therefore appears as a dark line between the epiphysis and the metaphysis. It may be an almost flat disk of cartilage, as at the proximal end of the femur, or have a much more complex curved shape at the distal end. The physeal cartilage is widest in the younger animals and narrows as adulthood is approached. On the epiphyseal side there is a narrow, dense region of bone which is thickest at its center. This is often retained in adult life, where it is referred to as the epiphyseal scar. Immediately on the metaphyseal side there is a region where there is no cortex and the outline may appear rather rough and irregular. This is the zone (the cut-back zone) where rapid remodeling of bone occurs during growth to reduce the diameter of the bone to its diaphyseal dimensions. The physis between the greater trochanter and the femoral diaphysis has a different appearance because at this site the mechanical force applied is distraction, as opposed to compression at most other physes. The border between the bone and cartilage at this physis is irregular and there is an area of dense bone on both sides of the physeal plate. Similar physes are found at other sites where a distraction force is applied, such as the olecranon and tibial tuberosity.

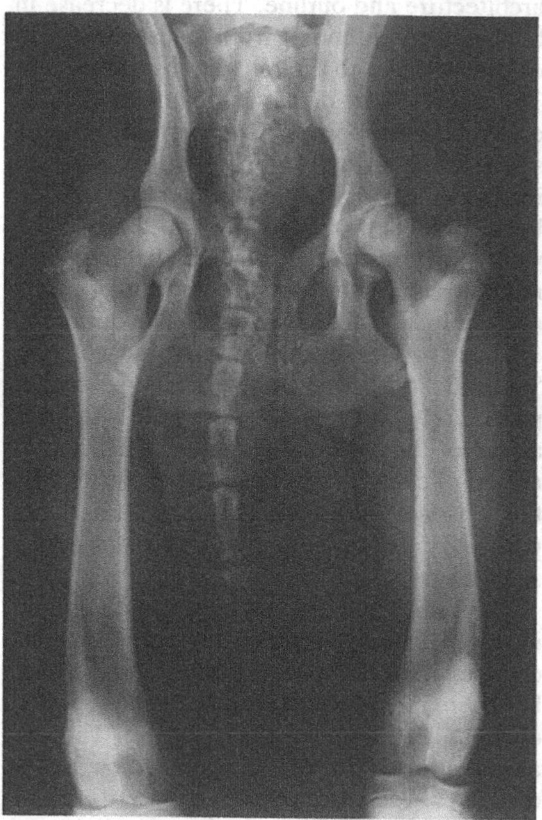

Fig. 12.13. Radiograph of a femur from a normal young dog. Note the characteristics of the proximal and distal femoral growth plates.

Radiographic diagnosis of bone disease

The aim of this section is to outline an approach to radiologic diagnosis illustrated by examples. Detailed information about specific diseases can be found in a number of definitive texts on veterinary radiology.

The radiographic image of a bone can change in only four basic ways. It can change in **shape, architecture, density** and **outline. Every disease of bone that can be detected radiologically will manifest itself by one or a combination of these basic changes.**

Reading a radiograph of bone involves seeking answers to a series of questions. The first and most important question is, are the bones normal or abnormal? As previously stated, the only way that this question can be answered is to be familiar with the normal appearance. If there is an abnormality, the next problem is whether the abnormality is confined to a single site or is apparent at many sites. To resolve this, it may be necessary to reexamine the patient or to look at additional radiographs. Defining the abnormality in terms of the four basic changes and finally more detailed characteristics within these should follow.

There is an almost infinite variety of ways in which the **shape of bone** can change, but an attempt can be made to categorize these. An alteration in shape may involve the entire bone or may be confined to a specific region. When an entire bone appears to have changed shape it should be determined whether the abnormality is caused by a change in the epiphysis or diaphysis. An increase or decrease in the size of the epiphysis relative to the diaphysis or a change in the shape of the epiphysis will alter the shape of the bone. Abnormally long or short diaphyses or bowing and twisting will also cause changes to the overall shape of the bone.

Change in the **density of bone** is related to alteration in the amount of mineral present. A decrease in the mineral content will result in decreased density of bone and this can be generalized, confined to one limb, to one bone, or to a localized area of a bone. Mineral may have been removed from bone by a variety of mechanisms, such as endocrine disorders, infection, destruction by a tumor, necrosis or disuse. Endocrine disorders result in a generalized demineralization, whereas, in disuse of a limb, only the bone in the disused limb will become demineralized. The pattern of demineralization will frequently give an indication of the cause. An increase in bone density is encountered much less commonly but can also occur as a generalized or localized condition. It can be the result of a number of pathologic processes. A generalized increase in bone density is seen as a congenital abnormality in a number of species. A localized increase in density can be caused by bone-forming tumors, some particular types of infection, particularly fungal infections, abnormal increased stress applied to a particular region, and by bone formation around an infected area.

Alterations to the normal architecture or structure of a bone can take many forms and are nearly always confined to localized areas. The most common and easily recognized change in architecture is seen when a bone is fractured, and, if correctly treated, goes through various healing stages till it gradually returns to its normal form. Architecture may be altered by infection, tumors or necrosis. These processes can cause destruction of an anatomical area such as the cortex or the trabecular pattern, or both.

A change in the **outline of a bone** will occur if there is any insult to the periosteum. The periosteum will form new bone if it has been disturbed from its normal position. The distribution and pattern of this new bone formation can frequently signal the underlying cause. There are a number of causes for the periosteum being moved from its normal intimate contact with the bone surface. The most common are trauma, avulsion of a ligament or muscle attachment, tumors and infection.

Examples of radiographic patterns in bone disease

The humerus of a dog illustrated in Fig. 12.14 shows a localized change in bone density, architecture and outline. There is decrease in bone density extending from the proximal end of the humerus to the mid diaphyseal region. The decrease in density is patchy, with no easily defined border. There are also some small localized and poorly defined areas of increased density. The architecture of the bone has been altered in that there is no visible cortex over the proximal diaphysis and metaphysis and the normal definition of cortex and medulla is no longer clear. The outline of the bone has also been changed by an irregular 'pallisade' type of periosteal reaction, which is particularly noticeable over the anterior aspect of the proximal diaphysis. This combination of changes and the patterns formed by the changes indicates that the underlying pathology is most likely to be an osteosarcoma.

The radius and ulna of a dog are illustrated in Fig. 12.15. In this plate there are also changes in density, architecture and outline, but there is a different pattern of change. The changes are limited to the diaphysis. There is an area of decreased density which has a distinct border, accompanied by an area of increased density. There is also a small area of increased density within the less dense zone. The architecture of the bone is changed by loss of part of the cortex and obliteration of part of the medullary canal. The outline of the bone has been altered by a periosteal reaction which has a rather indistinct outline. This pattern of combination of changes indicates that the underlying pathology is probably an osteomyelitis.

The radiograph of the pelvis and hip joint illustrated in Fig. 12.16 shows one normal and one abnormal femoral head. The abnormal femoral head shows changes in shape and density. The shape has changed from a regular semicircular shape to one that is flat and irregular and there are also irregularly shaped areas of decreased density. This pattern of changes indicates necrosis and remodeling of the femoral head. The lesion is long standing.

Radiographic patterns in joint disease

In a strict sense, the radiographic appearance of joint disease should be considered later in the chapter. However, because joint disease often has to be differentiated clinically from bone disease, it is placed here.

Synovial joints, composed of the articular cartilage, joint capsule and ligaments, have the same radiodensity as other soft tissues. They are therefore not readily seen on a radiograph. What are visible are the bones which underlie the cartilage. A joint appears as a radiolucent zone between two bones. This zone has often been wrongly referred to as the joint space, when in fact the true joint space is filled with synovial fluid and is extremely narrow. The apparent gap between the two

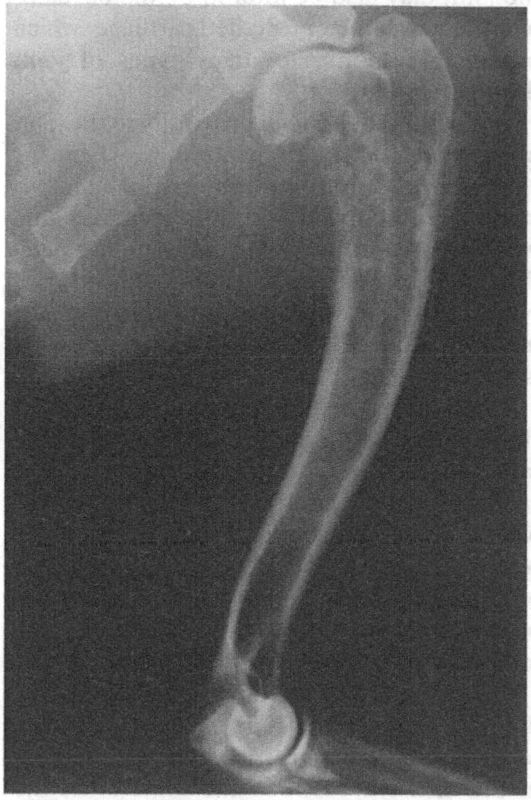

Fig. 12.14. Radiograph of the humerus of a dog showing localized changes in bone density, architecture and outline in the proximal humerus. The lesion is an osteosarcoma. For an explanation of the changes present, see the text.

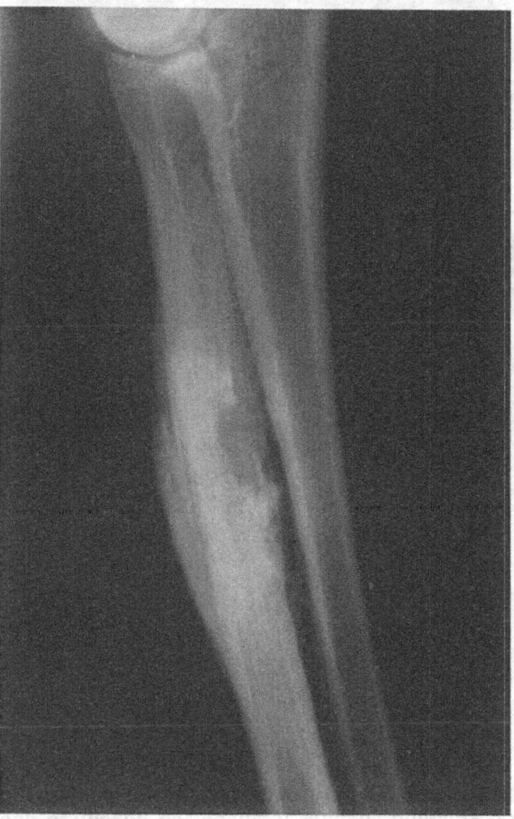

Fig. 12.15. Radiograph of the radius and ulna of a dog showing localized changes in bone density, architecture and outline. The lesion is an osteomyelitis.

bones is composed almost entirely of articular cartilage.

The bone immediately underlying the cartilage is dense and is referred to as the subchondral bone. It is this that appears to form the edges of the joint on the radiograph. Because the actual structures of the joint are not very radio-opaque, we have to rely largely on changes to the contiguous bone to indicate joint disease. This in turn means that radiography is not a very sensitive method of diagnosing joint disease and the disease is well advanced before radiographic changes can be seen.

The response of the bone in close proximity to a diseased joint is limited, but a number of patterns can be discerned. The distance between the bones on either side of the joint may change, the outline of the boney edges and subchondral bone may be altered, and the density of the bone in proximity to the joint may change.

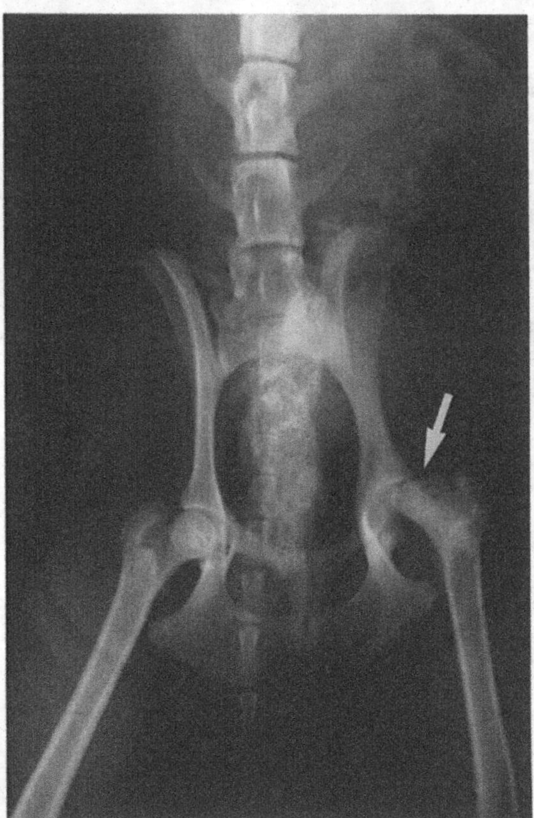

Fig. 12.16. Radiograph of the pelvis and hip joints of a dog showing changes in shape and density in one femoral head (white arrow). The changes include irregularity and flattening of the femoral head and focal loss of density.

The distance between the two bones will decrease if the articular cartilage is eroded. This occurs for example in degenerative arthropathy. Conversely, the distance will increase with, for example, severe synovial distension in the initial stages of some joint infections.

The outline of the subchondral bone can be altered by erosion following a break in the continuity of the articular cartilage. This is a feature of later stages of many joint diseases. The outline of the edges of the bone close to the joint may also be altered by the development of new bone or calcified cartilage, which occurs in response to many types of joint damage.

The density of bone in proximity to the joint may decrease either generally, when for example, infection spreads from a joint to cause an osteomyelitis, or in a localized area when an area of overlying cartilage has been damaged. Increase in density of subchondral bone will occur when the articular cartilage has been eroded and bone articulates directly with bone.

The humeral head illustrated in Fig. 12.17 shows alteration in outline and density of the bone underlying the articular cartilage. The caudal aspect of the humeral head has a 'flat' where some subchondral bone has been removed, and underlying this region is an area of decreased bone density where there has been some bone destruction. These changes indicate an osteochondrosis dissecans.

Figure 12.18 is of the tarsal (hock) joint of a dog. There are changes to the outline of the bone in proximity to the joints and a decrease in the distance between the bones. On the cranial and caudal aspect of the distal end of the tibia there is some new bone formation. This change in outline can also be seen on the tibial tarsal bone. The space between the central and third tarsal bone has decreased to the extent it has been almost obliterated, and there is new bone formation at the joint margins of the cranial aspect of both bones.

Pathologic features of bone disease

Now that the normal aspects of bone and the broad clinical and radiographic features of skeletal disease have been reviewed, the general responses and properties of bone in disease will be considered.

All the basic pathologic response patterns occur in bone, ranging from atrophy and inflammation to neoplasia. However, there are unique features of bone that influence the appearance and outcome of the pathologic processes.

One of the central features of bone is its ability to lose and replace substance. Bone also retains an almost embryonic capacity for repair. Rather like the repair of liver, new bone can form over a framework, but, unlike the liver, bone does not require the original scaffold. In areas devoid of bone, the provision of bone grafts or bone chips from another source will allow reconstitution. Indeed, if the gap between a fracture site is not too large, reforming of bone may take place without the provision of a scaffold. This capacity for repair is probably the property most frequently relied on in everyday practice.

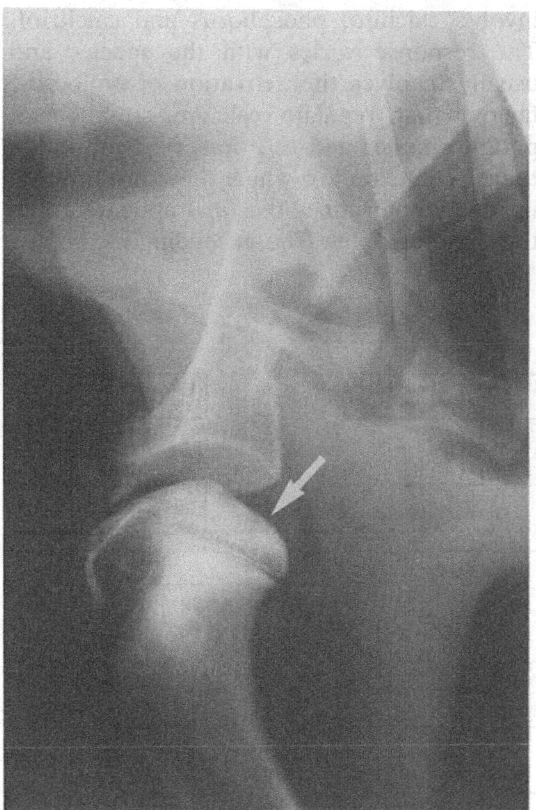

Fig. 12.17. Radiograph of the humeral head of a dog showing changes in outline and density of bone underlying the caudal aspect of the articular cartilage (white arrow).

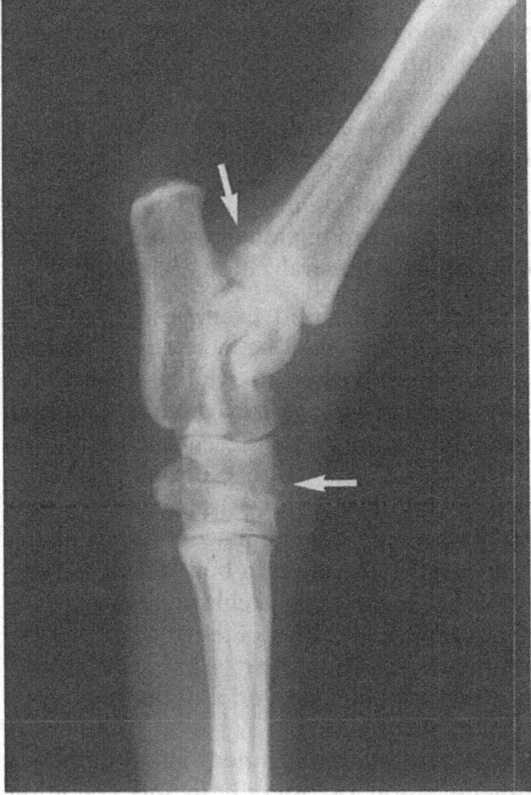

Fig. 12.18. Radiograph of the tarsal region of a dog showing changes in the outline of bones (white arrows). There is also a decrease in distance between the central and third tarsal bones.

Repair and remodeling is exploited in the repair of fractures, in the reconstitution of atrophied bone, and in osteomyelitis after the offending organism has been removed.

There are certain conditions that must be met before complete repair can occur. For example, the site of fracture must be immobile, the ends must be closely apposed, and the vascular supply must be adequate. The reader is encouraged to review surgical texts for detailed discussions of fracture repair.

Atrophy of bone

There will be some argument about the use of the term atrophy in the following discussion. Atrophy will be used to signify the reversible loss of one or all of the components of bone. For example, a failure to mineralize osteoid will be regarded as a form of atrophy. However, to avoid any confusion, the more commonly used terms osteoporosis, osteomalacia, rickets and fibrous osteodystrophy will be used where appropriate.

Broadly, bone atrophy may follow any of the three following processes:

1 An excessive rate of removal of bone.
2 Subnormal production of osteoid.
3 A defective mineralization of osteoid.

Even though florid examples of each of these processes can be seen in particular species, there is often an overlap of the three. Also, the same cause may produce different patterns of atrophy in different species.

The appearance of bone atrophy can take three forms. The first is **osteoporosis**, alternatively termed osteopenia. Osteoporosis is defined as a reduction in bone mass, but the bone that is present is architecturally normal. It may arise from either of the first two processes mentioned above; that is, an excessive rate of removal of bone, or a subnormal production of osteoid.

Osteomalacia is defined as a defective mineralization of osteoid. In this condition there is little effect on the production of osteoid. The juvenile form of osteomalacia is rickets. This disease has all the trademarks of osteomalacia, with the additional feature of defective mineralization of growth plate cartilage.

Fibrous osteodystrophy (osteodystrophia fibrosa, osteitis fibrosa cystica) is a complex form of atrophy, indicated by the terms used to describe it. The central features of the changes in bone are two-fold. The first is of bone atrophy, which may take the form of osteoporosis or osteomalacia, and the second, the presence of a variable amount of loose connective tissue between the remaining trabecular bone. There is a marked species variation in the gross and microscopic appearance of fibrous osteodystrophy. The features of bone atrophy are depicted in Fig. 12.19.

Apart from the specific diseases that regularly induce a particular pattern of atrophy, there are those metabolic diseases which involve calcium, phosphorus and calcitriol. The response varies with the species and usually involves the activation of endocrine factors that regulate calcium metabolism, parathormone, and to a lesser extent, calcitonin. The extent to which these are brought into play determines the final appearance of the bone atrophy. The exceedingly complex

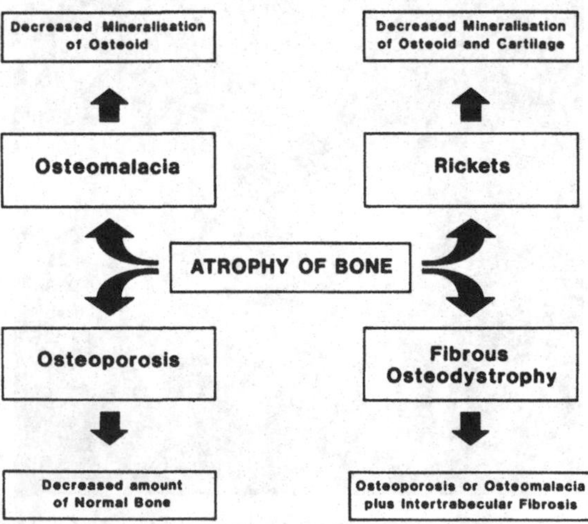

Fig. 12.19. The forms of bone atrophy and the major distinguishing features of each.

nature of metabolic bone disease is reflected in the literature on the subject.

Osteoporosis

The conversion of bone of normal mass to bone of lesser mass retaining normal architecture may be the result of increased bone resorption or decreased osteoid production. It may be localized or generalized.

Localized osteoporosis usually follows a lightening of the mechanical load normally placed on a bone, and is most commonly seen with the immobilization of a limb by plaster casting. As will be recalled, one of the major determinants of bone mass is the stress placed on that bone. An increase in the stress applied to a bone over that normally seen results in the deposition of bone, whereas a decrease results in the resorption of bone. The translation of variations in mechanical stress into deposition or resorption of bone is associated with changes in the electrical potential on bone surfaces, the so-called peizo-electric effect. The parathyroid gland has little influence on the development of localized osteoporosis. **Generalized osteoporosis** may develop after prolonged deficiencies of various nutrients such as dietary protein, copper, vitamin C or calcium. It can also be induced by high levels of corticosteroids and estrogens. That is not to say that osteoporosis is always the major feature of such aberrations. Calcium deficiency, particularly, may under certain circumstances produce either osteomalacia or fibrous osteodystrophy.

Osteomalacia and rickets

These two diseases are characterized by the presence of wide, unmineralized osteoid seams, and, in the case of rickets the additional feature of poor mineralization of growth plate cartilage. There is no defect in the production of osteoid, only in its subsequent mineralization. The major causes are a continued negative balance of vitamin D or of phosphorus. It may also be seen under certain conditions with calcium deficiency. Affected bones are of normal size, are easily cut with a knife and bend rather than break. In young animals the growth plates are widened and irregular. This is particularly so at the costochondrial junctions, producing the so-called rachitic rosary.

Phosphorus is an essential component of the mineral complex deposited on osteoid or cartilage. Dietary deficiencies of phosphorus are most commonly seen in herbivorous animals grazing on phosphorus-deficient pastures. There is a marked general effect on the rest of the body as well as the bones.

Vitamin D deficiency is not common in animals. It is characterized by hypocalcemia and hypophosphatemia, due mainly to lack of absorption of calcium and phosphorus from the intestine. Hypocalcemia leads to parathyroid gland hyperactivity, which increases bone resorption. Vitamin D is also required for mineralization of osteoid and cartilage, but the exact mechanism is unknown.

Fibrous osteodystrophy (hyperparathyroidism)

This is the most common metabolic bone disease and can be either of nutritional or renal origin. The changes in bone are due to a continued increase in circulating levels of parathormone. The gross and microscopic appearance of the disease varies from species to species. Where there is little osteoid production, it is termed **hypostotic** fibrous osteodystrophy. This type is observed mostly in cats. A **hyperostotic** form is characterized by abundant osteoid formation leading to an increased size of bones. A number of colloquialisms such as 'big head' or 'bran disease' are in common use for this disease of horses.

The disease is the result of either a nutritional imbalance of calcium and phosphorus, or of the imbalances in the renal regulation of calcium and phosphorus following renal failure. **Nutritional secondary hyperparathyroidism** occurs commonly in puppies or kittens fed on a whole meat diet. Meat contains little calcium and additionally contains relatively high levels of phosphorus. In herbivorous animals it is associated with cereal or bran diets low in calcium and high in phosphorus, or occurs if calcium is bound in the intestine by high levels of phytates from plants such as buffel grass (*Cenchus ciliaris*). In all these examples the absorption of phosphate leads to a hyperphosphatemia, which in turn induces hypocalcaemia.

Renal secondary hyperparathyroidism, as the name implies, is associated with chronic renal failure. It is rarely observed clinically, and is seen in only a few of the cases of chronic renal failure. The pathogenesis is associated with an inability of the kidney to excrete phosphorus. The inability to produce sufficient amounts of calcitriol, thus impairing the absorption of calcium from the intestine, may also be significant.

The trigger for continued release of parathormone is hypocalcemia induced by the hyperphosphatemia. The effect of parathormone on bone is increased osteoclasis and stimulation of osteoblasts, the result is either osteoporosis or osteomalacia accompanied by loose fibrosis of marrow spaces. There is a lack of mineralization of osteoid, but continuing mineralization of the cartilaginous growth plate in fibrous osteodystrophy. This phenomenon is yet to be fully explained. The mineralization of the growth plate appears affected only in vitamin D or phosphorus deficiency. Figures 12.20 and 12.21 demonstrate the pathogenesis of nutritional and renal secondary hyperparathyroidism.

It is worth returning to review the metabolic bone diseases. They are not easily separated, particularly with respect to the etiology, and, whenever calcium, phosphorus or vitamin D are implicated, there is always the specter of the transient or prolonged effects of parathormone lurking in the background. It is probably best to consider the metabolic bone diseases as a continuum, the histologic appearance depending to some extent on the phase of the disease process. It is wise to be familiar with the pattern commonly seen in different species, particularly in nutritional secondary hyperparathyroidism. Also worth re-emphasizing is the reversibility of the process occurring in bones, if the underlying etiology is removed.

Primary hyperparathyroidism and pseudohyperparathyroidism

Neoplasia of the parathyroid gland is of interest for two reasons. Firstly, because car-

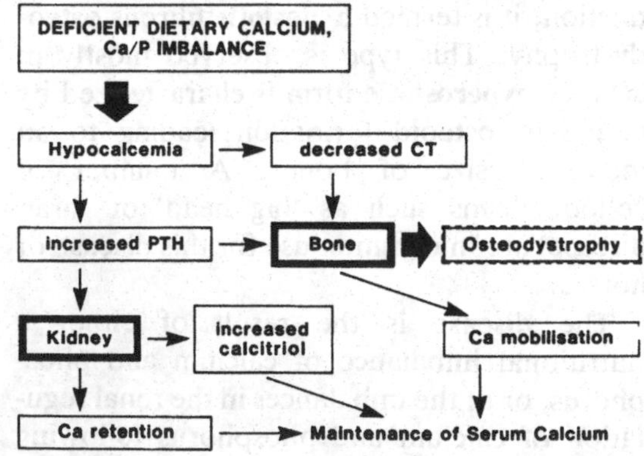

Fig. 12.20. The pathogenesis of nutritional secondary hyperparathyroidism. The major features are the response of bone and the kidney to elevated plasma levels of parathormone (PTH). Both favor the maintenance of serum calcium levels. CT, calcitonin.

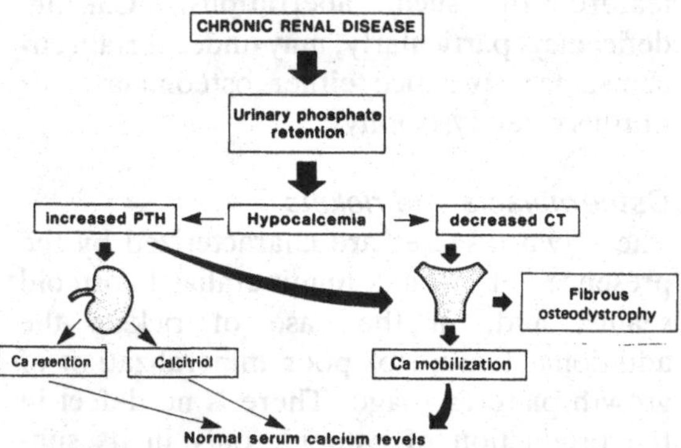

Fig. 12.21. The pathogenesis of renal secondary hyperparathyroidism. Increased plasma levels of parathormone (PTH) have little overall effect on the remaining renal tissue. However, the effect on bone is substantial. CT, calcitonin.

cinomas of this gland may invade and sometimes metastasize and, secondly, because some parathyroid adenomas and carcinomas are functional, producing a syndrome termed **primary hyperparathyroidism**. Functional parathyroid tumors release inordinate amounts of parathormone. The result is a condition closely resembling osteodystrophia fibrosa. The effect of excessive amounts of parathormone is two-fold, one is skeletal and the other results from prolonged hypercalcemia. The skeletal manifestations are lameness due to fractures and in some cases facial hyperostosis. Bones are thinned and fragile following excessive osteoclastic resorption. Prolonged hypercalcemia produces clinical signs of anorexia, depression, vomiting, polydipsia, polyuria and generalized muscle weakness.

Pseudohyperparathyroidism is a clinical syndrome with many features in common with primary hyperparathyroidism, as can probably be inferred from the title of the condition. Affected animals are hypercalcemic, with the attendant clinical signs of anorexia, depression, vomiting, polydipsia, polyuria and generalized muscle weakness. However, in contrast to primary hyperparathyroidism there is little if any radiographically observable change in bone structure. The cause of the hypercalcemia emanates from tumors of non-parathyroid gland origin. Those implicated to date in the dog have been tumors of the anal sac and more importantly lymphosarcomas. Neither secrete parathormone, calcitriol or prostaglandins, all of which are known to produce hypercalcemia. In the case of lymphosarcomas, hypercalcemia occurs only when the bone marrow is infiltrated. The substance released has been termed osteoclast-activating factor.

The clinical features of bone atrophy

Atrophy of bone, will almost by definition lead to structural weakness, and the clinical signs are usually referable to structural deficits. They may be frank abnormalities such as fractures, which should properly be termed pathologic fractures, but more often the clinical signs are of a shifting lameness associated with variable degrees of folding fractures. These folding fractures compress or expand the periosteum, causing pain. Because of the structural weakness there is often bending or bowing of long bones. There may also be marked deformity, particularly of the premaxillae and maxillae, in secondary hyperparathyroidism. This is due to extensive resorption and remodeling of these areas.

The search for the etiology in the absence of other primary organ disease, such as renal failure, depends on evaluation of the clinical and pathologic findings, combined with an analysis of dietary components, especially calcium, phosphorus and vitamin D.

Periosteal reactions

The periosteum is remarkably volatile. Any change to its well being will result in it responding angrily by laying down new bone. This type of reaction is seen in a variety of conditions, including infections, trauma, neoplasia, space-occupying lesions in the chest, and increased stress on the cortex. The periosteum may also respond by resorbing bone, particularly where pressure is applied locally.

The stimulus for periosteal new bone formation is not well defined, but there appear to be two general phenomena that lead to it. The first is periosteal lifting. Anything that lifts the periosteum, such as subperiosteal edema, hemorrhage, inflammatory exudate or neoplasia will, after a lag period, lead to periosteal new bone formation. The pattern observed can be variable in appearance. It usually appears as radiating spicules perpendicular to the outer cortical surface. It may be triangular in shape, as is seen in osteosarcoma for example, or it may be irregular and extend for some distance along the outer cortex, as in osteomyelitis. A word of caution: the periosteal reaction is often referred to as periostitis, which it is not as there is, in most cases, no accompanying inflammatory reaction.

The second stimulus to periosteal new bone formation is an increased stress placed on bone. The removal of bone from the endosteal surface and the deposition of bone on the periosteal surface is encompassed by the Flexure–Drift hypothesis discussed earlier.

Necrosis of bone

There are few specific examples of necrosis as the predominant feature of bone disease, with the exception of femoral head necrosis in small-breed dogs and pigs, and, metaphyseal osteopathy in large-breed dogs. Bone necrosis, however, may accompany a number of conditions. It may be observed in some cases of peripheral ischemia following thrombosis of the nutrient artery or following the intense vascoconstriction of ergot poisoning. There is also an element of necrosis of bone in every fracture and osteitis. In the latter two cases, if the fracture is stabilized, or the offending organism is removed, necrotic bone is resorbed by the usual process of osteoclasis. If the fragment of necrotic bone is too bulky, then it is sequestered.

Necrosis of the femoral head (Legg–Calvé–Perthes disease)

This disease is most frequently observed in the dog, but is sometimes seen in pigs and cats. In the dog, it is a condition of small breeds commencing between three and ten months of age, and is manifest initially as an intermittent failure of support, with some accompanying restriction of movement. This may progress to continuous carrying of the leg.

Radiologically, the initial sign is widening of the joint space, decreased density of the epiphysis and sclerosis of the femoral neck. This progresses to focal areas of radiolucency, flattening of the dorsal surface and eventual collapse of the femoral head. Secondary degenerative changes within the acetabulum are also evident (Fig. 12.16).

Histologically, the lesion is, as expected, necrosis of bone with areas of osteoclasis. It is thought that the disease is caused by interference to the blood supply to the femoral head, aided by the blood supply to the femoral head entering only through the synovial membrane.

Metaphyseal osteopathy (hypertrophic osteodystrophy)

This most enigmatic of diseases is observed in young, fast-growing, large-breed dogs. The onset is acute. The metaphyseal areas, particularly of long bones are hot, painful and swollen. The painful nature of the condition is amply demonstrated by many animals remaining in lateral recumbency and refusing to walk. Affected animals may be pyrexic. Many animals will recover completely. The lesions center on the metaphyseal area of all bones and are characterized by:

1 Widespread metaphyseal fractures.
2 Metaphyseal hemorrhage with neutrophil infiltration.
3 Sub- and extraperiosteal edema, hemorrhage and, later, new bone formation.

The radiographic reflections of the pathology are:

1 Metaphyseal density or lucency (Fig. 12.22).
2 Subperiosteal and extraperiosteal new bone formation subsequently develops. Hence the initial use of the term hypertrophic osteodystrophy.

The etiology is obscure and with obscurity comes speculation. Various suggestions have been oversupplementation with calcium, vitamin C deficiency and bacterial infections. There may be some truth in all these suggestions, but the inability regularly to reproduce the disease has not allowed critical examination to occur.

Osteomyelitis

Osteomyelitis may be localized or multifocal. Localized osteomyelitis is almost always bacterial and follows trauma, or the surgical exposure of bone to correct the effects of trauma. Multifocal osteomyelitis is usually associated with systemic or other organ infection, such as septicemia in young animals or in the debilitated with a chronic bacterial infection elsewhere in the body. Specific bacteria prone to affect bone are, for example, *Actinomyces bovis*, *Brucella canis* and, of the systemic mycoses, *Coccidiodes immitis* and *Blastomyces dermatitidis*.

Whatever the offender, bone responds to infections in predictable ways. There is an attempt to remove any necrotic bone induced either directly by the organism or indirectly by thrombosis. If the volume of bone is too large to be removed by osteoclasis, the necrotic bone is isolated by the production of new bone. The necrotic bone is termed a sequestrum, the new bone attempting to envelop the necrotic bone, an involucrum. If the volume of destroyed bone is limited, complete resolution can occur. However, if the infective focus or foci are not removed, the inflammatory reaction continues, and new bone is formed around the infected focus (Fig. 12.23).

The clinical signs associated with osteomyelitis vary with the site of infection. Vertebral osteomyelitis is usually manifest by varying degrees of paralysis and is seen mostly in economic animals. When long bones are involved there is lameness and sometimes discharging sinuses. The sinuses heal periodically, only to break down at a later time.

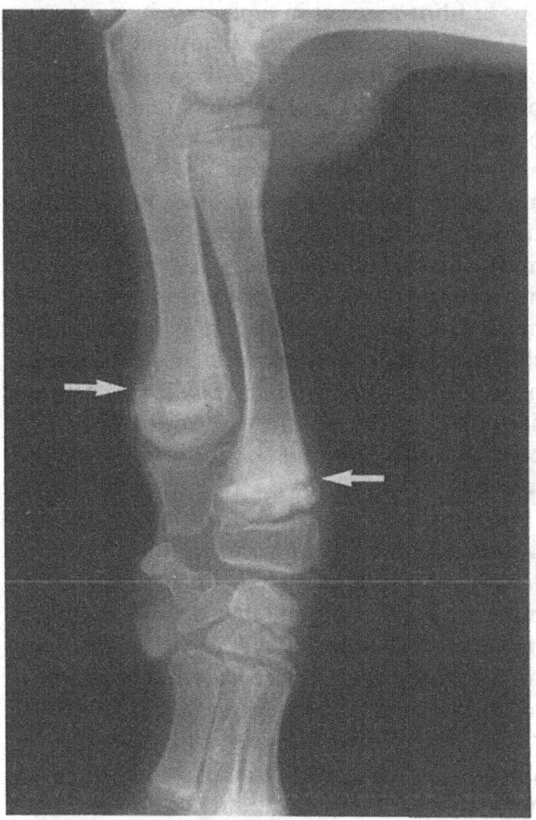

Fig. 12.22. Radiograph of the distal extremity of a four month old Great Dane with metaphyseal osteopathy. The changes are centered on the metaphyses and are particularly apparent in the distal radius and ulna (white arrows). Similar changes in density can be seen in the metacarpal bones.

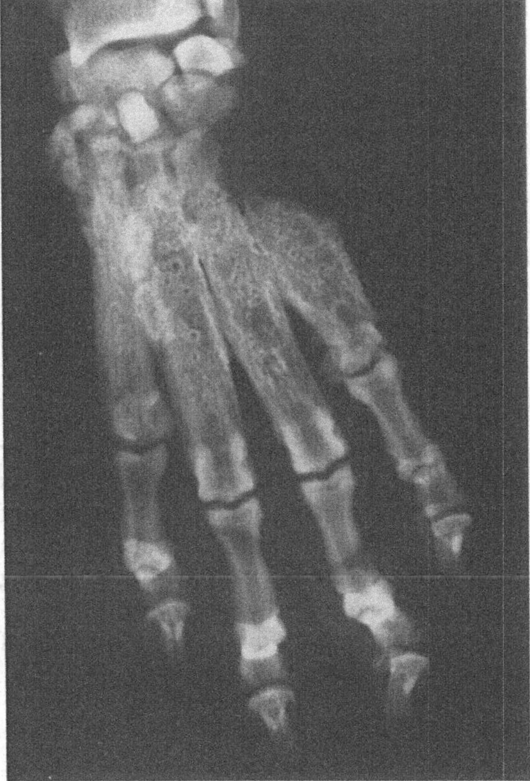

Fig. 12.23. Radiograph of the metacarpal bones of a dog with chronic osteomyelitis. There are changes in outline, architecture and density. There is extensive destruction and remodeling of cortical bone, with multifocal areas of lysis, which are surrounded by rings of new bone formation. A portion of one metacarpus was removed surgically.

Neoplasia

The importance of primary neoplasia of bone varies with the species of animal. Whilst malignant bone tumors in dogs, especially large dogs, are common, comparable tumors in farm animals are rare, with only occasional benign neoplasms seen. As with any other organ, the majority of tumors found reflect the differentiation or potential of the cells that comprise the tissue, the most common being of chondrogenic or osteogenic type. Most chondroid tumors are benign, and most osteogenic tumors, at least in the dog, are malignant. There are isolated examples of benign or malignant primary tumors arising from other cell types, such as endothelium and fibroblasts. The neoplasms arising from hematopoietic tissue are rare except for those in the cat.

Recognition and classification of neoplasia

The recognition, diagnosis and prognosis of tumors of bone depends on a knowledge of the clinical, radiographic and pathologic features of a particular tumor. In many cases, although one aspect may be suggestive of a particular type, in others a considered interpretation of all three is necessary.

Clinical features

Depending on the location of the mass, the animal will present with either lameness or swelling or both. The species affected is relevant, as discussed earlier. The location of the mass is of some assistance. For example, a mass in the distal or proximal areas of long bones in dogs is highly suggestive of osteosarcoma. Similarly, a mass on the rib cage of dogs is likely to be of chondroid origin.

Radiographic appearance

It is often difficult to assess the gross appearance of a bone tumor, but radiography reveals the fine detail of the mass that can usually be appreciated only histologically. Focal areas of lysis or sclerosis, the consequences of periosteal lifting, and the gradation between the tumor and normal bone is usually observable on radiography. The approach to, and interpretation of, radiography of bone has been dealt with earlier in the chapter.

Pathologic examination

The major difficulty that arises for the pathologist is the selection of the site for biopsy, which, in most cases, has to be taken by the clinician. It is imperative that a representative biopsy is taken. If taken improperly, only the new bone arising from the lifted periosteum may be present in the biopsy.

Benign tumors of bone

These types of tumors are uncommon in all species. They may be, osteogenic, cartilagenous or fibro-osseous in type.

Osteomas occur predominantly in the bones of the skull, mandible and maxilla. These progress slowly in size and may reach massive proportions, but are usually well demarcated and are composed of a mixture of cancellous and compact bone. **Chondromas** occur in the turbinates, flat bones, costal cartilage and limbs. **Ossifying fibroma** is a rare tumor of horses and ruminants found in either the maxilla or the mandible. It is a mass that slowly expands and destroys the surrounding bone and is composed of spicules of osteoid accompanied by a fibrous stroma. There is considerable argument about the nature of the lesion.

Malignant tumors of bone

Most malignant bone tumors arise from the mesenchymal components of bone and are named for the most differentiated tissue in the tumor. Once again, there is doubt with regard to the classification of some of them.

Osteosarcoma is the most common primary sarcoma of domestic animals, with the vast majority occurring in the dog. The histologic pattern of these varies in the dog and one classification is based on the variations in radiographic appearance, amount of matrix production, and the cell morphology. Osteosarcomas, according to this classification fall into three subclasses: simple, compound and pleomorphic. There is, as yet, no evidence to suggest that there are any clinical differences between these subclasses.

Osteosarcomas comprise more than 80% of all malignant bone tumors in dogs and cats. In dogs, there appear to be particular breeds predisposed to the development of osteosarcomas. Most are large or giant breeds of dogs and include Boxers, Great Danes, Saint Bernards, German Shepherds and Irish Setters.

The metaphyseal regions of the long bones are the areas most commonly affected, with clinical signs of lameness, swelling, pain or tenderness. The time from the appearance of the lesion to the death of the animal varies from one to eight months.

The sites of involvement for dogs vary for different breeds and weights. In general, the distal radius, tibia and femur and proximal humerus are preferred sites. The axial skeleton may be involved, particularly the skull and the ribs.

The radiographic pattern observed depends on the subclass of tumor involved. There is usually a combination of destruction of preexisting cortex manifested by areas of radiolucency, with areas of irregular radiodensity, which is the bone produced by the tumor. The lesion blends into the remaining normal bone and there is a regular, rather triangular area of subperiosteal new bone formation in response to lifting of the periosteum (Fig. 12.24). Some osteosarcomas may produce little bone and are observed as predominantly lytic lesions.

The pathologic appearance of osteosarcomas has already been covered under the radiographic appearance, where it was said that greater detail is manifest. There are two patterns, a lytic and a productive. Both destroy normal cortex, invade the medulla, but rarely cross joint spaces. There is a great variation in the microscopic appearance, from a simple osteoid-producing type to a very pleomorphic, highly cellular tumor.

Whenever the diagnosis of osteosarcoma is made, particularly in those originating in a metaphyseal area, a uniformly poor prognosis should be given. The tumor metastasizes frequently to the lungs with a prevalence probably approaching 100%. The prevalence is

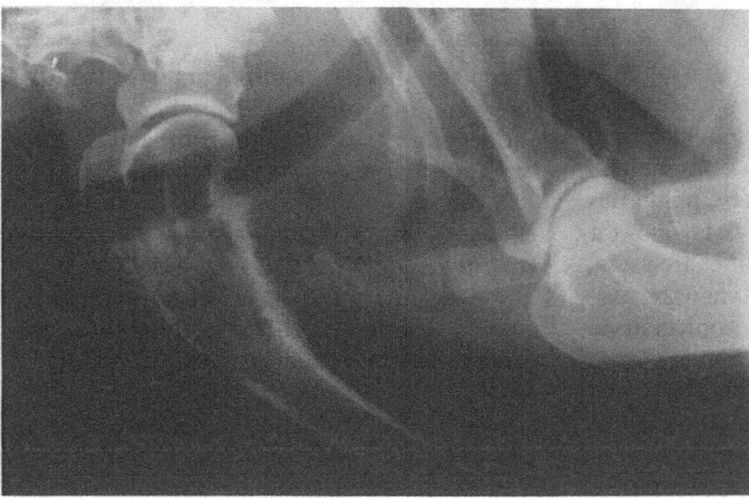

Fig. 12.24. Radiograph of an osteosarcoma involving the proximal humerus of a dog. There are changes in outline, architecture and density. A loss of cortical bone, focal areas of radiolucency and periosteal new bone formation is apparent.

somewhat lower in osteosarcomas that arise in the skull.

Diseases of growing bone

There are two major aspects of the growing skeleton that set it apart from an adult skeleton. Firstly, there is a need for the net accumulation of bone, whereas, in the adult, deposition equals resorption. Because of the need in growing animals for deposition to exceed resorption, there is a greater nutritional requirement. The substances that are most often in limited supply are calcium, phosphorus, vitamin D and protein. A lack of these nutrients affects adult bone also, but not as quickly or to the same extent. Therefore diseases characterized by atrophy, such as osteoporosis or fibrous osteodystrophy, are more commonly seen in the young. The circumstances leading to the development of metabolic bone disease have been discussed above. The second major difference between the immature and mature skeleton is disease associated with primary or secondary deficits in the cartilagenous growth plate. Occasionally there is also a defect associated with the development of osteoid.

Skeletal anomalies

Disturbances of skeletal development form a small but significant section of skeletal disease. There are, unfortunately, numerous isolated conditions and a battery of names for these conditions. Some of the names are useful and accurate, but others are somewhat misleading. An example of the confusion in terminology is the use of achondroplasia, chondrodystrophy, chondroid dysplasia and dyschondroplasia for the one condition of disproportionate dwarfism.

As with any disease grouping, it is helpful to use a framework which encompasses most, if not all, developmental disturbances. Specific disease entities may then be added to this framework.

Developmental skeletal disturbances may be classified under the general heading of **dysplasias** (literally, disturbances of growth). Such dysplasias result from abnormalities of the three major components of the embryologic or fetal skeleton, and they are the primitive mesenchyme, cartilage and osteoid. Few of the dysplasias observed in the newborn domestic animal have been adequately studied, mostly because of the complexity of skeletal development and the isolated nature of many of the conditions. The abnormality may be localized or generalized. Some dysplasias are undesirable states, especially in the context of not being sought after, but others, equally classifiable as disease processes, are thought to be desirable and are actively pursued. The latter is best exemplified by the chondroid dysplasias that characterize many breeds of dog such as the Basset Hound, Dachshund, Boxer, Boston Terrier and Bulldog.

The majority of skeletal dysplasias involve either the cartilagenous or the osteoid component of growing bone, with the chondroid dysplasias being the most important.

Chondroid dysplasias

These have been recognized to be the result of two basic abnormalities: primary disorders of endochondral cartilage growth and heterotypic proliferation of chrondroblasts.

Primary disorders of endochondral cartilage growth

These are the most important and most frequently diagnosed chondroid dysplasias. The generalized form occurs most commonly in cattle, but has been reported in dogs. Localized endochondral dysplasia is sought in a number of breeds of dog and cat. Boxers and Boston Terriers exhibit localized endochondral dysplasia of the basocranial area. The Dachshund and Basset Hound have dysplastic limbs. The types of localized abnormality, at least in dogs, are many and varied.

The external form of generalized endochondral dysplasia in any species is manifest by a short, squat, chunky animal with shortened, usually twisted, limbs. The mandible protrudes, the upper jaw is compressed and the forehead domed. In cattle, they are referred to as bulldog calves. Most of the well-recognized entities, particularly in cattle, are inherited. That is where simplicity ends and complexity begins. Generalized endochondral dysplasias are not uniform in type. At least in some breeds there is considerable phenotypic variability. The selection for short, compact animals particularly in beef breeds has led to an increased prevalence of dysplastic animals.

The morphology of the defect is identical for either localized or generalized endochondral dysplasia. The growth of bones from a cartilage model is accomplished by appositional and interstitial growth of cartilage. In endochondral dysplasia appositional growth is unaffected. It is a disturbance of interstitial growth. What growth there is, in addition to disturbed interstitial growth, is prematurely terminated. The microscopic appearance of the growth plate is characteristic. Instead of the normal linear arrangement of chondrocytes, these cells are irregularly arranged and are haphazardly placed. The normal process of degeneration, mineralization and resorption of cartilage occurs, but irregularly.

Heterotypic proliferation of chondroblasts

The presence of osteochondral proliferations, particularly in the periosteal area of the metaphyseal region of long bones, is rare. They are variously termed osteochondromas, multiple cartilagenous exostosis, dyschondroplasias or osteochondromatosis, and may be solitary (monostotic) or multiple (polyostotic). The mass (or masses) is characterized by a cartilage cap, below which there is trabecular bone. The mass grows whilst the bone grows and usually ceases when physis close. The lesions are considered to be ectopic cartilage from physeal plates and are probably hamartomas.

Osteoid dysplasias

Osteogenesis imperfecta is the name given to a disease in man which also occurs in Holsteins. There is inability to produce sufficient normal osteoid in this disease due to an abnormality in collagen.

Metaphyseal dysplasia or osteopetrosis is a lethal trait which is common in certain lines of Angus cattle. The diaphysial cortices are thin and the medullary cavity is obliterated by coarsely woven bone and cartilage. There is an exuberant osteoid formation or lack of removal of the primary spongiosa and secondary spongiosa. The names of the diseases imply the location of the abnormal osteoid.

Diaphyseal dysplasia refers to the deposition of radiating trabecular bone by the periosteum in newborn pigs. It is most commonly a lethal autosomal recessive trait, which usually involves one or both forelimbs. Occasionally the lesion regresses.

Miscellaneous developmental anomalies

There are numerous examples of the absence or smallness of various parts of the skeleton. These include absence of a limb or limbs (amelia), shortness or smallness of the limbs (micromelia), and absence of the distal limbs (peromelia). There are also examples of extra limbs and digits as well as fusion of the digits in most species.

Pituitary dwarfism

Although not a primary skeletal anomaly, a congenital pituitary abnormality may result in an animal that is dwarfed, but proportionate in stature. The best example of this type of dwarfism occurs in the German Shepherd breed. The pituitary is replaced by a craniopharyngeal cyst, with some isolated islands of anterior pituitary remaining. Many of these dwarfs are normal in all other respects, except for the retention of an infantile hair coat and, in some cases, the exhibition of hypothyroidism. Although not proven, the defect is probably a deficiency of growth hormone activity.

Joints

The overwhelming majority of joint affections fall into two categories, degenerative joint disease and infectious arthritis. Degenerative joint disease, either of a primary or secondary nature, affects all domestic species to a greater or lesser extent, and is seen most commonly in the young adult or aged animal. Infectious arthritis is usually polyarticular, and occurs with highest frequency in young farm animals following a septicemia.

Although strictly not confined to the joints, osteochondrosis dissecans (chondrosis) is also discussed because the clinical signs of the disease appear when the joint becomes involved.

The nature of joints

About half the joints allow growth and the other half allow movement. Those concerned with growth were discussed under bone. Here we will discuss only those that move. Moving joints normally function silently and efficiently for a lifetime. But when joints fail they cripple.

Stability versus mobility

As a joint is allowed more mobility, the problem of stabilizing it becomes greater. Mobile synovial joints like the stifle are often large and clumsy-looking because of their in-built stabilizing mechanisms. On the other hand, the stiff, barely moveable fibrous and cartilaginous joints, such as those between vertebral bodies, can have tough, simple unions to make them stable. Joint design reflects this compromise between stability and mobility.

Joints are best able to withstand heavy loads when they are in the so-called close-packed position. This is usually in full extension when ligaments are taut and articular cartilages are squeezed into maximum contact. In other positions, the ligaments slacken and the joint surfaces part a little. Synovia can flow between the cartilages and oil and nourish them. Joints are always vulnerable to injury when they are subjected to excess loads and sometimes when they are subjected to ordinary loads in these more lax positions.

Ligaments and joint capsule

The joint capsule is baggy to allow movement. Its fibrous outer coat becomes thickened or ligamentous where it is repeatedly under tension, such as in the sides of a hinge joint. Tension in the fibrous capsule is transmitted directly to the periosteum, with which it is continuous. Hence, a strained joint capsule is likely to induce periosteal new bone at the site of attachment. This is usually close to the rim of a hollow or female articular surface, but more remote from a rounded or male articular surface. The capsule frequently attaches close to the growth plate.

Joint movement

Joints which allow movement in any direction, like the ball-and-socket joints of hip and shoulder, cannot be stabilized with fibrous tissue. The joint capsule has to be roomy to allow rotation and any ligaments present must be slack for all but the limits of the range. Then the head is held in its socket not by fibrous tissue but by a sleeve of muscles. The hip and shoulder are thus particularly vulnerable to any generalized muscle weakness. Poor development of the muscles of the rump, for example, allows the femoral head to slip in and out of its socket, which may result in conformational changes in the joint surface, as it does in hip dysplasia in dogs.

Most other joints have a more restricted action and need fewer muscles to stabilize them. Just two collateral ligaments are needed to convert joints to simple hinges, which are the most common, the most efficient, lightest, and generally the least vulnerable of all synovial joints.

A few complicated joints allow movement in two planes but not a third. Both the elbow and stifle are of this type and allow flexion, extension and a little rotation, but no abduction or adduction. The solution at the elbow is to employ three bones, the ulna forming a simple hinge with the humerus, while the radius swivels against it. Having three independently growing long bones participating in a single joint, however, invites an unfair share of developmental anomalies to occur at the elbow. The solution at the stifle is provided by a flat tibial plateau. But this makes such a poor fit with the femur that fibrocartilaginous washers, the menisci, are required to clasp and support the rounded femoral condyles, and cruciate ligaments are needed to prevent backward or forward slipping. These additional stabilizing structures are not totally successful and may wear or rupture.

Two surfaces which rub against one another will ultimately wear out. Yet the bearing surfaces of joints have to last a lifetime, with the limb joints of a sheepdog, for example, having to sustain around a 100 million swings. The friction between them is less than ice sliding on melting ice so that wear is reduced to a minimum. Joints take a long time to wear, but once the process begins, it progresses rapidly. Friction rises, wear increases, friction rises even more and the process accelerates in a vicious circle.

Articular cartilage

This bearing surface is mostly of the hyaline variety, a name which reminds us of its glassy smoothness (Greek hyalos = glass). Its high water and chondroitin sulfate content makes it resilient, so that low surface elevations can be flattened. It contains arcades of collagen fibers which loop up from the subchondral bone, run tangential to the surface and then plunge deep again. When cartilage becomes worn, these surface layers of collagen are ruptured and the cartilage loses its smoothness; the deeper layers of matrix are exposed and the chondroitin sulfates, and proteoglycans are leached out.

Synovial fluid and synovial membrane

Synovial fluid nourishes and lubricates the articular cartilage. It is normally only present in small quantities so that it does not distend the joint capsule and limit movement and so that it maintains a high concentration gradient of nutrients. The fluid is principally a transudate of the capillaries, enriched by glycosaminoglycans. It is modified and absorbed by synovial cells, which line but do not form a continuous epithelium on the synovial membrane. Joints are unique in providing a large surface without an epithelium, where naked collagen may be exposed to the lumen. This has significance in the transport of nutrients, hormones, drugs, debris and microorganisms in to and out of the joint cavity.

Synovial fluid is squeezed out of the articular cartilage when it is loaded and lubricates the bearing surfaces. This weeping lubrication is supplemented by a type of boundary lubrication in which large organic molecules adhere to the surface of articular cartilage like seaweed on rocks. Moreover, hyaluronic acid, a long-chain polysaccharide found in relative abundance within synovial fluid contributes significantly to its lubricant properties. The chains entangle when joints are stationary, making the synovia viscous; they unravel and align themselves during motion to lower the viscosity.

Nerve supply

Articular cartilage is free of nerve endings and the synovial membrane is relatively free. Hence pain is never associated directly with a worn articular surface, and joints can sometimes creak without causing distress. The fibrous capsule and its associated ligaments, on the other hand, are richly innervated with the (so-called) proprioceptive triad of receptors which monitor movement and position and signal pain. These receptors guard the joint in the extremes of its range by inducing protective muscular contractions. They, above all, are the source of the pain of arthritis.

The response of joints to injury

Cartilage

Hyaline cartilage, at least in adult animals, has little capacity to repair or remodel deficits. Superficial cuts or abrasions of the articular surface do not heal, even after prolonged periods. This is probably associated with the shutdown of chondrocyte mitotic activity at the time of skeletal maturity. With superficial cuts, there is an initial limited mitotic activity, but this ceases after about seven days. However, the presence of superficial lesions does not necessarily lead to the development of perichondrial osteophyte formation. Indeed, experimental evidence suggests that this occurs rarely. Deep abrasion or lacerations of articular cartilage, that reach subchondral bone, fill with blood and are progressively organized until they become a fibrocartilagenous mass. The exposed subchondral cancellous bone becomes progressively thickened. Although healed in the sense of a scar, there is no restitution to the original state of hyaline cartilage. Most of the preceding information about cartilage healing comes from experimental models, but the same pattern occurs in disease, for example, when normal cartilage is removed as an osteochondral flap, or traumatically eroded or ulcerated. Similarly, erosion or ulceration of cartilage following the action of bacterial toxins or products of inflammation will result in the same lack of restitution.

Joint capsule

We are on firmer ground discussing the response of the joint capsule to injury mostly because it has a blood supply, and the full gamut of the inflammatory reaction and its consequences can be observed. The inflammatory response in all its forms is most commonly observed in infectious arthritis. Initially more subtle, but finally quite severe, changes can involve the joint capsule in degenerative joint disease, with little accompanying inflammatory response. It is the peculiarities of the synovial lining cells and the perichondrium that will be discussed. Synovial cells, as with any other cell, are sensitive to insult, and can demonstrate a range of degenerative changes. Synovial cells also have a remarkable capacity to undergo hypertrophy and hyperplasia. Any minor injury of any cause to either cartilage or joint capsule will lead to hypertrophy and hyperplasia of synovial cells (Fig. 12.25). Individual cells assume a cuboidal to low cuboidal appearance and the membrane is thrown up into folds. Instead of appearing as a flat, glistening white surface, the synovial membrane becomes either finely roughened or villous in appearance. On occasions, a fold of synovial membrane will adhere to eroded articular cartilage. This is termed a **pannus**.

A response, particularly with the degenerative arthropathies of either a primary or secondary nature, occurs in the perichondrial zone. This zone forms the boundary between the joint capsule, the articular cartilage and the periosteum. Under various forms of stress, this area initially undergoes fibrosis, which may be followed by either chondroid or osseus metaplasia, or both. This change can occur in the absence of significant damage to articular cartilage seen, for example, in some cases of joint instability following ligament rupture. The chondrophytes and osteophytes may occasionally break off and float freely in the joint. Once free, they are termed joint mice.

Synovial fluid

Alterations in synovial fluid reflect changes in the permeability of the vessels of the joint

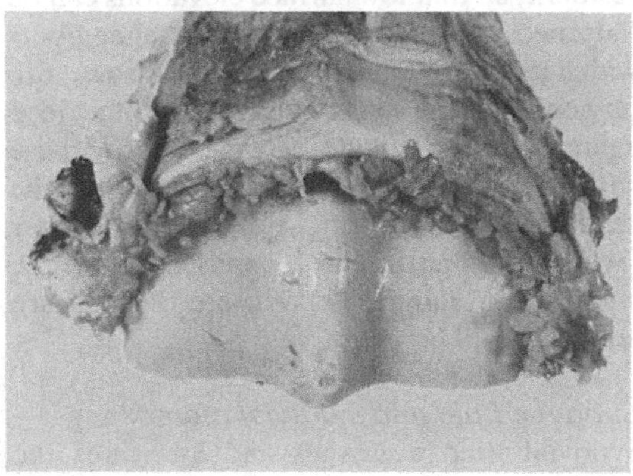

Fig. 12.25. Marked synovial hyperplasia in the metacarpal joint of a horse, so-called villonodular synovitis.

capsule, the function of the synovial lining cells, or changes in the articular cartilage. Depending on the disease process, the changes may include:

1 Migration of inflammatory cells of various types in to the fluid.
2 Diffusion of large molecules such as fibrin and gammaglobulin through more permeable capsular vessels.
3 A variation in the viscosity and proteoglycan content of the fluid.
4 Exfoliation of synovial lining cells into the fluid.

The differing proportions of these changes help to delineate the basic process occurring within the joint. Thus, in an acute bacterial arthritis, there are abundant neutrophils and the synovial viscosity will be reduced following the release of lysosomal enzymes from the neutrophils. Conversely, synovial fluid from a case of degenerative joint disease will be slightly turbid to clear, of normal viscosity, and may have moderate numbers of lymphocytes, macrophages and synovial cells. As with all things, there is overlap, but synovial fluid examination often helps in the characterization of joint disease.

The degenerative arthropathies

The term degenerative implies a slow inexorable path from health and fitness to debility and decay. Such is the lot of a joint that suffers from a degenerative disease. It has two origins: one, an inbuilt mechanism, the other, joint instability or repeated trauma. The former, without any identifiable antecedent, is regarded as primary, and the latter as secondary.

Primary degenerative arthropathy

All joints age, but some age more quickly than others. It is the premature development of the aging process, or a process of similar morphologic appearance that particularly afflicts certain lines of beef cattle. It affects dairy bulls also, but the lineage is by no means clear. Degenerative joint disease affects males of a number of beef breeds, including the Hereford, Aberdeen Angus, Galloway and Charolais in the United Kingdom, North America, and Australia. The incidence on particular farms may be sporadic or may reach epidemic proportions. The clinical onset of the disease varies from three months to two years of age. It is manifest by lameness, muscle wasting, crepitus on palpation of the greater trochanters, but there is variability in the presence of each sign. Affected bulls may stand with the hind feet placed forward, toes out and hocks inwardly rotated.

The lesions are limited to the hip joint and consist of a shallow acetabulum and cartilage erosion, which progresses to ulceration of the craniodorsal areas of the femoral head and acetabulum, the lesion commencing as a fine roughening and loss of lucency of the articular cartilage. Peripheral osteophytes develop. Although termed a primary degenerative arthropathy, some doubt has recently been cast on this view. It is alternatively thought to be secondary to a congenital shallowness of the acetabulum and so, in a sense, a hip or acetabular dysplasia. There appears to be little doubt that the condition is inherited and sex limited.

Secondary degenerative arthropathy

Whenever a joint is subjected to repeated minor stress, to a single major stress, or is abnormal in conformation, the development of secondary changes is inevitable. From the preceding, the range of species and joints involved can be quite varied. There are, however, some common examples such as hip dysplasia and cruciate ligament rupture in the dog, and bone spavin and ringbone in the horse. Many carpal swellings in the horse are degenerative arthropathies following repeated trauma. A similar process has been reported in dairy bulls, particularly those used at artificial insemination centers. In these, the lesions are centered on the vertebral column and the stifle joints.

Osteochondral disease

(Osteochondrosis, osteochondritis dissecans, chondrosis)

Initially considered to be a disease of limited importance, osteochonderosis dissecans is now recognized to be one of the most important joint diseases. This is particularly so in pigs, horses, cattle and dogs. Osteochondrosis dissecans appears to be a generalized process, but in each species particular joints are more commonly involved.

The basis of the lesion is a disturbance of endochondral ossification in rapidly growing animals. The resorption of cartilage on the diaphyseal side of growth plates ceases, probably due to the lack of provisional mineralization, leading to thickened areas of cartilage. The unresorbed cartilage then undergoes necrosis, due to impaired diffusion of nutrients leading in many instances to fissuring or collapse. There may also be collapse of the underlying subchondral bone. If the fissure in the cartilage extends to the joint surface, a cartilage flap may appear which can break off, exposing the underlying subchondral bone.

There is great variation in the gross appearance of the lesion, which is probably dependent on the varying stress placed on individual articular surfaces. The changes include a simple thickening of the articular cartilage to collapsing, folding and in some cases ulceration (Figs. 12.26 and 12.27). The particular joint affected varies with the species. In the dog it includes the humeral head, lateral condyle of the femur, and the ulna metaphysis. In horses, cattle and pigs it appears mostly to affect the femoral condyles.

Arthritis

Arthritis is inflammation of a joint. It may be acute, subacute or chronic. In domestic animals, arthritis is almost always infectious in origin. The infection is most commonly bacterial but may be viral, mycoplasmal, chlamydial, or occasionally fungal. Apart from an arthritis following local trauma or infection it is polyarticular. There are a number of circumstances common to many animals with arthritis:

1 Septicemia of variable bacterial genus, usually in young farm animals.
2 Specific bacteria that are prone to localize in joints such as *Erysipelas rhusiopathiae*, *Mycoplasma hyorhinis* or *Hemophilus suis*.
3 Viral diseases, such as ephemeral fever in cattle and retroviral arthritis in goats.

The morphologic type of arthritis varies, but is somewhat dependent on the inciting organisms. Streptococci are regularly associated with a predominantly fibrinous arthritis, staphylococci with a purulent arthritis. In the

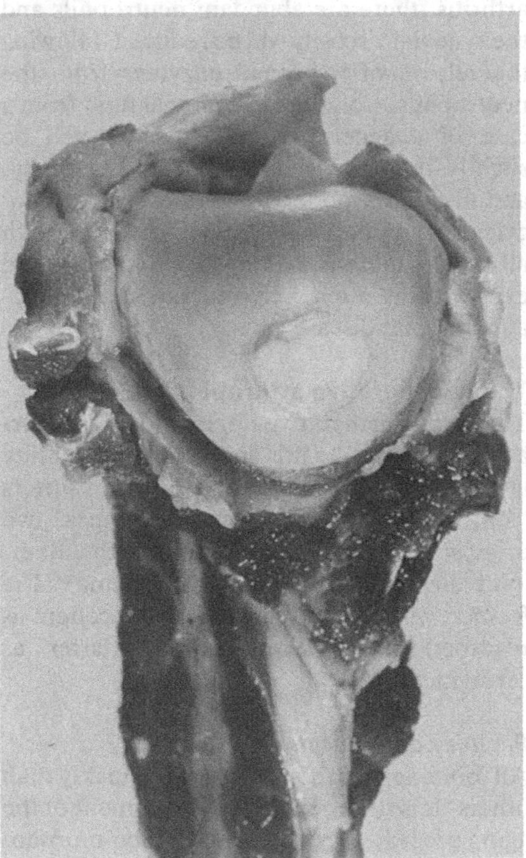

Fig. 12.26. Osteochondrosis of the humeral head from a dog. A circular area of articular cartilage is raised above the normal surface and is starting to break away.

acute phase of either infection, there are the trademarks of inflammation. The joints are hot, swollen and painful.

Resolution depends on the extent of damage wrought by the organism and the inflammatory response. In many cases of fibrinous arthritis, complete resolution is possible, but does not occur in chronic purulent arthritis, with its attendant architectural destruction. Articular cartilage is eroded, the joint capsule is thickened and fibrosed, and often the lining cells and subsynovium are necrotic. The best that can be achieved is an enlarged fibrosed joint of limited mobility.

Non-infectious arthritis

There are a number of arthritides where the etiology has not yet been elucidated, but they appear to have a substantial immune component. These have been divided into erosive and non-erosive types and include rheumatoid arthritis, feline progressive arthritis and a polyarthritis in Greyhounds. Although there are obvious differences between these types of arthritis, rheumatoid arthritis will serve as the example. It is well described for the dog, having necessarily many features in common with rheumatoid arthritis in man. One of the central features of the disease is the presence of a chronic polyarthritis characterized by synovial hyperplasia and a dense subsynovial infiltrate with lymphocytes and plasma cells. Additional features include a loss of bone density around the affected joints and focal radiolucent areas in subchondral bone and erosion of articular cartilage. A relatively constant finding is **rheumatoid factor**, an antibody directed against the endogenous immunoglobulin IgG. Rheumatoid factor may be either IgG or IgM. Much of the damage inflicted on the unsuspecting joint cartilage follows the initiation of inflammation by these immune reactions.

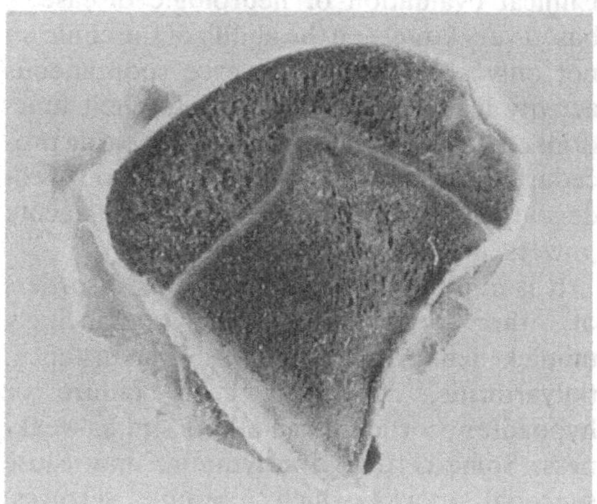

Fig. 12.27. Longitudinal section of the humeral head from the case shown in Fig. 12.26. The thickened flap of cartilage is separated from the underlying subchondral bone. There is also some increase in the density of trabecular bone immediately beneath the fissure.

Additional reading

Aegertes, E. and Kirkpatrick, J. A. (1975). *Orthopaedic Diseases*, 4th edn. Philadelphia, W. B. Saunders Co.

Banks, W. J. (1981). *Applied Veterinary Histology*. Baltimore, Williams and Wilkins.

Blood, D. C., Henderson, J. A. and Radostits, O. M. (1983). *Veterinary Medicine*, 6th edn. Bailliere Tindall, London.

Bojrab, M. J. (1981). *Pathophysiology in Small Animal Surgery*. Philadelphia, Lea and Febiger.

Bronner, F. and Coburn, J. W. (1981). *Disorders of Mineral Metabolism*, vol. 3 *Pathophysiology of Calcium, Phosphorus and Magnesium*. New York, Academic Press.

Bronner, F. and Coburn, J. W. (1982). *Disorders of Mineral Metabolism*, vol. 2 *Calcium Physiology*. New York, Academic Press.

Jubb, K. V. F., Kennedy, P. C. and Palmer, N. C. (1985). *Pathology of Domestic Animals*, 3rd edn. Orlando, FL, Academic Press.

Montgomery, D. A. D. and Welnourne, R. B. (1975). *Medical and Surgical Endocrinology*. Baltimore, The Williams and Wilkins Company.

Pool, R. (1978). In *Tumors of Domestic Animals*, 2nd edn, J. A. Moulton (ed.), pp. 89–147. Berkeley, University of California Press.

Sumner-Smith, G. (1982). *Bone in Clinical Orthopedics*. Philadelphia, W. B. Saunders Co.

Smith, L. H. and Thier, S. O. (1981). *Pathophysiology, The Biological Principles of Disease*. Philadelphia, W. B. Saunders Co.

Clive E. Eger, John McC. Howell
and Clive R. R. Huxtable

13 The nervous system

The nervous system is a communications network whose functions range from simple reflex activities to the complexities of awareness and conscious thought. Elements of the nervous system ramify to every part of the body, giving it a universal functional influence.

The functional integrity of the nervous system is directly related to the activity of neurons. When some noxious influence affects part of the nervous system, activity of the neurons in the area may cease, may be suppressed or may become excessive. In the central nervous system (CNS) these changes will inevitably affect other, interconnected, neurons and induce a change in their activity and coordination. The nervous system is arranged in a hierarchical manner so that dysfunction of one group of neurons is reflected as dysfunction of the subordinate neural pathways including those in the peripheral nervous system (PNS). The dysfunction may thus be projected widely throughout the system.

Fortunately for the clinician, the highly organized nature of the nervous system means that damage to a given area will always produce distinctively recognizable dysfunctions. With adequate clinical skills and an appreciation of the functional anatomy of the system, it is often possible to localize the seat of nervous dysfunction in an animal with neurologic disease.

The student is therefore advised to review the general anatomy of the nervous system along the lines indicated on p. 337, including the disposition of the meninges and the generation and flow-patterns of the cerebrospinal fluid.

The following discussion deals first with the clinical expression of nervous tissue disease and its anatomical correlations. This is followed by a review of the essential features of the neuron and its supporting cells, and then of the basic pathologic mechanisms involved in nervous tissue injury.

Principles of neurologic examination

Clinical evaluation of neurologic disease is based very largely on the ability of the clinician not only accurately to observe spontaneous activity in the patient but also to elicit functional deficits by means of specific testing procedures. The neurologist must rely on well-developed basic clinical skills and acute powers of observation.

It is important to remember that disorders of other body systems can convincingly mimick neurologic conditions. For instance, polyarthritis, congestive heart failure or hypoadrenocorticism can all present as weakness. Some cardiac dysrhythmias may cause syncopal attacks which resemble seizures, while orthopedic conditions such as bilateral cruciate ligament ruptures may be confused with spinal paralysis.

An overriding principle of clinical neurology is that damage to a given part of the nervous system will consistently produce the same clinical manifestation of neurologic dysfunction, *regardless* of the cause or nature of the damage.

Recognition of the dysfunction allows identification of its anatomical source and, therefore, the location of the lesion. When there seem to be exceptions to this rule, it is usually because other lesions are also present and other dysfunctions superimposed. For instance, it is very difficult to assess patients affected by sedatives or metabolic states which depress the nervous system.

Where a complex pathway involving long nerve tracts and numerous synapses is involved, a given dysfunction could be the result of lesions at one of several hypothetical locations along its length. In such cases, correlation with the other clinical dysfunctions will show that there is only one place where all the pathways controlling the deficient functions coincide, and that this must be the site of the lesion.

Should it not be possible to postulate a single anatomical site to account for all of the neurologic deficits, then it must be concluded that a **multifocal** or **diffuse** lesion is present.

It is vital that a systematic approach be taken to the neurologic examination. Many clinicians use a form or protocol such as that in Table 13.1 to ensure an orderly and complete examination.

Only when all sections of the nervous system have been assessed and the observations summarized can the abnormalities be correlated and some conclusions drawn as to the anatomical location of the lesion or lesions.

It is important to take time over the examination. A glance at the protocol (Table 13.1) will suggest that this is necessary and only the most obvious neural dysfunctions will be discovered during a standard consultation period. Ideally, the patient should be hospitalized to allow its initial agitation to subside and to permit it to be observed moving about at ease in a quiet, safe area with good footing. Subtle disturbances of gait, posture or behavior may then be observed, which would not be apparent with the animal restrained for physical examination. It is during this time also that an impression of the animal's mental status can be obtained from its interaction with the new environment and responses to strange sounds and sights.

The next stage is the 'hands on' part of the neurologic examination. The order in which the various parts of the examination are performed is a matter of individual preference. Some prefer to examine the cranial nerves first, if they suspect that the animal will have a short attention span and may become too agitated by the time the other tests have been done. In many cases, particularly with young animals and cats,rest periods are necessary during the examination, as these patients tend to become restless and impatient and a lack of cooperation quickly creates artefacts in the observations.

Once the anatomical diagnosis has been made it should be evident whether a focal lesion or a diffuse process is present. This will enable some ideas to be formulated as to the etiologic diagnosis. The identification of a focal lesion may suggest the possibility of a neoplasm, an abscess or a hematoma. Diffuse or multifocal lesions suggest the presence of inflammatory or degenerative processes with infectious, toxic or metabolic etiologies.

General clinical aspects of nervous dysfunction

Because a large proportion of veterinary patients are presented with ataxia, weakness, or some combination of both, it is useful to review the physiologic mechanisms responsible for the normal maintenance of coordinated movement, strength and muscle tone. This is encompassed in the concept of the lower motor neuron and the upper motor neuron.

The lower motor neuron

The spinal reflex arc is the fundamental unit, within the complex system of neuromuscular regulation, that permits the normal animal to

Table 13.1. *Sample neurologic examination form*

Date:

History:

Patient ID

General observations and physical examination

Mental status:

Gait and posture:

Cranial nerves:

II Menace –
Pupillary –
Ophthalmoscopic –
III Pupillary –
Strabismus –
IV Strabismus
V Motor: Mand –
Sensory: Ophth. – Palpebral –
Max. – Palpebral –
Corneal –
Mand. –
VI Strabismus –
Corneal –

VII Corneal, palpebral –
VIII Cochlear –
Vestibular –
Head tilt –
Nystagmus – Resting:
Vestibular:
Positional:
Post-Rotatory:
Strabismus
IX, X, XI
XII –

Spinal reflexes:

1 Flexor – LF RF
LH RH
2 Patellar – LH RH

3 Perineal –
4 Abnormal:
Crossed Extensor – F H
5 Triceps – Biceps –
6 Panniculus:

Attitudinal and postural reactions:

1 Tonic Neck and Eye –
2 Proprioceptive – LF RF
positioning LH RH
3 Placing – Optic –
– Tactile –
4 Extensor postural thrust –
LH RH

5 Wheelbarrow – LF RF
6 Hopping – LF RF
LH RH
7 Hemi-walks: L R
Hemi-stands: L R
8 Righting –

Pain perception –
Muscle tone –
Muscle atrophy –
Orthopedic examination –

Miscellaneous tests –

Summary of examination

Tentative diagnosis and prognosis –

ID, identification; Ophth., ophthalmic; Max., maxillary; Mand., mandibular; F, forelimb; H, hindlimb; R, right; L, left.

have precise control over voluntary movement.

The lower motor neuron (LMN) is the collective name given to the motor nerve cells in the ventral horns of the spinal cord gray matter, their axons extending out in the peripheral nerves, and the neuromuscular junctions where they terminate on the target muscles (Fig. 13.1). The LMN is, in the strictest sense, the 'final common pathway' by which all nervous impulses are transmitted to the skeletal muscles and is an intrinsic component of the spinal reflex arc.

Muscle tone is not an inherent property of the muscle but is maintained by reflex activity of the nervous system. The basic neural mechanism for maintaining muscle tone is the stretch reflex, which is also able to modify the tone so that voluntary movements can be superimposed. The LMN forms the motor arm (alpha motor neuron) of the stretch reflex (myotatic reflex) and it is by modification of the activity of this reflex that the higher centers in the brain (collectively called the upper motor neuron) can induce voluntary muscle movement.

Damage to the LMN results in the loss of normal resting muscle tone due to interruption of the myotatic reflex arc, loss of voluntary muscle movement because the 'final common pathway' is non-functional, and disruption of the reflex arc for tendon reflexes.

These deficits appear clinically as flaccidity (**hypotonia**) of the musculature, weakness and a reduction in reflex activity (**hyporeflexia**). In addition, the muscles rapidly undergo **atrophy** soon after disruption of LMN function. This denervation (neurogenic) atrophy is much more profound and occurs much more rapidly than the atrophy of disuse which occurs in muscles which have normal nerve supply. Its causes are poorly understood.

Neurogenic atrophy is also quite specific – only the muscle cells innervated by the damaged alpha motor neurons will atrophy. By contrast, disuse atrophy is a diffuse phenomenon occurring gradually and involving all the muscles of the affected limb.

In summary, LMN disease produces a group of signs resulting from damage to the pathway at any level from the ventral horn cell to the neuromuscular junction. These are:

1 Hypotonia/flaccidity,
2 Hyporeflexia/areflexia,
3 Weakness/paralysis,
4 Muscle atrophy – rapid, profound, specific.

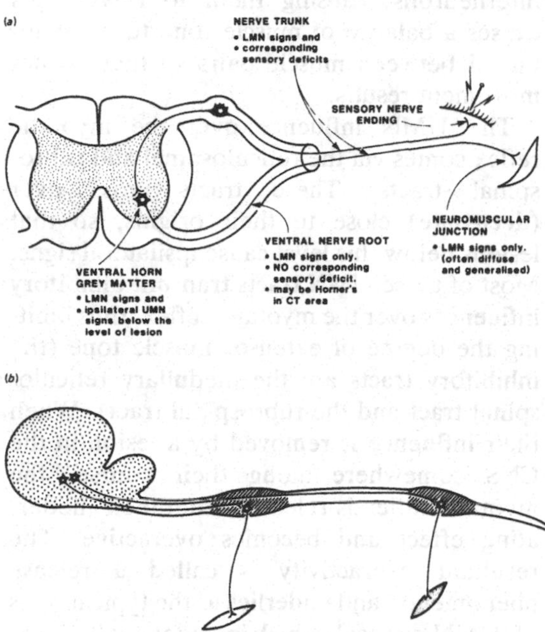

Fig. 13.1. (*a*) Schematic representation of the spinal reflex arc. The figure also indicates the typical signs caused by lesions located at various points on this neuroanatomical pathway. LMN, lower motor neuron; UMN, upper motor neuron; CT, cervicothoracic. (*b*) Schematic representation of the nervous system. UMN pathways are shown descending through the spinal cord to synapse with the LMN cell bodies in the gray matter of the brachial and lumbosacral intumescences (see also Fig. 13.6).

The upper motor neuron

The upper motor neuron (UMN) is the collective name given to all the motor systems within the CNS which are responsible for the initiation of voluntary movement, the maintenance of muscle tone for support of the body against gravity, and the regulation of posture to provide a stable background upon which voluntary activity can be superimposed.

Traditionally, the UMN is divided into the pyramidal and extrapyramidal systems. These two divisions are anatomically distinct but the pyramidal system, which is particularly concerned with finely skilled movements, is poorly developed in domestic animals and is therefore of little clinical relevance. The domestic animals depend much more, for control of their motor activity, on the other division of the UMN, known as the extrapyramidal system.

The **extrapyramidal system** comprises several groups of interconnected and functionally related structures that form a series of neurons in a multisynaptic pathway from the brain to the LMN. The UMN cell bodies are located in the cerebral cortex and in nuclei at all levels of the brain. Although they are all interrelated by feed-back loops, it is only the red nucleus and the reticular formation which give rise to tracts that actually descend the spinal cord and transmit the commands from the neuroanatomical bureaucracy down to the LMN (Fig. 13.1).

The **pyramidal system** (also known as the corticospinal system) is a monosynaptic system whose cell bodies are in the cerebral cortex and whose axons descend directly, via the pyramids of the medulla, to the spinal cord to terminate on interneurons in the ventral horn gray matter. Approximately 75% of corticospinal tract fibers cross in the pyramidal decussation and descend on the opposite side in the lateral corticospinal tract. Phylogenetically, the corticospinal system is a recent acquisition and therefore it is not surprising that its greatest physiologic expression is to be found in the higher primates. Its presence is related directly to the capacity of the animal to perform finely skilled movements. As may be expected, the clinical deficits resulting from lesions in this system produce much greater incapacitation in man than in subprimates, and the severe contralateral disabilities resulting from a cerebrovascular accident or 'stroke', which causes cerebrocortical damage in a human, are a good illustration of this. By contrast, similar lesions occurring spontaneously in cats cause minimal disturbances of gait and strength, although serious visual impairment is usual, due to damage to the occipital cortex.

The UMN modulates muscle tone and activity by its control over the myotatic reflex. As mentioned previously, this reflex utilizes information on muscle tension (from the stretch receptor muscle spindles) to induce more or less contraction in the extrafusal muscles via variations in the output of the alpha motor neuron (LMN). In this way a steady state of muscle tone is maintained. Information from the muscle spindles also passes to antagonistic muscles via inhibitory interneurons, causing them to relax. This causes a balance of muscle tone to be maintained between muscle pairs so that no net movement results.

The UMN influence over the myotatic reflex comes via the reticulospinal and rubrospinal tracts. These tracts cross over (decussate) close to their origins, so that lesions below this level cause ipsilateral signs. Most of these UMN tracts transmit inhibitory influences over the myotatic reflex, thus limiting the degree of extensor muscle tone (the inhibitory tracts are the medullary reticulospinal tract and the rubrospinal tract). When their influence is removed by a lesion in the CNS somewhere along their course, the myotatic reflex is released from their moderating effect and becomes overactive. The resultant overactivity is called a release phenomenon and underlies as the typical signs of 'UMN disease', which include:

- Spasticity/hypertonia of muscles, manifested as increased resistance to passive manipulation of the limbs.
- Hyperreflexia – as manifested by an exaggerated range and force of tendon reflex movements.

- Paresis – weakness due to disturbances in the mechanisms for initiating voluntary motor function.
- Abnormal reflexes – such as the crossed extensor reflex, described below.

The presence of spasticity, by giving the impression of rigid strength in the limbs, often masks the co-existing weakness in the early stages of UMN disease.

A facilitatory influence over extensor muscle groups is exerted by the pontine reticulospinal tract, and UMN lesions which damage this tract, but spare the inhibitory medullary reticulospinal and rubrospinal tracts, could give a clinical picture of hypotonia and hyporeflexia suggestive of LMN disease. This seems to occur in some cases of thoracolumbar intervertebral disk prolapse in dogs where, although the lesion is clearly suprasegmental (i.e. above the level at which the nerves to the hindlimbs arise), the hindlimbs are hypotonic and hyporeflexic. In such situations, other neurologic observations must be used to clarify the true location of the lesion.

Although occurring inconsistently, and being unreliable as diagnostic signs, **abnormal reflexes** are an additional manifestation of disruption of UMN influence.

The crossed extensor reflex is normally present in a standing animal. Its function is to increase muscle tone in the anti-gravity muscles of one limb to help take the extra weight when the opposite limb is lifted from the ground. The presence of this reflex in a recumbent animal (i.e. extension of one limb when the other is flexed in response to a noxious stimulus) is abnormal and is regarded as a sign of UMN disease.

Another abnormal reflex is the Babinski sign, which is a dorsal fanning of the toes in response to stroking the plantar aspect of the foot. The normal response would be for the toes to curl. In humans it is regarded as a sign of disturbance of the pyramidal component of the UMN system and as such should not occur in subprimates. It is, however, sometimes seen in dogs with cervical or thoracolumbar spinal cord lesions.

In summary, UMN disease produces a group of signs resulting from disruption of the influence of the UMN system on the LMN. These are:

1 Paresis,
2 Hyperreflexia – clonus.
3 Spasticity/hypertonia.
4 Occasionally, abnormal reflexes.

Common clinical syndromes

The great majority of animals with neurologic disease present with clinical signs that fall into the following categories. Although introduced here, these clinical categories are also discussed in relation to the various regions of the nervous system in the following section.

Seizures

A seizure or 'fit' is a paroxysmal, transient disturbance of consciousness, usually accompanied by abnormal somatic and visceral motor activity. Seizures are of sudden onset, cease spontaneously, and tend to occur repeatedly, often in 'clusters', with a variable time lapse within and between the episodes. The seizure episode itself is termed the ictus, while the time elapsing between seizures is called the interictus.

Many seizures take the form of violent convulsions and are known as epileptiform seizures, but this is not always the case. For instance, narcolepsy is a type of seizure characterized by sudden episodes of sleep.

Seizures originate as bursts of activity from neurons in the cerebrum, diencephalon or in the reticular formation of the brainstem and may be associated with organic lesions such as neoplasms or abscesses in these sites. They may, however, be induced by metabolic disturbance of structurally normal neurons, as in lead poisoning, ammonia poisoning or hypoglycemia. In idiopathic epilepsy and narcolepsy, there is intrinsic functional abnormality, causing spontaneous uncontrolled depolarization of the cell membranes of neurons.

Behavioral disturbances

Abnormal and sometimes bizarre behavior can be a manifestation of organic or metabolically induced brain disease. It may reflect structural lesions in the prefrontal lobe or diencephalon, or the influence of intoxication with agents such as lead, or the endogenous toxins that accumulate in renal or hepatic failure.

Many behavioral disturbances are psychologic in nature, however, and require a careful study of the animal and its environment, including the humans and other animals with which it interacts.

Cranial nerve deficits

Deficits in cranial nerve function are quickly appreciated because they involve the head, and attention is sought because of signs like a head tilt, blindness, or paralysis of the facial muscles.

In many cases, several cranial nerves are damaged simultaneously, particularly those arising from the pons and medulla, and adjacent nerve tracts are often involved as well, leading to coexistence of UMN deficits. The accurate identification of cranial nerve deficits occurring simultaneously with signs of dysfunction in other parts of the nervous system, particularly the spinal cord, is often the key to identifying the site of a brain lesion.

Incoordination of movement

The control of limbs and body movement involves the complex interaction of several parts of the CNS, including the cerebrum, midbrain, brainstem, cerebellum, vestibular system and spinal cord. Animals with lesions in these areas often have some disturbances in the control of voluntary movement and may appear ataxic, incoordinated, or weak, and have a tendency to walk in circles. The particular appearance of the abnormality shown by a patient may give vital clues to the location of the underlying lesion.

Weakness (paresis) and paralysis

Weakness and paralysis are very common presenting signs which can reflect the presence of lesions anywhere from the midbrain, through the spinal cord, to the peripheral motor nerves. It is in this context that an appreciation of the upper and lower motor neuron concept comes to the fore. A common locus of injury is the spinal cord, where focal or diffuse lesions occur frequently in many domestic animal species. Abscesses and intervertebral disk ruptures are good examples of focally acting injury, with ovine swayback and equine degenerative myelopathy as good examples of diffuse myelopathy.

Generalized interference with peripheral motor nerve function is often the result of electrolyte disturbances such as hypocalcemia or hypokalemia, or the action of toxins and venoms which depress the neuromuscular junction.

Horner's syndrome

This is a group of signs caused by disruption of the sympathetic innervation to the eye and the face and includes:

- Miosis – constriction of the pupil.
- Ptosis – drooping of the upper eyelid.
- Enophthalmos – retraction of the eyeball.
- Prolapse of the third eyelid.

The pathway by which the sympathetic fibers travel from the midbrain to the eye is very long and lesions in a variety of locations inside and outside the CNS can cause Horner's syndrome.

The coexistence of Horner's syndrome with other neurologic disturbances may aid greatly in the localization of the lesion underlying both problems. Horner's syndrome is discussed in detail together with the cranial nerves.

The anatomical basis of clinical neurologic diagnosis

The consequences of damage to the principal anatomical subdivisions of the nervous system will be described, along with appropriate test procedures used during clinical examination.

Telencephalon

The frontal lobe

The major section of the frontal lobe is the motor cortex which contains the cell bodies of the corticospinal (pyramidal) division of the UMN system, as well as the extrapyramidal neurons which project to the dorsal nuclei and other extrapyramidal nuclei in the brainstem. In animals, damage to the motor cortex does not result in severe paralysis, but learned and intricate movements may be lost, and there is mild contralateral hemiparesis. The relative mildness of clinical signs associated with damage to the motor cortex in domestic animals is in contrast to the results of similar lesions in humans and is a direct reflection of the predominant importance, in motor function, of the phylogenetically more primitive subcortical and brainstem structures (see The Upper Motor Neuron, above).

The prefrontal area of the frontal lobe, rostral to the motor cortex, is part of the limbic system and is therefore concerned with behavior. Signs of frontal lobe disturbance include compulsive walking and head pressing. It is common for such animals to turn their heads and wander in large circles towards the affected side when the lesion is unilateral or asymmetrical. This is known as the 'adversive' syndrome and is often seen in hydrocephalic dogs, and in cats with ischemic necrosis of the cerebral cortex (see p. 371).

The parietal lobe

The parietal lobe processes sensory information such as pain, touch and conscious proprioception. Damage to this area probably accounts for the deficits in postural reactions seen in animals with cerebrocortical lesions.

The occipital lobe

The occipital lobe is necessary for conscious perception of visual stimuli and lesions in this area (the visual cortex), although causing blindness, do not affect pupillary reflexes. However, the eye-preservation reaction ('menace response') would be absent (see Optic nerve, below).

The basal nuclei

The basal nuclei include the caudate nucleus, putamen and globus palidus and are part of the extrapyramidal UMN system. Lesions in this area cause contralateral weakness and UMN release phenomena. In practice, it is difficult to detect these signs in domestic animals unless the UMN system is affected more distally, in the midbrain or brainstem. Simultaneous injury to ascending sensory tracts often causes postural reaction deficits, which resemble weakness but are mainly due to deficiencies in proprioception.

Diencephalon

The diencephalon is the most rostral part of the brainstem and incorporates the hypothalamus and thalamus. Hypothalamic lesions may cause disturbances in autonomic functions such as appetite, thirst and temperature regulation. The hypothalamus is also a control center for most of the endocrine system so that damage here may be reflected as endocrine disturbances.

The thalamus is a major relay center for sensory information en route to the cerebral cortex, and thalamic lesions might be expected to cause contralateral deficits in functions such as vision, proprioception or pain perception. As the thalamus also relays motor information from the cerebellum and extrapyramidal nuclei to the telencephalon, contralateral dysmetria (disturbances of gait) may be observed. The thalamus also functions as part of the ascending reticular activating system and lesions may cause disturbances of consciousness, or convulsions.

The limbic system is a complex of neurons and tracts which forms two rings around the diencephalon and makes connections with nuclear groups in the rostral brainstem, diencephalon and cerebral cortex. The limbic system is involved with emotional and behavioral patterns and has many connections with the hypothalamus. This provides the mechanisms for the visceral component of many emotional reactions.

The key limbic structures involved in organic brain disease are the hypothalamus, hippocampus, amygdaloid body and the cingulate and septal areas of the cerebral cortex. Disorders of the limbic system cause emotional and behavioral disturbances and the disturbances of hypothalamic function already mentioned.

Seizures can have their origin in the hippocampus and amygdaloid body, and are generally characterized by bizarre behavioral manifestations. They are called psychomotor, or temporal lobe seizures because of the proximity of the temporal lobe.

Midbrain

The midbrain contains several important structures and severe neurologic deficits can therefore result from lesions in this area.

The **ascending reticular activating system** (ARAS) is part of the reticular formation which functions to arouse the cerebral cortex. It is responsible for maintaining wakefulness and is thought to be the seat of consciousness. Damage to it causes disturbances in consciousness, often resulting in a semi-comatose or comatose state.

The **oculomotor nucleus** lies in the midbrain adjacent to the midline and its axons pass ventrally through the reticular formation of the tegmentum, medial to the red nucleus and then through substantia nigra and crus cerebri, to emerge on the underside of the brainstem. (Oculomotor deficits are more fully discussed in the next section.)

Normal muscle tone is mainly influenced by two centers in the lower brainstem. The **pontine reticular formation** is a stimulator of extensor muscle tone via the pontine reticulospinal tract. Although it can function independently of higher center control, it is usually subject to a moderating influence from higher centers. The **medullary reticular formation** exerts an inhibitory influence over the extensor muscle tone via the medullary reticulospinal tract, but, to function, requires continual input from the cerebral cortex, basal nuclei and cerebellum. A lesion in the midbrain has the effect of inactivating the medullary reticular formation by removing the higher influences which activate it, at the same time releasing the potentially autonomous pontine reticular formation from the moderating influences of the cerebrum and basal nuclei. The result is predominance of facilitatory influences on the extensor reflexes (UMN release phenomenon), with the classical appearance of opisthotonus, or some lesser degree of extensor muscle hypertonia.

The red nucleus normally facilitates flexor muscle tone and inhibits extensor tone via the rubrospinal tract. Removal of this influence by a midbrain lesion adds to the predominance of extensor tone.

Lesions which cause swelling and increased intracranial pressure, such as trauma, neoplasia and polioencephalomalacia, can cause herniation of the occipital lobes ventral to the tentorium cerebelli. This directly compresses the midbrain and, if the compression is severe, unconsciousness (due to ARAS damage), opisthotonus (UMN release phenomenon), and ventrolateral strabismus with fixed, dilated pupils (oculomotor damage) will appear. Less severe midbrain damage will result in a less severe depression of consciousness, milder UMN signs, such as spastic tetraparesis or hemiparesis, and pupils which may be asymmetrical or bilaterally miotic.

Signs of midbrain involvement are therefore useful indicators of the severity of intracranial swelling, especially in trauma cases.

Medulla/pons

Lesions in this area cause UMN signs of ataxia, weakness and hyperflexia of all four limbs (ipsilateral signs if a unilateral lesion) and deficits involving cranial nerves V to XII.

The combination of cranial nerve deficits with UMN signs is a vital indicator of brainstem involvement and identification of the cranial nerves involved can allow accurate location of lesions within the brainstem.

In addition, the medullary and pontine areas of the brainstem contain control centers for respiration, blood pressure and heart rate and the ARAS. Severe lesions may therefore cause life-threatening depression of these vital functions.

Cerebellum

The cerebellum regulates, coordinates and 'smooths out' voluntary movements. It is positioned astride the brainstem and monitors information from various parts of the CNS concerning body position and movement. This it compares with information from UMN centers about intended movements, so that it can correct any mistakes or inaccuracies of muscle movement via its feedback circuits to the UMN centers. Without cerebellar function, voluntary movements can still occur, but they are jerky and lack fine control.

The cerebellum does not initiate movement and therefore cerebellar disease is characterized by **ataxia without weakness**. In severe cases, the patient may be incapacitated by the severity of the incoordination, but voluntary movements are elicited easily and have normal strength. The gait is forceful and ataxic and is characterized by an inability to regulate the rate, range and force of movements. This dysmetria usually manifests as **hypermetria** – an exaggeration of movements. The onset of voluntary movement is delayed and the response, once initiated, is exaggerated. Once the limb starts to move, it is raised too high and brought down too forcefully. The resting posture may show a broad-based stance on the thoracic limbs, and a **truncal ataxia**, a swaying of the body from side to side and forwards and backwards. A fine head tremor is also often evident which becomes exaggerated when voluntary movements of the head are initiated, particularly with precise maneuvers such as eating. This is known as an **intention tremor**. Muscle tone is usually increased and reflexes are normal or hyperactive. This is due to disruption of inhibitory cerebellar feedback to the UMN system.

In extreme cases, opisthotonus may occur, especially when the rostral lobe is involved. Unilateral cerebellar lesions produce ipsilateral signs.

A phenomenon yet to be explained fully is the absence of a menace response in animals with cerebellar cortical disease, but with normal vision and facial muscle control (see Optic and Facial nerves, below). Although usually associated with generalized lesions, this menace response deficit has been observed ipsilaterally with unilateral cerebellar lesions.

Because the cerebellum is closely integrated with the vestibular system (Fig. 13.2), cerebellar lesions can also cause vestibular signs, with loss of equilibrium, abnormal nystagmus, and a broad-based staggering gait with jerky movements. A 'paradoxical vestibular syndrome' occurs with unilateral lesions that affect the cerebellar peduncles and medulla. It is characterized by a head tilt to the side *opposite* to all of the other localizing signs, especially those UMN signs suggestive of brainstem involvement (see Vestibular nerve dysfunction, below).

Cranial nerves

The majority of cranial nerves are analogous to spinal nerves in that they have their origins in the gray matter of the CNS and extend beyond the brainstem towards their target structure. Others, such as the optic and olfactory nerves, are sensory extensions of the brain.

Many of the tests of cranial nerve function involve the observation of reflex activity. Each of these tests, therefore, involves a sensory and a motor pathway and often tests two cranial nerves simultaneously. Therefore, it becomes vital to correlate the results of all the cranial nerve tests in order to interpret correctly which nerves are normal and which are deficient in function.

Cranial nerve I: olfactory
The olfactory nerve is the sensory path for the conscious perception of smell, and loss of the perception of smell is called anosmia. Its function is difficult to assess. For instance, anosmic dogs will still show sniffing behavior when visually investigating their surroundings.

Behavioral responses to olfactory stimulation may be tested, using aromatic materials such as oil of cloves, xylol or fish oil. However, visually recognizable foodstuffs and irritants such as ammonia (smelling salts) should be avoided.

Anosmia is most commonly caused by rhinitis. Traumatic head injury may cause the olfactory nerves to be sheared off as they pass through the cribiform plate, but this is not common.

Cranial nerve II: optic
The optic nerve is the sensory path for vision and the pupillary light reflexes. Visual capacity is assessed by observation of the movements of an animal in unfamiliar surroundings or its response to the movement of objects nearby. The optic nerve can also be assessed by the eye-preservation and pupillary light reflexes.

The **eye-preservation** or **menace response** is tested by means of a threatening gesture made towards one eye. The normal response is for the eye to blink and the head to be averted. The operator must take care to avoid stimulation of the sensitive tactile hairs around the eye by direct contact or air turbulence. Some prefer to interpose a pane of glass, or use the blunt end of a pen or pencil for the menacing gesture. One eye should be tested, while the other is covered. Stoic, depressed or very young animals may need to be tapped gently near the eye a few times before they will begin to respond appropriately. The blink response requires intact facial nerve function, but, even without it, aversion of the head and retraction of the eyeball indicate intact visual perception. It has been observed that in some cases of cerebellar disease, the menace reaction is

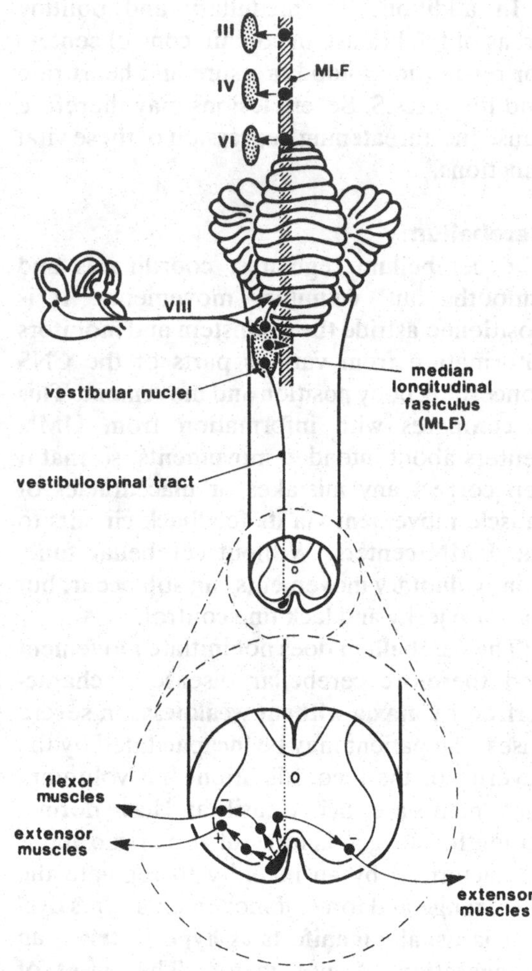

Fig. 13.2. The vestibular system showing the interconnections between the vestibular end-organ, the vestibular nuclei, the cerebellum and the nuclei of cranial nerves III, IV, and VI. The vestibulospinal tract is an upper motor neuron pathway exerting an inhibitory influence on extensor muscle tone on the contralateral side of the body and a facilitatory influence ipsilaterally.

deficient even though vision and facial nerve functions are normal.

In the **pupillary light reflex** (PLR), the optic nerve is the afferent arm of a reflex arc which passes via the optic chiasm to the optic tract and into the oculomotor nucleus in the midbrain (Fig. 13.3). The oculomotor parasympathetic fibers in the efferent arm of the reflex cause pupillary constriction in response to illumination of the retina. This is known as the **direct PLR**. There will also be a partial constriction of the pupil of the other eye due to some decussation in the optic chiasm and in the midbrain. This is known as the **indirect** or **consensual PLR**. The PLR is tested by shining a bright light into one eye and observing the brisk constriction of the pupil of that eye and the less complete consensual constriction in the other eye.

Combined assessment of the menace reaction with the PLR test allows the differentiation of lesions at various levels in the visual system (see Fig. 13.3 and Table 13.2).

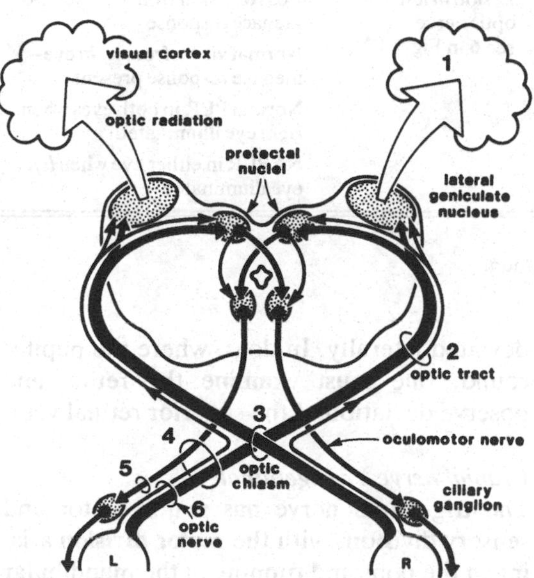

Fig. 13.3. The optic system, showing the pathways for visual perception and for reflex pupillary activity. The numbers refer to the locations of the hypothetical lesions discussed in Table 13.2. L, left; R, right.

Cranial nerve III: oculomotor

The oculomotor nerve arises in the midbrain and has two components:

1 Parasympathetic motor fibers for pupillary constriction.
2 General somatic efferent fibers for motor control of the extraocular muscles – dorsal, medial and ventral recti, and ventral oblique, and also for the motor pathway to the levator palpebrae muscle of the upper eyelid (see Fig. 13.4).

Testing of the parasympathetic division has already been discussed above in relation to the optic nerve. Unilateral dysfunction of the oculomotor nerve will cause the pupil to be constantly dilated relative to the normal eye. There will be absence of direct or consensual PLR on that side, although both will be present in the other eye provided both visual pathways are intact.

Lesions in the general somatic efferent division of the oculomotor nerve cause a fixed ventrolateral deviation of the eye (strabismus) and drooping of the upper eyelid (ptosis). The extraocular muscle paralysis can be confirmed by observing that the eye fails to move in coordination with the other eye as the head moves about.

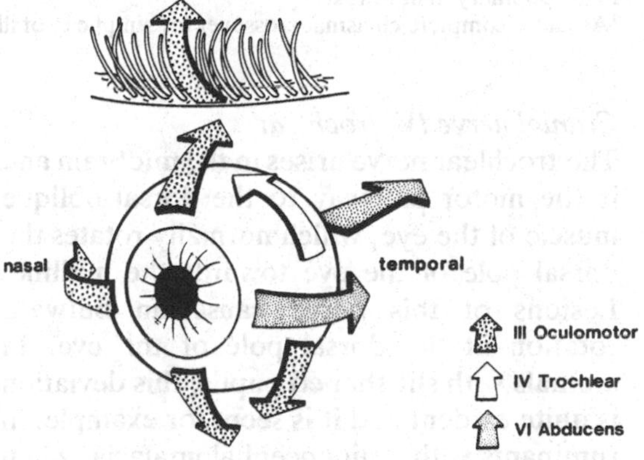

Fig. 13.4. Motor innervation of the extraocular muscles and upper eyelid.

Table 13.2. *Location of optic system lesions by combined assessment of visual responses and pupillary light reflexes*

Lesion in right visual cortex see **1** in Fig. 13.3	Loss of vision from *left* eye[a] – no menace response Normal[a] vision in *right* eye – menace response present Intact and normal PLR from both eyes because the lesion is above the level of the reflex circuit Both pupils have normal size	Lesion in left retrobular area affecting optic nerves see **4** in Fig. 13.3	Loss of vision from *left* eye – no menace response Normal vision in *right* eye – menace response present No PLR in either eye when *left* eye illuminated PLR in *right* eye only when *right* eye illuminated *Right* pupil of normal size *Left* pupil dilated, unreactive
Lesion in right optic tract see **2** in Fig. 13.3	Loss of vision from *left* eye[a] – no menace response Normal[a] vision in *right* eye – menace response intact Weak PLR in both eyes when *left* eye illuminated because the only input into the PLR reflex circuit is from uncrossed optic nerve fibres (temporal side of retina) Normal PLR in both eyes when *right* eye illuminated. Both pupils have normal size	Lesion in left oculomotor nerve see **5** in Fig. 13.3	Normal vision *both* eyes – menace responses intact PLR in *right* eye only, when either eye illuminated *Right* pupil of normal size *Left* pupil dilated, unreactive
Lesion at optic chiasm see **3** in Fig. 13.3	Loss of vision from *both* eyes – no menace responses No PLR from illumination of either eye Both pupils dilated	Lesion in left optic nerve see **6** in Fig. 13.1	Loss of vision from *left* eye – no menace response Normal vision from *right* eye – menace response present Normal PLR in both eyes when *right* eye illuminated No PLR in either eye when *left* eye illuminated

PLR, pupillary light reflex.
[a]Assumes complete chiasmal crossover for simplicity of illustration.

Cranial nerve IV: trochlear

The trochlear nerve arises in the midbrain and is the motor pathway to the dorsal oblique muscle of the eye, which normally rotates the dorsal pole of the eye towards the midline. Lesions of this nerve cause an outward rotation of the dorsal pole of the eye. In animals with slit-shaped pupils, this deviation is quite evident and it is seen, for example, in ruminants with polioencephalomalacia, when the medial aspect of the pupil is raised. In cats, the dorsal aspect of the pupil would be deviated laterally. In dogs, where the pupil is round, one must examine the retina and observe deviation of the superior retinal vein.

Cranial nerve V: trigeminal

The trigeminal nerve has major motor and sensory divisions with the **motor division** arising in the pons and running in the mandibular nerve to innervate the masseter, temporal, rostral digastric, pterygoid and mylohyoid muscles. Bilateral motor paralysis produces a dropped jaw that cannot be closed voluntarily.

There is drooling of saliva and considerable inability to prehend and chew. Unilateral lesions cause decreased jaw tone which can be assessed manually, but there is minimal disturbance of prehension or mastication.

In both bilateral and unilateral deficits, LMN **neurogenic atrophy** soon becomes evident. In the case of a unilateral deficit, this may be the first sign observed by the owner. Atrophy of the muscles of mastication causes the temporal area of the head to assume a hollow and wasted appearance and the zygomatic arch to become prominent. Because the masticatory muscles help to support the eyeball by forming the lateral and caudal sides of the orbit, their atrophy causes enophthalmos, which becomes even more obvious when the jaw is opened.

The **sensory division** feeds into nuclei dispersed through the mid- and hindbrain and is distributed through three main branches – hence the name of the nerve. The ophthalmic branch innervates the forehead, medial canthus and corneal surface. The maxillary branch innervates the upper lips, nose and lateral canthus and the mandibular branch provides sensory innervation for the lower jaw and cheeks.

Sensation should be tested over the distribution of all these branches. The **palpebral reflex** is a blink response, evoked by a touch to the eyelids. It is best performed with a small instrument brought up from behind the visual field to prevent the animal from seeing it and blinking in anticipation. Touching the medial canthus tests the ophthalmic branch; touching the lateral canthus tests the maxillary branch. It should be remembered that the palpebral reflex is dependent on intact facial nerve function. If facial paralysis is present, trigeminal nerve function must be assessed by means of the corneal reflex.

The **corneal reflex** tests the ophthalmic branch and is performed by holding the eyelids open and gently touching the surface of the eye with a clean wisp of cotton. Normally, the eyeball is retracted and third eyelid flicks passively across. This reaction is mediated by the abducens nerve, which innervates the retractor bulbi muscle.

If a light touch of the eyelids or lips does not elicit clear evidence of intact sensation, stronger stimulation may be employed by means of a pin-prick, or light squeeze with forceps. When there is no demonstrable response from these areas, stimulation of the sensitive mucosa of the nasal septum may be used.

Partial loss of facial sensation can be very difficult to demonstrate unless it is unilateral or asymmetrical. It is often necessary patiently to repeat the testing of both sides in order to gain a consistent impression of the sensory status.

Cranial nerve VI: abducens

The abducens nerve innervates the lateral rectus and retractor bulbi muscles of the eye. Dysfunction is manifested by a medial deviation (strabismus) of the eye, and suppression of the corneal reflex. Side-to-side movement of the head reveals an inability to abduct the eye.

Cranial nerve VII: facial

The facial nerve arises from a nucleus in the medulla and is the motor pathway to the muscles of facial expression. It is the LMN of some of the reflexes already described (e.g. menace, palpebral).

A unilateral deficit will result in hemiparesis or paralysis of the face, causing facial asymmetry. The lips and upper eyelid on the affected side tend to droop and the ear assumes a fixed, partially dropped position. The nose and upper lip will be drawn towards the normal side, and close inspection will reveal reduction or absence of the normal flaring of the nostril on inspiration. There will be deficiencies in the menace and palpebral reflexes, and, when the animal is alerted by sounds, the affected ear will not move together with the normal ear.

Prolonged facial paralysis often leads to the development of an involuntary periodic retraction of the eyeball, which serves to bring the third eyelid across the eye, moistening it. Without this mechanism the eye suffers from exposure and severe keratoconjunctivitis can develop.

As the facial nerve also provides sensory innervation from the rostral two-thirds of the tongue, this function can be tested by the application of a small amount of atropine to the affected side. Perception of the bitter taste will be delayed until it diffuses to the normal side of the tongue.

Cranial nerve VIII: vestibulocochlear

The vestibulocochlear nerve is linked to a complex of nuclei in the medulla and cerebellum and has two divisions. The **cochlear division** mediates hearing, but loss of hearing is very difficult to test clinically without sophisticated audiometric equipment. Unilateral deafness is virtually impossible to detect without audiometry. However, bilateral deafness may be fairly well proven by the observation of behavioral reactions to sound. Care must be taken to create a purely auditory stimulus so that the animal's attention is not drawn by visual stimuli or vibrations caused when the loud noise is created. When testing young dogs, it is often possible to catch them dozing and observe whether a loud sharp noise, such as made by striking together two stainless steel bowls, will rouse them.

Congenital deafness is not rare, and tends to be associated with white and merle coat colors, the defect often lying in the cochlear end organ. Deafness may be acquired, due to the toxic effects of some drugs such as aminoglycoside antibiotics, or as an ageing change.

The **vestibular division** provides information about the orientation of the head and its rate of change of movement and direction. The vestibular system is responsible for maintaining the position of the eyes, trunk and limbs with respect to the position or movement of the head at any time. Thus, with the aid of the visual system, and proprioceptive information from joints and tendons, it helps to maintain balance and orientation.

The receptor cells for the vestibular system are in the inner ear and their axons pass out of the petrous temporal bone as the vestibular division of the eighth nerve, entering the brainstem at the cerebellomedullary angle. Most of these axons pass to one or other of the four vestibular nuclei in the medulla, while some pass directly to the cerebellum.

The vestibular nuclei project axons to the spinal cord (via the vestibulospinal tract), the brainstem (via the medial longitudinal fasiculus), and the cerebellum (Fig. 13.2).

The **vestibulospinal tract** coordinates position and activity of the hindlimbs and trunk with movements of the head. It is a UMN tract and exerts a facilitatory (tonic) influence over the ipsilateral extensor muscles, and an inhibitory influence over the ipsilateral flexors and contralateral extensors.

The **median longitudinal fasiculus (MLF)** is a tract in the brainstem which connects with the nuclei of cranial nerves III, IV and VI. The vestibular nuclei project fibers into the MLF, linking all the nerves which control the extraocular muscles with the vestibular system to provide conjugate eye movements coordinated with movements of the head.

Vestibular information also passes up the brainstem to the **vomiting center**, providing conscious sensation of motion but also providing the underlying mechanism for motion sickness.

The cerebellum receives vestibular input as part of the battery of proprioceptive information it uses to regulate body movement. These inputs project to the flocculonodular lobe at the back of the cerebellum. Lesions in this area, for example, a neoplasm in the roof of the fourth ventricle, can cause severe signs of vestibular disturbance. Vestibular deficits are considered to be peripheral when the inner ear or the eighth nerve trunk are damaged and central when lesions involve the nuclei and tracts within the brain.

Clinical signs of vestibular disturbances are almost always related to **unilateral vestibular disease** and include the following.

- There is ataxia without weakness, due to loss of equilibrium. The animal may be reluctant to move and a broadbased stance is often observed as the animal attempts to compensate for the tendency to fall.

- There is tight circling towards the side of the lesion. In severe cases the animal will fall and roll to that side, especially when it shakes its head.
- Head tilt and postural abnormalities reflect loss of coordination between the head, trunk and limbs, and the head is often tilted towards the side of the lesion. There is a tendency to lean towards the side of the lesion and the trunk may be scoliotic, with the concavity directed towards the side of the lesion.
- Hypertonia and hyperreflexia occur on the side opposite to the lesion due to removal of the inhibitory contralateral UMN vestibulospinal influence. This imbalance in vestibulospinal influence may also explain the abnormalities in posture.

Another common manifestation of vestibular disease is abnormal nystagmus, which is best described after first reviewing normal **vestibular nystagmus**. This is an involuntary, rhythmic oscillation of the eyeball that occurs whenever the head moves. The gaze tends to remain fixed on an object so that, as the head moves, the eyes tend to 'lag behind' (slow phase). At some point the gaze is released and the eyes jerk towards the direction of head movement (quick phase) to 'catch up' with the new head posture. The plane of the nystagmus corresponds to the plane of the head movement and its function is to maintain visual fixation on stationary points as the body moves. For purposes of notation, the direction of such nystagmus is, by convention, ascribed to that of the quick phase (see Table 13.3).

Vestibular nystagmus can be demonstrated simply by moving the head slowly from side to side. The quick phase is towards the direction of movement. Immediately the head movement ceases, the oscillations of the eyeball stop. In all but rare cases, both eyes move in synchrony.

Abnormal nystagmus is that which occurs while the head is not moving. If it is seen when the head is in its normal extended position; it is called a **spontaneous** or **resting nystagmus**. If it is induced by holding the head fixed in a deviated position, such as with the chin elevated, it is called a **positional nystagmus**.

In peripheral vestibular receptor disease (in the inner ear) the nystagmus is either horizontal or rotary and always directed *away* from the side of the lesion. The direction of rotatory nystagmus is defined as 'clockwise' or 'anticlockwise' according to the change in the 12 o'clock position of the limbus during the quick phase.

In central vestibular disease (involving vestibular nuclei or cerebellar connections) the nystagmus may be horizontal, rotatory or vertical and may change direction.

Strabismus is another phenomenon seen sometimes in vestibular disease. In this case, it involves the ventral deviation of the eyeball on the affected side. This **vestibular strabismus** may be seen when the head is in a neutral position or may occur during postural reaction testing when the chin is lifted. This 'dropped eye' can be identified as being due to a vestibular disturbance if it can be corrected by altering the position of the head. In contrast, strabismus due to extraocular muscle paralysis would remain no matter what the orientation of the head.

Although most postural reactions are normal in vestibular disease, there is a deficit in the righting response. The animal may have difficulty rising from recumbency on the affected side.

In those rare cases of bilateral vestibular disease, there is complete loss of vestibular function, the ataxia is symmetrical and the patient tends to remain crouched on the ground with its limbs spread out. It may crawl along in this posture occasionally falling to one side or the other. A jerky, swaying, side-to-side head movement often occurs. There is no normal or abnormal nystagmus.

The **paradoxical vestibular syndrome** has already been mentioned in the context of cerebellar disorders. Some unilateral lesions in the cerebellar medulla and peduncles produce a head tilt and ataxia directed to the side

Table 13.3. *Clinical features useful in the distinction of central from peripheral vestibular disease*

	Peripheral vestibular disease	Central vestibular disease
Gait	Asymmetrical ataxia	Asymmetrical ataxia Hemiparesis
Head tilt	Always down on side of lesion	Usually down on side of lesion. NB: Paradoxical vestibular syndrome
Tone and reflexes	Contralateral hypertonia and hyperreflexia – usually mild	Ipsilateral hypertonia and hyperreflexia marked. Ipsilateral depression of conscious proprioception
Nystagmus	Always away from side of lesion Horizontal or rotatory	May change direction: horizontal, rotatory or vertical
Strabismus	Ipsilateral vestibular strabismus	Ipsilateral vestibular strabismus
Other	Horner's syndrome	Cranial nerve V – facial hypalgesia
	Cranial nerve VII – facial paralysis	Cranial nerve VII – facial paralysis Depression Head tremor, hypermetria Dysphagia

opposite the lesion. Such lesions also usually cause brainstem damage, which produces other cranial nerve and UMN signs and interferes with proprioception, thus allowing the true location of the lesion to be determined accurately.

The distinction between vestibular signs due to a peripheral lesion (inner ear) and those due to a central lesion (brainstem, cerebellum) is vital if one is to establish accurate diagnosis and prognosis. The prognostic implications for a peripheral lesion, often something like a middle ear infection, are clearly more favorable than those of a central lesion, which is likely to be a neoplasm. The diagnostic criteria for differentiating central from peripheral vestibular disease are shown in Table 13.3.

Cranial nerve IX: glossopharyngeal

The glossopharyngeal nerve is the motor pathway to the muscles of the pharynx along with some fibers from the vagus nerve. It also supplies parasympathetic motor fibers to the zygomatic and parotid salivary glands. It is the sensory pathway from the caudal one-third of the tongue and pharyngeal mucosa.

The signs of dysfunction include loss of gag reflex, loss of taste from back of tongue and dryness of the mouth due to lack of parotid saliva.

Function is assessed by observation of the palate and pharynx for asymmetry and by eliciting a gag or swallowing reflex by stimulating the pharynx with a tongue depressor or finger. In areas where rabies occurs, these tests should be performed with great care as dysphagia is one of the signs of the disease.

Cranial nerve X: vagus

The vagus nerve is the motor pathway to the pharynx, the larynx and the palate, and supplies parasympathetic motor fibers to the viscera of the body. The vagus nerve is the sensory pathway from the caudal pharynx, the larynx and the viscera of the body.

The signs of dysfunction include dysphagia,

megaesophagus, vocalization defects and tachycardia. Functional testing involves observation of the gag reflex and laryngeal movements.

Cranial nerve XI: spinal accessory
This nerve provides motor innervation to the trapezius muscle and parts of the sternocephalicus and brachiocephalicus muscles.

Signs of deficit are difficult to detect and are rarely recognized. Atrophy of the muscles may be detected but generally a lesion causing dysfunction of this nerve also causes other, more obvious and severe signs of brainstem or upper cervical damage.

Cranial nerve XII: hypoglossal
The hypoglossal nerve innervates the intrinsic and extrinsic muscles of the tongue, and the geniohyoideus muscle.

Signs of dysfunction include paralysis and atrophy of one or both sides of the tongue. If there is a unilateral lesion, the tongue will deviate towards the side of the lesion.

Functional testing involves observation of tongue movements, especially during drinking and eating. If the nose is moistened, many animals will lick at it, allowing tongue movement and symmetry to be observed.

Horner's syndrome

This refers to a group of signs caused by disruption of the sympathetic innervation to the eye and face. These are:

- Miosis (constriction of the pupil) – due to autonomic imbalance with predominance of parasympathetic influence over pupil.
- Ptosis (drooping of the eyelid) – due to reduced tone in the tarsal muscle of upper eyelid.
- Enophthalmos (retraction of the eyeball) – due to relaxation of smooth muscles of periorbita.
- Prolapse of third eyelid – occurs secondary to enophthalmos.
- Peripheral vasodilation – warmth and hyperemia of the ear and congestion of the nasal mucosa on the side of the lesion.
- Sweating – on the ipsilateral side of the face and upper neck in horses. (Sweating and ptosis are the main signs in the horse, miosis is not obvious.)

The presence of these disturbances is not particularly harmful in itself but it often serves to indicate the location of a lesion which may also be causing other, more severe, neurologic problems. An understanding of the anatomy of the sympathetic pathways to the eye can turn Horner's syndrome into a useful clinical tool for the accurate location of such lesions (Fig. 13.5).

The sympathetic innervation of the eyeball arises in the hypothalamus and pretectal

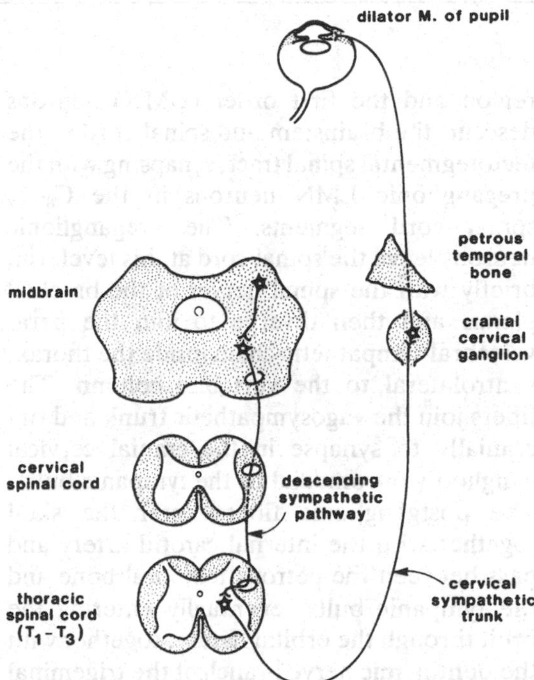

Fig. 13.5. Sympathetic innervation to the eye. Disruption of this pathway at any point can result in Horner's syndrome (Table 13.4), a distinctive combination of miosis, ptosis of the third eyelid, enophthalmos and, in horses, sweating, on the ipsilateral side. T_1–T_3 spinal cord segments, see Fig. 13.6; M., muscle.

Table 13.4. *Situations in which Horner's syndrome may occur*

Anatomical site of lesion	Concomitant signs that aid in location of lesion
Cervical spinal cord: C_1–C_7	
Not common; may occur with severe processes such as infarction	Tetraparesis – probably quite severe – with LMN signs to forelimb
Cervicothoracic spinal cord: C_8–T_2	
Usually due to injury with nerve root avulsion; neoplasia of nerve roots	LMN signs ipsilateral forelimb Ipsilateral hemiparesis Ipsilateral absence of panniculus reflex
Cranial thoracic sympathetic trunk	
Usually due to anterior mediastinal neoplasms, e.g. lymphosarcoma, thymoma	Regurgitation of food Dyspnea
Cervical vagosympathetic trunk	
Injuries to neck – trauma; surgical interference; jugular venepuncture	External signs of trauma, history of surgery
Middle ear	
Otitis media/interna	Vestibular signs, signs of facial nerve paralysis
Retrobulbar area	
Usually contusion or neoplasia	Proptosis Signs of periorbital trauma

region and the first order (UMN) neurons descend the brainstem and spinal cord in the tectotegmental spinal tract, synapsing with the preganglionic LMN neurons in the C_8–T_2 spinal cord segments. The preganglionic neurons leave the spinal cord at this level, run briefly with the spinal nerves of the brachial plexus and then diverge to join the paravertebral sympathetic chain inside the thorax, ventrolateral to the vertebral column. The fibers join the vagosympathetic trunk and run cranially to synapse in the cranial cervical ganglion ventromedial to the tympanic bulla. The postganglionic fibers enter the skull together with the internal carotid artery and pass between the petrous temporal bone and the tympanic bulla, eventually entering the orbit through the orbital fissure, together with the ophthalmic nerve branch of the trigeminal nerve.

Lesions at any one of several locations along this long and convoluted pathway could disrupt the sympathetic innervation to the eye and produce Horner's syndrome.

An example of the value of recognizing Horner's syndrome is in the analysis of brachial paralysis. Such a paralysis could be due to a lesion of the radial nerve in the antebrachium, a lesion of the brachial plexus in the axilla, or to traumatic avulsion of the brachial plexus nerve roots from the spinal cord. This last injury carries the worst prognosis for recovery and its recognition is important. When the nerve roots are avulsed in this way, there will also be damage to the sympathetic fibers as they emerge from the spinal cord and Horner's syndrome will be present. Damage to the nerves more distally, beyond the point at which the sympathetic fibers diverge to join the vagosympathetic trunk, will not cause Horner's syndrome. Thus the concurrent presence of Horner's syndrome with signs of brachial paralysis indicates the level at which the brachial plexus has been damaged and assists in determination of the prognosis.

Pharmacologic testing may be useful in locating the lesion responsible for Horner's syndrome. Such testing utilizes the phenomenon of 'denervation hypersensitivity', which makes smooth muscle, recently deprived of its innervation, supersensitive to the neurotransmitter, particularly when the lesion involves the postganglionic neuron. Differentiation of pre- and postganglionic lesions can be made in the first weeks of Horner's syndrome by instillation of 0.1 milliliter of 0.001% (w/v) epinephrine into the eye. Pupillary dilation will occur within 20 minutes with a postganglionic lesion, but may take up to 40 minutes with a preganglionic lesion. Common situations in which Horner's syndrome occurs are presented in Table 13.4.

Spinal cord

Clinical evaluation of spinal cord disorders is aimed at (1) location of the lesion and (2) estimation of the severity of the damage caused. Location of lesions is achieved by using the following: (*a*) observation and correlation of signs of UMN and LMN disturbances; (*b*) assessment of spinal reflexes and muscle tone; (*c*) observation of Panniculus reflex; (*d*) postural reaction testing; (*e*) detection and location of spinal pain; (*f*) presence of Schiff–Sherrington phenomenon.

Correlation of signs of UMN and LMN disturbances

For the purposes of the neurologic examination, the spinal cord can be visualized as having five zones, related to the distribution of major peripheral nerve systems (Fig. 13.6). These zones are defined in terms of spinal cord segments and it is important to realize that caudal to the thoracolumbar area there is a progressively greater discrepancy between the location of the spinal cord segments and the correspondingly named vertebrae. The cord segments become shorter towards the caudal end and the spinal cord is effectively compressed, so that, in the dog for instance, the L_4, L_5 and L_6 cord segments usually lie within the fourth lumbar vertebra. The practical relevance of this becomes evident when attempts are made to correlate the clinically determined location of a spinal cord lesion with radiographic changes in the spinal column.

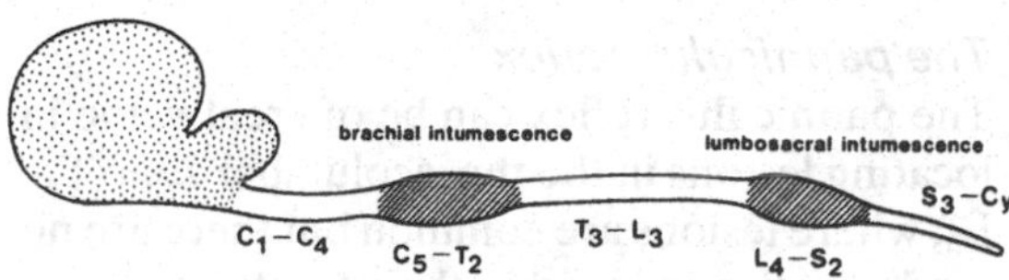

Fig. 13.6. The five functional zones of the spinal cord of the dog, defined according to the presence of major peripheral nerve origins.

The **upper cervical zone** comprises cord segments C_1 to C_4 and is anterior to the level of the brachial plexus. Extensive lesions within this area interfere with the UMN and sensory tracts of all four limbs and therefore cause ataxia, hypertonia, hyperreflexia, weakness (tetraparesis) and reduced conscious proprioception in all four limbs. The ataxia is often more severe in the hindlimbs and close examination of the forelimbs may be required to reveal the deficits there.

The **cervicothoracic** zone extends from C_5 to T_2 and includes the origins of the brachial plexus. Lesions in this area cause a combination of LMN signs in the forelegs and UMN signs in the hindlegs. Lesions confined to the anterior part of this zone may cause UMN signs in brachial plexus nerves which arise mainly from the more caudal segments.

Abnormalities in the forelimbs may include:

Hypotonia.
Paresis.
Hyporeflexia.
Neurogenic muscle atrophy.

Deficits in the hindlimbs include:

Hypertonia.
Paresis.
Hyperreflexia.
Reduced proprioception.

Lesions in the caudal part of this zone may also cause Horner's syndrome (see Fig. 13.5).

The **thoracolumbar zone** extends from T_3 to L_3 and therefore includes no major peripheral nerves. Lesions within this zone generally do not affect the forelimbs but cause UMN disturbances in the hind legs and also suppress pain perception. The typical deficits are:

Ataxia of the hindlimbs.
Paraparesis.
Hypertonia of the hindlimbs.
Deficient postural reactions in hindlimbs.
Reduced conscious proprioception in hindlimbs.
Reduced pain perception caudal to the lesion.

Location of lesions within this zone is often possible through observation of the panniculus reflex (Fig. 13.7). In addition, abnormal reflexes appear occasionally – notably the crossed extensor reflex.

Severe lesions in the anterior part of this zone can cause a clinical phenomenon, known as the Schiff–Sherrington phenomenon (see below), in which there is extensor hypertonia of the forelimbs.

The **lumbosacral zone** comprises cord segments L_4 to S_2 and is the origin of the major nerves to the hindlimbs, perineum and pelvic viscera. Lesions here cause LMN signs in the hindlimbs and perineum as well as adversely affecting bladder function (see Control of micturition).

Typical deficits include:

Hypotonia of anal and bladder sphincters.
Flaccid paraparesis or paraplegia.
Hyporeflexia.
Neurogenic muscle atrophy.
Diminished postural reactions.

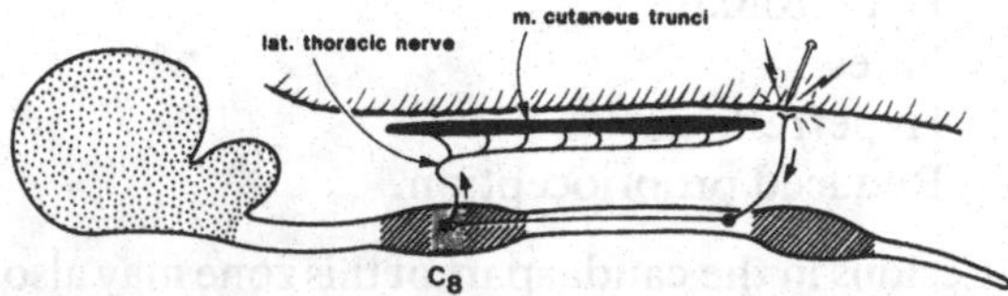

Fig. 13.7. The panniculus reflex. The sensory input to this reflex arc is segmental and a spinal cord lesion caudal to T_1 will disrupt the arc, resulting in an absence of the reflex from all dermatomes caudal to the affected cord segment. lat., lateral; m., muscle.

Reduced pain sensation from pelvic limbs, tail and perineum.

The **sacrococcygeal zone** is the terminal part of the spinal cord, S_3–Cy, and controls the tail. Hypotonia of the tail with reduced pain perception would be the main sign of a lesion confined to this zone.

Spinal reflexes and muscle tone

The testing of tendon reflexes is generally restricted to small animals as it requires the patient to be relaxed and in lateral recumbency. The tendon reflexes are a group of responses obtained by lightly tapping a muscle tendon and observing the immediate involuntary jerking response in the limb resulting from the reflex muscle contraction.

Reflex responses can be graded on a five-point scale:

0 Areflexia – no response
1 Hyporeflexia – subnormal response
2 Normal
3 Hyperreflexia – exaggerated response
4 Marked hyperreflexia; myoclonus

Myoclonus is the phenomenon of repeated muscle contractions in response to a single stimulus. Each reflex involves a circuit comprising a sensory nerve, a motor nerve and the spinal cord segments through which they communicate. Knowledge of the basic anatomy of each reflex will assist in the location of lesions where hyporeflexia or areflexia suggest LMN disease involving some components of a particular reflex arc. The anatomical and clinical features of the commonly examined spinal reflexes are summarized in Table 13.5.

The panniculus reflex

The panniculus reflex can be of great value in locating lesions in the thoracolumbar zone T_3–L_3, where lesions are common but there are no tendon reflexes to aid in their location.

The panniculus reflex is manifested as a twitch of the cutaneous trunci muscle in response to a cutaneous stimulation along the

Table 13.5. *Spinal reflexes*

	Sensory nerve	Spinal cord segments involved	Motor nerve	Technique	Observation
Biceps reflex	Musculocutaneous	$C_{6,7,8}$	Musculocutaneous	Percuss finger resting on tendon of insertion of biceps brachii muscle on flexor aspect of elbow	Flexion of elbow or palpable contraction of biceps muscle. *Often absent in normal animals*
Triceps reflex	Radial	$C_{7,8}\,T_1$	Radial	Percussion of triceps just above olecranon	Extension of elbow. *Often absent in normal animals*
Forelimb flexor reflex (withdrawal reflex)	Various nerves depending on area of limb stimulated (See Fig. 13.10)	$C_{6-8}\,T_{1-2}$	Axillary Musculocutaneous Median Ulnar	Cutaneous stimulation in area corresponding to sensory nerve being tested (Fig. 13.10)	Flexion of entire forelimb
Patellar reflex	Saphenous branch of femoral	$L_{4,5}$	Femoral	Percussion of patellar tendon of uppermost limb with patient relaxed and in lateral recumbency	Brisk extension of distal hindlimb – knee jerk. *Most reliable tendon reflex*
Hindlimb flexor reflex (withdrawal reflex)	Saphenous branch of femoral	$L_{4-7}\,S_1$	Sciatic	Pinch skin on medial aspect hock or metatarsus	Flexion of entire hindlimb
	Sciatic	$L_{6,7}\,S_1$	Sciatic	Pinch lateral toe near nailbed	(Flexion of entire hindlimb)
Perineal reflex	Pudendal	$S_{1,2,3}$	Pudendal	Stimulation of perineum	Anal sphincter contraction; tail flexion

back, somewhere between the scapulae and the tuber ischii. The cutaneus trunci muscle is innervated by the lateral thoracic nerve from cord segments C_8–T_1 (Fig. 13.7). The sensory nerves for the reflex are those from the specific dermatomes being stimulated. In the presence of a lesion in the thoracolumbar zone the sensory pathway from dermatomes caudal to the lesion will be disrupted. Hence, progressive skin stimulation moving cranially up the back will fail to elicit a twitch response until a dermatome above the level of the lesion is reached, where the reflex arc is intact.

Repetition of this procedure will allow a consistent 'sensory level' to be determined (i.e. level below which cutaneous stimulation fails to elicit a response). Projection to the corresponding cord segments will indicate where the lesion is located. Because the segmental nerves radiate caudally, the cord segments are usually several centimetres cranial to their corresponding dermatomes. In the dog, a useful guideline is a line drawn between the caudal borders of the costal arch which indicates approximately the L_1 level dermatome. The flank area between the costal arch and the front of the thigh can be divided into three equal bands which approximately represent the dermatomes of L_1, L_2 and L_3.

Postural reaction testing

Postural reactions are complex responses that occur whenever the animal shifts its weight or alters its posture. They require the integrated interaction of spinal reflexes with the sensory and motor systems of the brain and involve conscious recognition of what is happening. This is why they are called reactions rather than reflexes.

The purpose of testing the postural reactions is to reveal neurologic deficits which may be sufficiently mild, or so located as to cause minimal disturbance to gait or reflexes. Because the postural reactions involve many parts of the nervous system, lesions in any one of several areas could affect them. They, therefore, have little locatory value, but they form an essential part of the neurologic examination because they can reveal subtle deficits in strength and coordination when the gait appears normal.

By their nature, postural reaction tests are most readily applied to small animals. However, equivalent tests are available for use when examining horses. The tests involve the assessment of each limb individually and in combination so that an overall impression can be gained as to whether the deficits are more severe in the forelimbs, in the hindlimbs, on one side of the body or the other, or in a single limb. The tests should be performed on a surface which will give the animal good footing. It is often convenient to perform them in the sequence in which they are listed here.

Proprioceptive positioning: Proprioception is the perception of skeletal position and movement. Deficiency in proprioception can be a sensitive indicator of early or mild spinal cord disease and can aid in the differentiation of neurologic disease from orthopedic disease.

Much of the ataxia apparent during the early stages of progressive spinal cord disease is due to proprioceptive deficits. As the lesion progresses, however, paresis appears, due to motor nerve damage. The differentiation of the sensory component from the motor component of the ataxia is not important, since UMN and proprioceptive tracts accompany each other through most of the neuraxis and are simultaneously damaged in most cases.

Animals with proprioceptive deficiencies may be seen and heard to knuckle over or drag the feet while walking. There may be abnormal wear on the fronts of the hooves or toenails. Such animals tend to pivot on the deficient limb when turning sharply and may assume bizarre stances with the limbs crossed or widely abducted. The gait appears uncoordinated, especially when the animal is turning or changing pace.

Where such deficiencies are not immediately apparent, simple tests can be used to demonstrate them. With the animal standing, a paw is turned over so that weight is borne on the dorsum. Any delay in returning the foot to a normal plantigrade position suggests reduced proprioception. This test is far easier to perform on the hindlimb than on the forelimb. An alternative technique is to place the foot on a sheet of paper and gently abduct it by pulling the paper sideways. A normal animal will permit only a limited degree of displacement before moving the foot back towards the midline.

These tests are not appropriate for horses as it is normal for them to rest with one hoof 'knuckled over'. Proprioceptive deficits in this species must be inferred from observation of gait deficits, especially during turning, from any tendency to drag or scuff the feet, particularly on uneven ground, and from any tendency to adopt abnormal limb postures when standing.

It should be noted that depressed or sedated animals will often have unusual gait patterns and may tolerate the malpositioning of their feet for prolonged periods. Conversely, nervous or agitated dogs that are constantly shifting their feet may reposition a knuckled paw so that a proprioceptive deficit is masked.

Wheelbarrow test: This postural reaction tests the strength and coordination of the forelegs. The animal is lifted under the abdomen so that the hindlimbs are just off the ground and the weight is taken on the thoracic limbs. The animal's ability to support its weight evenly on the forelegs is assessed and it is then moved forwards to observe the forelimb gait. A normal animal will walk with strong symmetrical movements of both forelimbs and will hold its head extended. Animals with weakness or proprioceptive deficiencies will drag, knuckle or collapse on one or the other foreleg and may lower the head, perhaps even attempting to use the chin for support. If the gait appears normal, repetition of the maneuver with the chin elevated prevents visual compensation and may reveal previously inapparent deficits.

In horses, an equivalent test involves leading the animal down an incline and observing the competence of the forelimbs.

Forelimb hopping reaction: This maneuver tests each forelimb individually. One hand beneath the belly lifts the backlegs just off the ground while the other hand elevates one forefoot. The animal is gently pushed laterally towards the side of the remaining weight-bearing limb and should hop briskly and strongly with it. It is useful for the operator to stand astride the patient and move it through the same distance when testing each leg. In this way the number of hops on each side can be counted and compared, a reduced number indicating a gait deficit. Deficits may also show up as a tendency to drag the paw or for the onset of the hopping movement to be delayed. Hopping the animal medially is meaningless as most normal dogs will drag the paw when tested in this way.

Hindlimb hopping reaction: The forelimbs are elevated from the ground by supporting the animal under the sternum with the hand between the forelegs. One hindleg is flexed and the animal is forced to hop laterally towards the weight-bearing side. Both pelvic limbs are tested and the response compared.

This test is performed in horses by pulling laterally on the tail while the horse is being walked forwards. Any weakness will show as a tendency to stagger in the direction of the pull, whereas a normal horse will effectively resist.

Hemistanding and hemiwalking reactions: These tests are designed to reveal unilateral or asymmetrical deficits. With the animal standing squarely, both feet on one side are gently lifted off the ground. The animal's ability to support its weight on the remaining limbs is observed and then, by gently pushing it forwards or laterally, its ability to walk on that side is tested. This is also a way of testing hopping reactions on animals that are otherwise too large or uncooperative.

An equivalent test in horses is to push down on the withers while attempting to pull the animal laterally. This effectively tests the strength and coordination of the forelimb and, to a lesser extent, the ipsilateral hindlimb, therefore combining the hemiwalk and hemistand tests with the foreleg hopping test.

Extensor postural thrust: This is the hindlimb equivalent of the wheelbarrow test in the forelegs, in that it tests the strength and coordination of both hindlegs. The animal is lifted up and gently lowered to the ground. As the pelvic limbs touch the floor, they should extend to support the animal's weight and when the animal is moved backwards there should be a confident and symmetrical gait without any tendency to collapse on either leg or drag the feet.

In horses, the equivalent test involves backing the animal. Many normal horses resist this maneuver and the response must therefore be evaluated critically.

Placing reaction: This procedure tests forelimb proprioception and coordination. The animal is held off the ground and brought towards the edge of a tabletop. As the dorsum of the paws make contact, they should be lifted up immediately so as to bear weight on the surface. The test is performed with and without a blindfold (tactile placing reaction; visual placing reaction) as vision can help to compensate for a loss of proprioception.

The placing reaction test is prone to artefact and can be unreliable. Animals quickly learn where the tabletop is and begin to anticipate their response as the test is repeated. The operator should therefore move away from the table and move about before each repetition of the test. It is common, when the animal is supported on the operator's hip, for the paw on the side away from the operator to be lifted up more readily.

Tonic neck and eye reaction: This is a test of proprioception from the neck, and of forelimb strength and coordination. When the head and neck are extended so that the nose points dorsally, the normal reaction is for the forelegs to extend and the hindlimbs to flex. Incidental findings during this maneuver may be the presence of neck pain or the observation of vestibular strabismus.

Detection and localization of spinal lesions

Deep palpation of the spinal column and epaxial musculature may elicit a pain reaction from one particular area which helps to locate a spinal lesion. Cervical pain is assessed by gentle manipulation of the neck and observation of the symmetry and degree of muscular resistance.

It is important that this test be performed last as it can cause severe pain and markedly reduce the degree of cooperation by the patient. Manipulation of the spinal column is also contraindicated in trauma cases, especially if radiographs have not been taken.

Schiff–Sherrington phenomenon

The Schiff–Sherrington phenomenon is the clinical phenomenon of extensor hypertonia of the thoracic limbs and paralysis of the hindlimbs associated with acute thoracic spinal cord lesions in dogs.

There is no compromise of the descending UMN tracts to the forelegs and therefore gait and postural reactions are normal, in spite of the hypertonia.

The underlying mechanism relates to neurons ('border cells') in the lumbar spinal cord, especially in the L_2–L_4 region, with an inhibitory influence over the extensors of the forelimbs. Their axons run in the fasiculus proprius, a very deep tract which is affected only by severe spinal cord lesions. The observation of the Schiff–Sherrington phenomenon therefore has locatory value (indicating thoracic spinal lesion) and prognostic value (indicating severe spinal cord damage).

Initially, the hindlimbs are often hypotonic and hyporeflexic but this usually reverts within hours to the classical UMN signs that would be expected with a thoracic lesion. The hypertonia of the forelimbs may persist for weeks.

The Schiff–Sherrington phenomenon is seen frequently in very severe upper thoracic lesions, such as fractures with spinal cord transection. However, its presence is not necessarily a sign of a hopeless prognosis and other parameters such as deep pain perception should be assessed (see later). The foreleg hypertonia component of the syndrome has also been observed occasionally in less acute diseases, such as compression of the mid-thoracic spinal cord by an expanding tumor.

Assessment of the severity of spinal lesions

Proprioceptive testing, postural reactions and pain perception are utilized as indicators of the functional integrity of the cord. As the severity of spinal cord damage increases, the various functions are lost in a consistent sequence. The most vulnerable function is proprioception, followed by motor function and, finally, deep pain sensation distal to the lesion. This sequence of loss of function corresponds roughly to the sizes of the nerve fibers involved. Proprioception is carried by large fibers, which are sensitive to injury, and it is therefore a reliable indicator of early or mild spinal cord damage. At the other extreme, deep pain sensation is carried by small fibers, which are quite resistant to damage and whose dysfunction is therefore a critical indicator of severe cord damage and poor prognosis. In the resolution of spinal lesions, these neurologic functions return in the reverse order, proprioception being the last to reappear and often failing so to do if the lesion was severe.

The conscious perception of pain indicates the functional integrity of the peripheral nerves, their spinal cord segments, the ascending spinothalamic pathway in the spinal cord, and the related thalamocortical system in the brain. In the assessment of spinal lesions it is essential that spinal reflex movements are not confused with voluntary reactions in response to perceived pain.

The usual method of assessing pain perception is to apply a noxious stimulus such as a pinprick or toepinch with forceps. The severity of the stimulus should be steadily increased until a distinct pain reaction is obtained. The animal will turn around, attempt to bite the operator, or may try to move away. Remember that the limb withdrawal may be purely reflex and does not prove conscious perception of pain. A depressed or sedated animal, or one with severely painful injuries, may fail to respond to a toepinch stimulus, even though able to perceive it.

Control of micturition

The smooth muscle of the bladder wall, the detrusor muscle, is innervated by the parasympathetic pelvic nerve, which arises from the three sacral cord segments. Parasympathetic stimulation causes contraction of the detrusor muscle and evacuation of the bladder. Sensory neurons, responsive to stretching of the bladder wall, project axons via the pelvic nerve and sacral spinal nerve to the sacral cord segments. Some of these axons terminate on interneurons that complete the reflex arc with preganglionic parasympathetic neurons. Others ascend in pathways to the brain for central control functions.

The striated sphincter muscle of the urethra is innervated by the pudendal nerve; when the bladder contracts, the activity of the pudendal nerve is inhibited to allow urethral relaxation.

The detrusor muscle is also under the influence of sympathetic innervation arising in L_1–L_4 and passing, via the caudal mesenteric ganglion and hypogastric nerve, to the body and neck of the bladder. In the neck of the bladder, the sympathetic fibers terminate on α-receptors and increase sphincter muscle tone, aiding urine retention. In the wall of the bladder, the fibers terminate on β-receptors, and promote relaxation of the smooth muscle, allowing the bladder to fill. The stretch receptors also project via the pelvic nerve to the lumbar segments so that, as urine accumulates, sphincter tone increases and detrusor tone diminishes.

Reflex micturition requires facilitation of the sacral parasympathetic neurons (pelvic nerve) and inhibition of the sacral somatic neurons (pudendal nerve) and lumbar sympathetic neurons (hypogastric nerve). These reflex circuits are under the control of the higher centers in the brain. Ascending information from the stretch receptors informs the cerebral cortex of the distended bladder. Cortical influence over centers in the reticular formation, pons and medulla initiates voluntary urination and the information is carried by the tectospinal and reticulospinal UMN tracts. The basic aspects of the neural control of micturition and its disorders are illustrated in Fig. 13.8 and Tables 13.6 and 13.7.

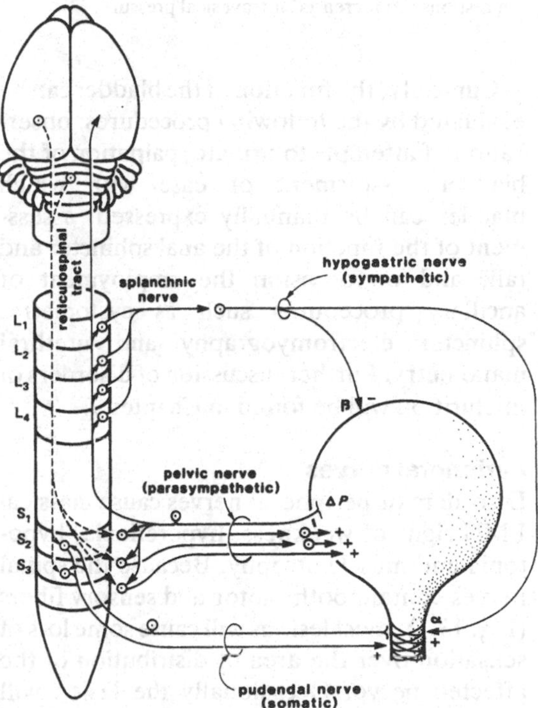

Fig. 13.8. Sensory and motor innervation of the urinary bladder and urethra during the filling phase (see also Table 13.7). α and β, indicate types of receptor; ΔP, change in bladder pressure.

Table 13.6. *Neural control of urinary bladder function*

Neural elements	Function	Status in filling	Status in micturition
Sacral somatic innervation	Sphincter contraction via pudendal N. to urethral striated muscle	Active	Inhibited
Lumbar sympathetic outflow (Hypogastric N.)	Contraction of urethral sphincter – α-adrenergic innervation via hypogastric N.	Active	Inhibited
	Relaxation of detrusor muscle – β-adrenergic innervation via hypogastric N.	Active	Inhibited
Sacral parasympathetic (pelvic N.)	Stimulates detrusor muscle via pelvic N.	Inhibited	Active
Sensory neurons of bladder wall	Activate different arm of reflex via pelvic N. Induces detrusor muscle relaxation[a] – via lumbar sympathetic efferent arm (hypogastric N.)	Active	Inhibited
Brainstem ponto-medullary centers cerebral cortex and cerebellum	Awareness of bladder filling, initiation and control of voluntary micturition		

N., nerve.
[a]in response to increased intravesical pressure.

Clinically, the function of the bladder can be evaluated by the following procedures: observation of attempts to urinate, palpation of the bladder, assessment of ease with which bladder can be manually expressed, assessment of the function of the anal sphincter and tail, and on occasion the employment of ancillary procedures such as cystometry, sphincter electromyography and urethral manometry. Further discussion of disorders of micturition will be found in Chapter 9.

Peripheral nerves

Disorders of peripheral nerves cause classical LMN signs of weakness, hyporeflexia, hypotonia and muscle atrophy. Because the spinal nerves contain both motor and sensory fibers (Fig. 13.1), most lesions will cause some loss of sensation over the area of distribution of the affected nerve. Occasionally the lesion will cause irritation of the nerve so that abnormal sensations are perceived (parasthesia) and self-mutilation may result.

It is important for the clinician to have a basic knowledge of the motor and sensory functions of each major peripheral nerve, as well as the spinal segments from which it arises, and its course as it passes from the spinal column down the limb.

By carefully mapping the motor and sensory deficits, this anatomical knowledge can be used to determine whether the lesion involves all or part of one nerve, a group of nerves, or possibly all the peripheral nerves. Where motor deficits are present without corresponding sensory deficits, a lesion of the gray matter of the cord or ventral nerve roots must be suspected (Fig. 13.1). Lesions within the spinal cord will often cause UMN signs to be present further down (Fig. 13.1).

When cutaneous sensation is assessed it is important to realize that there is considerable overlap between the areas innervated by adjacent nerves in the fore- and hindlimbs. It must be ensured that testing is confined to the area of skin innervated exclusively by a single nerve – the 'autonomous zone' of the nerve.

Figures 13.9 and 13.10 and Table 13.8 sum-

Table 13.7. *Mechanisms and manifestations of dysinervation of the urinary bladder*

Location of lesion (see Fig. 13.8)	Effects of lesion on control mechanisms	Clinical manifestations
Upper motor neuron		
(*a*) Thoracic or thoracolumbar	Ascending and descending UMN tracts disrupted	No voluntary micturition or sensation of filling
	Pudendal nerve influence over striated urethral sphincter muscle uninhibited	Hypertonia of sphincter causing resistance to manual expression of bladder, especially in males
	Reflex arc to detrusor muscle intact	Reflex urination may develop but small residual amount of urine will remain
(*b*) Lumbar lesion L_4–S_1	Sympathetic innervation (hypogastric nerve) to sphincter not disrupted, sphincter tone increased by combined UMN release phenomenon via pudendal nerve and normal α-adrenergic activity	Sphincter tone and resistance increased to greater degree than with thoracolumbar lesion, making manual expression even more difficult. Other manifestations similar to thoracolumbar lesion
Lower motor neuron		
(*a*) Sacral cord lesion	Sacral cord segments damaged, disrupting tonic efferent innervation to detrusor muscle and afferent innervation from pressure receptors in bladder wall (pelvic nerve.) Somatic innervation to sphincter muscles disrupted (pudendal nerve). Minimal sphincter tone from intact lumbar sympathetics	No voluntary micturition or conscious sensation of filling. Atonic, distended bladder with flaccid sphincter; overflow incontinence. Reflex micturition due to intrinsic contraction of detrusor muscle but evacuation weak and incomplete with large residual volume of urine
	Damage to other nerves arising from sacral cord segments, e.g. pudendal, coccygeal	LMN paralysis of tail and anal sphincter
(*b*) Pelvic nerve lesion	Tonic efferent innervation to detrusor muscle and afferent innervation from pressure receptors in bladder wall is disrupted	Atonic, distended bladder with no conscious sensation of filling
	Innervation to sphincter intact (hypogastric nerve, pudendal nerve)	Sphincter tone maintained. Bladder may be evacuated manually; α-blocker drugs help lower sphincter resistance
	Sacral segments not involved in lesion	Normal perineal sensation and sphincter function

LMN, lower motor neurone.

marize the sensory and motor functions of major peripheral nerves. More detailed descriptions can be found in standard anatomy texts.

Pathologic mechanisms of nervous disease

Pathobiology of the neuron

Basic nature of neurons

Neurons carry electrical impulses over long distances, connecting with one another at specialized points of contact – the synapses. All neurons have a cell body (soma or perikaryon) and a variable number of processes (Fig. 13.11) which may be few or numerous, and may be extremely long. For instance, a sensory neuron in a lumbar dorsal root ganglion of a horse may project its axon some 1.5 meters peripherally to terminate in the hind foot, and about the same distance centrally to terminate in the medulla. To maintain a cell structure of this size and complexity requires a prodigious amount of metabolic activity. The perikaryon is the metabolic factory for the whole cell and must sustain not only the cell body but the processes as well.

In clinicopathologic terms, the axon is the most significant extension of the cell. It is the channel for conduction of impulses *away* from the perikaryon. Although only a single axon leaves each cell body, many axons branch near their terminations.

The axon is devoid of ribosomes and is therefore totally dependent on the perikaryon for protein synthesis and almost totally dependent for breakdown and turnover of macromolecular components and organelles. This necessitates a bi-directional flow of molecules and organelles between perikaryon and axon which is carried out by the axonal transport mechanism. The cytoskeletal tubules and filaments (Fig. 13.11) form an important component of this active transport system, which is fueled by the consumption of energy in the perikaryon and along the axon.

The generation and propagation of an impulse by one neuron will cause the release of a transmitter at a synapse, which may initiate a response is another neuron. The response may be **inhibitory or excitatory** and the net effect will depend on the coordination of a pattern of inputs from several synapses.

The interconnection of neurons throughout the nervous system is highly complex and it is the stimulation and inhibition of chains and groups of neurons, organized into various pathways and systems, which underlies the diverse yet highly integrated functioning of the nervous system.

Neuronal injury

The general functional consequences of neuronal injury are illustrated in Fig. 13.12, which identifies the main functional regions of an individual nerve cell. Within the context of the entire nervous system, neuronal injury may be produced focally, or multifocally and may occur in a rather random fashion, as is the case in many traumatic injuries and microbial infections. Alternatively, particular groups or systems of neurons may be selectively and consistently injured by noxious agents. This applies particularly to chemical intoxications and to metabolic disturbances, which are often remarkably targeted to particular regions. Such disorders are often referred to as 'systemic', when a particular neuronal system

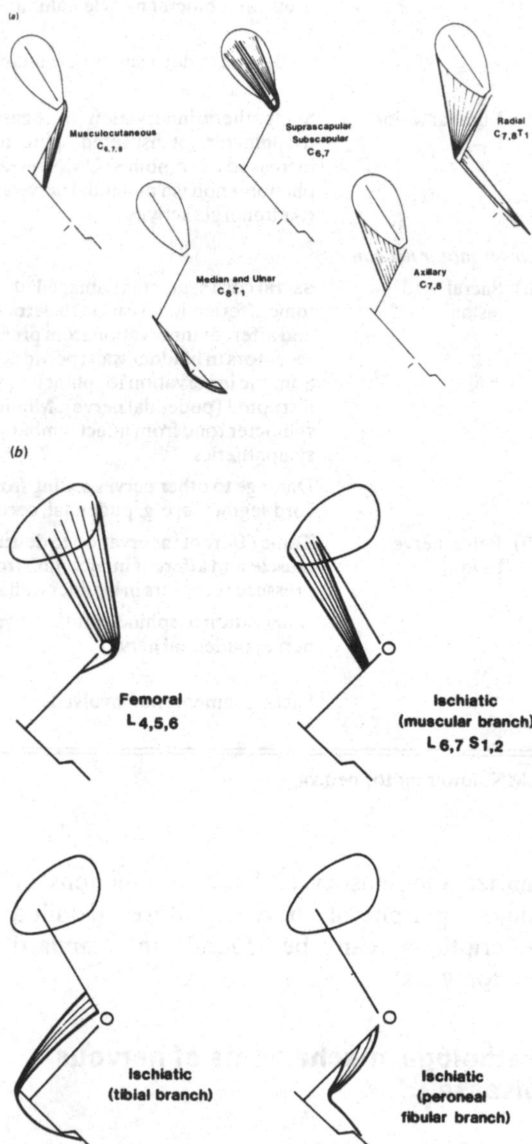

Fig. 13.9. Schematic representation of the motor innervation to the major muscle groups of the forelimb (*a*) and hindlimb (*b*).

is involved, for instance the optic tract or the spinal proprioceptive pathways.

Neuronal injury may be transient or permanent, and may involve a functional impairment with no structural change, or there may be both structural and functional damage. Necrosis and loss of neurons in one area may be compensated to a degree by the assumption of this activity by neurons in another area. Such adaptation can be remarkably effective, especially if the deficiency develops slowly over a long period of time, or in a young animal.

Neurons have such a high requirement for oxygen and glucose that deprivation of either will cause the cell to die in a very short time. In the case of cerebrocortical neurons, this is a matter of three to five minutes. Oxygen deficits may arise when there is inadequate respiration or circulation, or when there is a

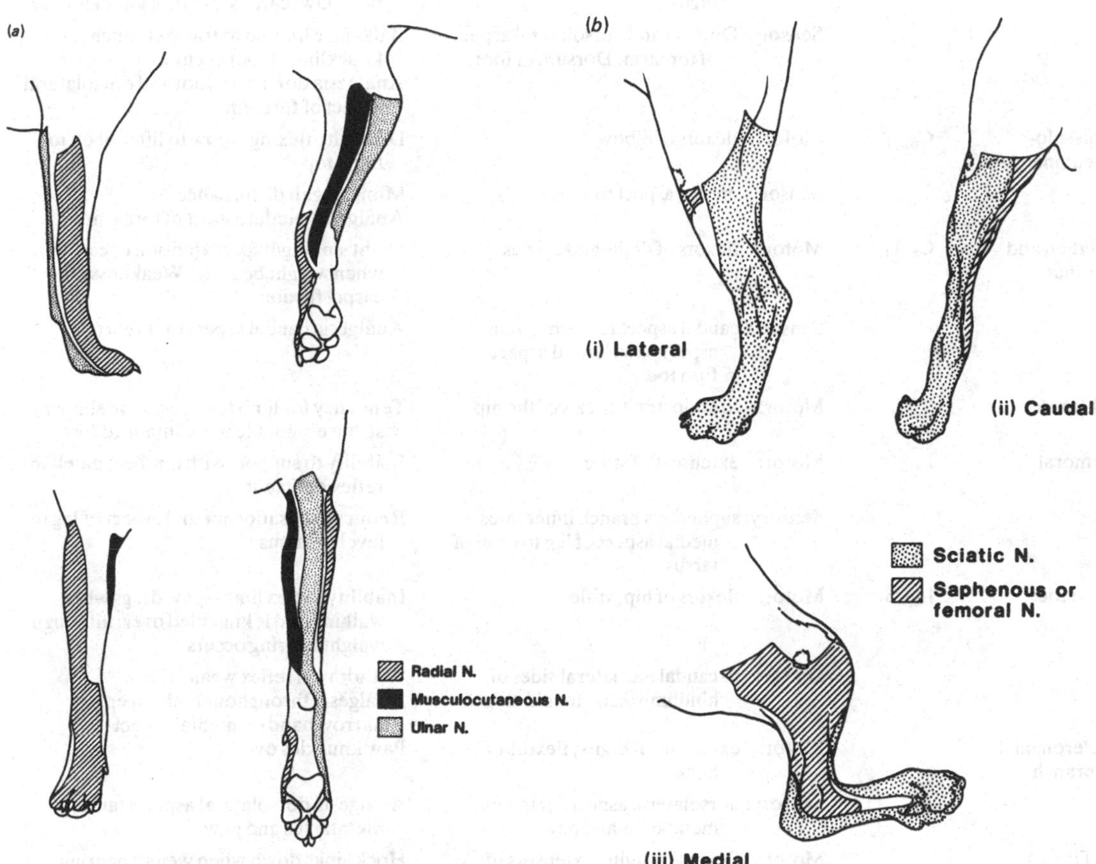

Fig. 13.10. (*a*) Zones of cutaneous innervation of the canine forelimb. Note that the shaded areas are those represented solely by the nerves indicated. The unshaded areas are zones of overlap between several nerves and should therefore be avoided when assessing nerve function clinically. N., nerve. (Redrawn after Bailey and Kitchell, 1984, see Additional reading.) (*b*) Zones of cutaneous innervation of the distal canine hindlimb. (Redrawn after Haghigi, 1900, see Additional reading.)

Table 13.8. *Fields of innervation of major peripheral nerves in the dog*

Nerve	Origin	Function	Signs of dysfunction
Suprascapular	$C_{6,7}$	Motor: extensors of shoulder	Atrophy with prominence of scapular spine Minimal gait disturbance
Axillary	$C_{(6),7,8}$	Motor: flexors of shoulder	Minimal gait disturbance; some weakening of shoulder flexion
		Sensory: narrow area lateral aspect of brachium	Atrophy Analgesia small area lateral aspect of brachium
Radial	$C_{7,8}, T_{1,2}$	Motor: extensors of elbow, carpus, digits	Inability to support – unable to extend or fix elbow; carpus and foot knuckle over
		Sensory: Dorsum and dorsolateral aspect of forearm. Dorsum of foot.	If damage limited to triceps branch, only knuckling of foot occurs Analgesia dorsum of foot and dorsolateral aspect of forearm
Musculo-cutaneous	$C_{6,7,8}$	Motor: flexors of elbow	Difficulty flexing elbow to lift foot on to table top
		Sensory: medial aspect forearm	Minimal gait disturbance Analgesia caudal aspect of forearm
Median and ulnar	$C_8, T_{1,2}$	Motor: flexors of digits and carpus	Slight sinking/hyperextension of carpus when weight bearing. Weakness of carpal flexion
		Sensory: caudal aspect forearm, volar aspect paw, lateral aspect fifth toe	Analgesia caudal aspect of forearm
Obturator	$L_{4,5,6}$	Motor: to adductor muscles of the hip	Tendency for limb to slide out on slippery surfaces, unable to maintain adduction
Femoral	$L_{4,5}$	Motor: extensors of stifle	Inability to support with hindleg; patellar reflex deficient
		Sensory: saphenous branch innervates medial aspect of leg to level of tarsus	Reduced sensation medial aspect of leg to level of tarsus
Ischiatic:	$L_{6,7}\ S_1$	Motor: flexors of hip, stifle	Inability to flex limb – paw drags when walking and is knuckled over, although weight bearing occurs
		Sensory: to caudal and lateral sides of hindlimb distal to mid-thigh	Withdrawal reflex weak Analgesia throughout limb except for narrow band on medial aspect
Peroneal branch		Motor: extensors of digits; flexors of hock	Paw knuckles over
		Sensory: dorsolateral aspects of tarsus, metatarsus and paw	Analgesia dorsolateral aspects tarsus, metatarsus and paw
Tibial		Motor: flexors of digits, extensors of hock	Hock sinks down when weight bearing
		Sensory: plantar aspect of paw	Analgesia plantar aspect of paw

toxic blockade of tissue respiration, as in cyanide poisoning. Energy metabolism may be compromised by hypoglycemia or more indirectly by the blockade of metabolic pathways. An example is the neuronal necrosis induced by inadequate thiamine availability, as seen in several species of domestic animals.

Impulses, transmitters and synapses

The ability of a neuron to generate and propagate an impulse depends on the electrochemical properties of the cell membrane. In a resting state, the interior of the cell has a negative electrical charge relative to the exterior and this is maintained by a state of dynamic equilibrium in the flux of ions across the membrane. Upon stimulation, there is an influx of sodium (Na^+), and the polarity across the membrane is reversed. The depolarization will spread from its point of origin, and, if a certain threshold is reached, the neuron will initiate an impulse that will travel along its axon as a 'propagated action potential'. When the impulse arrives at the end of the axon, it is usually passed on, not directly, but via the release of a *chemical transmitter*, which crosses the gap between the cells and binds to receptors on the membrane of the receiving cell. If a threshold depolarization is induced, this second neuron may then discharge and propagate an action potential. Some neurotransmitters are *inhibitory* and either stabilize, or provoke hyperpolarization, of the postsynaptic cell membrane.

A number of different chemical transmitters have been identified, and it appears that particular systems of neurons utilize a particular transmitter. The transmitters and/or their synthesizing enzymes are carried by 'fast flow'

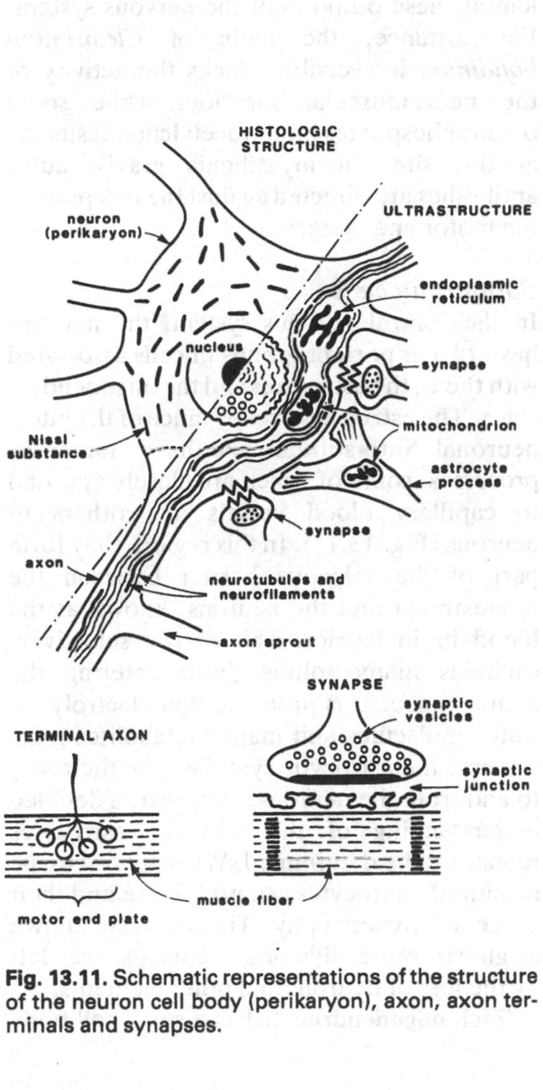

Fig. 13.11. Schematic representations of the structure of the neuron cell body (perikaryon), axon, axon terminals and synapses.

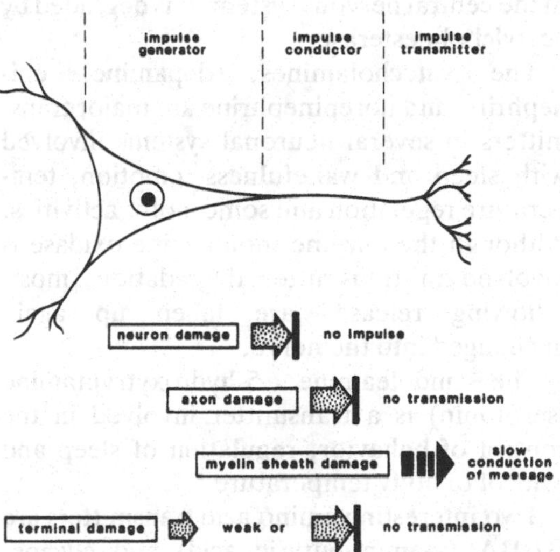

Fig. 13.12. This figure shows the major functional regions of a neuron and indicates the consequences of injury to any of these regions.

axonal transport from the neuronal perikaryon to the axon terminals. At the terminals, the transmitter is associated with vesicles. It has been suggested that transmitter is released when groups of vesicles fuse with the cell membrane, allowing their contents to diffuse into the intercellular cleft and bind with their receptors. This process involves entry of Ca^{2+} into the presynaptic fiber. Once the transmitter has performed its function, it is removed by enzymatic degradation or reuptake, and the receptors are made available for a new release of transmitter. The synapse is the specialized point of contact where these events occur and transmission across a synapse is uni-directional (Fig. 13.11). Any one neuron may contain thousands of synapses on its surface. Most are located on spinous projections from dendrites, many are on the cell body itself, and a few are on axons.

The most familiar chemical transmitter is acetylcholine, which is the major transmitter for the skeletal motor system. As such, it operates at the motor end plate, the special 'synapse' between motor axon and voluntary muscle and depolarization is passed on, not to another neuron, but to a muscle fiber. Acetylcholine is also a transmitter in many pathways in the central nervous system. It is degraded by acetylcholinesterase.

The catecholamines, dopamine, epinephrine and norepinephrine are major transmitters in several neuronal systems involved with sleep and wakefulness, emotion, temperature regulation and some motor activities. Although the enzyme monoamine oxidase is involved in transmitter degradation, most, following release, are taken up again unchanged into the nerve.

The indoleamine 5-hydroxytryptamine (serotonin) is a transmitter involved in the control of behavior, regulation of sleep and control of body temperature.

Two interesting amino acid transmitters are GABA (γ-aminobutyric acid) and glycine. GABA is the principal inhibitory transmitter in the brain, and glycine has a similar role in the spinal cord.

Neuroeffector and receptor structures

The neurons which are directly 'interfaced' with other organs form these connections by means of a variety of specialized structures. Motor endings are analogous to synapses, in that a chemical transmitter is released into a cleft to bind with a postsynaptic receptor on the muscle cell membrane. Sensory endings are transducers which convert mechanical, thermal, electromagnetic or chemical information into nerve impulses. Some are extremely complex, like the rods and cones of the retina; others are very simple, like the sensory endings lying between muscle fibers.

Some clinical disorders arise from a problem at these outposts of the nervous system. For instance, the toxin of *Clostridium botulinum* irreversibly blocks the activity of the neuromuscular junction, while some organophosphates inhibit acetylcholinesterase at this site. In myasthenia gravis, autoantibodies are directed against the receptors at the motor end plate.

Supporting cells

In the central nervous system, the neurons have two important groups of cells associated with them, the astrocytes and the oligodendrocytes. The **astrocytes** occupy much of the interneuronal 'interstitium' and have numerous processes, some of which are closely apposed to capillary blood vessels and others to neurons (Fig. 13.11). In this regard, they form part of the selective barrier between the bloodstream and the neurons, known as the blood–brain barrier. This barrier selectively excludes many solutes from entering the neural tissue. It is probable that electrolytes, water molecules and many metabolites must traverse the astrocytic cytoplasm on their way to and from the neurons. However, a detailed understanding of the role for astrocytes remains to be determined. When neural tissue is injured, astrocytes can proliferate and their processes hypertrophy. This process, known as gliosis, either fills or surrounds spaces left by the loss of neurons and other elements.

Each **oligodendrocyte** has a small cell body

and may have a dozen or so processes. Each process expands into a large sheet and forms a myelin sheath, which is wound concentrically and compacted around a segment of an axon over a distance of about 1 millimeter (Fig. 13.13). One oligodendrocyte may provide a segment of myelin sheath for each of several axons.

In the peripheral nervous system, the companions of the neuronal processes are the **Schwann cells**, which provide the myelin sheath for myelinated fibers, and a cytoplasmic investment for non-myelinated axons. In modified form, they become the capsule cell in peripheral ganglia, closely surrounding the cell bodies of ganglionic neurons.

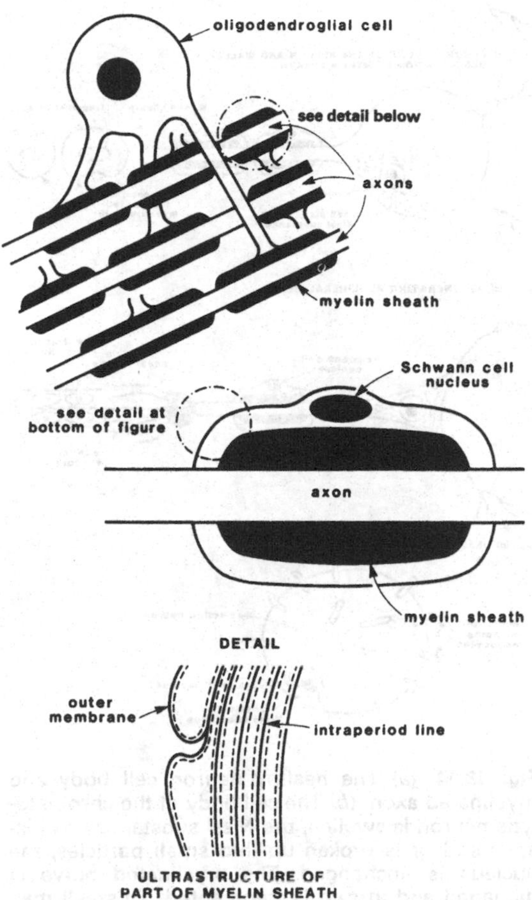

Fig. 13.13. Diagrammatic representation of the relationship to the axon of the oligodendroglial cell in the central nervous system and the Schwann cell in the peripheral nervous system. The compacted plasma membrane of these cells forms the myelin sheath.

Schwann cells produce myelin in the same general manner as do oligodendrocytes, but each cell makes myelin for one segment of one axon (Fig. 13.13). Schwann cells are fairly hardy and can indulge in phagocytosis of their own axon if it is damaged, and of their own myelin if necessary. They can proliferate vigorously, and are effective in the remyelination of regenerated peripheral axons.

Nature and function of the myelin sheath

Myelin is formed as a specialization in the process of oligodendrocytes or Schwann cells. The myelin itself is the result of chemical alteration and compaction of the cell membrane in a manner which obliterates both the internal cytoplasm and the extracellular space between the concentric turns (Fig. 13.13). The stability of the structure can be broken down by the action of specific anti-myelin antibodies, or toxins such as hexachlorophene and ammonia. These agents can cause the spiral turns of the sheath to unravel and the myelin to become vacuolated and disrupted. Although all myelin is basically a complex lipoprotein, central myelin is chemically and antigenically distinct from peripheral myelin. Larger axons, provided with a myelin sheath have an improved efficiency of impulse conduction. The sheath is interrupted at the nodes of Ranvier, leaving the axonal membranes exposed at these points, but insulated along the intervening internodes. Depolarization of the axon must consequently 'skip' from node to node in what is called 'saltatory' conduction. Its effectiveness can be judged by comparing the nerve conduction velocity in normal versus demylinated peripheral nerves. In a normal dog, the conduction velocity of myelinated nerves is about 50 metres/second; in a dog with demyelinating disease, this may fall to 5–10 meters/second, because of loss of myelin sheaths.

There is an intimate relationship between axons and their myelinating cells. Disintegration of an axon will immediately initiate removal of the overlying sheath. The myelinating cells themselves, particularly the Schwann cells, withdraw and digest much of their own myelin, but some is detached and ingested by macrophages. A myelinated axon, will tolerate loss of its coat for quite long periods, but will eventually degenerate if it is extensively denuded.

When myelin sheaths are destroyed, there is good potential for their regeneration. This is particularly so in the peripheral nervous system. In the CNS, remyelination may occur, with restoration of conduction to near normality, but in some lesions remyelination and persistent demyelination may occur side by side.

Regenerative ability

Neurons are one of the most specialized of mammalian cells and this has its price. After the proliferative flurry of the developmental period, final connections are formed and the power of cell replication is lost. Mature neuronal cell systems are permanent and each nerve cell potentially has a life span equal to that of the animal. Any mature neuron that dies is not replaced and is permanently lost.

There is, however, some ability to regenerate a lost axon if the parent perikaryon and the supporting cells are intact and healthy. The ability is maximal in the peripheral nerves following focal traumatic injury, provided the injury does not lead to marked separation of the ends of the damaged nerve fibers or cause death of the neuron. Axonal resprouting is a vigorous response to such injury. The axon segments distal to the point of injury will rapidly disintegrate and be removed along with the myelin sheath, in the process called **Wallerian degeneration** (Figs. 13.14 and 13.15). This leaves columns of Schwann cells and their associated basal lamina envelopes, which form conduits for the guidance of future axonal sprouts. Provided the new sprouts can easily re-enter these conduits, they will grow down the remaining endoneurium at a rate of 5–10 millimeters/day, may be remyelinated by Schwann cells, and will successfully establish specialized endings. If unable to re-enter the distal channels, the sprouts may form a disorderly tangle at the original site of injury.

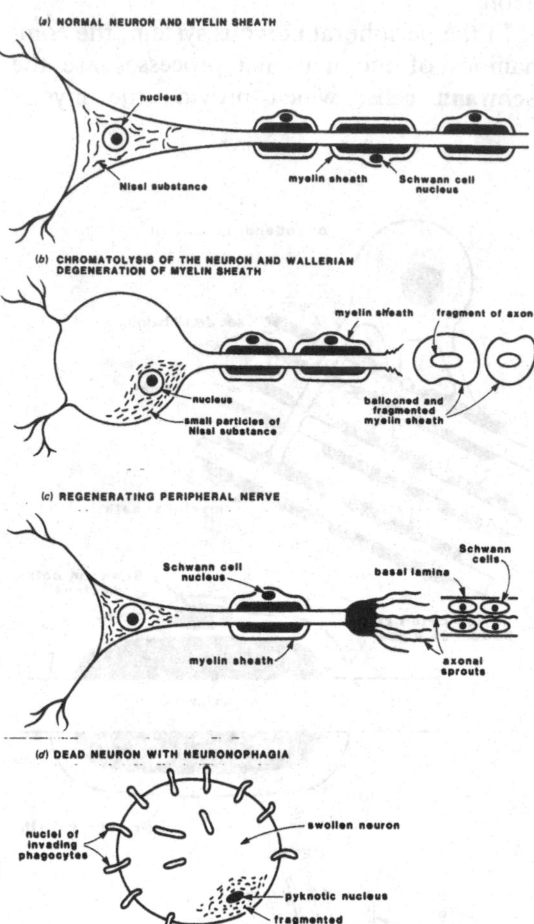

Fig. 13.14. (*a*) The healthy neuron cell body and myelinated axon. (*b*) The cell body of the chromatolytic neuron is swollen, the Nissl substance has disappeared or is broken up into small particles, the nucleus is unchanged. The myelinated nerve is damaged and in part is fragmented. This cell may return to normal. (*c*) The peripheral nerve is regenerating and axonal sprouts are migrating down the channels they formerly filled. They will make contact with proliferating Schwann cells and may remyelinate. (*d*) This depicts a dead neuron. The cell body is swollen, the pyknotic nucleus is shrunken and dark. Phagocytes are invading the cell.

In the CNS, the scope for axonal regeneration is extremely limited and the reasons for this are not well understood. The organization of the brain and cord white matter does not provide clear channels for axonal sprouts to follow. The more complex oligodendrocyte/axon relationship, and the lack of a stromal scaffold equivalent to the Schwann cell basement membrane, may disadvantage the central axon in comparison to the peripheral axon. Wallerian degeneration within the CNS means permanent loss of the affected axon.

Functional disorders of neurons

Neuronal activity may be severely compromised by agents which do not cause overt structural damage to the nerve cells. This category includes all the biochemical lesions that cause either a *blockade* or an *enhancement* of transmitter activity, or alternatively *interference* with maintenance of ion distribution across the membrane. The ability of certain agents to act reversibly and selectively in this way has resulted in their becoming useful analgesics, a fact we all appreciate when visiting the dentist.

The focus of attention here, however, is on the overtly neurotoxic agents. The discussion

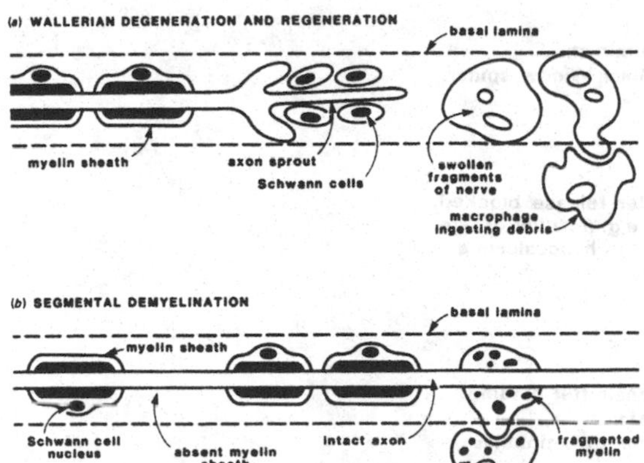

Fig. 13.15. (*a*) Wallerian degeneration and regeneration in a peripheral nerve. (*b*) Segmental demyelination in which individual Schwann cells and the myelin sheaths are affected.

will deal with the general principles; the interested reader can find the details in relevant texts.

If an **excitatory transmitter** has its effect nullified, the neuronal system involved will become inoperative. A number of examples will illustrate the key aspects. At cholinergic synapses and motor end plates the sequence of transmitter release and binding may be blocked at several points (Fig. 13.16). In the first place, the release of transmitter by the presynaptic membrane may be prevented. This is the mechanism of action of botulinus toxin, made especially serious because it is a prolonged effect. By contrast, a deficit in serum Ca^{2+}, as in bovine hypocalcemia, results in a similar problem at the synapse which is rapidly reversed by the provision of Ca^{2+} ions. In the second place the supply of vesicles at the synapse may be rapidly depleted by toxins, such as the venom of the black widow spider and some snakes. Such agents apparently promote a cascade of vesicle–membrane fusion, leading to a transmitter deficiency, which may arise quickly and be prolonged. Thirdly, the transmitter may be prevented from binding with the specific receptors on the postsynaptic membrane. This occurs when there is competitive binding of either toxins like α-bungarotoxin and tropane alkaloids, or of auto-antibodies, as in the disease myasthenia gravis.

In the cholinergic systems affected by such problems, the most common effect is paralysis of neuromuscular transmission. This will be of variable duration according to the basic cause. However, tropane alkaloids in high doses will also affect central synapses, and result in mental confusion or convulsions, in addition to other anti-muscarinic effects.

In regard to the blockade of **inhibitory transmitters**, a well-known example is strychnine poisoning. This alkaloid blocks glycine receptors on spinal interneurons. The escape from inhibitory stimulation by these neurons results in the violent spasms typical of the intoxication. Tetanospasmin (tetanus toxin) produces a similar effect by preventing the release of glycine at these synapses. Other drugs interfere selectively with the synaptic activity of GABA, and as it is the major inhibitory transmitter in the brain, the result is convulsions, rather than the tetanic spasms of the spinal lesion. Picrotoxin, derived from the fish berry plant (*Anamirta* sp.) probably acts in this way.

If we turn to the question of transmitter enhancement, there are fewer pathogenetic possibilities. A major mechanism is inhibition of the enzyme(s) responsible for transmitter breakdown. Another is the activity of 'false transmitters', which mimick the effect of the genuine article. The examples most relevant to veterinary medicine are the agents which inhibit acetylcholinesterase, and include some organophosphates and carbamates. The clinical effects reflect the excessive activity of the transmitter at central synapses, neuromuscular junctions and autonomic nerve endings. Similarly there are monoamine oxidase inhibitors which will result in the accumulation of norepinephrine at certain sites. Inhibition may be reversible or irreversible and this will affect the duration of the disturbance.

Some agents interfere severely with the depolarization of stimulated nerve cells and thereby put them out of action. Tetrodotoxin, from the puffer-fish for example, blocks Na^+ channels in the neuronal membrane and causes failure of impulse transmission. The effect is most marked in peripheral nerves and the dominant clinical effect is muscular weakness.

Primary neuronal degeneration

In some circumstances neurons degenerate or die, but the glial and vascular elements of the nervous tissue remain essentially intact. The major causes of such diseases are chemical toxins and acquired metabolic disturbances, but there are a number of genetically determined metabolic errors which give rise to abiotrophy (see below) or lysosomal storage disease. In diseases of this type the lesions are often restricted to specific neuronal systems and are bilaterally symmetrical. This reflects specific differences in metabolic activity between different populations of neurons. For instance, in organomercurial poisoning the neurons of the cerebellar cortex are vulnerable, whereas in organo-arsenical poisoning it

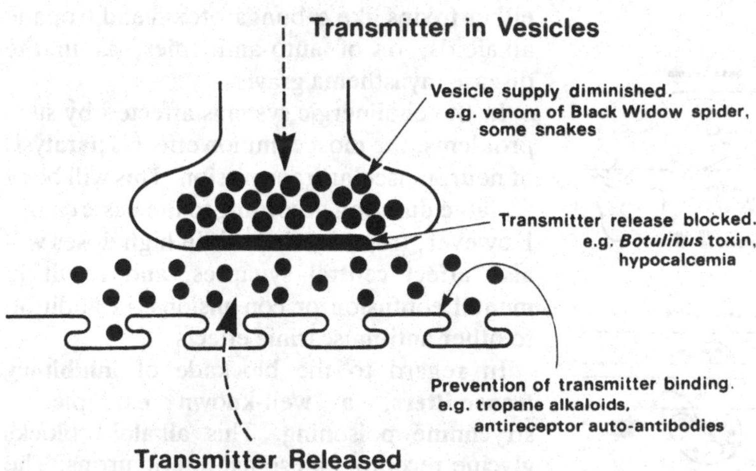

Fig. 13.16. Representation of the structure and function of a motor nerve ending in skeletal muscle and of the basic mechanism which may prevent normal function.

is the spinal motor neurons that are the prime target.

Neuronal necrosis

When a neuron dies, the cell body and all its processes fragment and are removed by phagocytosis (Fig. 13.17), leaving a permanent spatial and functional deficit. In the CNS, phagocytic removal is carried out by macrophages and astrocytes, and the structural repair is achieved by astrocytic proliferation and hypertrophy. In the peripheral nervous system, Schwann cells and macrophages remove debris and the Schwann cells also perform much of the reparative task.

Generally, a large number of neurons need to be lost before any gross changes become apparent in the CNS. When this does happen, affected areas may appear to have atrophied and this is most easily appreciated in superficial structures like the cerebral cortex or cerebellum.

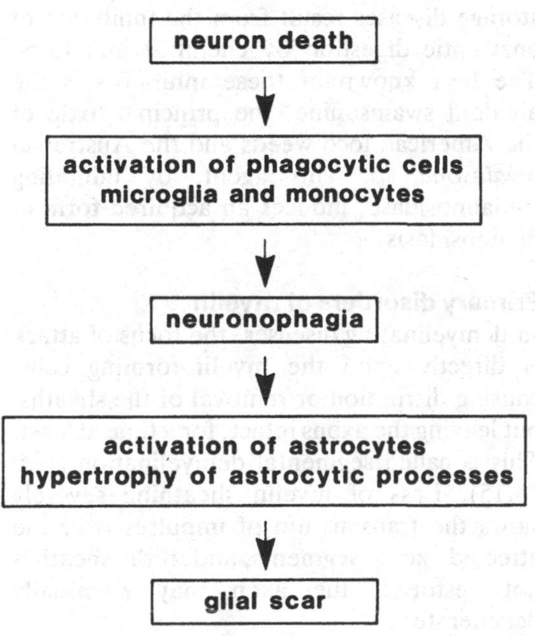

Fig. 13.17. This flow chart indicates that death of neurons activates phagocytes, which ingest the cell debris – neuronophagia. In turn, astrocytes are activated and the damaged area becomes filled by a glial scar.

Necrosis of neurons may occur as an acute episode, with a sudden onset of clinical signs, or it may be slow and insidious, with a matching chronic and progressive clinical syndrome. The delayed form of 'swayback' in young lambs and goat kids is an excellent example of neuronal necrosis. The disease is caused by a deficiency of copper, which results in degeneration and eventual necrosis of neurons, especially in the spinal cord. The clinical disease is dominated by indications of a diffuse myelopathy. Gross pathologic changes are not present, but histologically the degenerating neurons provide one of the best examples of chromatolysis in veterinary pathology (Fig. 13.14).

Another good example of neuronal degeneration is poisoning by dinotolmide, a coccidiostat used in domestic poultry. At toxic doses this agent causes necrosis of cerebellar Purkinje cells with subsequent neuronophagia (Fig. 13.14) and an acute clinical onset of cerebellar ataxia. In contrast, a slowly progressive loss of Purkinje cells occurs in dogs of the Kerry Blue Terrier breed, because of a genetic defect transmitted as an autosomal recessive trait. This disease can be considered as an abiotrophy – that is, a fault in the genetic program whereby certain cells undergo premature senescence and die well before the end of their expected life span.

Axonopathy

When neurons are sublethally injured by toxins or metabolic derangements, the frequent outcome is axonal degeneration or axonopathy. Although the perikaryon remains intact, the axon undergoes fragmentation and dissolution, and, if it has a myelin sheath, it too degenerates and is removed (see Wallerian degeneration, above). In many axonopathies, degeneration occurs primarily at the distal extremities and may extend in a retrograde manner towards the cell body. This has been called the 'dying-back' pattern of axonal degeneration. Some axonopathies are confined either to the peripheral nerves, or to the CNS, and are called, respectively, peripheral or central axonopathies. Others are both peripheral and central.

A typical example is the disease caused by long-term exposure to certain of the organophosphates. A severe central and peripheral axonopathy ocurs at the distal extremities of long axons in the peripheral nerves and spinal cord, causing a spectrum of clinical signs related to proprioceptive disturbance and muscle weakness. An interesting early sign is laryngeal paresis and vocalization defects, resulting from the vulnerability of the very long recurrent laryngeal nerves. In such chronic intoxication, regenerative attempts are usually abortive, even if exposure to the toxin ceases.

In German Shepherd dogs there is an example of a genetically determined progressive axonopathy known as giant axonal neuropathy. Clinical defects first appear at about 12 months of age and there is progressive deterioration, with impairment of proprioception and gait. Pathologically, the disease is characterized by large focal swellings, produced by accumulation of neurofilaments in central and peripheral axons.

Other well-known causes of axonopathy are the toxins found in plants of the cycad family, and in the buckthorn bush, *Karwinskia humboldtiana*.

Lysosomal storage diseases

The lysosomal storage diseases are a group of related conditions which result from deficiency in the activity of intracellular digestive enzymes. Each individual disease is based upon a deficiency of one particular lysosomal hydrolytic enzyme and is characterized by the progressive intracellular accumulation of molecules which the missing enzyme would normally digest. Neurons, being permanent cells, are very vulnerable to these disorders, because, for as long as material accumulates, each neuron has no option but to continue storing it in lysosomal vacuoles. Eventually the neurons become greatly swollen and distended.

Usually the vast majority of neurons are affected, and as all lysosomes are involved, glial cells, and many cells in other organ systems also show the changes. In most lysosomal storage diseases, neurologic signs are evident, predominantly expressed as disturbances of higher functions. However, the relationship between the storage process and neuronal dysfunction is not yet clear. Although in time, neuronal death may occur, recent research suggests that the storage condition may stimulate the formation of abnormal synaptic connections, particularly in the thalamus and cerebral hemispheres.

Most lysosomal storage diseases are due to genetically determined deficiencies of particular enzymes, are generally expressed early in life and are progressive and irreversible. Some examples are mannosidosis of cats, cattle and goats, glycogenosis type II of cattle, and fucosidosis of Spaniel dogs. Some lysosomal storage diseases result from the inhibition of enzymatic digestion by chemical inhibitors. The best known of these inhibitors is the alkaloid swainsonine, the principal toxin of the American loco weeds and the Australian *Swainsona* sp. This agent, by inhibiting α-mannosidase, induces an acquired form of mannosidosis.

Primary disorders of myelin

In demyelinating diseases, the focus of attack is directly upon the myelin-forming cells, causing disruption or removal of the sheaths, but leaving the axons intact, for a time at least. This is called segmental demyelination (Fig. 13.15). Loss of myelin sheathing severely slows the transmission of impulses over the affected axonal segments, and, if the sheath is not restored, the axon may eventually degenerate.

Loss of myelin can come about in several ways. Destruction of myelin-forming cells will result in fragmentation of the sheath and its removal by phagocytosis. More commonly, however, the sheath itself is attacked but its parent oligodendrocyte or Schwann cell is otherwise undamaged. The chief mechanism in this process is immune attack. Antibodies, lymphoid cells and macrophages can be directed against myelin, and their combined activities can remove it rapidly. This can be triggered by viral infections, the classic example being canine distemper, in which demyelination in the brain and cord is a major lesion, and no doubt contributes greatly to the functional disturbances. Immune-mediated demyelination is well recognized in the peripheral nervous system of dogs and results in stripping of the sheaths, particularly in motor nerve roots (see Peripheral neuritis, below). The myelin is removed from individual, but not necessarily contiguous, Schwann cell segments and is thus classified as segmental demyelination. In North American Coonhounds, the disease often follows racoon bites and is associated with an as yet unidentified factor in the racoon saliva. In other breeds, no specific causes or associations have yet been defined. If the immune-mediated attack subsides, remyelination can be completed in a matter of weeks, but, if it continues, a chronic progressive disease ensues. Tetraparesis and rapid muscle atrophy are the dominant clinical effects.

In Afghan hounds there is a genetically determined demyelinating disease which strikes at motor pathways in the spinal cord and causes irreversible tetraplegia. It is known as hereditary myelopathy of Afghan hounds.

Compression and swelling in the central nervous system

The CNS is firmly confined within the tough envelope of the dura mater, which is in turn encased by the rigid bones of the skull and vertebral column. In the cranium, the dura is in close proximity to the bones and it also has a fold, the tentorium cerebelli, which divides the cavity into anterior and posterior chambers. In the vertebral canal there is a space between the dura and the bone, which generally contains an insulating layer of fat. Because of this confinement, the brain in particular, is poorly placed to accommodate any increase in volume or any mechanical displacement. In either case, portions of the brain can suffer considerable secondary damage by compression and edema. Severe compression within the anterior compartment of the cranium may cause the occipital cortices to herniate under the tentorium cerebelli, and compression in the posterior compartment may force the cerebellar vermis to herniate through the foramen magnum into the spinal canal. In this way, pressure is exerted on vital centers in the medulla, and herniated tissue may suffer considerable damage. In the spinal canal there is more space available, but it is still limited, and the cord is particularly vulnerable to compression over the mobile vertebral junctions.

Compression may be produced by a single, space-occupying lesion, like an abscess, or neoplasm. In the brain, such a lesion may produce pressure which not only acts immediately around it, but is transmitted throughout the cranium to cause compression contralaterally (on the opposite side). The functional consequences of the lesion are greatly enhanced as a result. The single most important cause of localized spinal compression is prolapse of intervertebral disks. This is most commonly a problem in dogs and may occur at any level of the vertebral column. In some cases, the compression is so severe that extensive necrosis results in irreversible injury. In other instances only local edema is produced and functional recovery will follow successful surgical decompression. Spinal cord compression of intermediate severity will result in axonal degeneration, as exemplified by the 'Wobbler syndromes' in the dog and horse. Major causes of spinal compression are listed in Table 13.9.

Generalized swelling of the brain comes about when there is hydrocephalus and/or cerebral edema.

Hydrocephalus is a state resulting from increased pressure within the cerebrospinal fluid (CSF) compartments of the cranium. Under normal conditions, the total volume of CSF is turned over several times each day, as it

Table 13.9. *Causes of spinal cord compression*

1 Vertebral fracture
2 Vertebral malarticulation and malformation for example:
 atlantoaxial subluxation in miniature and toy dogs
 C_5, C_6 and C_7 malformation – malarticulation in Great Danes and Doberman Pinschers
 C_2–C_3, C_3–C_4 changes in Basset Hounds
 C_1–C_7 the wobbler syndrome in horses
3 Meningeal abscesses
4 Primary or secondary neoplasms or the meninges, cord, nerve roots or vertebrae
5 Dural ossification
6 Intervertebral disk protrusion

flows through the ventricular system of the brain and out into the subarachnoid space at the cerebellar–pontine angles. It is produced in the choroid plexuses by ultrafiltration of the plasma and re-enters the blood through the projection of the arachnoid villi into venous sinuses. It is known that 75% of the CSF re-enters the blood in the venous sinuses over the cerebrum.

If the free passage of the CSF is impeded at any point, the pressure will rise proximal to the obstruction, dilating the ventricular system and compressing the brain tissue. Localized inflammatory lesions, neoplasms and parasitic cysts may cause focal obstructions at the foramen of Munro, the aqueduct of Sylvius or at the cerebellar–pontine angles. In severe diffuse meningitis, extensive destruction of arachnoid villi over the cerebrum can block the re-entry of CSF into the venous sinuses.

The acute onset of hydrocephalus will provoke significant functional impairment and clinical neurologic disease. Slowly progressive hydrocephalus, however, although causing marked ventricular dilation and brain atrophy, can be compensated for to a remarkable degree and there may be no overt clinical deficits until the condition is well advanced. Clinical signs are often dominated by indications of prefrontal lobe injury, and an adversive syndrome.

Diffuse **cerebral edema** can follow severe traumatic head injury, decreased plasma osmotic pressure, or some intoxications. Poisoning with ammonia or hexachlorophene produces severe cerebral edema, but the classical example is porcine salt poisoning. The disease occurs when salt intake is excessive and water intake inadequate. Under these conditions cerebral sodium concentration becomes abnormally high and, if there is sudden access to unlimited water, a sudden massive increase in brain water content ensues. The resulting brain swelling causes compression and necrosis of the cerebral cortex and herniation of the cerebellar vermis.

Nervous tissue necrosis – malacia

In the CNS, necrosis involving all the tissue elements, gives rise initially to liquefaction, softening and focal hemorrhage in the affected area. This process is termed malacia – encephalomalacia when it occurs in the brain and myelomalacia when it occurs in the cord. It is further classified as polio- or leukomalacia, according to the involvement of gray and white matter, respectively. When extensive malacic lesions are provoked, there are usually significant concurrent edema and swelling which exacerbate the condition. If the patient survives the acute necrotic phase, the lesions resolve to leave permanent deformities within the tissue. Necrotic debris is removed by macrophages, and astroglial fibers and collagen derived from fibroblasts form a scar along the junction with normal surviving tissue. Areas of necrosis on the surface of the brain such as the cerebral cortex will eventually appear as depressed areas of 'atrophy' and surface deformity. Deeper lesions resolve into permanent cystic cavities.

Episodes of necrosis may occur acutely and be accompanied by the abrupt onset of clinical signs. It is fairly characteristic, however, for the clinical deficits rapidly to reach their maximal intensity and then stabilize. This is a useful generalization that aids in clinically differentiating this type of disease from inflammatory or neoplastic diseases, which are often progressive.

In veterinary medicine the most important causes of malacic lesions are intoxications, metabolic disturbances and traumatic injury. Infarction as a result of vascular disease is relatively uncommon, which is in marked contrast to the situation in man.

In the toxic and metabolic group of diseases the lesions are characteristically distributed in a bilaterally symmetrical pattern which reflects differing metabolic activities in various regions. The basis of this metabolic regionalism is not well understood and it is the focal symmetrical pattern of such lesions that has often brought them to notice.

The development of malacic lesions is an important aspect of the encephalopathy resulting from unavailability of thiamine in nervous tissue. In carnivores, this is usually due to a deficiency of dietary thiamine, while in ruminants it may also result from the action of thiaminases produced by rumen microorganisms. Ingested thiamine may be destroyed, or converted to analogs which compete with it. In the thiamine-related diseases there is necrosis of vulnerable areas, which is accompanied by the sudden onset of clinical neurologic disturbances. In ruminants, the extensive involvement of the dorsolateral aspects of the cerebral cortices has led to the specific name polioencephalomalacia for the condition. There are behavioral disturbances, blindness and often seizures. In carnivores the lesions are concentrated in nuclei in the thalamus and midbrainstem and are associated with rather characteristic seizures and blindness.

The epsilon toxin of *Clostridium perfringens* type D is a well known cause of malacia in lambs. On occasions, the absorption of this toxin from the gut, rather than causing the rapid death typical of enterotoxemia, causes symmetrical necrosis in the internal capsule, thalamus, the brainstem and white matter of the cerebellum. Hence, the condition is named focal symmetrical encephalomalacia and is characterized clinically by a 'dummy syndrome'. Similarly, in horses, a mycotoxin produced by a *Fusarium* sp. growing on moldy corn is the cause of a severe leukoencephalomalacia in the cerebrum. It is also expressed clinically as a 'dummy syndrome'.

Trauma as a cause of malacia is of most importance in the spinal cord, particularly in dogs with prolapse of intervertebral disks or vertebral fractures. The ventral spinal artery may be compromised and the resulting extensive necrosis and hemorrhage may involve large segments of the cord. Such devastation usually causes profound LMN deficits.

Infarction of the CNS is a feature of some vascular diseases, the best examples of which are porcine cerebrospinal angiopathy and equine herpesvirus myelopathy. In the former, regarded as a chronic form of edema disease, vascular damage with thrombosis produces randomly scattered small infarcts in the brain and cord. In the latter, inflammation of small spinal arteries results in multiple random infarction of the spinal cord. In both diseases, clinical signs depend upon the extent and severity of lesions. Extensive unilateral infarction of the cerebrum has also been recognized as an entity in the cat.

Infarction of the CNS secondary to vasculitis can also occur in bovine malignant catarrh and sporadic bovine encephalomyelitis.

In large-breed dogs, acute infarction of the spinal cord is the result of embolism by fragments of fibrocartilage which derive from intervertebral disks. It is not understood how these fragments gain access to the blood vessels, but it is presumed they are forced into the vasculature mechanically. The condition is referred to as fibrocartilagenous embolism of the spinal cord.

Inflammation in the central nervous system

Inflammatory lesions of the central nervous system are frequently a cause of clinical disease, with manifestations ranging over the entire spectrum of neurologic deficits.

Inflammatory processes may extensively involve the meninges, parenchyma and ventricular systems of the brain and cord, giving rise to diffuse **meningoencephalomyelitis**. Alternatively, lesions may be multifocal or even solitary. Obviously the more focal and restricted the tissue damage the more localized and restricted will be the functional deficits, unless secondary effects due to compression or swelling occur. In some diseases, vasculitis with accompanying thrombosis of arterioles or venules adds a significant extra dimension to the severity of any lesion. This is certainly true of several bacterial infections and a few of the viral infections. Diseases that are primarily multisystemic vasculitides, like bovine malignant catarrh and feline infectious peritonitis, can also cause significant damage to the CNS.

In many instances, the presence of inflammation can be detected clinically by changes in the CSF, because inflammatory cells and exuded protein alter its quality. However, diagnosis depends largely on the clinical findings and history. The process will be discussed in general terms on the basis of the major etiologic agents and the types of lesions they produce.

Bacterial infection

Many species of bacteria have been associated with neural inflammation, some causing specific disease entities. Bacteria may infect the nervous system by a number of routes. **Hematogenous infection** (Fig. 13.18) is likely to cause an acute diffuse or multifocal pattern of lesions, with extensive involvement of the brain and a strong likelihood of secondary brain swelling. The clinical picture is dominated by cerebral and brainstem dysfunction, with derangement, seizures, collapse and opisthotonous as common features. The reaction mostly produces a neutrophil-rich exudate, in some cases with suppuration.

Such bacterial meningoencephalomyelitis is most common in young animals and is well exemplified in *Escherichia coli* infection in calves and *Streptococcus suis* infection in growing pigs. As would be expected, the mortality rate is high and such is the extent of tissue destruction that complete recovery is a rarity. At the height of the disease, swelling of the brain results in herniation of the posterior cerebellar vermis through the foramen magnum, and the CSF is usually turbid or purulent. The leptomeninges may be thickened and hyperemic and the ventricular system dilated. Focal areas of hyperemia and softening may be scattered through the substance of the brain and cord.

Pyogenic bacteria carried in the bloodstream as organisms or as septic emboli may cause multiple abscess formation through the brain and cord, with relatively minor leptomeningeal involvement. In such cases there is a tendency for abscessation to occur in the cerebral cortex at the junction of the gray and white matter. Clinical disease in this instance may be of a less acute nature if abscesses are few in number and develop slowly.

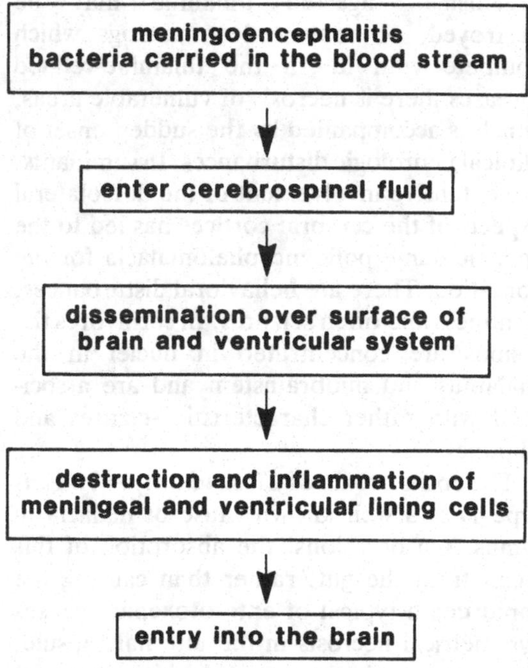

Fig. 13.18. This flow chart indicates the development of a meningoencephalitis as a result of the hematogenous spread of bacteria. The bacteria enter the cerebrospinal fluid to cause inflammation in the meninges and ventricles and from these sites enter the brain to produce an encephalitis.

Bacterial infection may also penetrate the CNS by *extension* from an adjacent structure. A common event is the spread of infection to the posterior brainstem from the middle ear. Another is the formation of an epidural spinal abscess by spread from vertebral osteomyelitis. Infection may also ascend spinal and cranial nerves, as in the case of *Listeria monocytogenes*. In most instances, infection by extension leads to anatomically localized and non-symmetrical tissue lesions. Clinical signs will reflect which anatomical sites are damaged. Listerial encephalitis is a good example. Organisms tend to enter the branches of the trigeminal nerve in the oro- and nasopharynx, and are transported via tissue spaces and lymphatics to the brainstem. In the brainstem, multiple small foci of necrosis and cellular infiltration are produced and a range of cranial nerve deficits expressed, although head tilt and circling are most prominent.

Viral infection

Many of the numerous viruses that cause nervous disease invade the CNS hematogenously, either being carried free in the plasma or in association with circulating lymphocytes. The rabies virus and some of the herpesviruses enter peripheral axons and are transported centrally by retrograde axonal flow. Other viruses may move from the nasal mucosa along the olfactory nerves. Once within the CNS, there may be extensive spread via the CSF. Viral infections tend to produce focally extensive to diffuse encephalomyelitis, with variable occurrence of moderate leptomeningitis. The lesions produced are generally characterized by a non-suppurative reaction, involving cells of the immune system, principally lymphoid cells and macrophages. In diseases such as canine distemper, the induced immune reactions can be responsible for much of the tissue damage. Viruses tend to be selectively destructive to the neurons and their myelinated axons, with glial elements surviving and actively taking part in the tissue response. It is for this reason that large-scale destruction of the neuropil does not often occur and lesions may not be detected by gross examination, even when they are very severe histologically.

Most viral encephalitides run an acute to subacute course and clinical neurologic deficits are generally widespread. An exception is the focal myelitis that often occurs in the caprine lentivirus infection ('arthritis–encephalomyelitis virus') producing clinical signs typical of a focal spinal cord lesion. In canine distemper, clinical signs may relate mainly to the cerebellum or basal ganglia.

Fungal infection

Fungal agents are a sporadic cause of CNS lesions and may reach the nervous tissues via the routes described for bacteria. Generally the lesions produced are subacute to chronic, focal or multifocal, granulomatous reactions. Vasculitis with thrombosis and secondary infarction may lead to extensive tissue destruction. The yeast-like organism *Cryptococcus neoformans* is a sporadic cause of diffuse leptomeningitis, mainly in the dog and cat.

Sporozoan infection

The major agent of this group is *Toxoplasma gondii*, which is an occasional cause of acute multifocal encephalomyelitis in many species. An unidentified sporozoan has also been found in a multifocal myelitis in the horse, and in other herbivores, sporozoan encephalomyelitis is assumed to be due to various of the *Sarcocystis* species.

Sporozoan agents infect the nervous system via the circulation of zooites in the bloodstream and produce a somewhat random multifocal pattern of myeloencephalitic lesions, and thus a variable set of clinical features. In general the diseases are of acute onset, and if lesions are widespread, progress rapidly to a fatal termination. In the case of toxoplasmosis, diagnosis in live animals may be aided by serologic tests. Cystic forms of the organism can be demonstrated histologically within, and at some distance from, lesions.

Sporozoan-induced lesions are focal areas of acute inflammation and necrosis, often with some hemorrhage present. Local tissue destruction is severe and lesions may be up to several centimeters in extent and visible to gross inspection.

Helminthic infestation

Helminthic agents migrating through the nervous tissues can result in extensive tissue destruction and a subacute to chronic granulomatous response. The agents include various nematode larvae, and cestode cysts and details can be found in appropriate reference texts.

Peripheral neuritis

In the peripheral nervous system, inflammatory lesions are rare, and are classified according to distribution. **Mononeuritis** involves isolated peripheral nerve trunks, whereas **polyneuritis** affects groups of similar nerve fibers innervating synergic groups of muscle fibers with bilaterally symmetrical distribution. **Polyradiculoneuritis** is similar, but additionally involves spinal nerve roots. The best-known examples of neuritis are acute idiopathic polyradiculoneuritis in the dog and equine polyneuritis. The former disease has already been mentioned in the context of immune-mediated primary segmental demyelination, and includes Coonhound paralysis. The inflammatory component of the lesion is provided by the infiltration of lymphoid cells and macrophages into the nerves. The equine disease predominantly affects the nerves of the cauda equina and involves a granulomatous reaction which may relate to immune responses to a persistent viral infection. There is a profound loss of function of the caudal nerves, with LMN deficits to the tail, perineum, rectum and bladder.

Primary neoplastic disease

Primary tumors of nervous tissue may arise from all structures associated with the nervous system, with the exception of mature neurons.

Secondary tumors may be deposited from any primary neoplasm arising outside the nervous system, but the most common non-nervous system-derived tumor is the lymphosarcoma, being part of the generalized lymphosarcoma syndrome.

The species most commonly involved in primary neoplasia of the nervous system is the dog and such tumors are most common in the brachycephalic breeds, e.g. Boxers, Pekes, Bulldogs.

Meningiomas are surface tumors and as space-occupying lesions they will depress the underlying brain. If they are large they may generate sufficient increased pressure to cause herniations. If they are small and have arisen in a non-vital area, they may be clinically silent and may be an incidental finding at necropsy. Clinical signs are associated with damage to an area such as the motor cortex or cerebellum.

Gliomas and **ependymomas** are much more likely to cause clinical signs than meningiomas. Ependymomas will be associated with the ventricles and are therefore likely to produce a blockage of the flow of CSF and induce hydrocephalus. The most common site of gliomas is the brainstem and they therefore interfere with the vital functions controlled by the brainstem nuclei and may stop the flow of messages to and from the higher centers.

Pituitary tumors grow within the confines of the sella tursica and cause much damage to the non-neoplastic pituitary tissue by pressure. Thus there may be endocrine effects which will vary in magnitude according to the age and stage of development of the animal, e.g. the difference produced by loss of growth hormone when it occurs (*a*) before and (*b*) after closure of the epiphyseal plates. Pituitary tumors also tend to grow out of the sella tursica into the pituitary stalk and involve the tissue around the third ventricle, thus destroying the hypothalamic nuclei and optic pathways. One of the major clinical effects of pituitary tumors is the production of diabetes insipidus.

Neurofibromas – affecting peripheral nerves

– will cause localized signs according to the site of the tumor. The brachial plexus is probably the most common site for these tumors.

Dysgenesis of the nervous system

A large number of neurologic disorders result from damage inflicted on the growing brain and spinal cord before, or shortly after, birth. In some instances there is a direct noxious influence on the nervous tissue, or it may be secondarily compromised by distortion of the skeletal investments. Damage inflicted at this time may have serious effects on the multiplication, migration or maturation of neurons, on the formation of myelin, or on the free flow of CSF.

In general, the consequences can include deficiency in nerve cell numbers, deficiency of myelin, dysplasia of nervous tissue, or hydrocephalus. Severe hydrocephalus in the immature animal can cause enlargement of the cranium, as the loosely united cranial bones allow for expansion in response to pressure.

The extent of the lesions can vary from localized to widespread, and both environmental and genetic defects are recognized as causes. Of the environmental agents, viruses and chemical toxins are the chief culprits. Agents of this type act as they do because of an affinity for cells that are in a state of intense metabolic activity or rapid proliferation. There are several good illustrations. Feline parvovirus requires rapidly proliferating cells in which to replicate. In young adult cats it attacks the cells of the bone marrow and intestinal crypt epithelium, causing the well-known infectious enteritis and panleukopenia. In the perinatal kitten, however, the virus finds a population of vulnerable cells in the cerebellar cortex. At this stage of growth there is intense proliferation of immature neurons in a layer along the cerebellar surface, called the external granular layer (Fig. 13.19). Normally cells from this layer migrate into the deeper tissue to form the granular layer of the cerebellum and make a great contribution to the total cerebellar mass. When large numbers of proliferating external granule neurons are destroyed by the virus, the cerebellum is left hypoplastic to varying degrees, and the kitten survives with cerebellar ataxia.

Several important general principles are well illustrated when fetal calves are infected with the Akabane or Aino arboviruses. The viruses destroy proliferating neurons on a much less selective basis than the feline parvovirus. The distribution of lesions depends upon which area of the CNS is in a phase of rapid development when the virus arrives on the scene. If infection occurs early in preg-

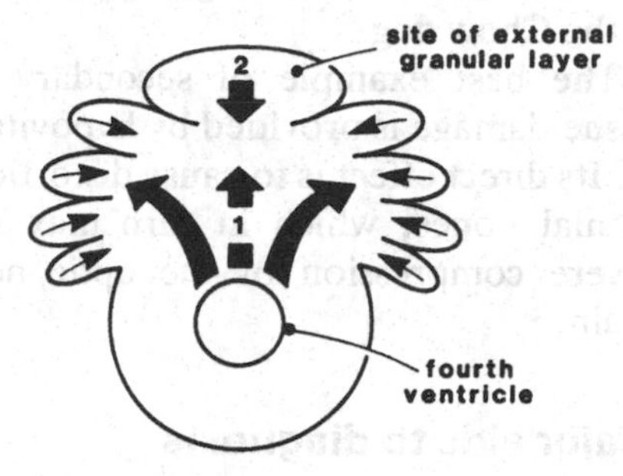

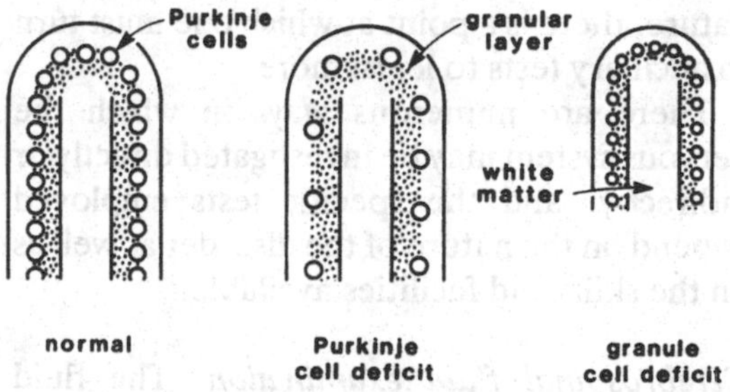

Fig. 13.19. In the developing brain, cells migrate out from the tissue adjacent to the fourth ventricle (1) and become Purkinje cells. Cells migrate in from the external granular layer (2) to become the neurons of the granular layer. These are shown diagrammatically in the transverse section of the cerebellum and pons. The three transverse sections of cerebellar folia show the normal situation, the effect of loss of Purkinje cells in abiotrophy, and the loss of granule layer cells in cerebellar hypoplasia.

nancy, there is total destruction of cells forming the cerebral cortical neuron population, and the result is hydranencephaly. If infection occurs just prior to birth, there is destruction of spinal motor neurons. Infection in between times overlaps the growth spurts of both areas, and both lesions occur.

In the pig, the virus of swine fever is capable of severely impeding the production of myelin in the developing CNS, leading to one form of congenital hypomyelinogenesis, and clinically a severe tremor syndrome. A similar disease occurs in lambs infected *in utero* with the bovine virus diarrhea togavirus.

Genetically determined disorders of myelin formation give rise to neonatal tremor syndromes and are well recognized in the pig and in the Chow dog.

The best example of secondary nervous tissue damage is provided by hypovitaminosis A. Its direct effect is to cause distortion of the cranial bones, which in turn may result in severe compression of the optic nerves or brain.

Major aids to diagnosis

The aim of the neurologic examination has been to achieve an 'anatomical diagnosis'. Although knowledge of the location and extent of the lesion give some indication of its nature, there is a point at which one must turn to ancillary tests to learn more.

There are numerous ways in which the nervous system may be investigated directly or indirectly, and the specific tests employed depend on the nature of the disorder as well as on the skills and facilities available.

Cerebrospinal fluid examination: The fluid which circulates within the CNS often reflects pathologic processes going on there. Changes in the number and type of cells in the fluid may indicate active inflammatory or degenerative disease and can be useful in differentiating viral from bacterial infections. The chemical constituents of the fluid also provide valuable diagnostic information. Cerebrospinal fluid is collected under general anesthesia and is drawn mostly from the cysterna magna at the base of the skull.

Radiography: This is another basic tool available in most practices. It is of little value in the diagnosis of most intracranial disease but may be useful in detecting hydrocephalus or lesions which have undergone mineralization or have destroyed bone. In the latter category would be neoplasms or chronic inflammatory lesions.

In spinal disorders, however, radiography is a vital diagnostic tool. Some lesions cause plainly visible alterations to the bony elements of the spine (intervertebral disk disease, vertebral infections and fractures) while others affect only the soft tissues and require contrast studies ('myelography') for their detection. Failure to demonstrate lesions by the use of these techniques may be a strong indicator that the pathology is occurring on a microscopic level (as in degenerative radiculomyelopathy).

Clinical chemistry: This is very useful in cases where biochemical lesions may underly the neurologic disturbances. For example hypokalemia may cause diffuse weakness, hypocalcemia tremors or tetany, and hypoglycemia seizures. Similarly, analysis of urine for the concentration of δ-aminolevulinic acid is a routine test for lead poisoning in the dog.

Electrocardiography: This is important when syncope is being considered as a cause of 'seizures'. It is possible to detect conduction problems that could be responsible for such attacks.

The encephalocardiograph can also provide a more sensitive indicator of potassium status than does serum chemistry. It is, therefore, valuable in the study of animals with diffuse weakness which might be due to hypokalemia or hyperkalemia.

Electroencephalography: This is a sophisticated tool for examining brainwave patterns. Certain abnormal waveforms can be indicative of different types of intracranial disease and, where the equipment is available, it has proven to be useful in some cases.

Eletrodiagnostics: The measurement of nerve conduction velocity (NCV) is routinely performed in many centers and gives valuable diagnostic information in cases with peripheral nerve involvement. Electromyography is often performed in conjunction with NCV studies and gives information about the neuromuscular junction and muscle cell function.

Computerized tomography: Although new and rather costly tools, computerized tomography scanners are in routine use in the larger veterinary institutions and prove especially valuable in the diagnosis of intracranial disease.

Tissue biopsy: Biopsy can be used as a diagnostic method where neuromuscular or peripheral nerve disease is suspected. Muscle biopsies can be taken surgically or with specially designed needles. Peripheral nerve biopsies can be taken from the ulnar and peroneal nerves and 30–50% of the nerve fascicles over a length of 2–2.5 centimeters can be excised safely without creating a functional deficit.

Immunology: Enzyme-linked immunosorbent assay (ELISA) tests for the presence of snake venom in blood or urine and testing for antibodies to acetylcholine receptors in myasthenic dogs are examples of useful applications of clinical immunology in neurologic diagnosis.

Additional reading

Adams, J. H., Corsellis, J. A. N. and Duchen, L. W. (1984). *Greenfield's Neuropathology*, 4th edn. London, Edward Arnold.

Adams, R. D. and Victor, M. (1981). *Principles of Neurology*, 2nd edn. New York, McGraw-Hill.

Bailey, C. S. and Kitchell, R. L. (1984). Clinical evaluation of the cutaneous innervation of the canine thoracic limb. *J. Am. Anim. Hosp. Assoc.* **20**: 939–50.

Chrisman, C. L. (1982). *Problems in Small Animal Neurology*. Philadelphia, Lea and Febiger.

Chusid, J. G. (1976). *Correlative Neuroanatomy and Functional Neurology*, 16th edn. Los Altos, CA, Lange Medical Publications.

De LaHunta, A. (1983). *Veterinary Neuroanatomy and Clinical Neurology*, 2nd edn. Philadelphia, W. B. Saunders Co.

Escourolle, R. and Poirier, J. (1978). *Manual of Basic Neuropathology*, trans. J. L. Rubinstein, 2nd edn, Philadelphia, W. B. Saunders Co.

Eyzaguiree, C. and Fidone, S. J. (1975). *Physiology of the Nervous System*, 2nd edn. Chicago, Year Book Medical Publishers.

Griffiths, I. (1982). Spinal disease in the dog. *In Practice* **4**: 44–52.

Haghigi, S. S. (1982). Electrophysiological studies of the cutaneous nerves of the pelvic limb of the dog. Ph.D. dissertation, University of California, Davis.

House, E. L. and Pansky, B. (1967). *A Functional Approach to Neuroanatomy*, 2nd edn. New York, McGraw-Hill.

Jenkins, T. W. (1978). *Functional Mammalian Neuroanatomy*, 2nd edn. Philadelphia, Lea and Febiger.

Mayhew, I. G., de Lahunta, A. Whitlock, R. H. Krook, L. and Tasker, J. B. (1978). Spinal cord disease in the horse. *Cornell Vet.* **68** (Suppl. 6).

Noback, C. R. and Demarest, R. J. (1977). *The Nervous System: Introduction and Review*, 2nd edn. New York, McGraw-Hill.

Oliver, J. E., Hoerlein, B. F. and Mayhew, I. G. (1986). *Veterinary Neurology*. Philadelphia, W. B. Saunders Co.

Oliver, J. E. and Lorenz, M. D. (1983). *Handbook of Veterinary Neurologic Diagnosis*. Philadelphia, W. B. Saunders Co.

Palmer, A. C. (1976). *Introduction to Animal Neurology*, 2nd edn. Oxford, Blackwell.

Sullivan, N. D. (1985). The nervous system. In *Pathology of Domestic Animals*, 3rd edn, K. V. Jubb, P. C. Kennedy and N. Palmer (eds.), pp. 202–338. Orlando, FL, Academic Press.

Sunderland, S. (1978). *Nerves and Nerve Injuries*, 2nd edn. Edinburgh, Churchill Livingstone.

Whitewell, K. E. (1980). Causes of ataxia in horses. *In Practice* **2**: 17–24.

Wayne F. Robinson

14 Muscle

The business of voluntary muscle is to contract, to exert an appropriate amount of force under the direction of the central nervous system. The coordination of muscle contraction and relaxation permits not only a wide range of movement, but also the maintenance of position in defiance of gravity.

The temporary or permanent loss of the ability to exert appropriate force characterizes most muscle disease. That is not to say that the defect is always within muscle fibers, as abnormalities of the tendon or tendon sheath are often the culprits.

The temporary or permanent loss of muscle function has a variety of causes which differ significantly between species. In athletic and companion animals, trauma is by far the most common cause, especially in those constantly stressed to perform. Resultant mechanical injury can vary from muscle or tendon sprains to, in the most severe cases, complete separations of the muscle–tendon unit. This is in contrast to economic animals – when the most important diseases primarily affect the muscle fiber itself, and range from the nutritionally based myopathies associated with deficiencies of selenium and vitamin E to the clostridial infections.

This chapter therefore has two major themes: those abnormalities of muscle and tendon that are of a traumatic nature and those that include degenerative and inflammatory diseases of muscle.

The nature of muscle

Individual muscles are composed of bundles of muscle fibers within a connective-tissue framework. The predominantly collagenous connective tissue may be subdivided into three basic levels. The epimysium, covering the muscle, the perimysium separating bundles of muscle fibers and endomysium investing each muscle fiber. Single muscle fibers, alternatively termed muscle cells or myofibers, are elongated multinucleate syncytia of variable length and width. Each muscle fiber runs the length of the muscle, from the origin to the insertional end. The combination of endomysial reticulin fibers and the basal lamina of the muscle fiber constitute the sarcolemma, a unit of particular importance when muscle regeneration occurs. The cell membrane of the muscle fiber is termed the plasmalemma.

Muscle fibers are replete with the contractile proteins myosin and actin, arranged in repeating units termed the sarcomere, and orchestrated to generate mechanical force. Most anatomical structures and biochemical pathways in muscle are directed towards assisting and regulating the contractile process. Such power generation requires the provision of large amounts of energy and in muscle this need is met by high-energy phosphates derived from abundant mitochondria and glycogen stores. A generous blood supply and liberal quantities of the oxygen-binding protein myoglobin optimize conditions for

Table 14.1. *Characteristics of muscle fiber subgroups*

	Fiber type		
Characteristic	1	Intermediate	2
Appearance	Red	Intermediate	White
Metabolism	Oxidative	Oxidative/glycolytic	Glycolytic
Contraction	Slow twitch	Fast twitch	Fast twitch
Myosin ATPase (pH = 10.4)	Low	Intermediate	High
Glycogen phosphorylase	Low	Intermediate	High
Myosin ATPase (pH = 4.35)	High	Low	Intermediate
Myoglobin	High	Intermediate	Low
NADH-tetrazolium reductase	High	Intermediate or high	Intermediate or low

aerobic metabolism. The sarcoplasmic reticulum, a relatively vast membranous network, regulates the amount of calcium present intracellularly and governs indirectly the extent to which actin and myosin are permitted to interact.

Although all muscle fibers contract, there are differences between them, particularly in their ability to continue to contract for prolonged periods of time. Indeed the difference between fibers is often visible to the naked eye. There are two general subtypes of muscle, those that are geared for short bursts of contraction (type 2, or white fibers) and those for sustained contractile efforts (type 1, or red fibers). These physiologic properties are reflected in the relative abundance or paucity of particular enzymes and organelles. For example, type 2 fibers contract more quickly, have less myoglobin and fewer mitochondria, but more glycogen than do type 1 fibers. There are also fibers that occupy an intermediate position and are classified as intermediate fibers. A particular muscle may be composed entirely of one fiber type or it may contain a mixture of fiber types. There is also considerable species variation (Table 14.1).

Clinical features of muscle disease

Clinically, disease of muscle or tendons is primarily seen as a diminution in, or a complete loss of, the ability to generate power. This may be due to an inherent weakness of the affected area or to pain associated with movement of the affected muscle or tendon. It may also be associated with physical restrictions of motion. It is appropriate at this point to emphasize that the abnormality may suddenly appear or be of gradual onset and the signs may involve a single muscle or tendon, a local group of muscles or all muscles.

With abnormalities of **sudden onset**, the history often includes abnormal motion and a reluctance to bear weight, both of which are referable to pain. Other observations include recumbency and, if the animal remains standing, fine to coarse trembling. There may also be an inability or difficulty with swallowing or eating and a change in the character of vocalization. These signs relate primarily to muscle weakness. The animal may also tire easily. With muscle and tendon abnormalities of sudden onset, the affected muscle or groups of muscles are often swollen. Physical examination may show hardness, pain on palpation and possibly heat. The abnormality may be asymmetrical or symmetrical in character. An additional finding in acute diffuse muscle disease is the presence of red urine (myoglobinuria).

Abnormalities of **gradual onset** may present with some of the features described earlier, such as abnormal motion and a reluctance to bear weight, recumbency and difficulty in rising, and difficulty with swallowing or eating. The animal may also tire easily and have difficulty bearing weight for long periods. Also prominent is symmetrical or asymmetrical muscle atrophy.

The foregoing is a broad attempt to highlight the most common clinical features, but it must be tempered with the fact that as compared with other organs, muscles and tendons are by their nature diverse, each with a specific function. The history and clinical signs observed depend on what muscles and tendons are involved. In fact, although there may be muscle groups that are obviously affected, in many cases other less obvious areas may be detected.

It should also be emphasized, particularly with animals that present for lameness, that consideration should be given to defects of bones, joints, central and peripheral nervous system and the integumentary system.

The laboratory confirmation of muscle disease

Once muscle disease is suspected, there are a number of ancillary tests such as serum enzymes, muscle biopsy and urinary myoglobin that can be carried out to confirm its presence. The first and most commonly used are assays for enzymes released into the circulation following acute muscle injury. Fortunately there are three such enzymes that are of use, one of which is considered to be specific for muscle.

Creatine kinase (CK) is present in abundance in both skeletal and cardiac muscle. It is a cytoplasmic enzyme which is liberated following even minor muscle damage. The half-life of CK in plasma is short and of the order of 2 to 4 hours. It reaches maximum serum concentration 6 to 12 hours after muscle injury, but is cleared in 24 to 48 hours after the muscle damage ceases. There are tissue-specific variants (isoenzymes) of CK, with CK_3 specific for skeletal muscle and CK_2 for cardiac muscle. They may be used if there is need to pinpoint the location of the muscle injury between the heart or skeletal muscle.

Although serum CK estimation is of great value, because of its cytoplasmic location and the liability to leak from minimally damaged cells, caution needs to be applied in the interpretation of rises in CK. For example, CK values rise in animals that are exercised or transported. However, the levels attained do not usually reach those seen with severe widespread injury to muscle.

Aspartate amino transferase (AST) is also of diagnostic use although it is not muscle specific. AST is also in high concentration in the liver. It is a mitochondrial enzyme released only after relatively severe cellular damage. It has a longer plasma half-life than CK, of the order of 12 to 18 hours. Used in conjunction with CK it can give an indication of the severity of muscle damage.

Lactate dehydrogenase (LDH) is also a useful indicator of muscle damage, although, as with AST, it is not generally muscle specific, being found in high concentration in the liver and red blood cells. However, there are isoenzymes of which one (LHD_5) is muscle specific.

Muscle biopsy is of great value in classifying the type of disease process. Biopsies are relatively easy to obtain under sedation and local anesthesia, but care must be taken to select the appropriate muscle and to fix the biopsied muscle under tension. A number of artefacts occur if it is not. There are a variety of muscle biopsy clamps available commercially to maintain muscle tension *in situ* before the biopsy is removed. Alternatively, the muscle biopsy may be tied to a wooden stick with suture material prior to removal.

Electromyography is a relatively specialized procedure and is only rarely carried out in everyday practice.

Tests for urinary **myoglobin** are also of value. It is important to distinguish between urinary hemoglobin and myoglobin, using appropriate methods. Myoglobin does not invariably appear in the urine following muscle damage. The damage must be widespread and it is usually only manifest in mature animals. Many young animals with widespread muscle disease do not exhibit myoglobinuria, presumably because of lower levels of myoglobin in muscle.

The adaptive response of muscle

Muscle is remarkably responsive to the workload placed on it, and that is so whether the workload is increased or decreased. Muscle atrophies when little used and hypertrophies when the workload is increased. It should be emphasized at this stage that the increase or decrease in mass is not due to a change in cell numbers, but to a change in cell size. There is, in the case of atrophy or hypertrophy, a change in many constituents of the muscle fiber, including contractile proteins and mitochondria. The metabolic mechanism for the translation of changes in muscle tension to changes in fiber size is not yet known.

While in many instances atrophy and hypertrophy are physiologic events, atrophy in particular is commonly the result of an interruption of the nerve supply to muscle and is termed **denervation atrophy**. The nerve fiber and the impulses traveling down it exert a trophic influence on muscle mass. Denervation atrophy may be localized or generalized depending on the initial insult. There are many examples of localized atrophy, the site of which depends on the particular nerve involved. Localized atrophy of muscles commonly follows radial nerve paralysis or intervertebral disk protrusion in the dog. Generalized neurogenic atrophy is less common but is seen in particular types of snake bite, where the venom produces prolonged blockade of neuromuscular transmission.

Disuse atrophy is another relatively common form of atrophy often associated with immobilization of a limb for prolonged periods. It should be remembered that atrophy is usually reversible providing the cause is removed.

Patterns of muscle disease and their etiologic implications

There are a wide variety of patterns of muscle disease, some primarily degenerative, some inflammatory and a few neoplastic. The degenerative pattern is by far the most common and the most important, and includes nutritional deficiencies, exercise-induced, toxic (e.g. snake venoms) and, rarely, inherited myopathies. Inflammatory muscle disease is often bacterial or parasitic in origin and sometimes immune mediated.

Degeneration (necrosis) and regeneration of muscle fibers

Because of the cellular differentiation of muscle fibers, the term degeneration is used in a special sense. The length of each muscle fiber and its multinucleate nature mitigates against necrosis of the whole muscle fiber. Most commonly, it is segmental in type and so the fiber is not in a strict sense necrotic, although segments may have lost all the characteristics of viability. Therefore, as most of the muscle fiber is intact and viable, the term degeneration is used.

There are also grades of severity of the degenerative process. At its least severe, only myofibrils and sacroplasm undergo degeneration. At its most severe not only are segments of the muscle fiber involved, but the endomyseal connective tissue and associated structures are also.

The most frequent mechanism associated with muscle degeneration in domestic animals is membrane injury. The loss of the ability of the membranous network of the muscle fiber (plasmalemma, T-tubules and sarcoplasmic reticulum) to regulate the movement of ions, particularly calcium, has lead to the suggestion that calcium has a central role in the degeneration of muscle fibers. Membrane injury leads to an overloading of muscle fibers with calcium. Mitochondria within the muscle fiber are particularly overloaded early in the degenerative process. The high intracellular calcium concentration leads to a state of prolonged contraction or hypercontraction of muscle fibers. Often the abundant mineral within the fiber is evident grossly. The causes of membrane damage leading to this train of events will be highlighted shortly.

The histologic appearance of muscle fiber degeneration is relatively uniform, giving little indication of the cause, but, in some cases, the pattern may be of sufficient specificity to suggest strongly an etiologic diagnosis. More helpful in this regard is the distribution of the degenerative change and the phase (age) of the lesion. Degenerative muscle lesions are often classified according to these two factors. The degeneration may firstly be monophasic and monofocal (at one stage and restricted to one muscle). Secondly, the pattern may be monophasic and polyfocal (a single-stage lesion present in a number of muscles), and finally the pattern may be polyphasic and polyfocal (lesions at different stages and present in a number of muscles).

Muscle retains an almost embryonic capacity for regeneration, but for effective regeneration to occur the retention of the scaffolding formed by the basal lamina and the surrounding connective tissues is essential. If this is the case, then regeneration occurs within the original sarcolemmal tubes. If the sarcolemmal tubes are destroyed, such as occurs with trauma and hemorrhage, infection or ischemia, regeneration is less orderly and less effective.

The mechanism of **regeneration** revolves around the removal of necrotic debris by macrophages followed by replacement of the necrotic segments by mononucleated myogenic stem cells (myoblasts). These arise from satellite cells, which normally lie between the basal lamina and the plasma membrane of the muscle fiber. They are a population of resting cells capable of undergoing mitosis when called on. After mitosis they migrate into the sarcolemmal tube and progressively increase in size, finally fusing with each other to restore the multinucleate syncytial nature of the muscle fiber. Regenerating muscle cells can be recognized histologically by the basophilia of the cytoplasm and the presence of large vesicular nuclei often arranged in rows. The cytoplasmic basophilia indicates active protein synthesis by rough endoplasmic reticulum.

The nutritional myopathies

In almost every country in the world there are instances of nutritional myopathies, and they occur most commonly in domestic ruminants, pigs and horses. While most have been documented to be the result of selenium or vitamin E deficiency or both, some, which are strongly suspected to be so are yet to be confirmed.

Selenium and vitamin E perform similar functions but apparently by different metabolic pathways. Selenium is an essential part of the membrane enzyme **glutathione peroxidase**, which is found in high levels in muscle and red cells, among other tissues. It acts to minimize lipoperoxidation by reducing free radicals and appears to act inside the cell. Vitamin E, traditionally known as an antioxidant, performs a similar function of minimizing lipoperoxidation, particularly on the outside of the cell membrane.

When there is either a deficiency of selenium or vitamin E, peroxidation of membranes occurs which alters membrane function sufficiently to impair the maintenance of the normally polarized status of the membrane. One of the major alterations which may not be the initial event is a massive influx of calcium into the cell. Mitochondria become loaded with calcium further depleting the ability of the muscle to produce high-energy phosphates. This is often referred to as mitochondrial calcium overload and the presence of mineral (calcium) in the muscle fiber gives the characteristic gross appearance to 'white muscle disease'.

Nutritional muscular degeneration is essentially a disease of young animals, usually from dams that have been on diets low in selenium or vitamin E. There are well-known areas in most countries where the soil is deficient in selenium and there are certain forage crops that reflect the soil selenium status. Vitamin E deficiency can be *absolute* where the level of the vitamin in the feed is inadequate or *relative* where feed levels of vitamin E are adequate, but contain high levels of unsaturated fatty acids. Because of their common properties, there appears to be some sparing effect of vitamin E on selenium activity and vice versa. Deficiency of either vitamin E or selenium does not necessarily mean clinical disease. There are often precipitating factors, such as unaccustomed exercise, that move the animal from a deficient state into clinical disease. Although skeletal muscle is affected in almost all cases, there are examples of myocardial involvement where sudden death is a prominent feature.

The exertional myopathies

Exercise is normally considered to be beneficial. It is part of the everyday activities of athletic animals and animals in the wild, but under certain circumstances, exercise and often mild exercise, becomes life threatening. Exertional myopathies have been recognized for at least the last 100 years in horses and more recently in the capture of wild animals. An especially important myopathy in pigs, porcine stress syndrome, is also included even though exercise may be of a very mild character. All are characterized by a severe degenerative myopathy similar in many respects, at least pathologically, to the nutritional myopathies.

Equine exertional myopathies

The most severe form is known under a number of titles such as azoturia, paralytic myoglobinuria, Monday morning disease and sacral paralysis. The less severe form comes under titles such as tying up, acute rhabdomyolysis (literally lysis of striated muscle) and transient exertional rhabdomyolysis.

There are conflicting theories to explain the rhabdomyolysis. Although the original theory was that of increased muscle lactate production leading to a metabolic acidosis, it was subsequently shown that less than 50% of affected horses have increased muscle lactate levels. Indeed in another study, six of seven horses had a metabolic alkalosis. It has been suggested that a variation in blood supply to type 1 and type 2 fibers results in a preferential hypoxia to the type 2 fibers. A further hypothesis proposed that local muscle ischemia follows a disturbance of electrolytes, particularly potassium ions, leading to increased neuromuscular irritability. There no doubt is some truth in each hypothesis, but the unifying concept has yet to be put forward.

Clinical signs of azoturia, the most severe form of the equine exertional myopathies, usually appear within one hour of a horse beginning work or training after a period of rest on full rations. Typically, affected animals are weak, particularly in the hindlimbs and are unable or reluctant to move. The animal may sweat and tremble. The affected muscles are hard and the animal usually has myoglobinuria, manifest as a dark red-brown urine. In the most severe cases, the horse may become recumbent and oliguric or anuric.

These clinical signs are the result of both the pain and swelling of affected muscles, the inability to maintain the normal stance from muscle weakness, with the myoglobinuria a result of the massive release of myoglobin into the blood, exceeding the renal threshold for myoglobin. Although myoglobin is not primarily nephrotoxic, the combined products of muscle necrosis lead to nephrosis. The combination of this and poor renal perfusion in severely affected animals leads to death following renal failure.

Capture myopathies

A severe degenerative myopathy sometimes occurs in wild animals that are either chased and subsequently captured or when they are transported. Clinically affected animals exhibit weakness and collapse, muscle tremors, dyspnea and hyperthermia. Often the animals die acutely. In those that survive for even a short period, there is myoglobinuria and in time there may be renal failure. The mechanism for the renal failure is probably as described for the horse. The pathogenesis is thought to be due to extensive muscle glycogen breakdown, but, if the comments about equine exertional myopathies are taken into account, then other mechanisms may well have to be considered.

Porcine stress syndrome

Although not strictly an exertional myopathy, porcine stress syndrome (PSS) is included in this group. It is alternatively termed malignant hyperthermia and in Europe *Hertztod*, and is recognized *post mortem* as pale, soft, exudative pork (PSE). PSS is a disease of great economic importance, occurring in pigs of particular genotypes and it is estimated that up to 30% of purebred breeding pigs are susceptible. The breeds affected include Landrace, Hampshire, Yorkshire and Pietrain. Crosses of these breeds are also susceptible.

The inherited defect appears to be in the uptake, storage and release of intracellular calcium leading to raised intracellular calcium levels and is exacerbated by excessive catecholamine release. It has been suggested that this uncouples oxidative phosphorylation, producing heat and the excessive consumption of muscle glycogen. The end result is high levels of localized acidosis because of high levels of muscle lactic acid. An extra and severe feature is the hypercontraction of muscle because of increased intracellular calcium levels.

The precipitating factors for the development of clinical signs include stressful procedures such as handling, transport and fighting. Affected pigs exhibit muscle rigidity and dyspnea. There is also tachycardia and hyperthermia. An interesting sidelight is that susceptible pigs can be identified by brief halothane anesthesia. There are, however, degrees of susceptibility to this provocation. Typical necropsy findings include the rapid development of rigor, pulmonary edema and hydropericardium. In some, epicardial and endocardial hemorrhages are prominent, as is myocardial pallor. The skeletal musculature is gray and edematous and the bundles easily separated. Histologic appearance of skeletal muscle varies from an irregularity of sarcomere length to segmental regions of hypercontraction. If the animal lives long enough, features of muscle fiber degeneration are more evident.

Muscular dystrophies

Truly dystrophic muscular disease is interesting, of some comparative importance, but distinctly uncommon. Muscular dystrophy is defined as a progressive hereditary degenerative disease of skeletal muscles. These are not to be confused with the nutritional myopathies which were often unfortunately referred to as dystrophies.

A number of histologic features separate dystrophies from other conditions and include a reduction in the number of muscle fibers, an increase in the number and size of nuclei and, most importantly, an ongoing segmental degeneration of fibers in the absence of effective regenerative efforts. Continued myofiber degeneration without effective regeneration is considered by some to be the fundamental feature of the dystrophic process, but there is some difference of opinion in this regard.

Muscular dystrophy has been reported in cattle (the Meuse-Rhine-Issel breed) in Merino sheep and sporadically in dogs. The best characterized is that in Merino sheep and will serve as the example.

Affected sheep usually begin to exhibit clinical signs from one to four months of age which include reduced growth rate and reduction of hindlimb flexion. The gait abnormalities progress in severity with exercise and age. By 6–18 months of age, most die of inanition or are killed. On necropsy, affected muscles are hard, gray and atrophic. Affected muscles include the vastus intermedius and the medial head of the triceps. Many other muscle groups are affected microscopically, and are characterized by progressive muscle fiber loss, degeneration with fibrosis and fat replacement. Fiber nuclei are prominent and there is a marked variation in fiber size. Often there is absence of myofibrils in the center and periphery of affected muscle fibers. The metabolic basis of this inherited defect is not known.

Myositis

Inflammation of muscle where the localizing clinical signs are limited to muscle is uncommon, whereas primary myositis accompanied by a significant and clinically overwhelming systemic component is met more often. It is also of great economic importance. Because of this, emphasis will be directed toward the latter group which are all infectious in origin, almost always bacterial and commonly clostridial.

Clostridial myositis

Clostridia are large, Gram-positive, spore-forming bacteria which are more or less anaerobic. They are widespread in soil and feces. Clostridia are potent exotoxin producers which have an effect both locally and systemically. Many are gas producers, the smell of which is familiar to most veterinarians. Only one of the clostridial species primarily affects muscle and that is *Cl. chauvoei*, the cause of blackleg. However, other clostridial species acting as opportunists may produce a myositis following wound infections.

Blackleg is a primary necrotizing myositis which affects both cattle and sheep. In cattle, blackleg originates from the spores of *Cl. chauvoei* which are already present in the tissues of the body, particularly muscle. The cause of the activation of the dormant spores is unknown, but in some cases it may be due to bruising. It usually affects cattle of 6 months to 2 years of age and is characterized by a very short period of clinical signs of 16–24 hours duration. Because of this, animals are often found dead. If found alive, the animals are dull, anorexic, pyrexic, with rapid respiration and pulse. They are severely lame, with the affected area swollen, painful and sometimes crepitant. The lesion may be found anywhere, but commonly involves the gluteals, shoulder area or ventral neck. Occasionally, the primary lesion may be in the tongue, diaphragm, heart or psoas muscles.

In contrast to cattle, blackleg in sheep is more frequently associated with wounds such as those from shearing, docking and crutching and wounds associated with fighting. It does, however, occur without any identifiable antecedent. If seen, clinical signs include pyrexia, anorexia and depression.

The lesion in muscle in both cattle and sheep is a necrotizing myositis following the elaboration of a hemolytic necrotizing exotoxin by *Cl. chauvoei*. The necrosis produced allows further proliferation of the organism and greater toxin production. Death is due to the systemic effects of the exotoxin. The affected area in muscle is dark red-brown to black and often contains small gas bubbles. There is often much edema fluid present at the periphery of the lesion.

Malignant edema and **gas gangrene** are usually mixed clostridial infections and although included are often diseases of subcutaneous tissues which extend into muscle. They are always associated with wounds of some type and the clostridia act as opportunists. Clostridial species involved include *Cl. septicum*, *Cl. perifringens* type A, *Cl. novyi* type A, *Cl. chauvoei* and uncommonly *Cl. sordelli*. Malignant edema occurs especially in ruminants, horses and pigs. Clinically, there is often local swelling of the infected wound, but no evidence of crepitus. The latter is seen in cases of gas gangrene. The animals are pyrexic, weak and depressed and usually die quickly of a profound toxemia.

Other types of myositis

As stated previously, this group is uncommon but well recognized. In most cases the cause and pathogenesis remain elusive but are suspected to be of immune mediation.

Eosinophilic and atrophic myositis of dogs

These may be different expressions of the one disease. They are, however, routinely considered separately as there are at least clinical distinctions between them. **Eosinophilic**

myositis is localized to the muscles of the head of young, large-breed dogs such as the German Shepherd and the Doberman Pinscher. Clinically, there is bilateral swelling of the masseters, temporalis and pterygoid muscles. The jaw hangs partially open, the dog is reluctant to eat and resists manipulation of the jaw. The acute episode subsides, only to recur, finally leading to atrophy and fibrosis of the affected muscles. There is in many cases a peripheral eosinophilia and, at least in the acute stage, a liberal supply of eosinophils in the affected muscles.

Atrophic myositis occurs in many breeds of dog, with no particular age preference; there is chronic progressive atrophy of muscles of the head without acute episodes. It is non-painful, and may be unilateral or bilateral, symmetrical or asymmetrical, resulting finally in an inability to open the mouth. In contrast to the eosinophilic variety, there is a low frequency of peripheral eosinophilia.

Both eosinophilic and atrophic myositis appear to be immune mediated and are responsive to corticosteroid administration. Pathologically, in the acute episodes of eosinophilic myositis, the affected muscles contain large numbers of eosinophils with some accompanying lymphocytes, plasma cells and neutrophils. In the chronic phase, the microscopic pattern is dominated by plasma cells and lymphocytes. Affected muscle fibers undergo atrophy and degeneration. The lesion in atrophic myositis is characterized by an infiltration with lymphocytes and plasma cells, with accompanying fiber atrophy and degeneration. In both, other muscles are affected microscopically, especially those of the neck. Eosinophilic myositis of cattle and sheep is a relatively rare condition usually discovered *post mortem*.

Idiopathic polymyositis occurs in large-breed adult dogs, when it is, once again, suspected to be of an immune basis. Typically, affected dogs exhibit exercise intolerance, episodic weakness, lameness, sometimes a stiffened gait with muscle pain. This progresses in time to generalized muscle atrophy. Initial clinical signs may appear acutely or may develop slowly. Both progress slowly over weeks to months. Affected muscles are heavily infiltrated with lymphocytes and there is progressive fiber atrophy and degeneration.

Parasitic myositis is of worldwide public health and economic significance, but is not often clinically apparent. The details of these diseases may be pursued in the appropriate texts.

Tendon, tendon sheath and muscle trauma

The shearing forces and strain applied to the muscle tendon unit may be sufficient to rupture either the muscle or its tendinous attachment. Less severe strain may wrench some of the fibers apart but the unit remains essentially intact, and is sprained or strained.

The clinical features of a sprain or rupture are those of either a depressed or a complete loss of function, which is usually accompanied by swelling and pain. When moving, which the affected animal is often reluctant to do, the pain is compensated for by placing as little weight or stress as possible on the injured area.

The fundamental question is not usually how it happened but what is the capacity for repair of the injured area. In this regard there are differences between tendons, ensheathed tendons and muscle.

Tendons possess great tensile strength, which is reflected in their structure. They are composed of dense connective tissue, which is predominantly collagen and is arranged linearly with a grossly visible periodicity. Although not readily apparent, the blood supply to tendons originates from different locations for differing parts of the tendon. The bundles of collagen fibers in tendons are held together by loose connective tissue termed the endotenon and the entire tendon is surrounded by a connective tissue sheath, the epitenon. A third structure, the paratenon abuts the outer surface of the epitenon.

The process of healing of sprained or ruptured tendons depends on a number of factors, including whether or not the tendon is naked, or encased in a synovial sheath. There is an added complication when healing of ensheathed tendons is considered. There is not only the repair of the tendon itself, but also of the tendon sheath. It is desirable to assist repair of the tendon by surgical apposition and the use of a stress-relieving suture pattern. Minimization of the formation of adhesions between the tendon and tendon sheath allowing the tendon to glide over contiguous structures is also desirable.

The process of healing of a paratenon-covered tendon, i.e. a tendon without a synovial sheath, seems to be agreed on. It is the migration of cells from the paratenon into the tendon proper that is critical. It appears that, if the paratenon is removed, then the reparative phase is minimal.

Other factors favoring tendon healing include secondary remodeling of the wound, which entails the orientation of newly formed collagen fibers along the lines of stress and the gradual introduction to weight bearing. Both lead to the restoration of tensile strength.

There remains some difference of opinion on the process of healing of a sheathed tendon. In contrast to tendons not covered by a synovial sheath, tendons within a synovial sheath do not have a paratenon. As mentioned previously, the paratenon is critical to healing of unsheathed tendons. While there is little agreement, it seems that ensheathed tendons heal from fibroblasts from the tendon itself. There is, however, almost equally compelling evidence suggesting that the tendon sheath must be intact for healing of the tendon to occur. Notwithstanding this difference of opinion, one of the problems associated with healing of tendons within sheaths is the formation of adhesions between the tendon and tendon sheath restricting or preventing movement.

Repair of traumatized muscle

Muscle trauma may follow violent or unaccustomed exertion, external trauma or bone fractures. The extent of muscle repair can be variable, from almost complete regeneration with little scar tissue to large-scale scar formation following extensive destruction of muscle architecture.

In traumatized muscle there is rupture of many muscle fibers and their attendant sarcolemmal sheaths. This is accompanied by intramuscular and intermuscular hemorrhage and edema. Once the ruptured fibers are pulled apart the resolution of the resultant gap is by fibrosis. The important factor therefore with muscle rupture is to reduce the gap and remove the attendant hemorrhage. This is usually accomplished surgically. In less severe trauma, when the muscle remains intact, healing is by a combination of muscle fiber regeneration and by fibrosis.

Additional reading

Adams, R. D. (1975). *Diseases of Muscle – A Study in Pathology*, 3rd edn. Hagerstown, MD, Harper and Rowe Inc.

Blood, D. C., Henderson, J. A. and Radostits, O. M. (1983). *Veterinary Medicine*, 6th edn. London, Bailliere-Tindall.

Bradley, R. (1980). Myopathies in animals. In *Scientific Foundations of Veterinary Medicine*, A. T. Phillipson, L. W. Hall and W. R. Pritchard (eds.), pp. 66–79. London, William Heinemann Medical Books Limited.

Chrisman, C. L. and Averill, D. R. (1983). Diseases of peripheral nerves and muscles. In *Textbook of Veterinary Internal Medicine*, 2nd edn, S. J. Ettinger (ed.), pp. 608–53. Philadelphia, W. B. Saunders Co.

Cullen, M. J. and Mastaglia, F. L. (1982). Pathologic reactions of skeletal muscle. In *Skeletal Muscle Pathology*, F. L. Mastaglia and Sir J. Walton (eds.), pp. 88–139. Edinburgh, Churchill Livingstone.

Earley, T. D. (1981). Tendon disorders. In *Pathophysiology in Small Animal Surgery*, M. J. Bojrab (ed.), pp. 851–66. Philadelphia, Lea and Febiger.

Hodgson, D. R. (1985). Myopathies in the athletic horse. *Comp. Cont. Educ. Pract. Vet.* **7**: 551–6.

Hulland, T. J. (1985). Muscles and tendons. In *Pathology of Domestic Animals*, 3rd edn,

K. V. F. Jubb, P. C. Kennedy and I. V. Palmer (eds.), pp. 140–99. Orlando, FL, Academic Press Inc.

Landon, D. N. (1982). Skeletal muscle – normal morphology, development and innervation. In *Skeletal Muscle Pathology*, I. L. Mastaglia and Sir J. Walton (eds.), pp. 1–87. Edinburgh, Churchill Livingstone.

David W. Pethick

15 Metabolic disease

The meaning of the term metabolic disease has always been a cause for discussion. Clearly all disease is associated ultimately with changes in the rates of various biochemical pathways and so in all diseases the metabolism of the animal is abnormal to a major or minor degree. Therefore, to call any animal disorder a metabolic disease would seem a hopelessly limitless definition. However, the term metabolic disease may be used to connote a biochemical change of non-infectious origin in an animal which has been subjected to 'apparently' optimal husbandry. Such a biochemical change results in an increase or decrease in a metabolite critical to the function of the animal. This metabolic change is induced by an imbalance in the input or output of the metabolite or a related metabolite. Therefore it is not surprising that metabolic disease is common in lactating or pregnant farm animals, where the rate of metabolite uptake from blood to the mammary gland or fetus is high. This observation has led to metabolic diseases in farm animals being known as production diseases. The implication is that these diseases are a failure to meet an excessive output associated with the continued desire of man to increase the production of livestock.

To clarify the understanding of metabolic disease four criteria have been used to limit the definition.

1 Metabolic disease is not infectious in origin. Some care is warranted, since infection may be a contributing factor. For example, primary spontaneous ketosis in lactating cows is distinguished from secondary ketosis, which may develop subsequent to infection. In this case secondary ketosis develops due to the inappetence associated with infection. Primary spontaneous ketosis is reserved for healthy cows which suddenly develop the disease.
2 Metabolic disease is not a specific genetic deficiency. Diseases due to the absence of a specific enzyme in a pathway (an inborn error of metabolism) are therefore not included. However, the overall genotype of the animal may be a factor, since beef breeds of cattle have a low incidence of metabolic disease.
3 Metabolic disease is not due to a simple dietary deficiency or overload. This is perhaps the most uncertain of the criteria put forward. For example, there is no doubt that an insufficient intake of digestible energy is a contributing factor to pregnancy toxemia of sheep, but it is not the only factor, since experimental starvation of pregnant ewes rarely produces the disease.
4 Metabolic disease results in a change in concentration of one or more blood metabolites. These changes explain the resultant clinical signs and are of diagnostic importance.

Examples of metabolic disease in economic

animals will now be discussed. In certain circumstances, the function of a specific organ or organs is impaired and so these diseases might well be considered under the appropriate organ systems, but to highlight the principles of metabolic disease they are considered here.

Parturient paresis (milk fever)

Prevalence and clinical signs

Milk fever is restricted mainly to dairy cattle with a prevalence of approximately 5%. It occurs less frequently in beef cattle, sheep and goats. Milk fever occurs in older cows immediately prior to or 2–3 days after calving. Indeed 75% of all cases occur within 24 hours of calving and cows in their sixth lactation are 50 times more likely to be affected. Clinical signs are initially excitement, tetany, fine muscle tremors and loss of appetite. Later the cow becomes paretic and eventually recumbent. If untreated, she will die.

A similar disease occurs most frequently in small, hyperactive breeds of dog and occasionally in cats (it is called puerperal tetany or eclampsia). The signs are similar to the ruminant disease and the main difference is that animals are most susceptible at peak lactation (one to three weeks *post partum*).

Etiology and pathogenesis

The primary abnormality is undoubtedly hypocalcemia. Total serum calcium concentration usually falls by 50% and therapy with calcium salts readily reverses the clinical signs. Hypocalcemia is responsible for the clinical signs because of its profound effects on the nervous system. The initial phase of excitement is probably the result of low calcium concentrations changing the permeability of neurons to potassium and sodium, making them more susceptible to depolarization. As the disease progresses the continued low calcium levels interfere with neurotransmitter release – particularly of acetylcholine at the neuromuscular junction, leading to paralysis and recumbency.

The critical question is: why does the calcium concentration fall so low and remain low? Plasma calcium is normally under strict homeostatic control involving parathormone, calcitonin and calcitriol. Recent work has shown that, in the transition from pregnancy to lactation, there is a rapid increase in the requirement for calcium due to the synthesis of colostrum, which in dairy cattle has a high calcium concentration. Because of inherent delays in homeostatic mechanisms, most cows have subclinical hypocalcemia at this time. In older cows consuming a high calcium diet (50–70 grams/day), it takes about one day for intestinal absorption to increase in response to the lowered plasma calcium. More importantly, bone resorption does not respond to lowered blood calcium for at least a week. These delays seem to be associated with a lack of responsiveness of the target organs (intestinal mucosal cells and osteoclasts), since plasma concentrations of parathormone, calcitonin and calcitriol respond appropriately and rapidly to hypocalcemia in cows with milk fever. Normally, cows will recover from subclinical hypocalcemia as the gut absorption of calcium homeostatically increases after a day or so. However, any decline in food intake will result in severe hypocalcemia because the cow is entirely dependent on dietary calcium during this time. Unfortunately, older cows often show a depressed appetite at parturition, possibly due to an elevated estrogen concentration. As lactation progresses into the first week and beyond, the calcium demand for milk increases further, but this has no effect on plasma calcium, since both intestinal absorption and bone resorption increase substantially to regulate calcium carefully. This correlates well with the observation that milk fever is rare in cows during full lactation.

The etiology of the disease in the dog and cat has not been studied in detail. However, it is suspected that bone resorption does not respond adequately to lowered blood calcium.

A seemingly paradoxical finding is that milk fever in cattle can be prevented by feeding low calcium diets in the last few weeks of pregnancy. This gives an adequate time for bone resorption to reach an optimal rate before parturition, when inappetence can be a problem. Consequently, the cow is no longer dependent on dietary calcium alone. Indeed, high calcium diets actually increase the incidence of milk fever. Thus, milk fever in ruminants would fit the definition of a true metabolic disease, with the prime lesion being the relatively slow response of bone resorption to hypocalcemia. The precise nature of this slow response has yet to be elucidated. The essential features of this disease are depicted in Fig. 15.1.

Hypomagnesemic tetany (grass tetany)

Prevalence and clinical signs

This disease occurs most commonly in lactating cattle although it can be seen in non-lactating animals including sheep. An overall prevalence of 1% has been reported in the Australian beef herd. Cases are largely seen in older lactating cattle which are grazing lush, grass-dominant, spring or fall pastures. Animals at greatest risk are those exposed to inclement weather or other stresses such as yarding and mustering. These stresses induce inappetance. As the common name grass tetany suggests, the characteristic feature of hypomagnesemia is tetanic contraction of muscle. Early clinical signs include excessive alertness, hyperirritability and fine muscular twitching. As the disease progresses intermittent muscular spasms, recumbency and paddling movements of the limbs can be observed usually in response to external stimuli. If left untreated the animal will die of respiratory failure. Death can often be within 24 hours after the initial onset of clinical signs.

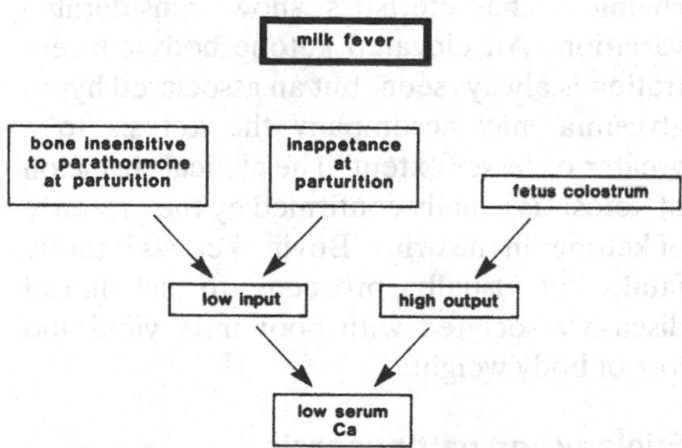

Fig. 15.1. Milk fever (parturient paresis) has as its central pathogenetic feature a high output of calcium to the fetus and into the colostrum. This, combined with a low input of calcium from bone (because of hormonal insensitivity) and inappetance at the time of parturition, leads to hypocalcemia.

Etiology and pathogenesis

Characteristically there is hypomagnesemia. Serum magnesium usually falls by 50% and there is an associated subclinical hypocalcemia. Sometimes serum magnesium is relatively normal. However, there is always a close correlation between a lowered magnesium concentration in the cerebrospinal fluid and clinical signs. Indeed, the disease is probably due to a dysfunction of the central nervous system, since ventricular infusion of a solution low in magnesium salts can induce the syndrome. Low extracellular magnesium (particularly a low Mg:Ca ratio) is known to increase the irritability of the central nervous system. In addition, the syndrome might also be due to low extracellular magnesium concentration at the neuromuscular junction, since low magnesium is known to potentiate the release of acetylcholine, resulting in the uncontrolled stimulation of muscle to contract.

The major question associated with the etiology is: why do rapidly growing, grass-dominant pastures induce hypomagnesemia? The most favored hypothesis is a decline in the amount of magnesium absorbed when animals graze such pastures. Implicit in this hypothesis is that there is little homeostatic control of blood magnesium and the normal concentration is determined mainly by the amount of magnesium absorbed from the gastrointestinal tract. Factors contributing to a fall in blood magnesium are:

1 Inappetance associated with inclement weather or faulty management practices.
2 High-risk pastures are often low in magnesium.
3 The proportion of magnesium absorbed from high-risk pastures is reduced. An important consideration here is that magnesium absorption occurs before the abomasum. Any dietary factors changing recitulorumen function may well have an effect on magnesium absorption. This may occur when animals eat high-risk pastures because of the following. (*a*) An **increased amount of magnesium is bound**, perhaps in microbial forms or by the chelation of ions by compounds such as transaconitate formed by the microflora. (*b*) **Protein is high** and rapidly fermented, leading to high rumen ammonia levels which tend to increase rumen pH. Under these situations, magnesium can form an unavailable magnesium ammonium phosphate complex. (*c*) The **potassium content is high**. Indeed, an accurate index of high-risk pastures is the ratio of K:(Mg + Ca). Elevated rumen potassium is known to reduce magnesium absorption, which occurs via an active sodium-linked mechanism and possibly high potassium levels antagonize the action of sodium. An elevated potassium intake could affect also the excretion of magnesium by the animal. A high potassium intake results in a homeostatic release of aldosterone, which increases both urinary potassium and magnesium excretion. Therefore, both a high urinary loss and low absorption of magnesium can be induced by high potassium concentration in the diet.

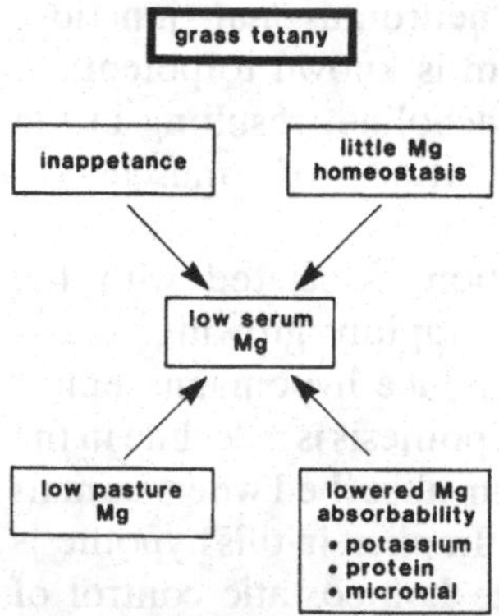

Fig. 15.2. The low serum magnesium (hypomagnesemia) seen in grass tetany is contributed to by a number of factors, including inappetance, low pasture magnesium, low absorbability of magnesium and poor homeostasis.

In conclusion, hypomagnesemia is more like a dietary deficiency than a metabolic disease. This is borne out in the treatment of the disease, which is usually by the administration of magnesium salts or by their addition to the pasture. In contrast to milk fever, a low magnesium diet aggravates rather than protects, and the syndrome is largely due to an inherently poor magnesium homeostasis (Fig. 15.2).

Bovine ketosis

Prevalence and clinical signs

This disease occurs mainly in high-yielding dairy cows during peak lactation. The prevalence is around 1–4% of the dairy herd. The clinical signs vary between two extremes: wasting and depression, and excitement and aggressive behavior. Similarly, the biochemical characteristics show considerable variation. An elevated ketone body concentration is always seen, but an associated hypoglycemia may accompany the ketosis to a greater or lesser extent. The clinical suspicion of ketosis is usually confirmed by the presence of ketones in the urine. Bovine ketosis is rarely fatal, but usually proceeds to subclinical disease associated with poor milk yield and loss of body weight.

Etiology and pathogenesis

To understand bovine ketosis, there is a need to appreciate the occurrence of several forms of the disease. The following classification is adapted from that presented by Kronfeld (1982; see Additional reading).

Alimentary ketosis

This occurs when the ration contains excessive amounts of ketone body-forming substrates. An example is the high butyric acid concentrations seen in some silages. The butyric acid is converted to 3-hydrody-D-butyric acid in the rumen epithelium and liver resulting in a rate of ketone body production exceeding the removal of ketones in other tissues such as skeletal muscle. This form of ketosis will not be further discussed.

Underfeeding ketosis

There is an insufficient intake of metabolizable energy (ME) due to insufficient or poorly metabolizable feed (primary underfeeding ketosis); or an insufficient intake of ME due to inappetence associated with another disease (secondary underfeeding ketosis).

This form of the disease is well understood and may be thought of as an accelerated response to undernutrition. For instance, when a lactating cow is fasted or underfed, hypoglycemia and ketonemia is greater than in a non-lactating cow.

Undernutrition reduces milk yield because less of the glucogenic precursors such as propionic acid and glucogenic amino acids are absorbed from the gastrointestinal tract. Therefore, glucose synthesis by the liver declines and consequently less glucose is delivered to the mammary gland for lactose synthesis (remembering that milk secretion is determined by the rate of lactose synthesis, since it is the primary osmotic component of milk – see Spontaneous ketosis, below). Nevertheless, the mammary gland still utilizes some glucose, and this initiates a hypoglycemia. To compensate, the lactating cow responds by accentuated changes in the ratio catabolic: anabolic hormones. Insulin levels decline, growth hormone is elevated and most probably the concentrations of adrenocorticotropic hormone (ACTH) and pancreatic glucagon rise. These changes result in the mobilization of body fat and protein in an attempt to maintain euglycemia and energy balance. Much of this mobilized fat, in the form of plasma non-esterified long-chain fatty acid (NEFA) is taken up by the liver. Due to the high value of the ratio catabolic: anabolic hormones and a low availability of glucogenic substrates, the liver is placed in a 'ketogenic setting' – that is, plasma NEFA taken up by the liver are directed to beta-oxidation rather than being esterified to glycerol to form triacylglycerol. The high rate of beta-oxidation results in acetyl CoA being only partly utilized by the tricarboxylic acid cycle, with the remainder converted to ketone bodies. Consequently, there is a high rate of ketogenesis. The peripheral tissues, particularly skeletal muscle, have a limited ability to utilize ketone bodies at high concentration and so the rate of ketone body production is greater than removal. **The end result is an elevated ketone body concentration**. This can lead to a mild keto-acidosis, which is detected as depression of the central nervous system. Ketone bodies or their decarboxylation products (acetone and isopropanol) may also have direct deleterious effects on the central nervous system at high concentration, but this is yet to be substantiated. In addition, the tendency for hypoglycemia may contribute to the nervous signs, although again no direct evidence supports this view.

The clinical signs of underfeeding ketosis are those of depression and wasting. Biochemically the cows show hypoglycemia and ketonemia.

Spontaneous ketosis

This type of ketosis develops when cows eat rations containing high levels of storage carbohydrate such as in rations high in cereal grains. These rations are considered nutritionally adequate on the basis of present knowledge and consequently this form of the disease is the most difficult to understand.

Spontaneous ketosis is exceedingly complex and its etiopathogenesis has always been controversial. The most convincing attempt to understand the disease has been made by Kronfeld (1982; see Additional reading). Below is this author's interpretation of Kronfeld's hypothesis. The primary cause of spontaneous bovine ketosis is thought to be a diet resulting in the absorption of an excess of glucogenic nutrients relative to lipogenic ones. The essential features of spontaneous ketosis are depicted in Fig. 15.3.

Glucogenic nutrients include propionic acid and glucogenic amino acids, while lipogenic nutrients include acetic and butyric acids, ketogenic amino acids and triacylglycerol. Implicit in the hypothesis is that milk yield is determined primarily by mammary weight and glucose supply. It is now clear that, in high

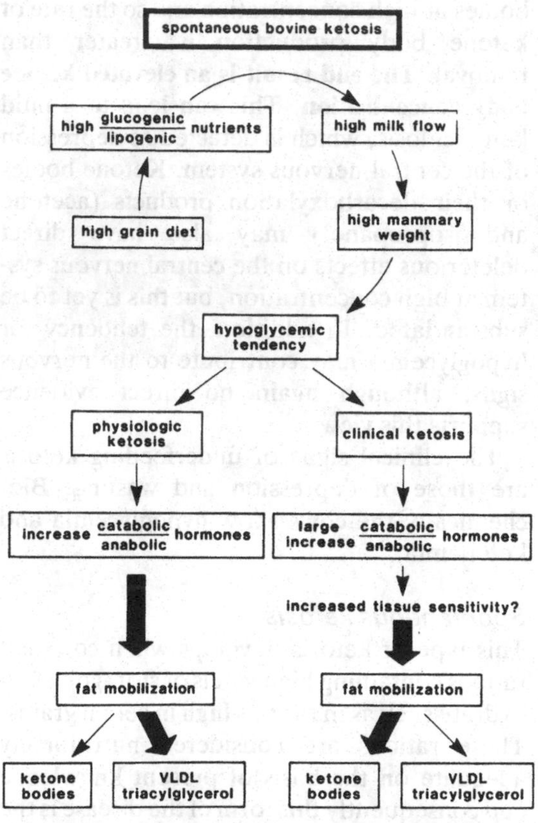

Fig. 15.3. Spontaneous bovine ketosis is seen in high-yielding cows which are often fed a high-grain diet. This promotes high milk yield and a tendency to hypoglycemia. Physiologic ketosis may arise with an increase in catabolic over anabolic hormones. Fat is mobilized leading to excess ketone body production. The more severe clinical ketosis appears to be an exaggerated form of physiologic ketosis. VLDL, very low density lipoproteins.

yielding cows, the daily milk volume is determined by the rate of lactose synthesis, which in turn is governed by glucose supply. Lactose is quantitatively the most important osmotic component of milk and so the more lactose produced the greater the volume of fluid flowing from the gland. Cows fed high cereal grain diets have high rates of propionic acid production (or amino acid production if high protein grains are used) in the rumen, which promote high rates of hepatic gluconeogenesis and so high rates of glucose supply to the mammary gland. This creates a demand for lipogenic nutrients to sustain milk-fat levels. A spectrum of outcomes is possible with three possible extremes, physiologic ketosis, low milk-fat syndrome and clinical ketosis.

Physiologic ketosis with ketonemia occurs in a mild form in most high-yielding cows. This group of cows shows a consistent response in early lactation, that is, a rise in the ratio catabolic:anabolic hormones, despite a diet yielding a high ratio glucogenic:lipogenic end-products of digestion. Consequently, body fat is mobilized. NEFA are removed by many organs of the body, including the mammary gland and liver. The liver converts the plasma NEFA to *both* triacylglycerol and ketone bodies. Importantly, the fatty acids taken up by the liver are not all converted to ketones, since much is esterified into triacylglycerol and redirected to the blood as lipoprotein-bound triacylglycerol. In lactating cows, lipoprotein triacylglycerols are an important source of milk fat. Therefore, a harmless or physiologic ketosis develops, where the concentration of ketone bodies in the blood is not sufficient to cause an acidosis. Indeed, a low to moderate rate of ketone body production is beneficial to an animal when body fat is mobilized. Ketone bodies are less toxic and more readily transported and metabolized than are the parent plasma NEFA.

Low milk-fat syndrome occurs in some cows which maintain low circulating levels of ketones and produce milk with a markedly reduced fat content. These animals respond to a feed promoting a high ratio glucogenic: lipogenic end products of digestion, by producing more insulin than 'normal' lactating dairy cows (i.e. a lower than normal ratio catabolic:anabolic hormones). Therefore less body fat is mobilized. In addition, dietary lipogenic precursors are limited, since feeds high in cereal grain do not contain a great deal of long-chain fat. Moreover these diets tend to encourage a greater ratio propionate:acetate absorption. Acetate is lipogenic but propionate is not. Consequently there are only limited substrates for milk-fat biosynthesis and a lowered milk fat develops. It is important to remember that glucose is not lipogenic in the mammary gland of the cow.

Clinical ketosis is the excessive or clearly pathologic state which develops in some cows. The reasons are unclear, but they are almost certainly multifactorial and there may be differences between animals. A possible explanation is that cows show either a marked increase in the ratio catabolic:anabolic hormones or a changed sensitivity to these hormones. Correspondingly, the rate of lipid mobilization is excessive and the liver is placed in a 'ketogenic setting'. Long-chain fatty acid taken up by the liver is diverted largely to ketone bodies and not to triacylglycerol (as in starvation ketosis). There is no direct evidence for this; indeed triacylglycerol formation must be active in hepatocytes because hepatic lipidosis is typical in ketosis. Further studies to assess the total balance of hepatic lipid metabolism is needed in conscious ketotic cows. The final scenario is for the **rate of ketogenesis to be greater than the rate of ketone removal.**

It is important to realize that spontaneous ketosis can rapidly progress to starvation ketosis. Initially, the clinical signs would be excitation and aggression, along with normal blood glucose and excessive ketonemia. The biochemical aberration responsible for these behavioral changes is not clear. Presumably it is not an acidosis associated with the elevated ketone body concentration, since accumulation of hydrogen ions usually causes central nervous system depression. Possibly it is a direct effect of ketones or their decarboxylation products on the central nervous system. Despite the mechanism, affected cows will have a depressed appetite and so a subclinical starvation ketosis would develop. Therefore, we have a complex interplay of different clinical signs, blood biochemistry and etiologies for apparently the same disease.

In summary, the precise mechanisms leading to spontaneous ketosis are still unknown. However, it seems that the nature of the hormonal response of cows in early lactation to diets which promote a high ratio glucogenic: lipogenic end products of digestion may be the key. This variant of the disease can be thought of as a true metabolic disease.

Treatment of bovine ketosis is traditionally aimed at starvation ketosis. Glucose is administered to correct hypoglycemia and insulin is used to inhibit fat mobilization. The response is usually encouraging but the rate of relapse is usually high. An alternative treatment is the injection of glucocorticoids. The most important effect is to inhibit milk yield and so prevent a hypoglycemia (starvation ketosis) or reduce the need for lipogenic precursors (spontaneous ketosis). Kronfeld has suggested that diets promoting both glucogenic and lipogenic end products of digestion should be used to prevent spontaneous ketosis. Such a diet would contain readily fermentable carbohydrate and long-chain lipid, the latter to be obtained from added long-chain fat or possibly from grains high in lipid. The hypothesis argues that the dietary fat is absorbed as chylomicron triacylglycerol, which:

(*a*) is utilized by the mammary gland for milk-fat synthesis;
(*b*) is poorly utilized by the liver and so does not promote rapid rates of ketogenesis;
(*c*) tends to prevent body fat mobilization and so decreases the concentration of ketone body precursors.

Ovine pregnancy toxemia

Prevalence and clinical signs

Pregnancy toxemia occurs most commonly in multigravid ewes during the last six weeks of pregnancy. The disease occurs sporadically and individual ewe flocks may have a high incidence. Affected animals are listless, unresponsive and lag behind the rest of the flock. Apparent blindness and ataxia develop such that ewes may be found wandering or propped against some object; at this stage anorexia is complete. Finally, ewes become comatose and death supervenes within 3–7 days. Two syndromes exist with a gradation of forms in between:

(*a*) Ewes with a history of undernutrition caused by such things as drought, footrot or insufficient trough room.
(*b*) Sudden stress such as mustering, transporting, shearing or snow fall.

In both cases, initially overfat ewes are more susceptible.

Etiology and pathogenesis

Hypoglycemia is considered to be the central feature. In the mid 1950s a correlation was shown between hypoglycemia and early clinical signs of pregnancy toxemia. Furthermore, the progression of clinical signs resembles those of insulin-induced hypoglycemia. However, it is not completely clear, since blood glucose can often be normal in ewes suffering the disease. In such cases it is argued that hypoglycemia-induced central nervous system damage has already occurred. Hyperketonemia is always a feature of pregnancy toxemia and in many cases a severe acidosis can eventuate. On necropsy, lipidosis of the liver and kidneys is typical.

Late pregnancy in sheep is associated with a marked increase in the demand for energy by the fetus. At term, the uterus and its contents remove 25% of the oxygen consumed by the mother. This energy is largely derived from maternal glucose in the fed state and accounts for 60–70% of the glucose synthesized by the ewe. This elevated glucose drain results in apparently well-fed ewes showing a biochemical pattern tending toward starvation. There is a high ratio of catabolic (glucagon, growth hormone and cortisol) to anabolic (insulin) hormones, which in turn results in fat mobilization and a mild ketosis. This is an example of physiologic ketosis. Indeed late pregnant twin-bearing ewes have been described as having a 'precarious' glucose balance. Despite this, it is not clear why hypoglycemia develops in ewes suffering pregnancy toxemia. However, the **inputs and outputs of glucose** in late pregnant ewes allow discussion of suggested mechanisms of hypoglycemia (Fig. 15.4).

Glucose input

In ruminants, glucose is derived mainly from hepatic gluconeogenesis. Substrates in fed animals are glucogenic end products of digestion, namely propionic acid (from the rumen) and glucogenic amino acids (from the small intestine). The extra demands of pregnancy require a greater feed intake to sustain a greater absorption of glucogenic substrates. Various factors can limit this adaptation. Environmental conditions (drought or snow fall) may limit food access. The feed may be of low quality (high roughage, low protein), resulting in diminished intake and a low yield of glucogenic end products of digestion per unit of metabolizable energy consumed. Even if adequate good-quality feed is available,

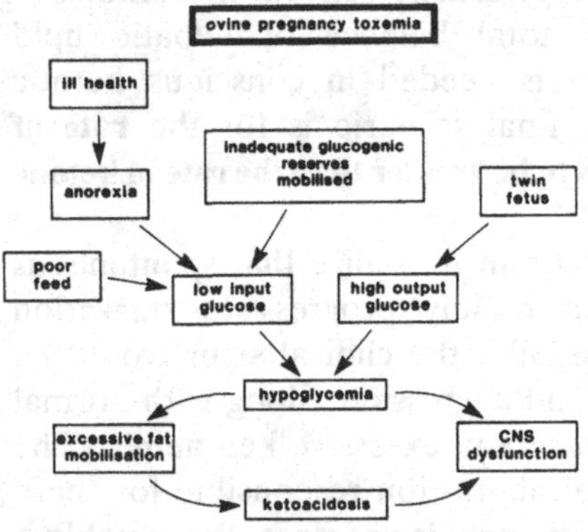

Fig. 15.4. The pathogenesis of ovine pregnancy toxemia revolves around a low glucose input combined with a high glucose output. Inadequate intake of glucogenic precursors and the drain of twin fetuses leads to hypoglycemia and ketoacidosis. CNS, central nervous system.

some twin-bearing ewes tend to show a decline in feed intake toward term. This may be related to abdominal limitations of space between the rumen, abdominal fat and the growing uterus. Any one of these factors will lead to a tendency for hypoglycemia, which will normally trigger the mobilization of body reserves to regain euglycemia. If, as in non-pregnant animals, the major tissue mobilized is fat, then only 5% of the energy mobilized is glucogenic (i.e. glycerol). An increase in the proportion of glucogenic precursor mobilized can only be achieved by mobilizing body protein. Normally this tends to be conserved while adequate depot fat is available, and this is more likely to be true the larger the fat reserves. The fat ewe may be unable to mobilize enough protein to supply the glucogenic needs of the pregnant uterus. The resultant rapid rate of fat mobilization can lead to problems including ketosis (see below) and fatty change in tissue. The latter has been demonstrated to lead to impaired liver and kidney function in affected ewes, by the use of dye-clearance tests. In conclusion, there are several mechanisms which may limit a pregnant, twin-bearing ewe from increasing her rate of glucose production to match the glucose requirement.

Glucose output

In undernourished pregnant, twin-bearing ewes, glucose is irreversibly lost from the circulation mainly via oxidation in the maternal central nervous system and the pregnant uterus. Other tissues, like skeletal muscle, oxidize only small amounts of glucose due to low insulin levels and a high rate of fat uptake (both plasma NEFA and ketone bodies). Both adaptations restrict glucose metabolism by skeletal muscle and almost certainly occur in all ewes. The glucose demand of the fetuses can adapt to maternal undernutrition: demand may be reduced by up to 40% and met partly by glucose from the maternal circulation (60%) and partly by fetal gluconeogenesis (40%). The degree to which fetal gluconeogenesis spares maternal glucose might be sufficiently variable that on some occasions maternal glucose may be the only source of glucose for the fetuses. In this situation, the ewe would be highly susceptible to hypoglycemia. The control of fetal gluconeogenesis, particularly in response to stress on the ewe, is not known, but it warrants further study, since it may help to explain the hypoglycemia of pregnancy toxemia in some cases. Finally, the maternal central nervous system has an obligatory glucose requirement, since the ovine brain, unlike the rat or human brain, cannot utilize ketone bodies as a fuel. This is undoubtedly an important reason why sheep suffer pregnancy toxemia.

The ketosis of pregnancy toxemia is often severe and reflects maximal ketogenesis along with underutilization of ketones by peripheral tissues. Hypoglycemia triggers fat mobilization. Added stress will further elevate the levels of plasma NEFA due to the lipolytic effect of cortisol and catecholamines. At peak NEFA levels, some 50% of the mobilized NEFA are converted to ketones in the liver and, since plasma NEFA are toxic at high concentration, this represents a detoxifying mechanism. The NEFA are converted to water-soluble substrates which can be utilized by many tissues such as muscle. The remaining NEFA are utilized directly by extrahepatic tissues. It is probable that at high NEFA concentration the enzymes which utilize ketone bodies become less active. So the ability of muscle to remove ketones decreases in ketotic sheep and ketone bodies accumulate to levels that can depress blood pH.

The success in treating ovine pregnancy toxemia relies on early detection of the syndrome. At this stage, highly palatable concentrates or carbohydrate therapy will often lead to recovery. Treatments include administration of glycerol, glucose, glucose plus insulin, propylene glycol or a concentrated oral rehydration solution. As the disease progresses, such treatments are less effective, presumably because the hypoglycemic encephalopathy becomes irreversible. An alternative aproach is to induce parturition. In one study, this resulted in a 96% recovery rate and highlights the central role of the fetus in the etiology of the disease. Equally, cesarian section may be considered in some circumstances.

In conclusion, the hypoglycemia of ovine pregnancy toxemia may arise either because the pregnant ewe can not produce enough glucose for the fetus or because the fetus is unable in part to sustain its own glucose requirement.

Additional reading

Allcroft, R. and Burns, K. N. (1968). Hypomagnesemia in cattle. *N. Z. Vet. J.* **16**: 109–28.

Baird, G. D. (1982). Primary ketosis in the high-producing dairy cow: clinical and subclinical disorders, treatment, prevention and outlook. *J. Dairy Sci.* **65**: 1–10.

Blood, D. C., Henderson, J. A. and Radostits, O. M. (1983). Metabolic diseases. In *Veterinary Medicine*, 6th edn, pp. 970–1014. London, Bailliere Tindall.

Buswell, J. C., Haddy, J. P. and Bywater, R. J. (1986). Treatment of pregnancy toxaemia in sheep using a concentrated oral rehydration solution. *Vet. Rec.* **118**: 208–9.

Ettinger, S. J. (1983). Puerperal tetany in the bitch. In *Textbook of Veterinary Internal Medicine, Diseases of the Dog and Cat*, 2nd edn, pp. 1574–6. Philadelphia, W. B. Saunders Co.

Hay, W. W., Sparks, J. W., Wilkening, R. B., Battaglia, F. C. and Meschia, G. (1984). Fetal glucose uptake and utilisation as functions of maternal glucose concentration. *Am. J. Physiol.* **246**: E237–E242.

Hunt, E. R. (1976). Treatment of pregnancy toxaemia in ewes by induction of parturition. *Aust. Vet. J.* **52**: 338–9.

Kronfeld, D. S. (1972). Ketosis in pregnant sheep and lactating cows. *Aust. Vet. J.* **48**: 680–7.

Kronfeld, D. S. (1982). Major metabolic determinants of milk volume, mammary efficiency, and spontaneous ketosis in dairycows. *J. Dairy Sci.* **65**: 2204–12.

Lindsay, D. B. and Pethick, D. W. (1983). Adaptation of metabolism to various conditions: metabolic disorders. In *Dynamic Biochemistry of Animal Production*, P. M. Riis (ed.), pp. 431–80. Amsterdam, Elsevier.

Martens, H. and Rayssiguier, Y. (1980). Magnesium metabolism and hypomagnesaemia. In *Digestive Physiology and Metabolism in Ruminants*, Y. Ruckebusch and P. Thivend (eds.), pp. 447–66. Lancaster, England, MTP Press Ltd.

McClymont, G. L. and Setchell, B. P. (1955). Ovine pregnancy toxemia. I. Tentative identification as a hypoglycemic encephalopathy. *Aust. Vet. J.* **31**: 170–4.

Ramberg, C. F. Jr, Johnson, E. K., Fargo, R. D. and Kronfeld, D. S. (1984). Calcium homeostasis in cows, with special reference to parturient hypocalcemia. *Am. J. Physiol.* **246**: 698–704.

Peter E. Williamson

16 The reproductive system

When an animal suffers a reproductive problem the presenting signs may be dramatic, as in abortion or acute pyometra, but more often the animal is presented for examination simply because attempts at breeding have been unsuccessful. Furthermore, the underlying cause of infertility may no longer be present, even though the effects are still apparent. So one of the most important facets of investigating a reproductive problem is to obtain an accurate breeding history for the animal. Understanding the causes of infertility also requires an appreciation that the reproductive system is a dynamic system dependent on the finely tuned interaction of pituitary and gonadal hormones. To this end the reproductive anatomy, physiology and endocrinology of the domestic species will be reviewed briefly, before any discussion of breeding disorders.

Abnormalities of reproductive function generally follow disturbances of the normal hormonal and physiologic patterns. Such disturbances can arise from endocrine dysfunction, from invasion of the reproductive tract by pathogens or tumors, or from abnormalities of development. A number of reproductive disorders will be discussed, but the list of diseases will not be exhaustive. The object is to illustrate in principle the major dysfunctions of the system.

From a clinical standpoint the female reproductive system is intrinsically more complex and interesting, in that diseases can often be treated, or aberrant cyclic patterns modified, by the administration of exogenous hormones. In the case of the infertile male, the role of the veterinarian is usually confined to diagnosis and management advice, as treatment is often ineffective.

Development of the reproductive tract

The genital organs of both sexes arise from the same embryonic tissue, with divergence occurring after the establishment of the basic structures. In both sexes, the gonads first appear as a ridge (the genital ridge) which protrudes from the dorsal wall into the coelom (body cavity) of the embryo, adjacent to the developing excretory organ (mesonephros). If the gonad is to develop into a testis, the epithelium lining the coelom grows inward to produce solid cords which envelope the germ cells and their supporting cells (Sertoli cells). These cords remain solid until after birth and in some species even up until puberty. Connective-tissue compartments are formed between the developing seminiferous tubules, within which the hormone-producing interstital cells of Leydig arise.

In ovarian development, differentiation may proceed in a similar manner, with the germ cells enclosed in cords connected to the mesonephros. Alternatively, germ cells may proliferate to form initially a localized mass without cord formation, depending upon the species. The central cells then degenerate, leaving a sterile medulla and a cortex rich in germ cells.

In the male, the mesonephric (Wolffian) duct abutts the testes for delivery of sperm to the exterior. The mesonephric tubules connect with the rete testis to form the efferent duct system, located near the dorsal pole of the testis. The epididymis, which joins this efferent duct system, is formed by the coiling of the proximal part of the mesonephric duct, while the vas deferens and ampulla are formed by its distal part. The other accessory sex glands in the male (prostate and bulbourethral glands) arise from the urethra. In the female embryo the mesonephric duct disappears and the uterus and oviduct develop from the paramesonephric (Müllerian) duct, which is formed from connective tissue adjacent to the degenerate mesonephric duct.

Early in embryonic life the urogenital sinus develops from the division of the cloaca. Following closure of the urachus (which connects the bladder and the allantois), there is expansion of the urogenital sinus and, in the male, the ureter and vas deferens open separately through the wall of the sinus. At the same time the testes descend towards the inguinal canal from their abdominal location near the kidneys. In the female, with growth of the urogenital sinus, the paramesonephric ducts develop and enter (fused or paired to varying degrees, depending on species) into the caudal part of the sinus, which becomes the vagina. The preceding developmental process is depicted in Fig. 16.1.

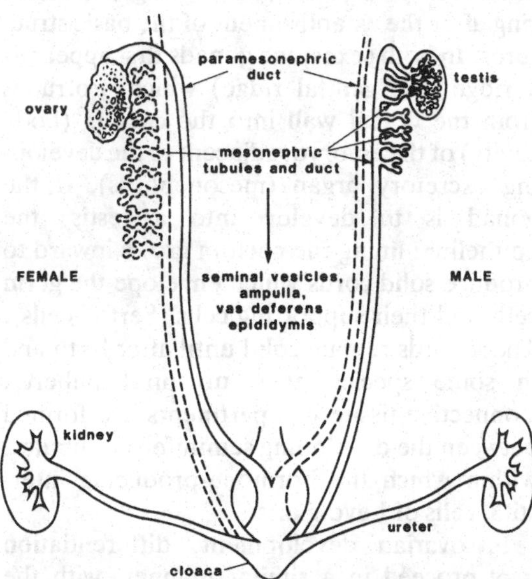

Fig. 16.1. The development of the male and female reproductive tracts from common embryonic structures. The uterus and fallopian tubes arise from the paramesonephric duct, whereas the epididymis and vas deferens arise from the mesonephric duct.

The development of the external genitalia is essentially similar in both sexes. The genital tubicle becomes the penis in the male and the clitoris in the female. In the male, the genital folds form the prepuce, while the more laterally situated genital swellings continue to develop as the scrotum. In the female the folds and swellings give rise to the vulva. Separation of the penis from the prepuce may not be complete until puberty.

Female reproduction: anatomy and physiology

The female reproductive tract consists of paired ovaries for the production of gametes, and a tubular duct system to facilitate fertilization of ova and to support pregnancy (Fig. 16.2).

The ovaries and duct system

The paired ovaries are generally ovoid in shape and held within the body cavity near the kidneys. In most species, the central medulla is surrounded by a cortex which contains the germinal epithelium with oogonia. In the mare, this pattern is reversed, with a central 'cortical' region surrounded by medullary tissue. As a consequence, mares ovulate into the hilus of the ovary, through the ovulation fossa, whereas, in other species, follicles rupture to the outer surface of the ovary, and the expelled ovum is trapped within the ovarian bursa and directed into the oviduct by the waving fimbria.

Gamete production in the ovary proceeds by mitosis within the germinal epithelium prior to birth. No further oogonia are produced after birth and if postnatal destruction or loss of the gametes occurs the animal will be infertile. At birth, mitosis ceases and meiotic division occurs up to the dictyotine stage, at which point the ova rest until puberty. These early meiotic divisions occur in primordial follicles, in which the developing oocytes are surrounded by a layer of cells and a basement membrane. Follicular growth does not resume until puberty, and then only a few of the primordial follicles develop at each successive estrous cycle, to form antral and then Graafian follicles, which eventually rupture and shed their ova (Fig. 16.3).

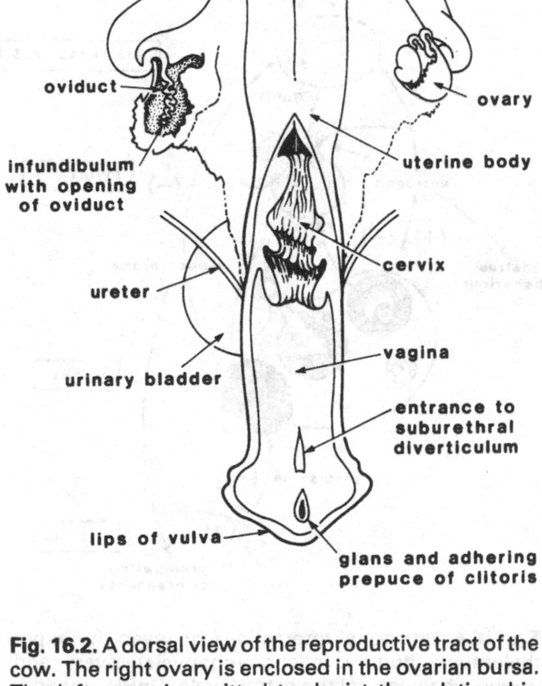

Fig. 16.2. A dorsal view of the reproductive tract of the cow. The right ovary is enclosed in the ovarian bursa. The left ovary is omitted to depict the relationship between the ovarian bursa, the infundibulum and oviduct (fallopian tube).

The female duct system consists of five regions:

1 The external genitalia.
2 The vagina.
3 The cervix.
4 The uterus, usually with a body and two horns.
5 The oviducts, which lead from the ovarian horns to the ovaries.

Fertilization occurs in the oviducts and the fertilized ovum passes into the uterus, where implantation and development of the embryo occur.

The estrous cycle

Reproductive patterns vary between species. Some animals, such as cows and sows, display estrous cycles throughout the year, while mares and ewes exhibit definite breeding seasons controlled primarily by changing day length. Most domestic species show recurring periods of sexual receptivity (estrous cycles) throughout their breeding season unless they become pregnant, whereas others such as the bitch have only one breeding period with each season. Estrous cycles are characterized by periods of ovarian follicular development coincident with sexual activity (estrus), which alternate with periods when the corpus luteum

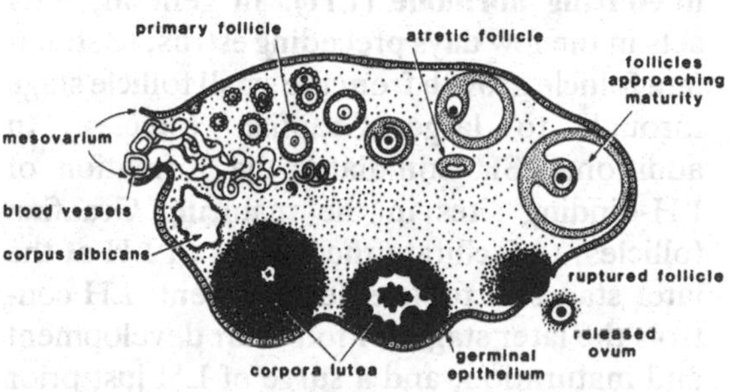

Fig. 16.3. The follicular cycle. During the estrous cycle the follicles enlarge, mature and ovulate. The corpus luteum develops following follicular rupture. The end-stage follicle is often termed a Graafian follicle.

is active (diestrus). In the few days preceding estrus, follicle(s) grow and mature in the ovary, a process which culminates in ovulation. The luteinization of granulosa cells in the ruptured follicles results in the formation of the corpus luteum, the dominant structure on the ovary between successive estrous periods.

Hormonal control of the estrous cycle

The ovarian cycle and uterine development during pregnancy are controlled by a complex interaction of hormones; the gonadotropins released from the adenohypophysis (anterior pituitary gland) and the steroid hormones released from the ovaries. General information on the structure and function of hormones is contained in Chapter 10. The principles described there can be applied to the reproductive hormones.

The **gonadotropins** are glycoprotein hormones involved in the regulation of ovarian function. Secretion of gonadotropins is controlled by gonadotropin-releasing hormone (GnRH), itself released by the hypothalamus. GnRH is released in pulses, and the amplitude and frequency of pulses determines the level of gonadotropin release. Furthermore, the sensitivity of the adenohypophysis to GnRH varies with the season and stage of the cycle, providing an additional regulatory mechanism.

The hormones of principal concern are **follicle-stimulating hormone** (FSH) and **luteinizing hormone** (LH). In general, FSH acts in the few days preceding estrus, to stimulate follicle growth from the small follicle stage through to large Graafian follicles. In addition, FSH stimulates the formation of LH-binding sites in the maturing Graafian follicles, to facilitate the actions of LH at the later stages of follicle development. LH controls the later stages of follicular development and maturation, and a surge of LH just prior to, or during, estrus causes the rupture of the mature follicle(s). In addition, LH causes luteinization of the granulosa cells of the ruptured follicle to produce the corpus luteum, which secretes **progesterone** between successive estrous periods or through the early stages of pregnancy. Progesterone, acting in concert with low mid-cycle estrogen levels, blocks release of the gonadotropins from the adenohypophysis by a negative feedback mechanism which persists until the corpus luteum is destroyed (Fig. 16.4).

As follicles develop in the ovary under the influence of the gonadotropins, the ovary secretes increasing amounts of the steroid hormone **estrogen**. This is a generic term for a group of compounds, the principal secreted forms of which are 17β-estradiol and estrone, produced by an interaction of the theca interna and granulosa cells of the developing follicle. Estrogens influence many diverse body functions, some of the more important

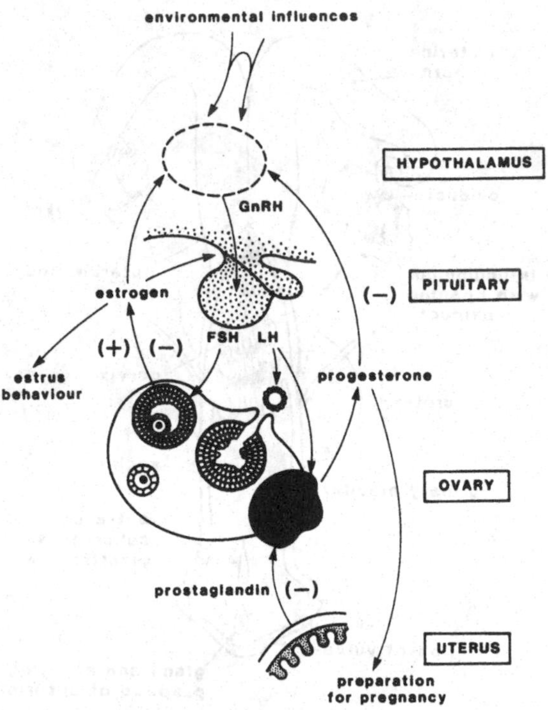

Fig. 16.4. Hormonal interactions between the hypothalamus, pituitary (adenohypophysis), ovary and uterus whch control the estrus cycle. GnRH, gonadotropin-releasing hormone; FSH, follicle-stimulating hormone; LH, luteinizing hormone.

being the control and development of female secondary sexual characteristics, the stimulation of estral behavior, the physiologic changes in the reproductive tract associated with estrus, and the regulation of the rate of passage of ova through the oviducts. The low level of estrogens present in the mid-cycle period, in concert with progesterone produced by the mid-cycle corpus luteum, inhibits the release of LH. However, as the Graafian follicles undergo rapid growth in the few days before ovulation, they produce large amounts of estrogens and at this point the effect of estrogens on LH release is reversed and they precipitate the surge of LH which causes ovulation.

After ovulation, the corpus luteum replaces the ruptured follicle and produces large amounts of progesterone, the most important hormone of pregnancy. It prepares the uterus to receive the embryo by stimulating both the development of complex endometrial glands and thickening of the uterine wall. Implantation cannot occur in the absence of progesterone, and it also inhibits gonadotropin secretion from the adenohypophysis during the mid-cycle period or in pregnancy. In the dog, cat, swine and goat, the corpus luteum is the main source of progesterone throughout pregnancy, but in other species the placenta becomes the principal source within a few months. The period of the cycle dominated by an active corpus luteum is called the **luteal phase**. During both the luteal phase and in pregnancy the dominance of progesterone as the circulating steroid hormone has importance in the etiology of a number of diseases of the reproductive tract. The high concentration of progesterone in the reproductive tract has an inihibiting effect on the local immune system, making the tract more susceptible to infection.

If fertilization and subsequent implantation in the uterus do not occur following ovulation, destruction of the corpus luteum ensues, thus ending progesterone secretion. Recent research indicates that this destruction is precipitated by the production of oxytocin from the corpus luteum itself, which in turn stimulates the endometrium in most domestic species to synthesize and release a luteolytic hormone, identified as **prostaglandin $F_2\alpha$**. The prostaglandin is then carried in the bloodstream to the corpus luteum, where it causes its destruction. The demise of the corpus luteum leads to a dramatic decrease in progesterone production and a release of the negative feedback on gonadotropin secretion from the adenohypophysis. Gonadotropins are again released and initiate a new wave of follicle development. If the animal becomes pregnant, the synthesis and secretion of prostaglandin $F_2\alpha$ is prevented by the implanting embryo and the corpus luteum is maintained into pregnancy. In this event estrous cycles cease.

The growth and disappearance of structures in the ovary over the period of an estrous cycle are illustrated in Fig. 16.5, and the interactions of the hormones from the adenohypophysis and the ovary which control the reproductive cycle, as previously discussed, are illustrated in Fig. 16.4. A disturbance in this delicate hormonal balance will generally result in disruption of the normal reproductive patterns, a lack of estrus, or the failure of ovulation and persistence of large follicles on the ovary. Prostaglandins may be released from the

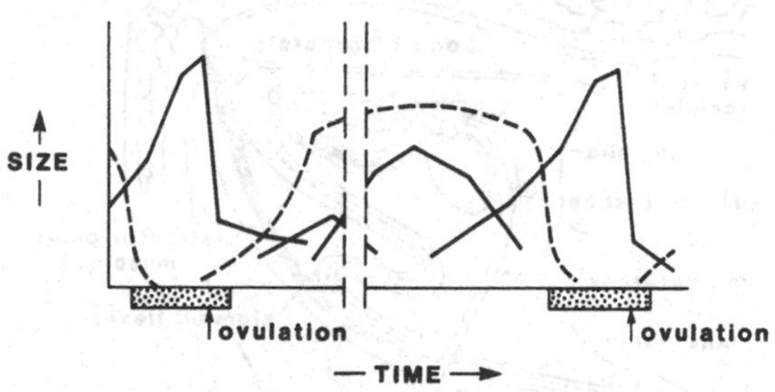

Fig. 16.5. The growth and demise of follicles (solid lines) and the corpus luteum (dashed lines) over one period of an estrous cycle in the mare. The shaded bars represent the period of estrus. Note that follicular development continues through diestrus, but these follicles regress because gonadotropin levels are not suitable.

uterus as a result of acute inflammation or irritation of the endometrium, and can cause destruction of the corpus luteum in mid-cycle or in early pregnancy. These abnormal conditions will be dealt with as the diseases of the reproductive tract are discussed.

Male reproduction: anatomy and physiology

The reproductive tract of the male consists of a penis, a scrotum enclosing the paired testes, the efferent tubules, and a duct system which leads from the testes to the penis. The duct system includes the epididymides, the vas deferens and several accessory glands, the ampulla, prostate, seminal vesicles and bulbourethral glands (Fig. 16.6). The Leydig (interstitial) cells found in the connective tissue of the testes between the seminiferous tubules secrete **testosterone** under the influence of LH. The Sertoli cells within the seminiferous tubules support the production of germ cells. Spermatozoa formed in the seminiferous tubules pass into the epididymis, undergoing maturation during passage from the head to the tail. At ejaculation the secretions of the seminal vesicles, prostate and bulbourethral glands are added to the spermatozoa to form the ejaculate.

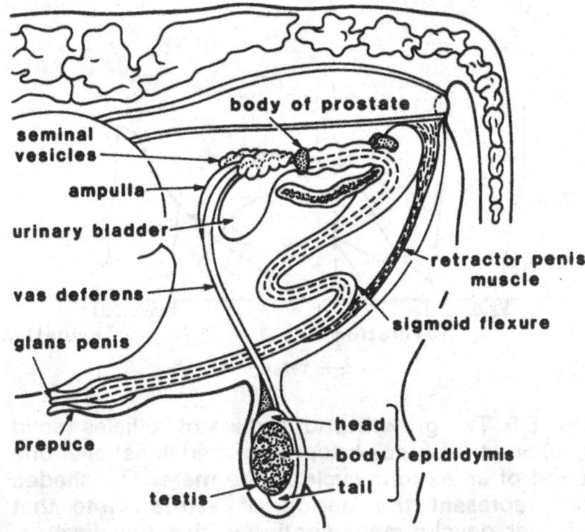

Fig. 16.6. The reproductive system of the bull.

The germinal epithelium is sensitive to many adverse influences, with the major cause of disruption being increased testicular temperatures. In terms of disease processes, their location outside of the body cavity is of paramount importance. Most domestic animals have testes located externally, in scrotal sacs, to enable them to be maintained at a temperature below that of the rest of the body. This is achieved by a number of mechanisms. The arterial blood is cooled prior to entry into the testicles by a counter-current heat exchange arrangement, including the pampiniform plexus which surrounds the spermatic cord. The cooler venous blood in the plexus allows heat exchange and cooling of the blood entering the testes. As a further regulatory mechanism the cremaster muscles, which are attached to the testes, can either be contracted as the environmental temperature decreases to allow testicular warming, or relaxed in high temperatures to allow more cooling. The scrotal skin also has a profusion of sweat glands which allow evaporative cooling in warmer weather. The testicular temperature is maintained 4–7 deg. C below rectal temperature in mild weather, but in very hot weather this temperature differential can be reduced to 2 or 3 deg. C.

Developmental anomalies

Freemartinism

Anomalies of development of the reproductive system occur quite frequently in the different species, and amongst these probably the freemartin is the most common. It is a sterile female calf born twin of a male and affects 95% of females born as twins with males. It is characterized by the suppression of the development of the female genitalia and the development of vestiges of the male reproductive system. When twin embryos are present in the bovine uterus, anastomoses of blood vessels occur at around 28 days, and the twins share a common blood circulatory system. Sexual differentiation does not occur until 40 or 50 days of embryonic development and factors are exchanged in the common blood supply, causing disruption of normal development of the reproductive tract in the female. The female twin has a hypoplastic non-patent vagina, a hypoplastic vulva and an enlarged clitoris. The tubular genitalia may vary from vestigial remnants of the Müllerian ducts through to well-developed uterine horns. Vestigial seminal vesicles are always present and communication between the uterus and vagina always absent. Vestigial ovaries are present, often containing sterile spermatic cords surrounded by Leydig cells.

The basis of the condition is fusion of the chorioallantoic circulations of the twins, allowing the interchange of cells and hormones. The sexual organogenesis in the female is modified as a result of the transfer of some agent, or agents, from the male co-sibling. The sharing of hormones and cells in the crossed circulation result in the production of genetic mosaics (chimeras). Cells from both the male and female fetus can be identified in the tissues of the calves born co-sibling. While this exchange of cellular tissue and genetic material is a two-way event, no definite abnormalities of the male sexual organs have been identified and neither has there been a clear demonstration that fertility of the male is reduced.

Intersexes

Hermaphrodites are congenitally malformed animals which are bisexual, with gonads of both sexes present. Pseudohermaphrodites have the external genitalia of one sex but the gonads of the other. Recent studies on sex ratios, sex chromatin and cytogenetics provide evidence that the majority of hermaphrodites are genetic females. This type of abnormality is most common in goats, especially polled goats, pigs, dogs and horses.

A number of additional abnormalities occur in development in both sexes. These abnormalities are usually manifest by hypoplasia or segmental aplasia of the reproductive tract.

Ovarian dysgenesis

Ovarian hypoplasia is one of the most frequently observed congenital malformations in cattle and in some breeds it is genetically determined. In this condition the primary abnormality is a reduced number of germ cells in the ovary and thus there is often a reduction in, or absence of, development of follicles, with the resultant absence of estrogens and other steroidal hormones from the ovaries. In extreme cases the ovaries may not develop, with only a fibrous thickening found on palpation. There are varying degrees of cyclic abnormalities, ranging from anestrus to shallow, erratic estrous periods. Accompanying this will be varying degrees of underdevelopment of the secondary sexual characteristics of the animal.

Ovarian dysgenesis has been recognized in mares, the primary problem being an absence of germ cells in the ovary, which is generally due to an abnormal chromosome complement. The mares are phenotypically female but are infertile and generally have weak estrous patterns or are anestrous. They are generally not responsive to the stallion even if displaying the shallow estrous signs. Clinically the external genitalia appear normal, though small, and the uterus and cervix are small and flaccid, with the cervix generally dilated. The ovaries are small, smooth, firm and have no palpable follicles.

Segmental aplasia (female)

A number of conditions are recognized in which segmental aplasia (under- or non-development) of the oviducts and uterus occur, but in which the ovaries are generally normal. The degree of aplasia of the uterus varies markedly, with the severe condition being common in a number of cattle breeds. The effect of segmental aplasia will depend on its location. If the caudal portion of a uterine horn is aplastic, its cranial portion may be normal and functional, with the endometrial glands producing normal secretions. Because of the aplasia in the base of the horn, the remaining horn is a blind-ending sac, which generally becomes dilated with the uterine secretions. If the aplastic segment is at the tip of the horn no indication of blockage of the uterine horn will be evident on palpation, but the uterus is asymmetrical, with a cord-like thickening in the region of the uterotubal junction. Depending on where the aplastic segment occurs, and whether it is unilateral or bilateral, the effects on fertility of the animal can be quite severe. As these abnormalities are common to certain breeds, implying a genetic cause, a great deal of care should be taken, where this condition is known to exist, in selecting cattle for breeding programs.

Testicular hypoplasia

Testicular hypoplasia is perhaps the most common developmental anomaly of the male, affecting all species, being especially common in the horse and in some breeds of cattle, where it is considered to have an hereditary basis. As testicular hypoplasia is frequently unilateral and may not be complete, spermatogenesis and semen quality will vary, depending on the severity of the condition. Hypoplastic testes are apparently more sensitive to degeneration than testes in normal animals, and the inherent reduction in fertility may be compounded by superimposed degenerative changes. Epididymal hypoplasia generally accompanies hypoplasia of the testes.

Cryptorchidism

Cryptorchidism, the incomplete descent of the testes from the body cavity into the scrotum, occurs in all domestic species. The condition is caused by a dominant gene in the stallion, and an autosomal sex-linked recessive gene in other species. Because of the requirement for the testes to be held at a temperature several degrees below normal body temperature, severe testicular degeneration occurs in cryptorchid animals whose testes have not descended at puberty, when spermatogenesis is in full flight. Where one testis has descended, the animal is described as a unilateral cryptorchid or monorchid. The descended testis usually produces some normal sperm, but there are low sperm numbers in the ejaculate. In the dog, there is some evidence that the cryptorchid testis is predisposed to neoplasia and, in the stallion, cryptorchidism leads to the development of aggressive and sometimes vicious behavioral characteristics as the animal matures. This is probably due to the excessive production of androgens from the retained testis.

Segmental aplasia (male)

Other male developmental anomalies occur in the epididymis and efferent tubules, where segmental aplasia causes blockage leading to the impaction of sperm and cystic dilation of the ducts. This condition is an inherited trait in rams and goats, occurring mainly in the head of the epididymis. The basement membrane of the epididymal tubule eventually degenerates due to pressure atrophy of the tubule wall, and sperm leak into the interstitial tissues of the testicle or epididymis. The sperm, being 'foreign protein' at this site, provoke a chronic granulomatous inflammation, which can be palpated as a firm swelling.

Hormonal activity and abnormal estrous cycles

Females of each domestic species have defined cycle lengths which are characteristic for the species; some cycle regularly throughout the year, while others cycle only at particular times of the year. Deviations from normal occur under various circumstances, some being the result of a disease-state and others secondary to external influences. Apart from the congenital ovarian disorders, which have been discussed previously, all are thought to be the consequence of disturbed hormonal balance. The primary source may be the ovary itself, such as with estrogen-producing tumors, or the ovary may be the victim of inappropriate influences arising from higher centers, such as the pituitary and hypothalamus. These in turn may be influenced by other regions of the brain, or they may be affected by the presence or absence of hormonal feedback from the uterus.

The consequences of such hormonal disturbances are variations in the regular pattern of the estrous cycle. Cycle abnormalities cause the females to remain either in a sexually quiescent state termed anestrus or in a state of prolonged sexual receptivity termed nymphomania.

The least-defined group of cycle disturbances is associated with metabolic stressors of various kinds, such as lactational stress in immature first-calf heifers or a poor nutritional state in any species. The result is anestrus, which in most cases is probably mediated by suppression of the release of FSH and LH by higher centers.

Cystic ovarian disease

This subgroup encompasses a multiplicity of conditions expressed clinically either as anestrus or nymphomania. Cystic ovarian disease has been reported to occur in many domestic species but is most common in high-yielding dairy cows, and sows. It is encountered to a lesser extent in mares and does.

In this syndrome, the normal sequence of events is disturbed at some point during the development and rupture of the follicle and the subsequent formation of the corpus luteum. For instance, the follicle may develop, fail to rupture, and then may or may not undergo partial luteinization. A follicle that fails to rupture, but remains unluteinized is termed a **follicular cyst**, while a follicle that undergoes partial luteinization is termed a **luteal cyst**. LH is at the center of the pathogenesis of follicular or luteal cysts, and two possible mechanisms have been proposed. The first is primary deficiency of LH release at the time of ovulation, and the second is failure of adequate development of LH binding sites in the maturing follicles, making them refractory to the LH surge which precipitates ovulation. This may imply an inadequate response to FSH by granulosa or theca cells. A follicular cyst represents a profound deficiency of LH or LH receptors, whereas a luteal cyst reflects a less severe deficiency. Follicular cysts produce little or no progesterone and as a result additional follicles grow under the influence of continuing gonadotropin release.

Cystic ovarian disease in the cow is associated with nymphomania or anestrus and, in many cases, nymphomania followed by anestrus. It is most common in mature high-yielding dairy animals in the early *post-partum* period.

The clinical signs observed are the manifestation of excessive levels of estrogen and are related initially to the presence of several large, thin-walled cysts in the ovary which remain unluteinized or partially luteinized. At this stage, affected cows are often detected because they exhibit estrous behavior for a prolonged period, mounting other cows and standing to be mounted. The vulva is swollen, the uterus edematous and the cervix enlarged, edematous and patent. There is also a copious discharge of vaginal mucus. In severe cases the sacroiliac ligaments relax and milk production drops.

As the condition progresses, the ova degenerate in the follicles along with the granulosa cells, and the cysts become atrophic and retreat into the ovary. The cow then moves into a clinical anestral phase. The uterus becomes thin-walled and the cervix small.

At this stage, the uterine and ovarian changes are permanent and the affected cows remain anestrous. The progression from follicular to atrophic cysts produces conditions in the uterus that predispose to infection. In the initial estrogenic phase, the proplastic stimulation of endometrial glands tends to lead to their obstruction and cystic dilation, and in addition excessive mucus accumulates in the uterine lumen.

In the case of luteal cysts, there is sufficient progesterone secreted by the partially luteinized follicles to block the development of further follicles. Cows affected with luteal cysts are anestrous, but most will recover spontaneously after a prolonged period. On rectal examination there are one to two thick-walled ovarian cysts.

The most subtle disorder in this group does not strictly speaking include the formation of an ovarian cyst, but is clearly part of the disease spectrum. It occurs when there is failure to ovulate, but there is complete luteinization and formation of a corpus luteum. In spite of failure to ovulate, the cow's cycle will be normal. This is an occasional event which will not affect the same cow repeatedly. The variation in LH levels or effectiveness is necessarily minimal. Fig. 16.7 depicts the events associated with bovine cystic ovarian disease.

Cystic ovarian disease in sows was recognized initially by a seasonal syndrome known as 'summer infertility'. However, it is now evident that there is a problem throughout the year with a peak in the summer months. It is seen mainly as a problem of intensive husbandry, probably highlighted in recent times because of the accurate recording systems employed in large piggeries. The infertility is considered to arise following temperature stress, overcrowding and less than optimal management. The stress-induced release of adrenocortical steroids results in depression of LH secretion (Fig. 16.8). The syndrome is characterized by failure of ovulation and/or follicular luteinization, resulting in either regression of the follicles and ovarian inactivity, or persistence of luteinized follicles. In either case the sow remains anestrous. As the condition generally arises after mating of the sow, it is often assumed that she is pregnant. Some sows may return to estrus between 40 and 120 days post-mating.

In contrast to the situation in other species, '**cystic follicles**' in some mares is a 'normal' condition, occurring at the beginning of the breeding season. It is commonly referred to as spring heats, most frequently involving mares in poor condition or young maiden mares, and is characterized by a prolonged estrus lasting 10 days to two months. Follicles develop in successive waves, but fail to mature. On palpation, multiple follicles are present on one or

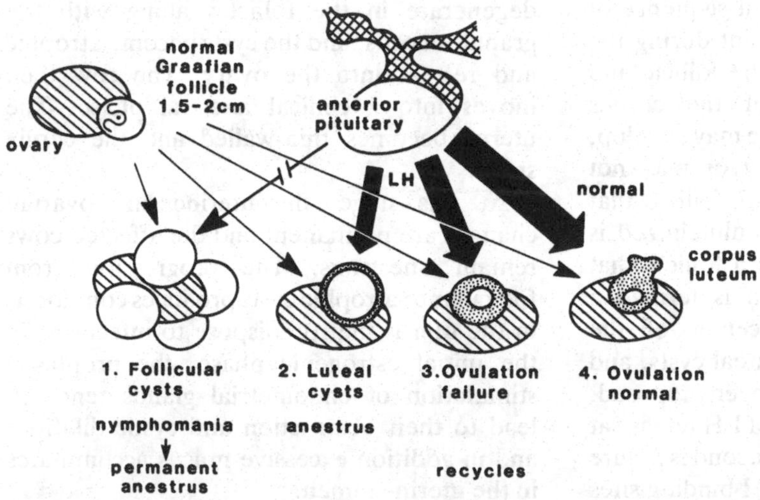

Fig. 16.7. Cystic ovarian disease in the cow. The nature of the cyst is governed by the extent of deficiency of luteinizing hormone (LH), or receptors for it.

both ovaries which are sometimes referred to as 'bunch of grape' ovaries. The etiology of the condition probably involves a slow release of low levels of gonadotropins early in the breeding season rather than the normal cyclic release of both FSH and LH. There is no permanent ovarian damage.

Uterine abnormalities

The functional relationship between the uterus and the ovaries is exemplified when the uterus is distended for reasons other than pregnancy. For instance, in the cow, fetal death and subsequent mummification or uterine infection result in the continuing distension of the uterus, persistence of the corpus luteum, and a state of anestrus. In mares, persistence of the corpus luteum occurs without uterine distension. This condition is called prolonged spontaneous diestrus and follows failure of the uterus to release prostaglandin and is usually successfully corrected by prostaglandin therapy.

Functional ovarian tumors

Ovarian tumors are relatively common in animals, but, as might be expected, only functional tumors affect the cycle length.

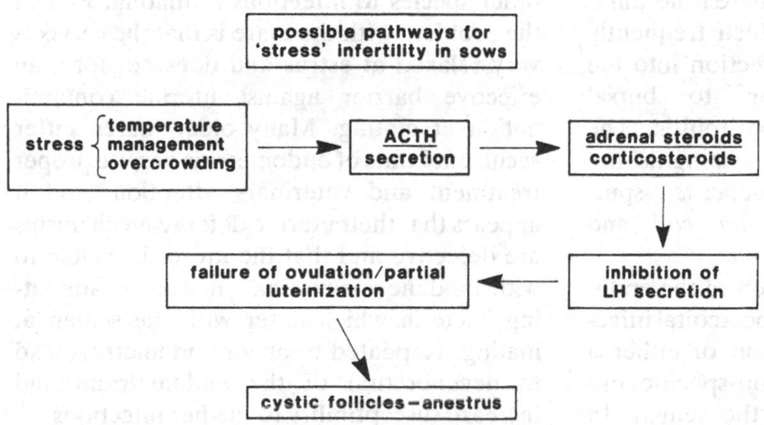

Fig. 16.8. Suggested pathogenesis of cystic ovarian disease in sows associated with the so-called summer or 'stress' infertility syndrome. ACTH, corticotropin; LH, luteinizing hormone.

The most frequently observed is the equine granulosa/theca cell tumor. Although granulosa cells in the normal female produce estrogen, functional granulosa cell tumors often result in elevated levels of testosterone as well as estrogen. In the mare, functional granulosa cell tumors produce three distinct clinical patterns; anestrus, nymphomania (continuous or intermittent estrus) and stallion-like behavior, the last mentioned being associated with particularly high levels of testosterone. As granulosa/theca cell tumors are unilateral and often benign they are amenable to surgical removal.

Inflammation of the female genital tract

The major cause of inflammation in the female genital tract is bacterial infection and a number of generalizations can be made about such infections.

- The majority of infections occur at estrus or following parturition.
- All segments of the tract are subject to infection.
- The non-pregnant uterus is generally very

resistant to infection, due to the enhancement of uterine defences under the influence of estrogens. The pregnant or mid-cycle uterus is far more susceptible to infection.

– Endometritis is common after abnormal parturition where excessive trauma or bruising occurs, or when the placenta is retained. Similarly the likelihood of infection is enhanced by delayed involution of the uterus, a frequent sequel to the retention of fetal membranes.

Salpingitis, or inflammation of the fallopian tubes, occurs in 70–75% of cases of metritis because of ascending infection from the uterus, but sterile salpingitis can result from the trauma of manual manipulation during rectal examination of the reproductive tract. In either case the epithelial cells of the oviduct are readily desquamated, especially at the apex of the ridges, and denuded ridges fuse to form fibrous transductal adhesions leading to ablation of the tubes. This, not unexpectedly, can cause blockage of the oviducts, failure of passage and fertilization of ova, and hence sterility. In the absence of infection, obstructed oviducts may become thin-walled tortuous tubes, distended with clear secretory fluids, a state termed **hydrosalpinx**. When infection has occurred there is accumulation of pus (**pyosalpinx**), which frequently leads to the spread of the infection into the ovarian bursa, predisposing to bursal adhesions and occasionally peritonitis. The principal infective agents in salpingitis are bacteria including *Staphylococcus* spp., *Streptococcus* spp., *Escherichia coli* and *Actinomyces pyogenes*.

Endometritis, or inflammation of the endometrium, may be produced by postcoital infection, reflecting the transmission of either a specific venereal disease or a non-specific contaminant, usually present in the semen. In some animals the establishment of a non-specific infection is probably allowed when the resistance of the uterus has been reduced by traumatic injury or degenerative changes in the endometrium.

Specific venereal disease in cattle can be caused by a number of organisms including *Campylobacter fetus* and *Tritrichomonas fetus*. These diseases are spread rapidly amongst susceptible herds by venereal contact and cause inflammatory changes in the uterus which inhibit fertilization or implantation of the embryo. Similar venereally transmitted endometritis is recorded in other species, for example that resulting from infection of the mare with certain capsule types of *Klebsiella* sp. and *Taylorella equigenitalia*. These diseases are well documented and can be recognized from a combination of clinical signs, the isolation of the organisms and the epidemiology of the spread of the disease through a herd or group of animals.

In the cow, endometritis following the introduction of non-specific semen contaminants at mating is generally caused by coliform organisms or *Actinomyces pyogenes*. In the mare, a wider range of 'non-specific' organisms, including *Streptococcus zooepidemicus*, coliforms, *Pseudomonas* spp. and *Corynebacterium equi*, have been linked, along with numerous others, with endometritis and infertility.

The mare seems to be more susceptible than other species to infections at mating. Part of the problem with the mare is that the cervix is very relaxed at estrus and does not form an effective barrier against uterine contamination at mating. Many older mares suffer recurrent bouts of endometritis despite proper treatment and veterinary attention, and it appears that their uterine defense mechanisms are defective and that the uterus is unable to withstand the normal challenge of contaminating bacteria which enter with the semen at mating. Repeated bouts of endometritis lead to degeneration of the endometrium and increase susceptibility to further infections.

Infection of the uterus in other species is more likely in the luteal phase of the estrous cycle and may arise either through hematogenous spread of organisms or from an ascending infection of the lower reproductive tract. This increased susceptibility to infection during the luteal phase is of special significance in embryo transfer programs where the collection and transfer of embryos to and from the uterus generally occurs during the luteal phase of the estrous cycle. Careful attention must be paid to cleanliness and to the sterility of catheters used to collect and deposit ova, as even the introduction of normally non-pathogenic organisms at this stage of the cycle can lead to endometritis, a hostile environment and likely death for the transplanted embryo.

Post-partum infection of the uterus

In cows, endometritis is most often observed in the *post-partum* period. Organisms introduced from the environment may persist when an abnormal or traumatic parturition has occurred, and may cause a serious and persistent endometritis. Bacteria known to cause endometritis can generally be cultured from the uterus in more than 90% of cows two weeks after calving, whereas they are recovered from only 5–9% of cows after 60 days. In addition, the uterine flora fluctuates in the first 50 post-calving days, with a variety of bacteria involved. It is therefore very difficult to judge the significance of cultured organisms if swabs are taken in the early *post-partum* period.

The most common organisms known to cause endometritis during this period are *Actinomyces pyogenes*, *E. coli* and *Streptococcus* spp. Infection with *A. pyogenes* generally causes more purulent discharge and greater damage to the endometrium than the other bacteria. Organisms such as *Brucella abortus* or *Campylobacter fetus* may cause endometritis in cows, after having primarily infected the pregnant uterus and caused abortion.

In dairy cows, the principal clinical signs associated with *post-partum* endometritis are a drop in food intake and milk output, and infertility characterized by a long period between calving and conception.

Pyometra

Bacterial invasion of the uterus, either as an ascending invasion or through hematogenous infection, may lead to a condition termed pyometra. In this condition the uterus becomes distended with pus and debris from the inflammatory reaction, especially if the cervix is closed. Pyometra occurs mainly in the cow and the bitch, and a closed cervix is a characteristic feature. It can also occur in the mare, where the conformation of the pelvis and uterus may allow gravity to cause retention of the inflammatory debris. In the cow and the bitch, mechanical obstruction of the cervix may be the result of either chronic fibrosing cervicitis, or, a segmental aplasia which leads to pooling of fluids in the isolated segment of the uterus.

Functional closure of the cervix may also occur because of high levels of circulating progesterone. If a corpus luteum is present in the ovary at the onset of pyometra, uterine distension will lead to its persistence, because of decreased synthesis or failure of release of prostaglandins from the endometrium. Progesterone also depresses uterine defenses and decreases myometrial contractility.

The cystic hyperplasia–pyometra complex in the bitch and cat

In the bitch and queen, the hormonal fluctuations of successive estrous cycles can result in certain individuals developing cystic, endometrial hyperplasia. During diestrus, excessive secretions from the hyperplastic endometrial glands accumulate within the uterus, as the dominant influence of progesterone causes the cervix to be closed. The pooled secretions within the uterus provide an ideal medium for bacterial growth. Occasionally pathogenic bacteria arrive via the bloodstream and initiate pyometra, the most common organisms being *E. coli* and *Proteus* spp. The clinical onset of the disease invariably occurs a few weeks after an estrous period and it often terminates in severe toxemia and death.

The prevalence of pyometra in the bitch is not high, but the disease has been reproduced experimentally by imposing hormone regimes which alternate progesterone and estrogen dominance. From such evidence it has been concluded that naturally occurring estrous and diestrous periods over a number of breeding seasons initiate the predisposing proliferative endometrial changes.

Clinically the animals are depressed and anorexic, with frequent vomiting, polyuria and polydypsia. Usually there is a vaginal discharge present, but if the cervix is closed this may be absent. The uterus becomes distended, thin-walled and fragile, so extreme care must be taken when palpating the abdomen. Polyuria and polydypsia are the result of impaired urinary concentrating mechanisms (see Chapter 9).

The preceding examples demonstrate how hormones circulating at various times in the reproductive cycle can influence the onset and course of inflammatory diseases in the uterus. The principal hormone involved is progesterone, which depresses the defense mechanisms of the uterus, as well as closing the cervix and reducing the contractility of the uterine muscles. Thus, infection or mechanical intervention during the progestational phase is more likely to lead to infection and pyometra.

Fetal injury, death and abortion

Death of the embryo or the fetus is a major cause of reproductive wastage in most domestic species. Between 20% and 40% of conceptuses are lost at the preimplantation or early postimplantation stages of development, the loss varying between species and between animals of the same species. Most of these losses occur in the first month of pregnancy and the effects on subsequent cycling and fertility of the animal will depend on the stage at which the conceptus dies. If death occurs before implantation, the conceptus will be **resorbed or expelled** at the next estrus with little effect on the cycle. As pregnancy progresses, death of the fetus will generally result in prolonged inter-estrous periods with an irregular return to estrus.

In addition to resorption there are other possible consequences of the death of the embryo or fetus.

- **Mummification** may occur when fetal death takes place after the development of the skeleton. The process involves the slow resorption of the fetal and surrounding fluids, leaving the skin and bones of the fetus intact. It can only occur if there is no bacterial infection concurrent with, or causing, the death of the fetus. The mummified fetus may be spontaneously expelled, or, if a small number of fetuses die in a multiparous species such as the pig, may be delivered with the live fetuses at the time of normal parturition.
- **Maceration** involves infection of the uterus, either causing or following fetal death, and resulting in putrefaction of the soft tissues of the fetus. There is concurrent metritis, which can result in endometrial damage.
- **Abortion** is defined as the expulsion of the fetus before the period of viability.
- **Stillbirth** is the expulsion of the dead fetus within the period of viability, close to the time of normal parturition.

During the period that the corpus luteum is active, the effect of fetal death is unpredictable. With fetal death, the corpus luteum may degenerate and thereby promote the subsequent expulsion of the dead fetus. Alternatively, persistence of the corpus luteum will promote resorption or mummification of the fetus. During the last third of pregnancy some hormonal support from the fetoplacental unit is essential and fetal death will generally result in expulsion of the fetus within a few days.

The **mechanisms of abortion** are not clearly understood, but are probably similar to the mechanisms involved in normal parturition (Fig. 16.9). Disruption of the placenta, or other factors causing fetal stress stimulate the release of cortisol from the fetal adrenals, the very process which normally triggers the onset of parturition. The high levels of cortisol act on the placenta, stimulating the enzyme system which increases the synthesis of estrogen and decreases the production of progesterone. This changing hormone balance promotes the production of prostglandin in the endometrium and increases the sensitivity of the uterine muscles (myometrium) to oxytocin, a neurohypophyseal hormone, causing them to undergo powerful contractions. Prostaglandins are also released when the placenta is inflamed, destroying the corpus luteum if it is still active and relaxing the cervix. Prostaglandins also cause powerful contractions of the myometrium. In some cases the disruption of the placenta is thought to remove the immunologic barrier between the fetus and dam, causing a reaction by the dam against the fetus. The net effect of these reactions is the expulsion of the fetus.

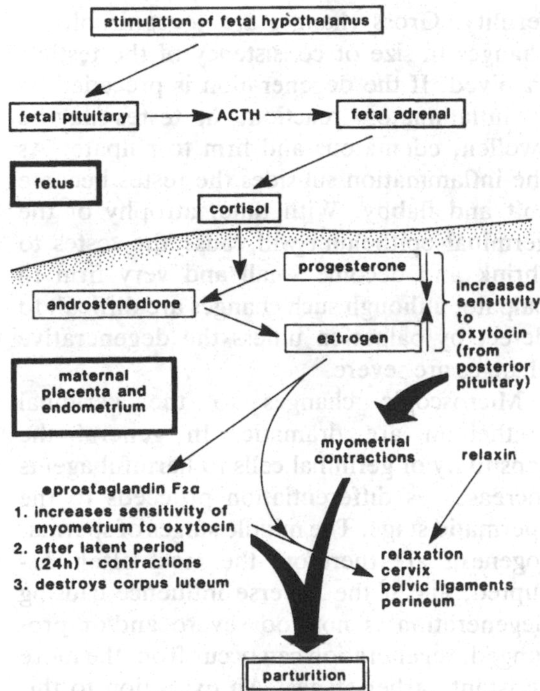

Fig. 16.9. Hormonal mechanisms controlling parturition. The important point to note is the initiating stimulus from the fetus. ACTH, corticotropin.

In contrast to this response to acute fetal stress or death, chronic fetal disease will often lead to the premature birth of weak and diseased offspring.

A wide range of agents can cause abortion in the various domestic species and to help with an understanding of the pathogenesis it is convenient to place the agents into three broad groups. Although non-infectious agents can be involved they are of minor significance.

1 Infectious agents which cause abortion as a major part of their pathogenic effect, such as *Brucella abortus*, infectious bovine rhinotracheitis virus (IBR), *Campylobacter fetus*, *Tritrichomonas fetus* and numerous fungi.
2 Infectious agents which lead to sporadic abortion in a non-specific manner when affecting the uterus, such as *Salmonella* spp., or *Actinomyces pyogenes*.
3 Infections or other agents causing systemic distress in which abortion is one of the possible manifestations.

In most cases, the agent causes death of the fetus or damage to the placenta, with inflammatory or degenerative changes which precipitate the abortion. The lesions in the placenta and fetus depend upon the route of entry of the infective agent into the uterus. There may be hematogenous infection from the dam, as in listeriosis, IBR infection and certain fungal infections in cattle. Alternatively there may be infection of the uterus prior to conception, such as occurs in infection with *T. fetus*, *C. fetus* or *A. pyogenes*. Finally, infection may spread from the cervix into the pregnant uterus, as seen in *A. pyogenes*, and fungal infections in the mare.

Once within the uterus and placenta, the agent can spread to the fetus directly across fetal fluids or via the umbilical vein. The method of infection will to some extent determine the site of lesions in the fetus. For example, spread of *Listeria* down the umbilical vein will produce extensive lesions in the liver, whereas spread of *B. abortus* or fungi from the placenta across the allantoic and amniotic fluids produces lesions on the fetal skin and, as amniotic fluid is swallowed and aspirated, in the fetal lungs.

The pathologic reactions within the infected fetus and placenta depend on three major factors. First, the **nature of the infective agent** may dictate whether there is principally a necrotizing reaction as in toxoplasmosis, or a vigorous chronic inflammatory response, as in brucellosis. Second, **the stage of immunologic development** of the fetus determines its capacity to mount an immune response to infection. This capacity is acquired during gestation at different stages depending on the species of animal and the particular antigen provoking the response. In general, it is operative in cattle by 150–170 days, in sheep by 75–80 days and in the pig by 56–70 days.

The nature of the reaction to a particular agent can be greatly influenced by whether or not the fetus is mature enough to mount an immune response at the time of infection. For instance, if the bluetongue virus infects a pregnant ewe before approximately day 75 of gestation, the fetus suffers extensive necrosis of the brain, and develops porencephaly. If infection occurs after 75 days, the induction of inflammation results in meningoencephalitis. Similarly, if a sow becomes infected with porcine parvovirus before 60 days of gestation, the principal lesion in the affected fetuses is massive necrosis of tissues. Should infection occur between days 60 and 80, the fetuses will develop inflammatory lesions, and after 80 days, when immunocompetence is fully established, no fetal lesions will occur.

The third major factor is **the stage of organ development** at the time of infection. At certain stages of organogenesis, cells in some organs are highly susceptible to particular viral agents. Good examples are the bovine viral diarrhea virus (BVD) and hog cholera virus. BVD infection earlier than 100 days may give rise to fetal death, mummification and abortion. However, if infection occurs between 100 and 170 days, developing cells in the cerebellum are destroyed, leading to cerebellar degeneration and cerebellar hypoplasia at birth. There may also be defects in other organs. Infection between 90 and 200 days also finds developing hair follicles vulnerable and calves may be born with partial or total absence of hair in localized areas.

Pathology of the male genital system

Testicular degeneration

The testicular germinal epithelium is sensitive to many adverse influences and its degeneration is the most frequent cause of infertility in male animals. Progressive, irreversible degeneration occurs fairly frequently in all species, often without any indication of primary cause or pathogenesis.

Testicular degeneration may be unilateral or bilateral depending on whether it is caused by a systemically acting agent or a specific, localized infection, and it invariably reduces fertility. Gross changes are recognizable by changes in size or consistency of the testicle involved. If the degeneration is preceded by an inflammatory reaction, the testes may be swollen, edematous and firm to palpate. As the inflammation subsides the testes become soft and flabby. With time, atrophy of the germinal epithelium will cause the testes to shrink and become small and very firm to palpate, although such changes are difficult to detect by palpation unless the degenerative changes are severe.

Microscopic changes to the germinal epithelium are dramatic. In general the sensitivity of germinal cells to harmful agents increases as differentiation proceeds to the spermatid stage. The middle stages of spermatogenesis are therefore the most often disrupted, and if the adverse influence causing degeneration is not too severe and/or prolonged, regeneration can occur from the more resistant earlier stages. An exception to this generalization is if irradiation has been the primary cause of testicular degeneration. In this case the spermatogonia, the base cells from which all other stages of sperm arise, are more susceptible. Damage by irradiation therefore can permanently destroy the germinal epithelium.

As has been previously discussed, by far the most common cause of testicular degeneration is elevation of testicular temperature. There are numerous situations when testicular temperature may be elevated by disease or environmental conditions. These include cryptorchidism, excess fat in the scrotum in bulls and rams prepared for showing, high environmental temperatures to which the animal does not adapt, and edema and hematoma of the scrotum, caused by trauma or scrotal dermatitis. Similarly in periorchitis, inflammation extends from the body cavity to the tissues surrounding the testes.

Several hypotheses have been advanced to explain the effects of high temperature on the germinal epithelium. The two most widely accepted are the occurrence of tissue hypoxia and the lack of glucose substrate for tissue metabolism. Tissue hypoxia is thought possible because it has been shown in rams that there is no compensatory increase in blood flow to the testes when the testicular temperature is elevated. With increased testicular temperature, the metabolic rate also increases and creates an increased demand for oxygen, which cannot be met, as the blood flow remains constant. Similarly it is thought that the increase in metabolic rate of the germinal epithelium induced by the elevation in temperature leads to an increased demand for glucose substrate, which cannot be met without an increased blood supply. Alternatively, it is thought that the elevated temperatures may inhibit the action of enzymes essential for the phosphorylation of glucose during its normal metabolism. This would have the effect of reducing the availability of glucose to the testicular epithelium.

Other adverse influences which may lead to testicular degeneration include extreme cold leading to scrotal frostbite, localized systemic infections, poor nutrition, torsions or compressions of the spermatic cord, obstructive lesions of the efferent tubules and autoimmunity. In addition, intoxication with agents such as iron, thallium, and lead salts or an excess or deficiency of hormones may be involved. The last of these is most commonly due to neoplasms of the pituitary and hypothalamus, or to the therapeutic administration of exogenous hormones, especially testosterone.

The most important thing to remember in predicting the outcome of seminal degeneration is the time course of the degeneration. The appearance of abnormal sperm in the ejaculate generally reflects the stage of spermatogenesis affected by the degenerative influence. The semen picture in the ejaculate depends on the cell types affected, the duration of spermatogenesis and the duration of epididymal transport. For example, when elevated temperature produces mild degenerative changes in the germinal epithelium, spermatocytes and spermatids are the stages most affected. The time course of spermatogenesis means that these cells do not reach the stage of spermatozoa for around 20–35 days after the damage has occurred and then require a further 15–20 days for epididymal passage. The defective cells will not appear in the ejaculate until 40 or 50 days after being damaged. It is therefore important to realize, when examining an ejaculate to establish whether seminal degeneration has occurred, that the semen picture in the ejaculate reflects time past. It is necessary to take serial samples over quite a wide time span to establish if degenerative changes have occurred and if the changes are likely to be prolonged or transient in nature. When gross lesions or easily recognizable changes to the testes are not apparent, it is imperative that the ejaculate be monitored over several months before a reasonable prognosis can be made. If the adverse influence has been mild and of short duration, spermatogenesis may return to normal following regeneration from the undamaged spermatogonia. If, however, the adverse influence is severe or acts for a prolonged period, the damage will be much more severe and the chances of normal spermatogenesis resuming are far less likely.

Inflammation of the penis and prepuce (balanitis and posthitis)

Clinically significant inflammation of the penis and prepuce (balanoposthitis) occurs infrequently in most domestic species. A large population of microorganisms resides in the prepuce of all males. If the balance of normal flora is upset by disease, the introduction of a specific pathogen, or by intervention, such as the use of antiseptic solutions to wash the penis, inflammation will become established and can cause severe ulcerative lesions. In the bull, amongst the most common agents causing balanoposthitis is bovine herpesvirus I the cause of the well-recognized female condition, infectious pustular vulvovaginitis. The signs of this infection in the bull are a thin watery discharge from the prepuce and the presence of numerous small gray-white pustules on the preputial mucosa. These pustules ulcerate in severe cases, and coalescence of the ulcers can give rise to extensive lesions, with edema and swelling of the penis and prepuce. If the disease is uncomplicated, the lesions will begin to disappear in around 8–10 days and healing should be complete in around two weeks. Equine herpesvirus III causes the lesions characteristic of equine coital exanthema, which occur on the vulva of the mare and on the prepuce and penis of the stallion. Canine herpesvirus can cause similar lymphoid nodules, especially at the base of the penis. Lesions in both of these conditions generally resolve over a short period if secondary infection does not occur. In the sheep, infection with *Corynebacterium renale* gives rise to severe ulcerative lesions on the prepuce, with the bacteria hydrolyzing urea contained in urine. The condition occurs frequently in sheep grazing high-protein or leguminous pastures. There also appears to be a hormone interaction involved, as wethers (castrated male sheep) suffer the condition more than entire rams, and administration of testosterone preparations greatly reduces the risk.

Orchitis

Inflammation of the testes (orchitis) is seen in most species with many infectious agents isolated. The pathology of orchitis involves non-specific inflammatory changes in the testes, generally accompanied by testicular degeneration. In bulls, localization of virulent *Brucella abortus* or the live vaccine strain 19 of the organism can cause a severe orchitis and testicular degeneration. It is important to remember that bulls must not be vaccinated with the strain 19 when vaccination programs for brucellosis are being undertaken. In stallions, orchitis is an infrequent condition, generally associated with vasculitis. Poor semen quality generally follows even a mild orchitis in the stallion. Sometimes abscess-forming organisms, for example *Salmonella abortus equi*, infect the stallion's testes, generally causing severe degenerative changes in the infected testes. In dogs, orchitis is a common disease, and is generally accompanied by epididymitis. In bacterial orchitis the most common route of infection is via reflux from the bladder or vas deferens. In these cases *E. coli* and *Proteus vulgaris* are the most commonly isolated organisms. *Brucella canis* can also cause a severe orchitis in dogs.

Epididymitis

Epididymitis occurs more frequently than orchitis in domestic species, but the two are generally concurrent. Epididymitis may be caused by spermatic granulomas resulting from duct anomalies during development, trauma or infection. While lesions of the epididymis occur in most domestic species the most commonly observed lesions are in the ram. Two specific organisms are responsible for most cases of epididymitis recorded in the ram. *Brucella ovis* causes large palpable swellings, generally in the tail of the epididymis. The lesions may be unilateral or bilateral and cause blockage of the duct system and edema and degeneration of the testes concerned. *Actinobacillus seminis* is another organism found commonly causing epididymitis in rams and lesions may involve the tail or the head of the epididymis and the seminal vesicles and prostate may also be infected.

In male dogs *Brucella canis* can cause a severe epididymitis which will also often involve the prostate glands and will cause testicular atrophy.

Inflammation of the accessory sex glands

Seminal vesiculitis is common in bulls and palpable changes occur to the seminal vesicles, with a swelling of the organ and a smoothing of the normally lobulated structure. Firmness may also be palpated due to fibrosis, which occurs after prolonged infection. Semen quality will be reduced and the fertility of the bull is affected. The organisms involved in seminal vesiculitis have not yet been defined although *A. pyogenes* has been shown to cause a chronic interstitial inflammation of the seminal vesicles and *Brucella abortus* causes a characteristic fibrinopurulent discharge in the semen. Numerous viruses have been suggested as agents and are thought to give mild transient signs following infection.

Prostatitis is common in old dogs, often following primary prostatic hyperplasia. The cause of the hyperplasia is most probably hormonal as castration prevents the development of hyperplasia and often reduces the size of the already hyperplastic prostate. Prostatitis is invariably bacterial, with the most common organisms being *E. coli*, *Proteus vulgaris*, streptococci and staphylococci. Clinical signs vary, but included are hematuria, pyuria, dysuria and incontinence. Acute prostatitis is usually diffuse, but as the disease progresses there is a tendency toward abscess formation.

Testicular tumors

Tumors of the testicle are most common in older dogs and occur to a lesser extent in older bulls. Of the factors predisposing to these tumors, cryptochidism is of importance, with a many times higher probability of tumors occurring in cryptorchids.

The tumors of most clinical significance originate from the interstitial cell (Leydig cell), the Sertoli cell and the spermatogonia. **Interstitial cell tumors** are the most common testicular tumor in the dog, are invariably benign and often discovered as an incidental finding on necropsy. **Sertoli cell tumors** are of more than usual interest because a minority of them are active hormonally. In such cases there are widespread systemic effects of the inappropriate synthesis and secretion of estrogens. Clinical signs observed include a bilaterally symmetrical alopecia, gynecomastia, pendulous prepuce and prostatic enlargement. In both functional and non-functional Sertoli cell tumors, the locatory sign is testicular enlargement. The **seminoma** arises from spermatogonia and is common in cryptorchid testes. When present in descended testes the tumor is suspected when comparatively sudden enlargement occurs. Of the discussed tumors, the seminoma most frequently metastasizes.

The author thanks Dr David Pass for his contribution, especially concerning infectious infertility.

Additional reading

Jones, D. E. and Joshua, J. O. (1982). *Reproductive Clinical Problems in the Dog*. Bristol, P. S. G. Wright.

Jubb, K. V. F., Kennedy, P. C. and Palmer, N. (1985). The female genital system. In *Pathology of Domestic Animals*, 3rd edn, pp. 365–92. Orlando, FL, Academic Press.

Ladds, P. W. (1985). The male genital system. In *Pathology of Domestic Animals*, 3rd edn, K. V. F. Jubb, P. C. Kennedy, N. Palmer (eds.), pp. 393–449. Orlando, FO, Academic Press.

Lien, D. H. Pyometritis in the bitch and queen. In *Current Veterinary Therapy*, vol. VII *Small Animal Practice*, R. W. Kirk (ed.), pp. 925–32. Philadelphia, W. B. Saunders Co.

McDonald, L. W. (1980). *Veterinary Endocrinology and Reproduction*, 3rd edn. Philadelphia, Lea and Febiger.

Morrow, D. A. (1986). *Current Therapy in Theriogenology*, 2nd edn. Philadelphia, W. B. Saunders Co.

Patten, B. M. (1948). *Embryology of the Pig*, 3rd edn. New York, McGraw-Hill Book Company.

Roberts, S. J. (1986). *Veterinary Obstetrics and Genital Diseases (Theriogenology)*. Pomfret, David and Charles Inc.

Index

Printed in the United States
By Bookmasters